Simplicity and Complexity in Proteins and Nucleic Acids

Goal of this Dahlem Workshop:

To examine the structural, dynamic, and functional features of proteins and nucleic acids that lead to an apparent simplicity of many functions, but that show a surprising complexity when studied in detail.

Report of the 83rd Dahlem Workshop on
Simplicity and Complexity in Proteins and Nucleic Acids
Berlin, May 17–22, 1998

Simplicity and Complexity in Proteins and Nucleic Acids

Edited by

H. Frauenfelder, J. Deisenhofer,
and P.G. Wolynes

Program Advisory Committee:
H. Frauenfelder, J. Deisenhofer, P.G. Wolynes,
F.G. Parak, W. Saenger, and P. Schuster

DAHLEM UNIVERSITY PRESS
Berlin

E. Löschhorn
Kaiserswerther Str. 16–18
D–14195 Berlin, Germany

Tel: +49 (30) 838 73404
Fax: +49 (30) 838 73444

Orders & customer service enquiries:

D. Hildebrand-Che
email: aaa@fu-berlin.de

Visit our Home Page at
http://www.fu-berlin.de/dahlem/

Die Deutsche Bibliothek CIP-Einheitsaufnahme
Ein Titeldatensatz für diese Publikation ist bei der Deutschen Bibliothek erhältlich

ISBN 3–934504–01–9

Typeset in 10/12 pt Times by Dahlem Konferenzen

Printed and bound in Germany by Alphabet KG

Contents

The Dahlem Konferenzen ix

List of Participants xi

1 **Introduction** 1
H. Frauenfelder, J. Deisenhofer, and P.G. Wolynes

2 **What Is an Energy Landscape and Why Do We Need to Know?** 5
P.G. Wolynes

3 **How Do We Explore the Energy Landscape for Folding?** 15
C.M. Dobson

4 **How Do We Explore the Energy Landscape for Function?** 39
G.U. Nienhaus

5 **Information Contained in Protein Structures Determined by NMR in Solution or by X-ray Diffraction in Crystals** 59
M. Geyer and I. Schlichting

6 **Group Report: What Do We Need to Know about Proteins and Nucleic Acids?** 81
K.A. Dill, Rapporteur
J. Deisenhofer, G.R. Fleming, H. Frauenfelder, K. Gerwert, J.A. McCammon, and H. Michel

7 **Is Complexity of Biological Macromolecules a Result of Randomness or Not?** 107
T. Garel and H. Orland

8 **Hemoglobin and Allostery: Is It Understood?** 117
W.A. Eaton

9 **Proteins Reactions and Conformational Change: Simple, Complex, or Both?** **139**
M. Karplus

10 **Group Report: Are There Common Themes in the Function, Energy Landscape, and Dynamics of Proteins?** **179**
W.A. Eaton, Rapporteur
R.H. Austin, C.M. Dobson, B. Halle, M. Karplus, H. Orland, F.M. Richards, I. Schlichting, and P.G. Wolynes

11 **Are Molecular Motions Important for Function?** **193**
J.A. McCammon

12 **What Is the Relationship between Dynamical Effects and Biological Functions?** **199**
A. Warshel

13 **Averaging versus Zooming-in Complexity: Contraria sunt Complementa?** **213**
F.G. Parak and M.U. Palma

14 **From Simplicity to Complexity and Back: Function, Architecture, and Mechanism of Light-harvesting Systems in Photosynthetic Bacteria** **227**
K. Schulten

15 **Group Report: How Does Complexity Lead to an Apparently Simple Function?** **255**
K. Moffat, Rapporteur
J.-P. Changeux, D.M. Crothers, H. Grubmüller, G.U. Nienhaus, M.U. Palma, F.G. Parak, K. Schulten, and A. Warshel

16 **Why Structure Changes More Slowly than Sequence in Protein Evolution** **281**
C. Chothia, A.M. Lesk, I. Gelfand, and A. Kister

17 **How Much Sequence Variation Can the Functions of Biological Molecules Tolerate?** **297**
W.A. Baase, N.C. Gassner, X.-J. Zhang, R. Kuroki, L.H. Weaver, D.E. Tronrud, and B.W. Matthews

18 **Is Preservation of Diversity a Constraint on Molecular Evolution?** **313**
P. Schuster

19 Chance and Necessity in Biomolecular Chemistry — Is Life as We Know It Universal? 339
S.A. Benner and C.Y. Switzer

20 Group Report: What Are the Constraints on the Evolution and Diversity of Biopolymers? 365
S.G. Boxer, Rapporteur
S.A. Benner, M. Brunori, C. Chothia, T. Kiefhaber, Z. Luthey-Schulten, B.W. Matthews, J.N. Onuchic, R. Rigler, and P. Schuster

Author Index 383

Subject Index 385

The Dahlem Konferenzen

In 1974, the Stifterverband für die Deutsche Wissenschaft[1] in cooperation with the Deutsche Forschungsgemeinschaft[2] founded the *Dahlem Konferenzen*. It was created to promote an interdisciplinary exchange of scientific ideas as well as to stimulate cooperation in research among international scientists. Dahlem Konferenzen proved itself to be an invaluable tool for communication in science, and so, to secure its long-term future, it was integrated into the Freie Universität Berlin in January, 1990. As has been evident over recent years, scientific research has become highly interdisciplinary. Now, before real progress can be made in any one field, the concepts, methods, and strategies of related fields must be understood and able to be applied. Coordinated research efforts, scientific cooperation, and basic communication between the disciplines and the scientists themselves must be promoted in order for science to advance.

To meet these demands, Dahlem Konferenzen created a special type of forum for communication, now internationally recognized as the *Dahlem Workshop Model*. These workshops are the framework in which coherent discussions between the disciplines take place and are focused around a topic of high priority interest to the disciplines concerned. At a Dahlem Workshop, scientists are able to pose questions and solicit alternative opinions on contentious issues from colleagues from related fields. The overall goal of a workshop is not necessarily to reach a consensus but rather to identify gaps in knowledge, to find new ways of approaching controversial issues, and to define priorities for future research. This philosophy is implemented at every stage of a workshop: from the selection of the theme to its breakdown in the discussion groups, from the writing of the background papers to the formulation of the group reports.

Workshop topics are proposed by leading scientists and are approved by a scientific board, which is advised by qualified referees. Once a topic has been approved, a Program Advisory Committee of scientists meets approximately one year before the workshop to delineate the scientific parameters of the meeting, select participants, and

---

1 The Donors' Association for the Promotion of Sciences and Humanities, a foundation created in 1921 in Berlin and supported by German trade and industry to fund basic research in the sciences.

2 German Science Foundation.

assign them their tasks. Participants are invited on the basis of their scientific standing alone.

Each workshop is organized around four key questions, which are addressed by four discussion groups of approximately ten participants. Lectures or formal presentations are taboo at Dahlem. Instead, concentrated discussion — within a group and between groups — is the means by which maximum communication is achieved. To facilitate this discussion, participants prepare the workshop theme prior to the meeting through the "background papers," the themes and authors of which are chosen by the Program Advisory Committee. These papers specifically review a particular aspect of the group's discussion topic as well as function as a springboard to the group discussion, by introducing controversies or unresolved problem areas.

During the workshop week, each group sets its own agenda to cover the discussion topic. Cross-fertilization between groups is both stressed and encouraged. By the end of the week, in a collective effort, each group has prepared a report reflecting the ideas, opinions, and contentious issues of the group as well as identifying directions for future research and problem areas still in need of resolution.

A Dahlem Workshop initiates and facilitates discussion between a certain number — necessarily restricted — of scientists. Because it is imperative that the discussion and communication should continue after a workshop, we present the results to the scientific community at large in the form of this published volume. In it you will find the revised background papers and group reports, as well as an introduction to the workshop theme itself.

The difference between proceedings of many conventional meetings and this workshop report will be easily discernable. Here, the background papers have not only been reviewed by formal referees, they have been revised according to the many comments and suggestions made by *all* participants. In this sense, they are reviewed more thoroughly than scientific articles in most archival journals. In addition, an extensive editorial procedure ensures a coherent volume. I am sure that you, too, will appreciate the tireless efforts of the many reviewers, authors, and editors.

On their behalf, I sincerely hope that the spirit of this workshop as well as the ideas and controversies raised will stimulate you in your work and future endeavors.

Wedigo de Vivanco, Director
Dahlem Konferenzen der Freien Universität Berlin
Thielallee 66, D–14195 Berlin, Germany

List of Participants with Fields of Research

ROBERT H. AUSTIN Dept. of Physics, 122 Jedwin Hall, Princeton University, Princeton, NJ 08544, U.S.A.

Protein dynamics, single molecule manipulation (DNA), cell dynamics and sorting

STEVEN A. BENNER Depts. of Chemistry and Anatomy and Cell Biology, P.O. Box 117200, University of Florida, Gainesville, FL 32611–7200, U.S.A.

Chemistry, biology, and evolution of proteins and nucleic acids

STEVEN G. BOXER Dept. of Chemistry, Stanford University, Stanford, CA 94305–5080, U.S.A.

Protein electrostatics and dynamics; membrane assembly, function, and manipulation; photosynthesis

MAURIZIO BRUNORI Dip. di Scienze Biochimiche, Università di Roma "La Sapienza," Piazzale A. Moro 5, I–00185 Rome, Italy

Structural dynamics and folding of heme proteins

JEAN-PIERRE CHANGEUX Molecular Neurobiology Laboratory, Institut Pasteur, 25 rue du Dr. Roux, F–75724 Paris Cedex 15, France

Allosteric proteins, membrane receptors, ligand-gated ion channel, nicotinic receptor, signal transduction

CYRUS CHOTHIA MRC Laboratory of Molecular Biology, Hills Road, GB–Cambridge CB2 2QH, U.K.

Principles that determine the structure and evolution of proteins; mechanism of the conformational changes in proteins that are related to function

DONALD M. CROTHERS Dept. of Chemistry, Yale University, P.O. Box 208107, New Haven, CT 06520–8107, U.S.A.

Nucleic acid structure, interactions, and dynamics

JOHANN DEISENHOFER Howard Hughes Medical Institute, University of Texas Southwestern Medical Center, 5323 Harry Hines Blvd., Dallas, TX 75235–9050, U.S.A.

Structural biology, X-ray crystallography, membrane protein structure

KEN A. DILL Dept. of Pharmaceutical Chemistry, University of California, 3333 California Street, Suite 102, San Francisco, CA 94118, U.S.A.

Statistical mechanical theory and modeling of proteins and RNA

CHRISTOPHER M. DOBSON Oxford Centre for Molecular Sciences, University of Oxford, New Chemistry Laboratory, South Parks Road, GB–Oxford OX1 3QT, U.K.

Protein structure, dynamics, and folding; biophysical techniques

WILLIAM A. EATON Laboratory of Chemical Physics, Bldg. 5, National Institute of Diabetes and Digestive and Kidney Disease, National Institutes of Health, Bethesda, MD 20892–0520, U.S.A.

Biophysical chemistry: protein folding, dynamics, and self-assembly

GRAHAM R. FLEMING Dept. of Chemistry, University of California, Berkeley, CA 94720–1460, U.S.A.

Ultrafast spectroscopy applied to dynamics in liquids, solutions, and proteins

HANS FRAUENFELDER Center for Nonlinear Studies, MS B 258, Los Alamos National Laboratory, Los Alamos, NM 87545, U.S.A.

Biological physics, protein dynamics

KLAUS GERWERT Ruhr-Universität Bochum, Institut für Biophysik, Gebäude ND/04, D–44780 Bochum, Germany

Molecular reaction mechanisms of proteins, bacteriorhodopsin, photosynthetic reaction center, trans p21, protein folding, time-resolved spectroscopy, FTIR, FT Raman, photolysis

HELMUT GRUBMÜLLER Max-Planck-Institut für biophysikalische Chemie, Arbeitsgruppe für theoretische molekulare Biophysik, Am Fassberg 11, D–37077 Göttingen, Germany

Theoretical biophysics, protein dynamics, conformational heterogeneity and dynamics of proteins, effective models for protein dynamics

BERTIL HALLE Condensed Matter Magnetic Resonance Group, Dept. of Chemistry, Lund University, P.O. Box 124, S–22100 Lund, Sweden

Water in biological systems, amphiphilic liquid crystals, nuclear spin relaxation

MARTIN KARPLUS Dept. of Chemistry and Chemical Biology, 12 Oxford Street, Harvard University, Cambridge, MA 02138, U.S.A., and
Laboratoire de Chimie Biophysique, ISIS, Université Louis Pasteur, F–67000 Strasbourg, France

Theoretical chemistry

THOMAS KIEFHABER Abt. Biophysikalische Chemie, Biozentrum der Universität Basel, Klingelbergstr. 70, CH–4056 Basel, Switzerland

Protein folding and protein stability; protein folding pathways and reaction kinetics

ZAIDA LUTHEY-SCHULTEN Dept. of Chemistry, A554 CLSL, Box 19-6, University of Illinois, 600 South Mathews Avenue, Urbana, IL 61801, U.S.A.

Protein folding and protein structure prediction using statistical mechanical approaches; structural genomics

BRIAN W. MATTHEWS Institute of Molecular Biology, Howard Hughes Medical Institute, and Dept. of Physics, University of Oregon, Eugene, OR 97403, U.S.A.

Protein structure, protein stability, protein folding and design, protein–DNA interactions

J. ANDREW MCCAMMON Dept. of Chemistry and Biochemistry, Dept. Of Pharmacology, University of California at San Diego, La Jolla, CA 92093–0365, U.S.A.

Theoretical and computer simulation studies of protein dynamics and function

HARTMUT MICHEL Max-Planck-Institut für Biophysik, Heinrich-Hoffmann-Str. 7, D–60258 Frankfurt/Main, Germany

Structure and mechanism of membrane proteins, especially crystallization and X-ray structure analysis of, e.g., photosynthetic reaction centers and cytochrome oxidase

KEITH MOFFAT Dept. of Biochemistry and Molecular Biology, University of Chicago, 920 East 58th Street, Chicago, IL 60637, U.S.A.

Time-resolved macromolecular crystallography, applications of synchrotron radiation in structural biology

G. ULRICH NIENHAUS Universität Ulm, Abt. Biophysik, D–89069 Ulm, Germany, and Dept. of Physics, University of Illinois, Urbana, IL 61801, U.S.A.

Biological physics, protein dynamics

JOSÉ N. ONUCHIC Dept. of Physics, Mail Code 0319, University of California at San Diego, 9500 Gilman Drive, La Jolla, CA 92093–0319, U.S.A.

Protein folding and dynamics; "electron transfer reactions in protein"

HENRI ORLAND Service de Physique Théorique, CEA Saclay, F–91191 Gif-sur-Yvette, France

Statistical physics of disordered systems, soft condensed matter, and proteins

M. UGO PALMA Dip. di Scienze Fisiche ed Astronomiche, Università di Palermo, Via Archirafi 36, I–90144 Palermo, Italy

Functional aspects of water–biomolecule interactions; kinetically interacting processes in self-assembly of extended biomolecular structures

FRITZ G. PARAK Physik-Department E17, Technische Universität München, D–85748 Garching, Germany

Biophysics: structure and dynamics of proteins; electronic configuration of metals in metalloproteins

FREDERIC M. RICHARDS Dept. of Molecular Biophysics and Biochemistry, Yale University, P.O. Box 208114, 260 Whitney Avenue, New Haven, CT 06520–8114, U.S.A.

General aspects of the structure of proteins as studied by both experimental and theoretical approaches. Principal focus of present work (as yet at a very preliminary stage): the experimental determination of solvent-accessible, macromolecular surface areas

RUDOLF RIGLER Dept. of Medical Biophysics, MBB, Karolinska Institutet, S–17177 Stockholm, Sweden

Thermodynamic fluctuations and single molecule spectroscopy

ILME SCHLICHTING Max-Planck-Institut für molekulare Physiologie, Abt. Physikalische Biochemie, Rheinlanddamm 201, D–44139 Dortmund, Germany

Protein crystallography, emphasis on enzyme mechanisms, kinetic crystallography

KLAUS SCHULTEN Beckman Institute, University of Illinois, 405 North Mathews Avenue, Urbana, IL 61801, U.S.A.

Molecular biophysics: general theory, structure analysis, computational modeling

PETER SCHUSTER Institut für Theoretische Chemie und Molekulare Strukturbiologie der Universität Wien, Währingerstr. 17, A–1090 Vienna, Austria

Molecular evolution of RNA

ARIEH WARSHEL Dept. of Chemistry, University Park, University of Southern California, Los Angeles, CA 90089–1062, U.S.A.

Computer simulation of biological reactions

PETER G. WOLYNES Dept. of Chemistry, CLSL, Rm. A544, Box 16-6, University of Illinois, 600 South Mathews Avenue, Urbana, IL 61801, U.S.A.

Theoretical chemistry, protein folding

1

Introduction

H. FRAUENFELDER[1], J. DEISENHOFER[2], and P.G. WOLYNES[3]

[1]Center for Nonlinear Studies, MS B 258, Los Alamos National Laboratory, Los Alamos, NM 87545, U.S.A.
[2]Howard Hughes Medical Institute, University of Texas Southwestern Medical Center, 5323 Harry Hines Blvd., Dallas, TX 75235–9050, U.S.A.
[3]Dept. of Chemistry, University of Illinois, Urbana, IL 61801, U.S.A.

Goal of the 83rd Dahlem Workshop: To examine the structural, dynamic, and functional features of proteins and nucleic acids that lead to an apparent simplicity of many functions, but that show a surprising complexity when studied in detail.

Many biomolecules show at first sight a simple behavior that can be expressed by a one-line equation. Despite nearly a century of dedicated work by legions of scientists, however, the function of not even one biological macromolecule is fully understood in microscopic terms. The overwhelming weight of biomolecular research is directed at understanding the structure and function of individual proteins and nucleic acids. Studies dedicated to find the universal characteristics of biological macromolecules are rare. Proteins and nucleic acids are also an exciting playground for physicists and theoretical chemists as prototypes of complex systems. Within the past two decades, the study of complex systems has become an active field. Much of the work is theoretical; well-characterized experimental studies on complex systems are difficult. Here is where biomolecules can be valuable. They have been selected by more than 3 billion years of evolution. Their complexity is consequently not random, as in a glass, but has a purpose. The exploration of the energy landscape and dynamics of proteins and nucleic acids may therefore shed more light on complexity.

The functions of some biomolecules appear at first sight to be extremely simple. Consider as an example myoglobin, the "hydrogen atom of biology." Textbooks declare that its main function is storage of dioxygen in muscles. Thirty years ago, the

Simplicity and Complexity in Proteins and Nucleic Acids
Edited by H. Frauenfelder, J. Deisenhofer, and P.G. Wolynes

binding process, $Mb + O_2 \rightleftarrows MbO_2$, was considered to be adequately described as a simple one-step process. Three decades of work by many groups have shown that the "simple" function is exquisitely complex and is still not fully understood.

Biological macromolecules are the nexus of many different disciplines: biology, for their importance for the evolution and processes of life; biochemistry and biophysics, for their structure, dynamics, and function; chemistry, for their motions and reaction mechanisms; computer sciences and mathematics for their challenging problems; materials science, for their superb nanostructures and their properties; medicine, for their importance in health; physics, for their complexity. In some respects proteins and nucleic acids appear as simple building blocks of the cell, with well-defined properties. In other respects, the same molecule emerges as a truly complex system itself with emergent properties. Close collaboration and interaction among the various disciplines could result in major progress; however, a number of stumbling blocks lie on the path to ideal collaborations. Language is one problem: sometimes the same word means different things in different disciplines while, on the other hand, identical aspects are characterized by different words in different fields. Culture is another issue: the disciplines differ considerably in their experimental and theoretical approaches and, more importantly, in their meaning of "understanding." Some biochemists and molecular biologists are satisfied if they know in what reaction or binding process a protein participates, while most physicists strive for a mathematical description with quantitative predictive power.

This Dahlem Workshop was organized to expose the different approaches and languages, and to pave the way for a closer interaction among the interested disciplines. The participants at the workshop were challenged to find answers to four questions:

1. What do we need to know about proteins and nucleic acids? The different disciplines require different information in order to build an integrated picture. Biology, biochemistry, biophysics, chemistry, computer science, and physics each have different pictures of proteins, different goals, and search for different answers. What are the different approaches and cultures, and how can they be unified and made mutually relevant? What do the different disciplines mean by "understanding"? At present, the interaction between the biological and the physical sciences is to a large extent accidental.
2. What are the common structural and dynamical properties of biological macromolecules? Few proteins have been as extensively studied as myoglobin. Are these studies overkill or do they contain valuable lessons for all biomolecules? Can we already sketch a picture of a universal biomolecule and of its structural features, its energy landscape, and its dynamics? Which properties are essential for function and which ones are irrelevant? Does the structure determine the energy landscape, the dynamics, and the function? What do we have to learn to make the connection quantitative and to be able to predict the energy landscape, the dynamics, and the function from the structure? How does

the environment (the solvent, allosteric effectors, neighboring biomolecules) affect the structure, the energy landscape, and the dynamics?

3. How does complexity lead to an apparently simple function? The function of a protein in the cell can often be described simply. Is this property always true? Is it possible to understand, based on a knowledge of the structure, the energy landscape, and the dynamics of how the simplicity emerges? How much more would we have to know and understand about the fundamental aspects to be able to predict the function? What do we mean by "simple" and "complex" in biomolecules and in the sciences in general? What are the main characteristics of complexity, and why is complexity needed in proteins?
4. What determines the diversity of proteins and nucleic acids? How do the energy landscapes of related proteins fit the respective biological roles? How perfect are proteins and nucleic acids for their functions? Is preservation of diversity a constraint on molecular evolution? Were present-day proteins and nucleic acids built from smaller units in a hierarchical manner? Are the current building blocks — amino acids, nucleotides, and saccharides — uniquely suited for making functional polymers?

The questions listed here are clearly more that can be solved in a workshop, but the well-established approach and culture of the Dahlem Workshops helped to make a beginning at answering at least some of the questions, preparing a path into the future, and establishing or strengthening links among the various disciplines. The steering committee collected a group of about forty scientists, from biologists to theoretical physicists. About ten participants were assigned to each topic listed above and individual authors, or small groups, were requested to prepare survey or background papers before the workshop. These papers are reproduced here. The entire workshop then took place in Berlin. The four groups met in various combinations to clarify and discuss the issues. No talks were given! The discussions were often heated. The results of the deliberations of the four groups led to the four group reports also reproduced here. Obviously, it is not possible to summarize the many hours of discussions and writings. The papers speak for themselves. We believe that the meeting was very successful and that it was indeed a step in the right direction, namely to increase collaboration among the disciplines and to narrow the gaps in language and culture.

We remember the workshop with great pleasure. Taking place in the vibrant city of Berlin, culminating with listening to Beethoven's violin concerto played by Gil Shaham, the workshop created a feeling of unity. The excellent help and support of Professor Klaus Roth and Mrs. Julia Lupp helped to make the workshop a pleasant experience for organizers and participants.

2

What Is an Energy Landscape and Why Do We Need to Know?

P.G. WOLYNES

Dept. of Chemistry, University of Illinois, Urbana, IL 61801, U.S.A.

Keywords: energy landscapes, glasses, protein folding

> "Landscape painting, and fresh and vivid descriptions of nature, alike, conduce to heighten the charm emanating from the study of the external world, which is shown us in all its diversity of form by both, while both are alike capable in a greater or lesser degree, according to the success of the attempt, to combine the visible and invisible in our contemplation of nature."
>
> — Alexander von Humboldt, "Cosmos," translated by E.C. Otte 1850

Since Newton's work in mechanics, the dominant approach in science has been to describe nature by postulating a universal law of development and then to describe its unfolding consequences. Later in chemistry, a somewhat different description emerged in which a system is said to exist potentially in a few states and then to make, at random, transitions between these states. This has also proved a natural and compelling language. The "chemical" description of the universe is not inconsistent with classical mechanics, but in a sense it supersedes it and remains useful in the quantum theory. Much effort and progress has been made in connecting mechanical and chemical descriptions of nature. In the study of several complex material systems such as glasses, the usual chemical phenomenology has been strained to its limit. Most important for us at this meeting, the same stresses appear when we try to understand biomolecular function. To cope with these problems the idea of an energy landscape has been introduced. The notion of an energy landscape is an attempt to provide an overview of not only the possible states of the complex system but also their energies and their dynamical connectivities using statistical ideas appropriate to such large systems. The goal of using energy landscapes is to provide a richer phenomenology that

Simplicity and Complexity in Proteins and Nucleic Acids
Edited by H. Frauenfelder, J. Deisenhofer, and P.G. Wolynes

can encompass all experiments on biomolecules under all thermodynamic conditions and, at the same time, to provide an idea of when simplifications can be made so that the behavior of a system can be understood and computed, not merely described as in an historical narrative.

Before describing what is new about energy landscapes, it is important to present more explicitly the traditional viewpoints used to describe biomolecular behavior. The setting for a classical biochemical description of a process is considered complete when all of the "species" in the system have been classified. A species is thought to have a "molecular structure." Next, transitions between species are described through chemical reaction rate coefficients in simple rate laws. Each such transition is thought of, at least provisionally, as an elementary event. Thus we describe a biochemical process by a "mechanism," which is a shorthand for a set of differential equations for species populations. An example is the set of reactions of the citric acid cycle, which many of us suffered through memorizing in school. The other, now traditional, description of a biomolecular event is provided by molecular dynamics. This brings the Newtonian paradigm into the biomolecular realm. An energy function is postulated on the basis of the (approximately) known interactions between the constituent parts of a macromolecule. Newton's laws (which for many of the processes we discuss are a very good approximation) are then integrated much as they would be in astronomy. This leads to a trajectory that describes where all the atoms go in a particular instance. This trajectory can then be made into a movie or described by the investigator using all his or her powers of narration. Much like watching a wildlife program on television or reading Kipling, such films or narrations can be very stimulating. Unfortunately, while publishable, such stories rarely yield crisply testable scientific propositions. To relate trajectories to experiment requires much more. It is necessary to follow the history of not one molecule in one particular circumstance, but to view many such trajectories and examine their properties with the quantitative power of statistics for us to gain practical insight.

The reason that the traditional chemical description and mechanical description of a molecular system can coexist happily is that the dynamics of many systems exhibits a considerable time scale separation between the vibrational motions most directly related to the forces and those "conformational" motions that sensibly change the properties of the system. Viewed in terms of the motion of the atoms then, there are dynamical basins of attraction. When a molecule starts out in a configuration belonging to one of these basins, it remains in that vicinity for many vibrational periods. The application of traditional chemical phenomenology relies on their being just a few basins of attraction, which can be classified and specifically named. Such a small number of basins will occur naturally when the scale of energy differences between structures is much larger than the thermal energy. Thus, at room temperature covalent chemistry is described by a few basins or chemical "species" that are related to minima of the energy. On the other hand, if configurational energy differences involve only weak van der Waals interactions, at normal temperatures no individual interaction will be sufficient to stabilize the system into a clearly defined dynamical basin. Only for a large enough system will the collective effect of many simultaneous weak interactions lead to sufficient separation of time scales to allow such dynamical basins to appear. For a macroscopically large system, the most familiar situation is that a system can remain in one of these basins for very long times. We refer to a system as being in one of its "phases." At a sharply defined temperature, a phase transition occurs and a system can live in either basin (say liquid or crystal) forever. At other temperatures, one of the alternate phases is "metastable" and while the system can remain in the metastable basin for quite some time, it will eventually find the stable one. For a big but still microscopic system, such as a cluster, these "phases" describe basins with long but not infinite interconversion times that can exist in chemical equilibrium over a small temperature range. When a mesoscopic system is collectively trapped by this mechanism, we should speak not of energy basins but of free-energy basins because the competition between energy and entropy, reflecting the number of microstates in a basin, becomes critically important. For biomolecules, like large molecular clusters, when only a few classified dynamical basins of attraction of the free energy arise from these collective interactions, much of the phenomenology familiar from covalent chemistry can be still used. However, the basis of the phenomenology is very different. For example, ordinary rate laws will apply. Transitions between the few collective basins can be described by exponential relaxation. In a macroscopic system, nucleation from a metastable phase is a Poisson process like radioactive decay or a first-order reaction. Studies of the nucleation of small homogeneous drops show the probability per unit time of seeing a transition interval scales with system size. On the other hand, in nucleation the free-energy barriers have large variable entropic components so the rate coefficients in the rate laws do not follow the Arrhenius relationship as in covalent chemistry. The same is true for many biomolecular transformations. It is still possible that specific configurations will act as the dynamical bottlenecks for moving from one basin to another. If so, the Arrhenius law will work.

When the interactions stabilizing the system are numerous and weak as in biomolecules, however, the emergence of classical phenomenology with a few species is by no means immediately guaranteed. Dynamics of large, weakly bound systems may still be considerably slower than the vibrations, but clearly classifiable dynamical basins of attraction may not be apparent. This is most familiar in the case of liquids becoming glasses, a very nearly universal phenomenon. The glass transition occurs for all known substances when crystallization is prevented, with the exception of the quantum mechanical liquid, helium. What is seen is that as a highly viscous liquid approaches the glassy state, one finds a continuum of motions which are considerably slower than vibrations. These motions exist over a wide range of time scales. Measured relaxation dynamics of many quantities is nonexponential. Classical chemical kinetic phenomenology completely fails. Individual, Poisson-distributed transitions between a few states do not describe the motions. If the slowness is to be accounted for by a picture of dynamical basins, these basins must be so numerous that they cannot be easily classified. For this reason, when a substance becomes a glass by being cooled below the glass transition temperature, its properties depend on many aspects of its history. The “age” of the glass can be inferred by measuring its current properties in some cases. The idea of a few species fails, but what replaces it? Clearly, for this sort of system, an overview of all possible states of the system must be developed that can deal with their extremely large number. Since there are so many possible microstates, this description must be statistical in nature. Even at the gross level of fitting rate data one must consider a distribution of relaxation times. Simultaneously, this distribution must be connected to a distribution of many other characteristics of such a glassy system.

While it is in the context of understanding the properties of glassy systems that the notion of energy landscapes first came to the fore, experiments on the dynamics of proteins reveals that at low temperature proteins also cannot be described by the classical chemical approach using a small number of species. At physiological temperatures some processes in proteins can be described with simple rate laws, involving only a few conformational species; however, as more probes are used, the necessary number of species usually proliferates. Frauenfelder’s seminal experiments on myoglobin at low temperatures showed that natural proteins behave like liquids that become glasses when cooled. While these observations might be dismissed as “artificial,” since organisms do not live under these conditions of temperature, nevertheless, they immediately raise the question as to why the simple behavior should ever be seen. If we must invoke physiology to explain the validity of simple phenomenology then somehow evolution is involved! Even under physiological conditions it is not clear whether all important processes can be described by the classical chemical phenomenology. Recent single-molecule experiments dramatically reveal many slowly converting states of enzymes. If simple rate laws are a result of evolution, what happens when systems are perturbed by mutation? In the mutant protein some processes are turned off while others are turned on. Are the new processes described by the traditional phenomenology? This has been seen in the dynamics of folded proteins carrying out functional motions. Dynamical complexity is seen even more often in the study of protein folding where intermediates are observed under some conditions but not under others. The study of energy landscapes has proved to be particularly fruitful in understanding the folding process. Folding is often the essential first step of function. Thus we see consideration of complex energy landscapes must enter the description of the evolution of protein and biomolecule properties where both the functional molecules in surviving species and dysfunctional molecules in less fit, now extinct individuals must be considered.

For complex and large biomolecular systems the energy landscape approach might sometimes be avoided. However, if the phenomenology of glasses emerges under some conditions, the statistical energy landscape approach cannot be avoided. For such systems the landscape idea is of computational use even when the chemical picture with few species suffices as a phenomenological framework. Systems with simple phenomenology have remnants of the statistical parts of the energy landscape. These remnants influence the *values* of the coefficients in the simple rate laws observed macroscopically. Knowledge of the statistical energy landscape theories, at the very minimum, can be helpful in appreciating the design of the simple system. How has the complex behavior been avoided by the natural system? (What is the arrangement of an energy landscape that allows there sometimes to be a few dynamical basins, while in other cases there are many?)

How can we picture the arrangement of the energy states of a system? One mode of description appropriate for smaller systems is to *globally* define the energy states. This is, perhaps, the best place to use the term "energy landscape," since everything is taken in at once. For a macroscopic system the landscape characteristics must be associated with some length scale and, therefore, should be considered "local," i.e., two parts of the system sufficiently distant from each other have no direct influence. (Parenthetically we can argue it is the growth of the "global" from the "local" that is the main puzzle of macroscopic glasses: Why is one end of a macroscopic glass rod moved by pushing on the other? Notice the similarity to the problem of allostery!) To define a "local landscape," some correlation length scales are needed. Recent experiments on viscous liquids suggest that in harmony with theory such length scales do indeed exist.

The theory of *globally* defined landscapes is much more advanced than locally defined landscapes. Two extreme models of globally defined energy landscapes are useful in considering real biomolecules (Figure 2.1). One extreme model energy landscape is the completely rugged or random one. Each configuration is assigned an energy from a random distribution. This is known as the Random Energy Model. The other limiting global landscape is a smooth, funneled landscape (Figure 2.1a). For the completely rugged landscape the chemical phenomenologies break down dramatically. However, for the funneled landscape much of classical kinetic phenomenology can be used. How rates emerge from molecular details for funneled landscapes is quite different from how they emerge in traditional covalent chemistry: The $H + H_2$ paradigm is very misleading. Proteins are probably small enough to be described by global landscapes. Real protein landscapes are in between the two extremes of ruggedness and partake of characteristics of both.

What do we mean by a rugged energy landscape and why do such landscapes occur? A rugged energy landscape is one with a lot of local minima. These minima must be similar to each other in energy so they can be populated at reasonable temperature and transitions between them are active. On the other hand, there must be substantial barriers between the minima. Two mechanisms have been found for such a multiplicity of states: frustration and symmetry. The existence of many distinct minima can be caused by the conflict between different interactions that can potentially stabilize the system. This is the frustration mechanism for getting a complex landscape. When there is "frustration" or conflicting interactions, it will be difficult to find a single configuration that minimizes simultaneously each of the contributions to the energy by themselves. The energy of any particular configuration is then a sum of randomly positive and negative terms, so it is a nearly random quantity. Thus many states exist whose structures are globally different, but have similar energies and may be thermally occupied. From the mathematical viewpoint, the most studied frustrated system is that of a spin glass with very long range interactions, for which the so-called mean-field theory is exact. For this model it has been shown rigorously that there is a transition

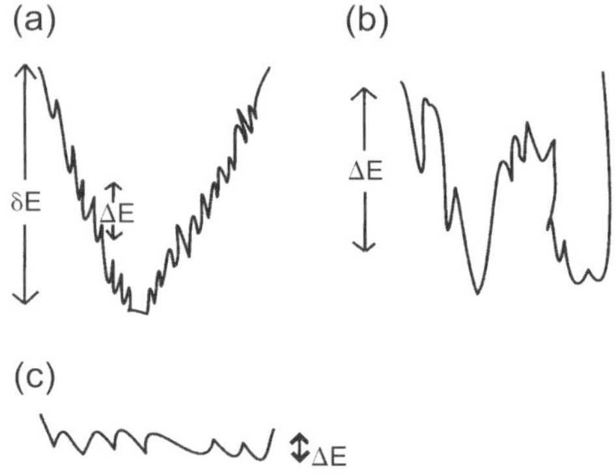

Figure 2.1 Three energy landscapes. The energy is sketched as a function of many coordinates with the *rule,* however, that nearby minima are near in structure. All such landscapes need to be quantified by the number of states and a range of correlation of structure with energy. (a) The funneled landscape has two energy scales: the depth of the main funnel δE and the variation of energies ΔE. (b) A rough landscape is statistically similar everywhere. Only one energy scale is needed to describe it. (c) A smooth landscape on one scale may still be rough, as looked at on another. While ΔE may be small in absolute terms, it is large compared to $k_B T$.

temperature where complex behavior emerges. Below this temperature, in the thermodynamic limit, many states coexist with very high barriers between them. In addition, these frozen states are shown to be very different in structure from each other. This is the sign of ruggedness. This same sort of organization for local energy minima was postulated by Bryngelson and Wolynes for random heteropolymers on the basis of the universality shown to occur in the randomly frustrated spin systems. By rewriting the mean field arguments for the case of a polymer this universal arrangement of states, like that of the random energy model, was also confirmed by detailed analytical calculations.

Ironically, extreme symmetry of interaction can lead to a multiplicity of states where the landscape concepts can also be used. Consider a homopolymer: because all of the interactions are the same, the energy of two different topologies of the polymer will be very close — only a small surface term distinguishes the energy of different topologies of a long chain. We might say the landscape is "smooth" on the interaction energy scale but in fact at temperatures capable of distinguishing between different topologies there will be significant barriers between these states that are hard to surmount. A consequence is that crystallization of a long homopolymer is usually kinetically controlled.

The other limiting energy landscape using the global language is a smooth landscape with a gradient or "funnel." Despite the newness of this term in the biomolecular context, this sort of energy landscape is the one most familiar in traditional physics. On a funnel landscape, one global minimum clearly dominates over the local minima. The lower energy local minima are formed by introducing defects or excitations into this global minimum structure. This is essentially the classical description of a slightly imperfect crystal. When a sufficiently large number of interacting defects is introduced, the resulting structures again may not be recognizable as being derived from the ground state. However, in a funnel landscape these local minima will be sufficiently shallow so that they pose no problem for the system to move. We do not need to worry about being frozen or trapped in such states with different structures. Real energy landscapes for a folding protein are thought to be intermediate between the completely funneled ideal landscape and the rugged landscape of the random heteropolymer.

Because both aspects coexist in the folding landscape, folding will sometimes follow the usual kinetic laws of the funneled landscape while under other conditions it will follow the complex dynamics of rugged landscapes. An important use of the landscape theory is to quantify the thermodynamic conditions under which these different behaviors emerge. This is easiest to do if a complete model of the system can be simulated on the computer. To make landscape theory useful outside of the virtual world of computers, one must examine the consequences of the different forms of energy landscape and compare results with experiment. This is an ongoing activity. The use of *ab initio* atomistic dynamics calculations also can help in this characterization.

Why are protein landscapes, considered from the global perspective, intermediate between the two extreme limits? *We have to bear in mind that proteins are finite systems.* The distinction between funnel and rugged landscapes can only be made sharply in the large-system limit. For any finite system, the number of different minima, especially the deep ones, is finite. Again, for smaller systems it may be possible to classify these basins taxonomically even when a much bigger system with similar input parameters would have to be analyzed statistically. The role of the finite-size effects in the glass transitions aspects has been especially confusing to some workers in the field. It has led to many of the controversies about the energy landscape perspective in the biomolecular context. Hopefully, these issues will be clarified as the theory of *locally* defined energy landscapes is further developed. In a system with short-range forces and correlations we can apply the landscape ideas to a limited region of the system at a time (see Figure 2.2). There will be interface between different regions, each part being in a low-energy state, but with strain at the junction. The glassy phenomenology of liquids has been explained using this idea which leads to scaling concepts but only the first steps have been made for numerical computations. For a local funneled landscape, the corresponding theory is the analysis of nucleation based on the capillarity model.

Proteins are big in atomic terms but mesoscopic when considered as condensed matter systems, creating the conceptual difficulties. An additional complexity is that chain connectivity can introduce long-range correlations because of topology and excluded volume effects. Considering this, one possible answer to why landscapes are both funneled and rugged is that proteins are so small that we cannot clearly make a distinction between funneled and rugged, global and local, much as was the case in the old controversy between the adherents of molecular orbital and valence bond theories

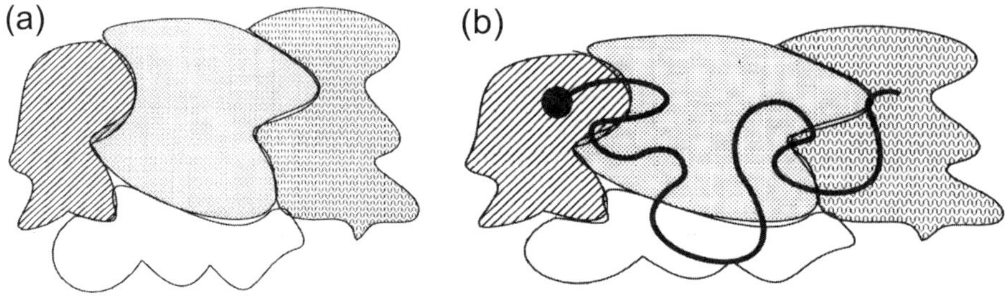

Figure 2.2 (a) With finite range interactions a system naturally breaks up into *regions*, each of which can be described by global landscape models. The size of these regions is important in determining whether relations such as the Arrhenius law will hold. (b) For polymers, one region may have its dynamics entangled with another.

of the chemical bond. If so, both limits have elements of correctness and the problem is to learn how to do quantitative calculations.

Nevertheless, it is difficult to escape the notion that proteins that evolved today are too big to have these intermediate style landscapes by accident. To use the term of David Sherrington, they correspond to "sculpted" landscapes. This sculpture occurred by the process of evolution and selection. A slightly different alternative is that the funneledness arises because the big proteins are made by the concatenations of smaller objects each with a rough landscape. Modular structures can only have *locally* rough landscapes and must be globally funneled. Understanding the nature of the energy landscape therefore tells us about the constraints of evolution or protein biogenesis. Many aspects of protein taxonomy can be understood using the idea that the landscapes must be sculpted in order to avoid dynamical difficulties in folding. These include the prevalence of certain superfamilies of structures which are highly overrepresented in the known universe of proteins. Also, the known mutation stability of proteins can only be understood by the assumption that landscapes are partially sculpted by the evolutionary process.

The most important problems where we need to know about how energy landscapes work occur when we create landscapes ourselves. Protein structure prediction usually involves setting up in the computer an artificial system which is meant to mimic, in some respect, the way actual proteins behave. A simulation with this artificial landscape must lead to a folded protein. Structure prediction, therefore, involves creating energy landscapes. It follows that it is important that we know the quantitative statistical characteristics of landscapes in order to compare different structure prediction schemes. The theory can be used to guide our search for energy landscapes that predict structures well. Several practical protein structure schemes use the explicit energy landscape approach.

Finally, landscape ideas can be used to improve the design of protein sequences that will fold. An example of this is our emerging understanding of why the simplest designs fail to give truly foldable molecules.

REFERENCES

There are so many works using landscapes that a real bibliography is hard to fit in this short space. The reader is directed to two reviews which aim at covering the aspects involved in biomolecules. These are:

Frauenfelder, H., S. Sligar, and P.G. Wolynes. 1991. The energy landscapes and motions of proteins. *Science* **254**:1598–1603.

Onuchic, J.N., Z. Luthey-Schulten, and P.G. Wolynes. 1997. Theory of protein folding: The energy landscape perspective. *Ann. Rev. Phys. Chem.* **48**:539–594.

3

How Do We Explore the Energy Landscape for Folding?

C.M. DOBSON

Oxford Centre for Molecular Sciences, University of Oxford, New Chemistry Laboratory, South Parks Road, Oxford OX1 3QT, U.K.

Keywords: protein folding, biophysical methods, aggregation, amyloid

ABSTRACT

Many aspects of the processes by which proteins fold differ significantly from the ways in which simple molecules undergo chemical reactions. Just as new methods are needed to develop a framework for a theoretical description of the folding process, new approaches are required to extend experimental strategies from those appropriate for studying simpler systems. The energy surfaces or landscapes on which proteins fold are distinct in a number of fundamental ways from those which determine the reactions of small molecules. Their characteristics influence the design and interpretation of experimental procedures required to characterize the structural changes that occur during folding, and hence to elucidate the underlying mechanism of the folding process. Specific examples from recent studies, both of normal folding and of aberrant folding and aggregation associated with the onset of disease, are discussed in this article to illustrate recent progress in "mapping" by experiment the energy landscape for folding.

INTRODUCTION

In a living cell, polypeptide chains are synthesized on ribosomes from the message contained within the cellular DNA. For these polypeptides to have biological activity, it is usually necessary for them to fold to specific compact structures. This folding process begins while the developing chains are still attached to the ribosome. It is thought, however, that the nascent chains become associated with helper proteins (molecular chaperones) whose tasks include protecting them from premature folding and

Simplicity and Complexity in Proteins and Nucleic Acids
Edited by H. Frauenfelder, J. Deisenhofer, and P.G. Wolynes

unwanted intermolecular interactions (Hartl 1996). After detachment from the ribosome, the folding process takes place in a molecularly crowded environment, and other chaperones are on hand to aid in efficient conversion into correctly folded proteins. In complex organisms, many processes are often necessary prior to complete folding, including translocation across membranes and correct localization in cellular compartments. These processes are central to the correct development and functioning of the organism, and failure of the correct folding process, or the occurrence subsequently of unfolding or misfolding events, is increasingly linked to the onset of disease (Thomas et al. 1995; Sunde et al. 1997).

Despite the importance of the cellular control and localization mechanisms, it is clear that the code for the folding of all, or at least most, proteins is contained within their amino acid sequence, and that the fundamental nature of the folding process can be established by studies of isolated proteins in simple laboratory environments. This is the framework for structural studies of folding reactions. The starting point of these studies is therefore not the ribosome but a denatured polypeptide chain such as can be produced, for example, by addition of high concentrations of chemical species such as urea to solutions containing a protein. The extreme case of such a denatured chain is a random coil, a state in which no nonlocal interactions are present. The first step in understanding folding requires an understanding of this state of a protein. Recently, there has been much interest in random coils because of their importance in folding and because of the information that their study, particularly by NMR spectroscopy, can elicit relating to intrinsic local conformational preferences of different sequences (Smith et al. 1996).

From the point of view of the folding reaction, an important aspect of the random coil state is that it is extremely heterogeneous, both in terms of its global properties and of its local ones (Figure 3.1). Even if only two conformations were to exist for each residue in a chain of 100 residues (e.g., if an individual peptide bond could be in the α- or β-region of the Ramachandran surface) there would be 10^{30} possible different conformations. This is the origin of the Levinthal Paradox (a systematic search of this number of conformations would take at least 10^{15} s), but it also illustrates a crucial feature of the experimental study of folding. Given that a typical experimental sample will contain at most 10^{18} molecules, the number of possible structures of similar energy is so large for a random coil that every molecule is likely to have a different conformation. The process of protein folding involves converting all of these into the unique native state; all polypeptide chains (which will differ substantially in their molecular dimensions in the random coil state) must undergo collapse to a highly compact state in which all (or nearly all) side chains are in specific conformations rather than being disordered as in the random coil state.

An essential characteristic of this process arises because an ensemble of protein molecules has a very large number of distinct conformations at the initiation of the folding reaction, and because the stability of the native state relative to others is the result of a very large number of weak interactions. This characteristic is that the series of events describing the process of folding of distinct individual protein molecules can be

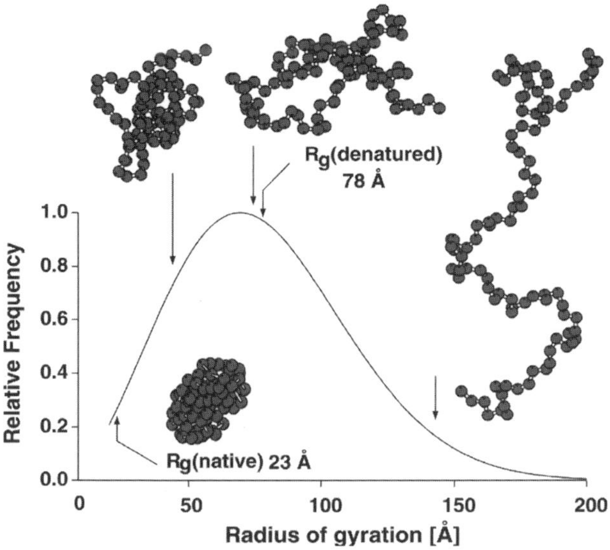

Figure 3.1 Distribution of the radius of gyration (R_g) calculated for a model of a random coil polypeptide of 300 residues simulated by a freely jointed chain of 100 spheres. Representative structures having different values of R_g are indicated. The effective R_g values for the native and denatured states are indicated. The diagram illustrates that the ensemble of random coil conformers contains both highly compact and highly extended states. The local conformational distributions of the main chain torsion angles are defined by the familiar Ramanchandran surfaces. From Smith et al. (1996).

very different. Individual molecules can form the interactions characteristic of the native state in a different order as a result of the stochastic nature of the folding process. On the energy landscape of folding there will be, therefore, a very large number of trajectories describing the behavior of an ensemble of molecules. Any individual molecule will sample very few of the very large number of possible conformations, although the whole ensemble of molecules will sample a large fraction of the accessible surface. This is the essence of the avoidance of the Levinthal Paradox and arises directly from the physicochemical properties of the polypeptide chain.

This characteristic of folding represents a fundamental difference from more familiar chemical reactions, where the trajectories of different molecules are usually extremely similar, reflecting the much greater role of entropy in folding than in the reactions of small molecules. Given that present structural methods detect the properties of large ensembles of molecules (cf. below) this difference is of paramount importance. The structural heterogeneity of the molecular ensemble prior to the formation of the native state generates a major challenge to the experimentalist (Dobson et al. 1998). As well as requiring that we think in terms of distributions of structures at any stage of the reaction, this heterogeneity raises the question of whether a conventional description of a reaction, in terms of simple kinetic parameters relating to well-defined reaction coordinates and unique transition states, is appropriate to folding. As we shall see, although some aspects of the folding process may approximate to that expected from simple reaction theory, others will not and must be analyzed in a manner that takes into account more explicitly the characteristics of the molecular ensemble.

TIME-RESOLVED STRUCTURAL MEASUREMENTS

These challenges to experimental studies of folding are compounded by the fact that, under conditions appropriate to their study, most proteins fold rapidly (often in seconds or less) requiring the utilization of fast reaction methods. One approach to dealing with these complexities is to apply a battery of biophysical techniques, each of which is able to probe at least one of the characteristics of a native protein as the folding process occurs (Plaxco and Dobson 1996). Thus, for example, stopped flow far UV circular dichroism (CD) can monitor the average content of secondary structure present as a function of time after initiation of folding, for example by dilution of a protein in concentrated denaturant solution in a rapid mixing device. If, however, near UV circular dichroism is detected, the incorporation of aromatic residues in close-packed environments can be monitored as a function of time. An increasing range of techniques is being developed for such studies, and Figure 3.2 summarizes some of these in schematic form. Among the most interesting is the utilization of small angle X-ray scattering in time-resolved mode to monitor molecular dimensions and shape in real time during folding. Importantly, recent developments in mixing devices and in novel ways of initiating folding are reducing the time scales on which many of these experiments can be carried out (Eaton et al. 1997).

Application of this general strategy to an increasing range of systems is revealing a number of important features of folding reactions (Miranker and Dobson 1996). One class of behavior is typical of the results of studies of a number of small proteins with simple folds. Here, the various characteristics of native structure all develop, to within experimental error, at the same rate and with approximately exponential kinetics. This is indicative of "two-state" behavior, i.e., that within the ensemble of structures present at any given time there are essentially only unfolded or fully folded molecules; there is no accumulation of intermediate partially folded species. By contrast, studies of larger and more complex proteins typically show that different characteristics of the native structure appear at different times after the initiation of folding, indicating that significant populations of partially folded molecules are present during the folding reaction. This shows that different aspects of the folding process can be "decoupled" from each other and provides the opportunity to probe the factors that encode the different features of the overall structure. A particular characteristic of many of these studies is that the data indicate the formation of compact structures often with a near native-like complement of secondary (main-chain) structure prior to the formation of the structural rigidity of the side chains characteristic of the fully formed tertiary structure.

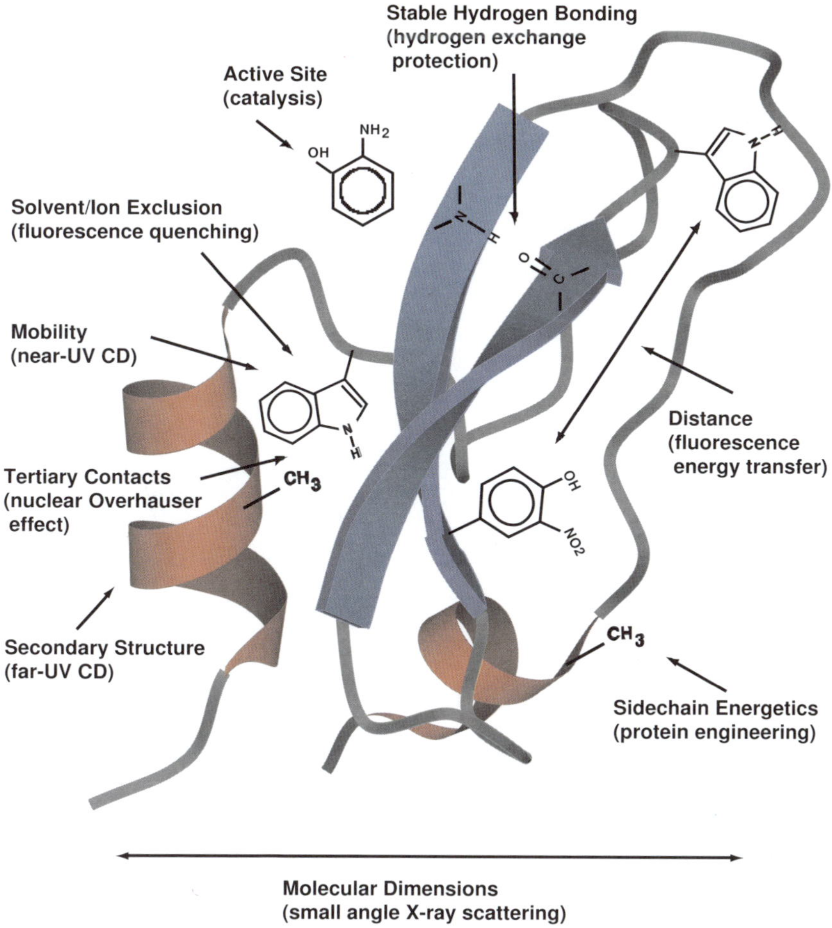

Figure 3.2 Schematic representation of some of the characteristics of a globular protein that can be followed during folding by biophysical techniques (stopped- or quenched-flow) with time resolution in the second to millisecond range (from Dobson et al. 1998; adapted from Plaxco and Dobson 1996). Examples of specific techniques able to probe the different features are indicated. Other properties that can be monitored but are not indicated include the existence of hydrophobic cavities and overall thermodynamic stability. Although no individual probe can monitor all of the structural changes during folding, the use of a set of complementary probes can provide a detailed picture of the distributions of conformations that are sampled at different stages of folding.

One of the most intensively studied families of proteins showing complex folding behavior is that of the c-type lysozymes and their structural homologues, the α-lactalbumins. These proteins have a nonsequential fold, two structural subdomains (one largely helical and the other whose secondary structure is largely β-sheet) and

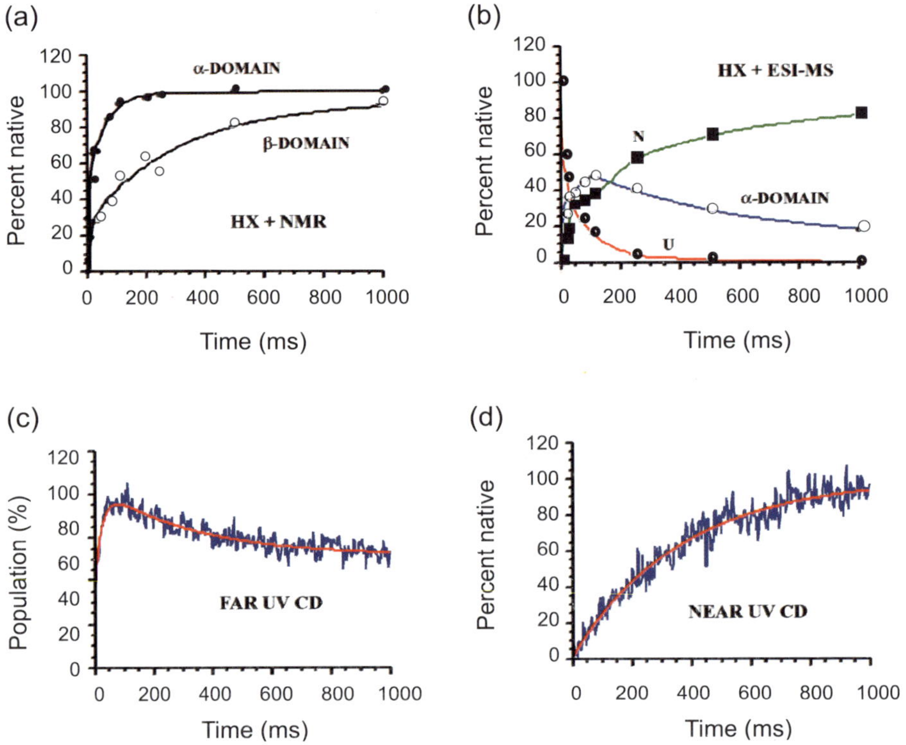

Figure 3.3 The folding kinetics of hen lysozyme initiated by 11-fold dilution (pH 5.5) from 6 M guanidine hydrochloride at 20°C. (a) Hydrogen exchange labeling and two-dimensional NMR; average protection of amides in the α- (•) and β-domains (o) is shown. (b) Hydrogen exchange labeling and ESI-MS; unprotected state (•); species substantially protected only in the α-domain (o); species with native-like protection (n). (c) Far UV CD (225 nm). (d) Near UV CD (289 nm). *(Continued on next page.)*

four disulfide bridges which are left intact in the experiments of interest to us here. Figure 3.3 shows some of the results of an extensive series of stopped and quenched flow experiments with one member of this family, hen lysozyme (Dobson et al. 1994). The experimental data show clearly that different characteristics of the native fold (see Figure 3.2) are formed with very different kinetics. Thus a hydrophobic core and extensive secondary structure develop in a few milliseconds, whereas close packing and the ability to bind specific inhibitors develop at least 100 times more slowly. Importantly, also, the kinetics at which stable hydrogen-bonded structure develops, monitored by the protection against solvent exchange of amide hydrogens, has been found to differ between different subdomains of the protein, showing the existence of species having relatively native-like structure in some regions of the polypeptide chain but not in others. In addition, the development of hydrogen exchange protection of individual amides occurs with nonexponential kinetics. This has been shown to arise from

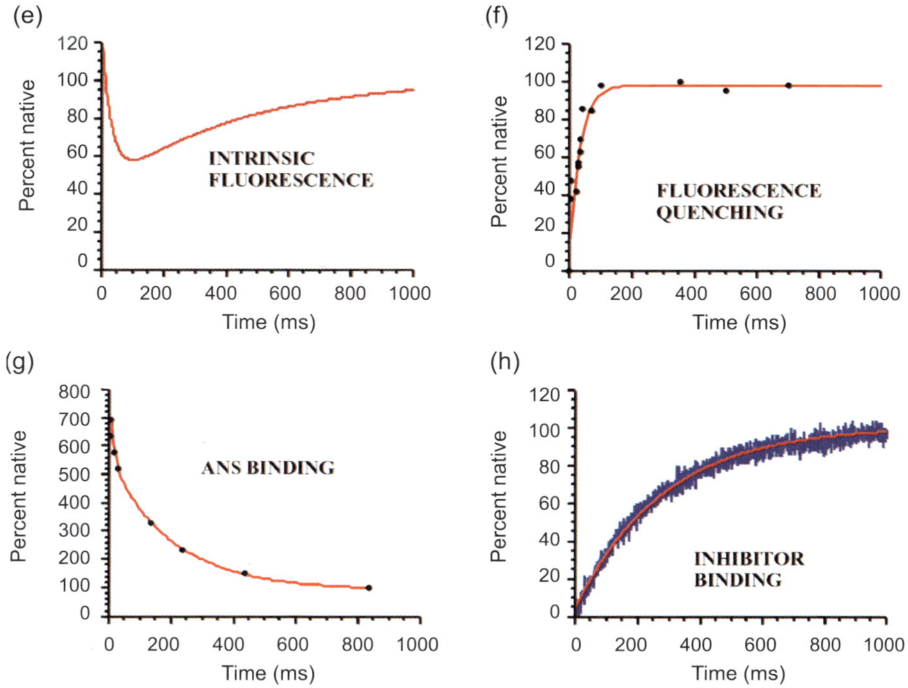

Figure 3.3 (*continued*) (e) Intrinsic tryptophan fluorescence. (f) Quenching of fluorescence with iodide ions. (g) Binding of ANS. (h) Binding of a fluorescent inhibitor (4-methyl umbelliferyl- N,N′diacetyl-β-chitobiose). Values are expressed as a percentage of the value of the native state which is arbitrarily defined as 100, except for (b) which represents the percentage of each species, given a total population of 100%. The contribution of about 20% of very slow-folding molecules, which probably arise as a consequence of *cis/trans* proline isomerism, has been ignored. From Dobson et al. 1994.

heterogeneity at early stages of the folding process resulting in the kinetics of formation of the same elements of structure being different in different populations of molecules (Dobson et al. 1994). This has been supported by experiments which indicate heterogeneity in the formation of fully stable and functionally competent native molecules (Kiefhaber 1995; Matagne et al. 1997).

Experimental data relating to folding have traditionally been interpreted using the language of conventional reaction rate theory, particularly in terms of intermediates and folding pathways. Such approaches suggest that there are marked differences in the fundamentals of the folding behavior of different proteins. Recently, however, we have begun to compare the results of experiments with the concepts emerging from theoretical models of energy landscapes and the predictions of dynamics simulations (Dobson et al. 1998). This approach has not only given new insights into the way in which folding occurs, but has highlighted the underlying similarity in the folding mechanisms for different proteins, the differences in experimental behavior chiefly reflecting the different interplay within the ensemble of molecules of the energy and entropy terms in driving the folding reaction. As discussed earlier, recent advances are not simply in the description of folding in terms of energy surfaces. All chemical reactions can be described as occurring on a surface that relates the energy of the system to parameters defining the atomic positions. The important question is the nature of such surfaces, the topologies of the energy landscapes for given proteins under specific conditions, and how this relates to the sequences of the polypeptides and the environments in which folding takes place. It is in this area that major progress has been made (Bryngelson et al. 1995; Dill and Chan 1997; Dobson et al. 1998). To relate the ideas of energy surfaces to experimental observables, it is desirable to represent the folding process by some analogue of the reaction coordinate of a simple reaction. Given the large number of atoms in a protein, this is not a trivial task. One approach is to use a "progress variable" to describe the reaction. A possible candidate, particularly appropriate to experimental structural studies, is Q, the fraction of native contacts present at a given stage of folding. Landscapes can then be represented in terms of the energies of different conformations as a function of this progress variable.

LANDSCAPES FOR FOLDING REACTIONS

The existence of "two-state" behavior suggests that the folding of some proteins can be represented by a simple surface or landscape, where molecules do not accumulate in intermediate regions and where a single variable is adequate to describe the progress of the reaction. Such a landscape must, however, have a barrier or "bottleneck," so that partially folded states do not contribute significantly to the ensemble average (Dobson et al. 1998). The bottleneck indicates that structures with intermediate values of Q have relatively high free energies. One possible reason for this, indicated by the results of simple lattice model calculations, is that the barrier is largely entropic, i.e., is associated with the high chain entropy of these intermediate states. This fits in very well with the results of protein engineering studies, where certain residues have been found by systematic site-directed mutagenesis experiments to be particularly important for rapid folding (Fersht 1997). This implies that these residues must be structured in the folding "transition state" (on the landscape model there will be an ensemble region of conformational space); without these contacts, the barrier height is significantly greater. The bottleneck may thus represent the need to search to form this "nucleus" around which the rest of the chain can coalesce. The results of protein engineering studies suggest that this folding nucleus can involve a relatively small number of residues whose contacts bring key regions of the polypeptide chain together (Fersht 1997). Experiments that probe the characteristics of the transition state regions of the energy surfaces of proteins exhibiting apparent two-state behavior are of particular importance to investigate the nature of the processes that can be rate-limiting in folding reactions.

The folding landscapes of proteins showing complex kinetic behavior clearly must differ significantly from those for proteins which show, at least to a good approximation, two-state kinetics. In the case of proteins where chain collapse occurs early in folding, and prior to the formation of extensive native-like tertiary interactions, the energy falls rapidly as a function of Q and the folding process can be represented as a search within a disordered globular state for stabilizing native contacts. Relatively simple surfaces representing such a situation emerge from simulations and can be related to the experimental behavior of some proteins that do not show two-state kinetics (Dobson et al. 1998). In the case of a protein such as lysozyme, it is clear from the type of behavior shown in Figure 3.3 that the folding reaction will require at least one additional variable for an adequate description of its progress. To visualize this, we can generate semi-empirical surfaces to represent the experimental behavior of more complex proteins in terms of features observed in a variety of simpler simulations.

A highly schematic representation of a surface appropriate for more complex folding behavior, such as that observed in the lysozyme experiment, is shown in Figure 3.4. No attempt has been made to indicate the "roughness" of the surface that undoubtedly exists, nor to define quantitatively the axes on the figure. The surface does, however, illustrate some important characteristics that relate to experimental findings for the folding of this protein. First, the effective energy drops rapidly after initiation of folding, and we associate this with the formation of an ensemble of compact molecules with extensive secondary structure but disordered tertiary interactions. There is a region where the surface has a deep local minimum, which would explain the existence of a substantial population of intermediate partially folded species during folding. A ridge separates this from a region of the surface where such a deep minimum does not exist; this permits heterogeneous kinetic behavior. On such a surface one can envisage that for certain conformations the manner in which stabilizing native contacts are formed results in fast trajectories to the native state. For others, the trajectories result in species that have high barriers to the native state, associated perhaps with the existence of nonnative structure, or with species where the probability of forming the critical nuclei for folding is relatively low. These barriers can either be surmounted or, alternatively, avoided by unfolding steps that enable the molecules to join that part of the landscape on which folding is fast. In either case, these molecules will fold slower than those whose trajectories do not lead into these "folding traps."

If Figure 3.4 represents, even in outline, a surface on which the folding of a protein such as lysozyme can be represented, then we can begin to explore methods by which sequence-specific structural features can be related to progress variables describing folding. One approach is to look in more detail at kinetic data of the type summarized in Figure 3.3. Of particular interest are hydrogen exchange pulse-labeling data because they can provide, through the site specificity of NMR measurements, information about the development of persistent structure in different regions of the protein. Thus, for example, it is possible to identify that the major intermediate populated during lysozyme folding has more persistent native-like structure in the sub-domain with helical character than in the other subdomain whose principal secondary structure is

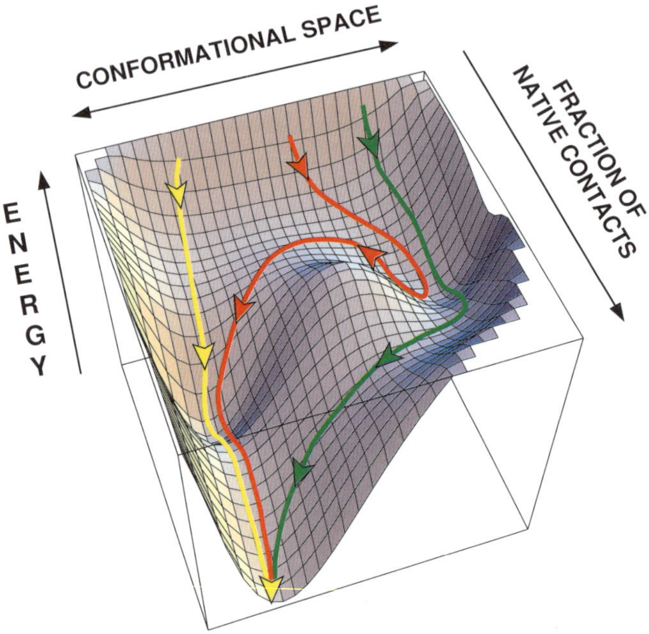

Figure 3.4 Schematic average effective energy surface based on the experimental data shown in Figure 3.3. Several possible folding trajectories are shown. One is for fast folding (left), one for slow folding that crosses a high-energy barrier (right), and one for slow folding that returns to a less folded state and then follows the valley corresponding to the trajectory for fast folding. The effective energy represents all the contributions to the free energy except the configurational entropy. The latter is proportional to the steepness of the cross-sections at specific values of Q, a progress variable representing the fraction of native contacts formed during folding. Full details, along with representation of "real" surfaces calculated from Monte Carlo simulations using a 27-mer lattice model, are given in Dobson et al. (1998).

sheet. In studies of the most highly protected amides during kinetic refolding of the equine variant, and its relationship to stable partially folded species formed at low pH (see below), it has been possible to identify a core of residues that appear to stabilize this intermediate (see Figure 3.5) (Morozova et al. 1995). Like the folding nuclei of two-state proteins, this core involves a relatively small number of residues that appear to bring together, into native-like spatial proximity, regions of secondary structure that are distant in sequence. This may well be a key element of the mechanism by which the native fold forms relative to alternatives. This point will be taken up again later in this chapter.

Such experiments, despite their value for speculating on the nature of interactions within partially folded states, do not provide direct information about the specific residue contacts that characterize the progress variable of folding. In principle, NMR spectroscopy can provide such information, through nuclear Overhauser effects

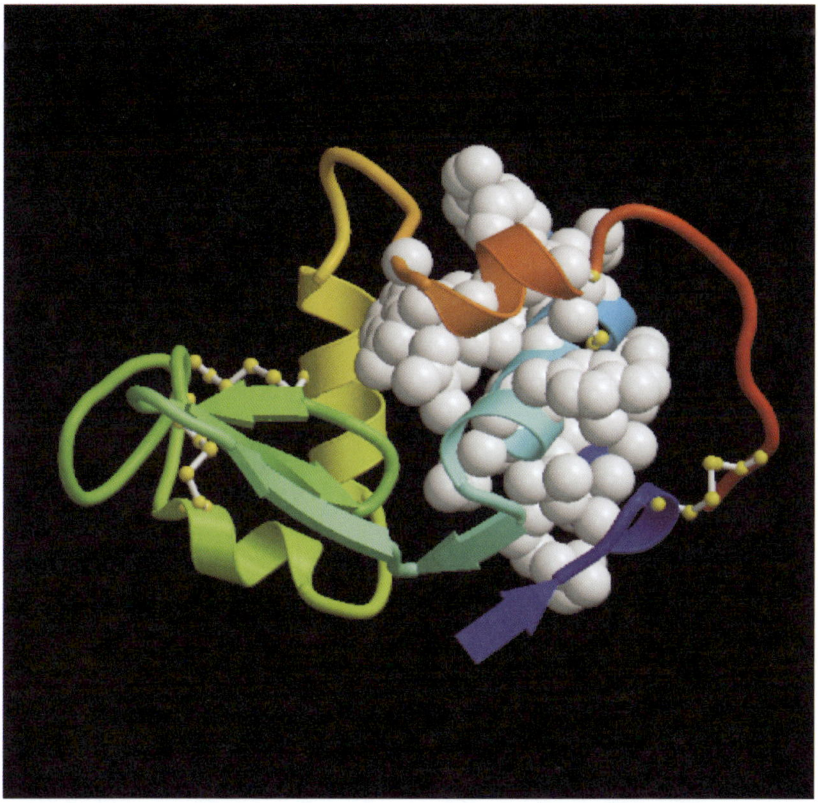

Figure 3.5 Results of NMR measurements of the protection against solvent exchange of amide hydrogens of equine lysozyme at pH 2.0. The residues illustrated in space-filling representation have high protection factors and form a hydrophobic "core" bringing together different regions of the polypeptide chain in the α-domain of the structure. Adapted from Morozova et al. (1995). Recent experiments have indicated that these residues also develop protection most rapidly during the kinetic folding of the protein (Morozova-Roche et al. 1999).

(NOEs). These distance-dependent effects arise from through-space dipolar coupling between nuclei and can be detected in appropriately designed NMR experiments. This is the information through which the solution structures of native proteins are determined by NMR. Moreover, in the event of a distribution of contacts, such as anticipated for species formed during the folding of an ensemble of molecules, the observed NOEs will reflect an average over this distribution. If sufficiently detailed measurements could be made, the distributions of contacts between the different residues in the protein could in principle be defined. To achieve this, methods need to be developed to allow NMR data to be collected as a protein folds in the same way that, for example, CD data are able to probe structural transitions in a more general manner. The challenge is that NMR spectroscopy is a demanding technique lacking high sensitivity and requiring measurements to be made in extremely homogeneous high magnetic fields. Despite these restrictions, major progress has recently been made in initiating reactions within NMR probes, enabling a wide range of techniques to be applied to characterize the folding reactions of proteins (Hore and Dobson 1998). Indeed, experiments are beginning to be developed to address the objective of mapping the landscape at the level of individual residues. One particularly promising approach is to monitor NOEs at different stages of folding by transferring them to the well-resolved spectrum of the native state in a rapid double mix experiment. This "rf pulse labeling" experiment has been demonstrated in principle and, with improvements in its selectivity, should enable detailed information to be extracted (Balbach et al. 1997). It is clear, therefore, that the folding process can be monitored effectively by NMR, and strategies to enable NOEs to be measured even for transient species have been developed, at least in outline. This is likely to be one of the most fruitful areas of research in the quest to define atomic level models for folding.

PROBING THE STABILITY OF RESIDUE CONTACTS

Kinetic studies monitor the ensemble of molecular trajectories on an energy surface. In Figure 3.4 it can be seen that these trajectories do not necessarily relate in a simple way to a progress variable for folding. Thus although on average the trajectories move to increasing values of Q, representing the stabilizing nature of native contacts, even on a relatively smooth surface some regions of trajectories represent moves to lower Q. To test a surface of the type shown in Figure 3.4, it might be possible to compute the kinetic behavior that would result from the surface for comparison with experiment. Such an approach would enable, at least in principle, refinement of an initial energy surface from kinetic data. However, a complementary approach is to try to measure the relative stabilities of individual contacts within the protein and to construct the surface from such information. One strategy towards this end is to attempt to monitor the ways in which the different interactions within the protein structure respond to conditions under which the overall stability of the protein is altered. Although the overall energy surface will be changed by such conditions, it may be possible to reconstruct that appropriate for the protein in the absence of the perturbing influence.

One example of this general approach involves the use of site-directed mutagenesis to probe the roles of individual residues in stabilizing a protein (Fersht 1997). As well as the transition states discussed above, this approach has been used to characterize intermediates populated during folding reactions by mapping the regions of their structure that are important for stability. It has provided unprecedented detail concerning the likely structures of such species and their role in folding. A complementary approach utilizes NMR spectroscopy to monitor the stepwise loss of structure of a protein during progressive unfolding with denaturants. If such experiments are initiated for the native state, the stability of this state is such, relative to any intermediates, that unfolding is usually highly cooperative and effectively two-state. However, certain proteins can be denatured under relatively mild solution conditions (e.g., at low pH) in such a way that they remain in a compact partially folded state, often referred to as a "molten globule." The absence of close-packed tertiary interactions within such states can allow them to be progressively unfolded by addition of increasing concentrations of chemical denaturants (Schulman et al. 1997).

One of the most highly studied of these compact denatured states is the A-state formed by α-lactalbumins. These proteins are close structural homologues of the c-type lysozymes, which were discussed above but which do not (except for the Ca^{2+}-binding equine variant, see Figure 3.5) form such stable species under these conditions. At pH 2, α-lactalbumins retain a native-like far UV CD signal, indicative of extensive native-like secondary structure, but have much reduced ellipticity in the near UV region, showing that the close-packed persistent side-chain interactions characteristic of the native state have largely been lost. The low pH or A-state is thus similar at least qualitatively to species formed in the early stages of the kinetic folding of lysozymes discussed above, and indeed which are also known to form during the kinetic refolding of α-lactalbumins themselves. Intensive studies of the A-state have been carried out using a variety of approaches in order to probe its structural properties (Kuwajima 1996). Experiments have shown that the effective hydrodynamic radius characterizing the A-state is within about 10% of that of the native state. NMR experiments have provided evidence for extensive side-chain disorder, while hydrogen exchange experiments have indicated the existence of relatively persistent secondary structure in regions of the α-domain of the molecule. Measurements of the propensity for the formation of different disulfide bond pairings have indicated a native-like architecture of the overall polypeptide chain. We therefore believe that the A-state represents an ensemble of predominantly compact structures, the members of which have a largely native-like fold but which do not have specific side-chain conformations. An important conclusion of these studies is that the overall architecture of the protein is likely to be defined by the overall pattern of residues in the sequences (e.g., hydrophobic vs. hydrophilic character) rather than by the specific close-packing of side chains.

To define in more detail the structural properties of such species, NMR studies are an obvious next step. NMR approaches have proved highly successful when applied to peptides, unfolded proteins, and some partially structured species notably of myoglobin (Shortle 1996; Smith et al. 1996; Eliezer et al. 1998). However, the spectra of α-lactalbumins at low pH (and indeed many partially folded species) are characterized by extremely broad resonances that do not give rise to resolved lines in the heteronuclear multidimensional experiments needed for these studies. The reason for this reflects important characteristics of the folding process. Except for the native state, the ensemble of structures characterizing the protein is highly heterogeneous. In the case of highly unfolded "random coil" states, the interconversion of the contributing conformations is sufficiently rapid to give extensive motional averaging and narrow resonance lines. In highly compact denatured states, however, the barriers to reorganization are significantly greater, and this can result in relatively slow interconversion between different conformations. This is a manifestation of the "roughness" of the energy surfaces revealed in theoretical studies. Because of the large number of states and the range of time scales involved, it appears that essentially every resonance of residues incorporated into the compact structure of α-lactalbumin is subject to exchange broadening.

This apparent limitation, however, provides an exciting strategy for probing the interactions present within such an ensemble, and hence to begin to map the energy landscape for folding (Schulman et al. 1997). Addition of chemical denaturants destabilizes the interactions and ultimately results at high concentrations in a state that approximates a random coil, both in terms of its global dimensions and local conformational preferences. Experiments, however, show that the loss of structure in the A-state is not uniform as the denaturant concentration is increased. Rather, as the members of the ensemble become on average increasingly less compact, certain regions of the average structure become differentially unfolded. When the degree of unfolding is sufficient to give rise to rapid motional averaging, the NMR resonances narrow and become visible. By assignment of the resonances to individual residues, it becomes possible to map out the regions of the polypeptide chain in terms of their stability within the compact conformations and, conversely, to observe which regions of the structure remain compact even when others are unfolded.

Analysis of the results of experiments of this type have shown the power of this approach. At pH 2, only the three residues at the N-terminus of human α-lactalbumin are sufficiently unfolded in the A-state to give narrow resonances; the rest are involved in compact structure and therefore give rise to broad signals. As the concentration of urea is increased, additional well-resolved resonances close to random coil chemical shift positions become visible. These resonances appear at different urea concentrations and enable the unfolding of different regions of the polypeptide chain to be monitored (Figure 3.6). The most readily unfolded regions of the protein are in the β-domain; however, both within this domain and within the α-domain, it is clear that loops and the termini of secondary structure units unfold before the core regions of structure (Schulman et al. 1997). In addition, certain helices are less stable than others and can be unfolded without disrupting completely the compact α-domain core formed by the remainder. Interestingly, at high denaturant concentrations, this core region appears to remain even when CD indicates the absence of significant secondary structure. Moreover, recent experiments with human α-lactalbumin, in which disulfide bonds have been selectively removed by mutagenesis experiments, indicate the important result that the compact regions of structure are not dictated by these cross links, but by the nature of the noncovalent interactions (Redfield et al. 1999).

These experiments provide an experimental means of monitoring the change in the energy surface in response to changes in the relative stability of the compact and highly unfolded states. In the case of the A-state of α-lactalbumin, there is remarkable correlation between the results of these experiments and the known architecture of the native state, strongly supporting the contention that native-like interactions are generally more stabilizing than alternative nonnative ones. There is also excellent

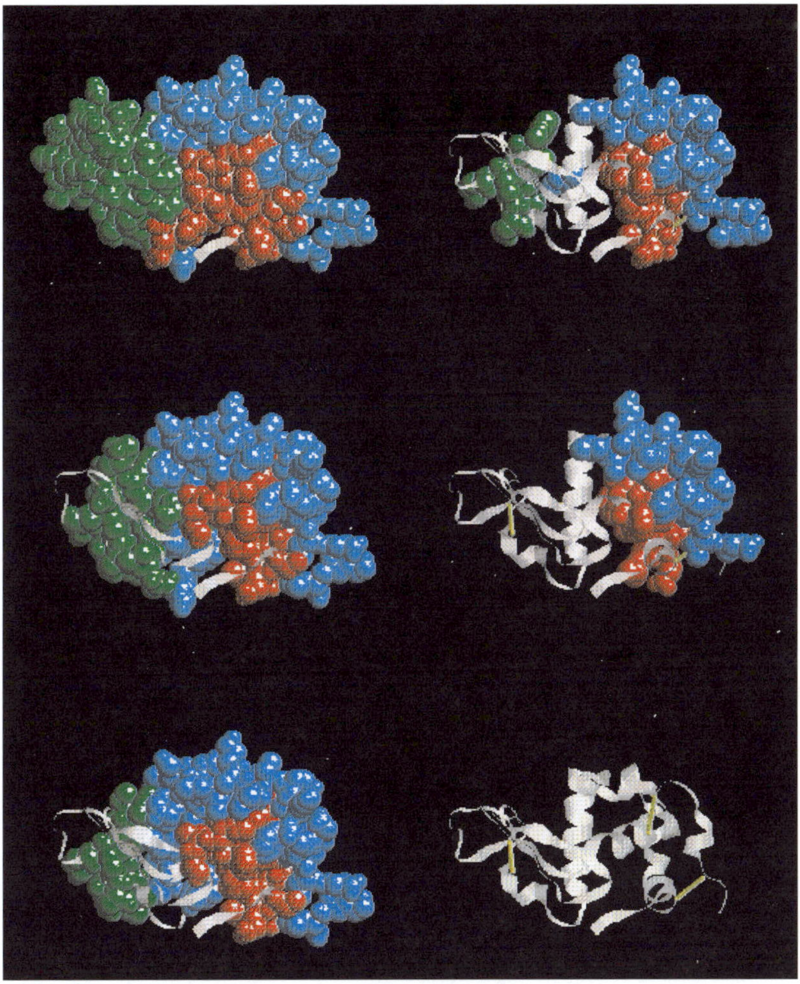

Figure 3.6 Progressive unfolding of the A-state of human α-lactalbumin (α-LA). The residues that are not visible in the NMR spectrum at a given concentration of urea, and thus are presumed to remain collapsed, are highlighted on the structure of native human α-LA. These residues are indicated with a space filling representation for the data at (a) 0 M urea at 20°C, (b) 3 M urea at 20°C, (c) 6 M urea at 20°C, (d) 10 M urea at 20°C, (e) 10 M at 50°C, (f) 8 M guanidine hydrochloride at 50°C. The backbone is represented by a gray ribbon. From Schulman et al. (1997).

agreement with complementary probes of the relative involvement of different residues in intact structure. For example, the sites of proteolytic cleavage in the A-state are located in regions that are most readily unfolded in urea. Similarly, the most highly protected amides are located within the most stable core regions. Both findings suggest that local fluctuations giving rise to accessible regions of structure correlate with the measured stabilities in the urea denaturation experiments. It seems possible, therefore, that these experiments can indicate the regions of the protein that are likely to be most stable at different stages of folding, i.e., to add residue-specific contact information to a landscape of the type indicated in Figure 3.4. Moreover, from cases where equilibrium and kinetic experiments of partially folded states of the same protein can be compared, there is evidence of a direct similarity between at least their major characteristics (Jennings et al. 1993). It may be that in folding, the direct dependence of the trajectories on the stability of the interactions defining the energy landscape results in a closer correlation of kinetic properties with thermodynamic ones than is found in simpler reactions. The next stage is to try to bring together experiments of the type described in this section with the NMR strategies described in the previous one, for example, by carrying out kinetic experiments as a function of denaturant concentration.

INTERMOLECULAR INTERACTIONS IN PROTEIN FOLDING

Until now, the emphasis in this chapter has been entirely on intramolecular interactions. However, intermolecular interactions can also be important for folding, both *in vitro* and *in vivo*. In some cases this is because the native state of a protein involves more than one polypeptide chain, e.g., it is a dimer or higher oligomer. In other cases, however, it is a result of unwanted aggregation processes that can compete with the folding of isolated molecules. One of the roles of molecular chaperones is to prevent aggregation occurring in the complex cellular environment, which could result in irreversible precipitation of the incompletely folded polypeptide chains (Hartl 1996). It is now clear, however, that there are diseases associated with the failure to prevent such events, for example, as a result in genetic mutations that result in a greater tendency of specific proteins to aggregate (Thomas et al. 1995). Other diseases are associated with intermolecular interactions that occur subsequent to the normal folding reaction and result in insoluble amyloid fibrils or plaques (Sunde et al. 1997). These include sporadic and genetic conditions such as Alzheimer's disease, and also infectious diseases such as the transmissible spongiform encephalopathies including some forms of Creutzfeldt-Jakob disease.

One of the features of the amyloid fibrils associated with all these diseases is that they are remarkably similar in their morphology (Sunde et al. 1997). They are typically long and unbranched, with cross sections of ca. 100 Å. X-ray fiber diffraction suggests that they have a cross-β structure with strands running perpendicular to the fibril axis. One of these amyloid diseases turns out to be associated with single amino acid variants of lysozyme (Pepys et al. 1993), the protein whose wild-type folding properties have been discussed in detail above. This provides an opportunity to use similar experimental methods to map the intermolecular energy landscape for a relatively simple well-defined case. These variant proteins have been expressed, and under many solution conditions they are soluble and stable. Crystallographic and NMR studies show that under these conditions the proteins are folded and their structures are closely similar to that of the wild-type protein. Under other conditions, such as elevated temperatures, the molecules undergo a structural transition to fibrils which show the characteristics of the disease-related *in vivo* material. The ready formation of the fibrils *in vitro* has allowed investigation of this aggregation phenomenon using biophysical methods of the type discussed above. Initial results of these studies suggest in outline how these fibrils might form and allow speculation concerning the molecular nature of the aggregation phenomenon (Booth et al. 1997).

Studies of the behavior of the amyloidogenic variants using the approaches described earlier for the wild-type protein have revealed a number of important characteristics (Booth et al. 1997). In particular, thermal unfolding monitored prior to aggregation shows that both variants are less stable than the wild-type protein. More importantly, these studies show that, unlike the cooperative unfolding of the wild-type protein, the unfolding transition of the amyloidogenic variants involves the formation of intermediates with many characteristics similar to those observed during the kinetic refolding of wild-type lysozyme discussed earlier. Although these intermediates have yet to be fully characterized, their existence can be rationalized in terms of the results discussed earlier. This is because the locations of both amino-acid substitutions are in regions that are important for the stability of the native structure, but not in the regions of structure forming the "core" of residues needed to stabilize the partially structured species. In effect, the mutations destabilize the native state to a greater extent than the intermediates, resulting in an increase in the population of partially unfolded species for the amyloidogenic variants compared with the wild-type protein. One can speculate that the relatively unfolded strands in the β-domain of these species could then form intermolecular instead of intramolecular contacts. It is possible that the relatively structured α-domain could act as a scaffold for this process giving it specificity. Since the fold lacks high cooperativity in the partially structured state, as shown above for the A-state of α-lactalbumin, it would be possible to recruit more of the chain into the developing aggregate to generate an increasingly stable state that goes on to form fibrils. One can imagine that such a process takes the polypeptide chain into a region of the energy landscape for the intermolecular process that is more stable than the landscape for intramolecular folding at the concentrations of the proteins involved, and which has such high barriers that it is essentially impossible to return even under other conditions.

This partial unfolding process occurs, however, *in vitro* on heating. Why should the fibrils form *in vivo*? A clue to this comes from a study of the hydrogen exchange properties of the variant proteins under conditions where the native state is stable. Compared to the wild-type protein, hydrogen exchange of even internal amide hydrogens is fast. This suggests that under physiological conditions there could be a sufficient concentration of partially folded species to allow slow aggregation. As with processes that give ordered crystals, it is probably essential for aggregation, which like crystallization is strongly dependent on nucleation phenomena, to occur slowly under near-equilibrium conditions in order to obtain ordered fibrils (rather than an amorphous precipitate). Perhaps amyloidogenic variants are those that achieve a delicate balance between rapid disordered precipitation and complete solubility under conditions found in the living system. Only then can the well-ordered array of strong intermolecular hydrogen bonds form which can match in energy the ordered intramolecular secondary structure of globular proteins. Less specific aggregation probably involves weaker and less stable interactions between surfaces of partially structured globules where many intramolecular interactions are preserved; such aggregates may simply redissolve under changing solution conditions or be susceptible (unlike amyloid) to proteolytic degradation. They would therefore not persist to cause long-term pathological conditions. This can perhaps best be thought of in the context of the varied states available to a protein molecule under different circumstances, as illustrated schematically in Figure 3.7.

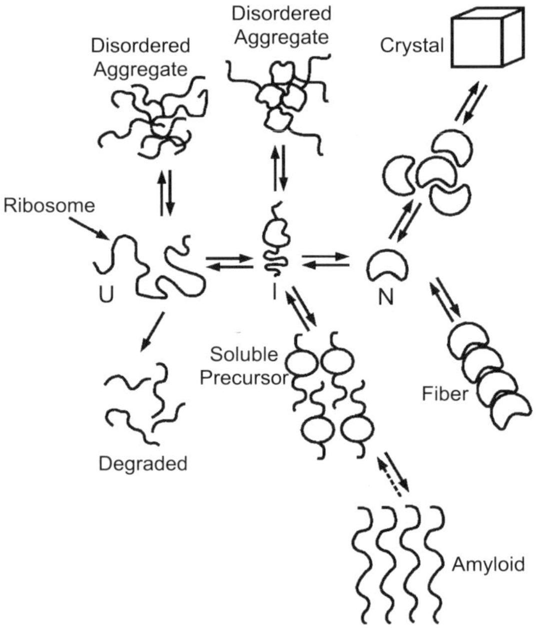

Figure 3.7 Overall representation of some of the multiple states available to a polypeptide chain. Intermolecular interactions can be ordered (e.g., crystals, fibers, or amyloid) or disordered, as in many aggregates whether of unfolded, partially folded, or native states. Some intermolecular associations will involve weak or nonspecific interactions and are readily reversed. Others, such as amyloid, are characterized by strong intermolecular hydrogen bonds which compete effectively with intramolecular interactions in the folded state. This renders formation of this material an effectively irreversible process under physiological conditions. The state that a given polypeptide chain will find itself in under given conditions will be governed by both thermodynamics and kinetics. The latter will be determined by a range of factors including nucleation phenomena.

If this argument is correct, one might anticipate that many, or indeed all, proteins could form amyloid fibrils if the appropriate conditions could be found. There is increasing evidence that this is the case. We have shown for example, that an SH3 domain, one of the most common motifs in protein structures, will form extremely well-ordered amyloid fibrils if exposed to low pH for periods of hours or days (Guijarro et al. 1998). Interestingly, under these conditions, the protein in its soluble monomeric state is denatured. However, NMR studies have shown that the ensemble of unfolded molecules is significantly more compact than that anticipated for a random coil, suggesting that partially structured members of this ensemble could be responsible for the formation of initial aggregates that result ultimately in ordered amyloid structure. This complements the lysozyme example where the partially structured species are in equilibrium with highly folded, rather than highly unfolded, species. It suggests that amyloid formation may be a generic property of polypeptide chains and that systematic studies of a range of proteins of different character will reveal the key factors determining its formation. Although amyloid represents a kinetic or thermodynamic "sink" in the energy landscape, the growth of this ordered material is slow and often will not compete successfully with other processes such as normal folding. The diseases with which it is associated, however, are known to develop slowly. Some involve genetic mutations, such as seen in lysozyme, which might give the native protein a predisposition to form partially structured intermediates that are precursors to the formation of fibrils. Some are related to aging phenomena, where proteins may be damaged or exposed for relatively prolonged periods to conditions in which they are less stable. Others are associated with infection, where an external agent or stimulus could promote the formation of the precursor species perhaps by seeding or otherwise aiding nucleation processes. The family of amyloid diseases includes examples where all these factors could play a role (Sunde et al. 1997).

FUTURE OBJECTIVES

In this chapter I have outlined a number of approaches to map by experiment the energy surfaces or landscapes on which folding takes place, and discussed some of the results such studies have achieved. Although many of the results of these experiments could be described in terms of the conventional views of small molecule reactions, I believe that substantial additional insights emerge from their analysis in terms of the "new view" of folding. The concepts of specific pathways and well-defined intermediates are seen to be specific cases of a more general picture. Ideas of "on" or "off" pathway species are redundant and can be replaced with ideas of efficient or less efficient folding routes, as Figure 3.4 suggests. Perhaps the most fundamental issue is that concerned with the existence of the massive conformational heterogeneity associated with the folding process, particularly in its earliest stages, and the profound implications this has for experimental studies. It is essential to continue development of structural methods, such as NMR, that have the potential to define at an atomic level the nature of such ensembles. Once an outline landscape has been constructed, it can be used to simulate the kinetic parameters for comparison with experiment. This should allow refinement of such surfaces. A particularly exciting additional approach is suggested by recent progress in the development of a variety of approaches able to study the properties of single molecules (Eigen and Rigler 1994). Experiments of this type have the potential to characterize directly the distribution of rates associated with different aspects of the folding reaction. Not only will this allow a more direct test of theoretical predictions, it will provide a framework for the more rigorous interpretation of experimental measurements of ensemble properties.

In the case of "two-state" folding reactions, the rate-determining step is likely to be the search for sufficient contacts to stabilize a compact state despite the much greater configurational entropy of the unfolded state. For the small proteins characterized by this type of folding kinetics, it is likely that the value of Q will need to be relatively high for a compact state to be stable; hence once these contacts are formed, the native state will result rapidly, giving rise to the observed kinetic behavior in which folding appears to be highly cooperative. Characterization of these systems requires defining the largely unfolded ensemble in which the search takes place, mapping the global properties of the transition state region (e.g., by observing the effect on the folding and unfolding rates of changes in solution conditions and variables such as temperature and pressure), and investigating the role of individual residues through site-specific mutagenesis experiments.

In the case of larger proteins, with substantial numbers of hydrophobic residues, stable collapsed states will be formed even in the absence of favorable native contacts, i.e., nonspecific hydrophobic interactions will drive collapse to disorganized globules having low values of Q. Here the search for the native contacts will be within compact molecules, and energetic barriers are likely to arise largely from the need to reorganize regions of native or nonnative-like structure. Investigation of species "trapped" in this way, for example, using the strategies outlined here, will provide information about the relationship between sequence and structure. A particularly interesting aspect of the folding of these molecules relates to the limiting rates at which the collapse process can occur (Eaton et al. 1997). It seems likely that these may be as high as 10^5–10^6 s^{-1}. For events on these time scales, equilibration of conformers within the ensemble may not be fast relative to the collapse process (Dobson et al. 1998). If folding is initiated from a random coil state, such as illustrated in Figure 3.1, those molecules that are substantially extended at the instant folding is triggered will take longer to reach the fully compact state than those that are relatively collapsed at this instant. The conversion of the initial wide distribution of molecular dimensions to a narrower one representing a compact ensemble is therefore likely to deviate substantially from simple exponential kinetics, and interpretation of experimental measurements of the rates of such events will require explicit consideration of the nature of the ensemble average as a function of time.

One of the intriguing issues concerning intermolecular interactions is why amyloid fibrils from different precursor proteins appear so similar. The explanation is likely to be that the fundamental nature of the fibrils is determined by the physicochemical properties of polypeptide chains in general, rather than the specific sequences of side chains that determine the structures of globular proteins. Thus, the dimensions of the structures might be determined by factors such as the persistence length of unfolded polypeptide chains or statistical probabilities of the lengths of strands or the occurrence of turns. The significance of partially folded states in aggregation may be primarily because these represent species in which the cooperative native fold has been lost under solution conditions which still favor the formation of stable noncovalent bonds, including hydrogen bonds, which are required for aggregation to occur. These are just the sorts of problems that can be addressed by a combination of experiments and simple theoretical models such as have had such a revolutionary impact in the field of the intramolecular folding of proteins. Mapping the landscapes for intermolecular interactions has the prospect of widening the perspective of protein structure to include states of proteins that would a few years ago have seemed to be of esoteric interest, but which are now set to become center-stage. As with the folding of isolated molecules, a wide range of techniques will be needed for these studies (Fink et al. 1998). It is likely that techniques able to study structural features in noncrystalline solids will be particularly important, including IR, electron microscopy, and solid state NMR, along with methods such as mass spectrometry, which can detect and characterize oligomeric species in solution.

The ultimate objective of "mapping the energy landscape of folding," whether intra- or intermolecular, is to understand the manner in which the physiochemical properties of the polymer chain and the amino acid sequence code for the structure. This will be achieved if the specific nature of the contacts formed between residues during folding can be defined by experiment and then simulated by theory. In this chapter, I have attempted to discuss progress towards this objective and strategies by which further progress might be made. This progress will not only result in the intellectual satisfaction of understanding in detail one of the central features of biological molecular design, but will have many practical consequences. These range from prediction of structures from sequence, an essential aspect of the efficient exploitation of information from the various genome projects, to novel therapeutic strategies to combat a variety of debilitating human diseases.

ACKNOWLEDGMENTS

This is a contribution from the Oxford Centre for Molecular Sciences which is supported by the U.K. BBSRC, EPSRC, and MRC. The research described here has also been funded by other bodies notably the Wellcome Trust, the Howard Hughes Medical Institute, and the European Commission. Valuable contributions to the work discussed in this paper were made by a large number of colleagues, many of whom are cited in the references. I wish to acknowledge particularly, however, Martin Karplus for many stimulating and lively discussions on the topic of this paper, and Andrej Salí, Christina Redfield, and Lorna Smith for help with producing diagrams.

REFERENCES

Note: Additional material relevant to this chapter has emerged since its submission and is reviewed in:

Dobson, C.M. 1999. Protein misfolding, evolution and disease. *Trends Biochem. Sci.* **24**:329–332.

Dobson, C.M., and M. Karplus. 1999. The fundamentals of protein folding: Bringing together theory and experiment. *Curr. Opin. Struct. Biol.* **9**:92–101.

* * *

Cited references:

Balbach, J., V. Forge, W.S. Lau, J. Jones, N.A.J. van Nuland, and C.M. Dobson. 1997. Detection of residue contacts in a folding intermediate. *Proc. Natl. Acad. Sci. USA* **94**:7182–7185.

Booth, D.R., M. Sunde, V. Bellotti, C.V. Robinson, W.L. Hutchinson, P.E. Fraser, P.N. Hawkins, C.M. Dobson, S.E. Radford, C.C.F. Blake, and M.B. Pepys. 1997. Instability, unfolding and aggregation of human lysozyme variants underlying amyloid fibrillogenesis. *Nature* **385**:787–793.

Bryngelson, J.D., J.N. Onuchic, N.D. Socci, and P.G. Wolynes. 1995. Funnels, pathways and the energy landscape of protein folding: A synthesis. *Proteins* **21**:167–195.

Dill, K.A., and H.S. Chan. 1997. Nucleation mechanisms in protein folding. *Nature Struct. Biol.* **4**:10–19.

Dobson, C.M., P.A. Evans, and S.E. Radford. 1994. Understanding protein folding — The lysozyme story so far. *Trends Biochem. Sci.* **19**:31–37.

Dobson, C.M., and P.J. Hore. 1998. Kinetic studies of protein folding using NMR spectroscopy. *Nature Struct. Biol.* **5**:492–495.

Dobson, C.M., A. Salí, and M. Karplus. 1998. Protein folding: A perspective from theory and experiment. *Angew. Chem. Intl. Ed.* **110**:4000–4025.

Eaton, W.A., V. Munoz, P.A. Thompson, C.K. Chan, and J. Hofrichter. 1997. Submillisecond kinetics of protein folding. *Curr. Opin. Struct. Biol.* **7**:10–14.

Eigen, M., and R. Rigler. 1994. Sorting single molecules: Application to diagnostics and evolutionary biotechnology. *Proc. Natl. Acad. Sci. USA* **91**:5740–5747.

Eliezer, D., J. Yao, H.J. Dyson, and P.E. Wright. 1998. Structural and dynamic characterization of partially folded states of apomyoglobin and implications for protein folding. *Nature Struct. Biol.* **5**:148–155.

Fersht, A.R. 1997. Nucleation mechanisms in protein folding. *Curr. Opin. Struct. Biol.* **7**:3–9.

Fink, A.L. 1998. Protein aggregation, inclusion bodies and amyloid. *Folding and Design* **1**:R9–R23.

Guijarro, J.I., M. Sunde, J.A. Jones, I.D. Campbell, and C.M. Dobson. 1998. Amyloid fibril formation by an SH3 domain, *Proc. Natl. Acad. Sci. USA* **95**:4224–4228.

Hartl, F.U. 1996. Molecular chaperones in cellular protein folding. *Nature* **381**:571–580.

Jennings, P.A., and P.E. Wright. 1993. Formation of a molten globule intermediate early in the kinetic folding pathway of apomyoglobin. *Science* **262**:892–896.

Kiefhaber, T. 1995. Kinetic traps in lysozyme folding. *Proc. Natl. Acad. Sci. USA* **92**: 9029–9033.

Kuwajima, K. 1996. The molten globule state of α-lactalbumin. *FASEB J.* **10**:102–109.

Matagne, A., S.E. Radford, and C.M..Dobson. 1997. Fast and slow tracks in lysozyme folding: Insight into the role of domains in the folding process. *J. Mol. Biol.* **267**: 1068–1074.

Miranker, A.D., and C.M. Dobson. 1996. Collapse and cooperativity in protein folding. *Curr. Opin. Struct. Biol.* **6**:31–42.

Morozova, L.A., D.T. Haynie, C. Arico-Muendel, H. Van Dael, and C.M. Dobson. 1995. Structural basis of the stability of a lysozyme molten globule. *Nature Struct. Biol.* **2**:871–875.

Morozova, L.A., J.A. Jones, W. Nuppe, and C.M. Dobson. 1999. Independent nucleation and heterogeneous assembly of structure during folding of equine lysozyme. *J. Mol. Biol.* **289**:1055–1073.

Pepys, M.B., P.N. Hawkins, D.R. Booth, D.M. Vigushin, G.A. Tennent, A.K Soutar, N. Totty, O. Nguyen, C.C.F. Blake, C.J. Terry, T.G. Feest, A.M. Zalin, and J.J. Hsuan. 1993. Human lysozyme gene mutations cause hereditary systemic amyloidosis. *Nature* **362**:553–557.

Plaxco, K., and C.M. Dobson. 1996. Time-resolved biophysical methods in the study of protein folding. *Curr. Opin. Struct. Biol.* **6**:630–636.

Redfield, C., B.A. Schulman, M.A. Millhollen, P.S. Kim, and C.M. Dobson. 1999. α-lactalbumin forms a compact native-like molten globule in the absence of disulfide bonds. *Nature Struct. Biol.* **10**:948–952.

Schulman, B., P.S. Kim, C.M. Dobson, and C. Redfield. 1997. A residue-specific NMR view of the non-cooperative unfolding of a molten globule. *Nature Struct. Biol.* **4**:630–634.

Shortle, D.R. 1996. Structural analysis of non-native states of proteins by NMR. *Curr. Opin. Struct. Biol.* **6**:24–30.

Smith, L.J., K.M. Fiebig, H. Schwalbe, and C.M. Dobson. 1996. The concept of a random coil. Residual structure in peptides and denatured proteins. *Folding and Design* **1**: 95–106.

Sunde, M., L.C. Serpell, M. Bartlam, P.E. Fraser, M.B. Pepys, and C.C.F. Blake. 1997. Common core structure of amyloid fibrils by synchrotron X-ray diffraction. *J. Mol. Biol.* **273**:729–739.

Thomas, P.J., B.-H. Qu, and P.L. Pederson. 1995. Defective protein-folding as a basis of human disease. *Trends Biochem. Sci.* **20**:456–459.

4

How Do We Explore the Energy Landscape for Function?

G.U. NIENHAUS

Abt. Biophysik, Universität Ulm, 89069 Ulm, Germany, and
Dept. of Physics, University of Illinois at Urbana-Champaign, Urbana, IL 61801, U.S.A

Keywords: myoglobin, protein structure, dynamics and function, conformational substates, energy landscape, fluctuations, relaxations

ABSTRACT

Within the properly folded, native state, proteins possess a complex energy landscape with a large number of local minima, called conformational substates (CS). The heights of the energy barriers separating CS differ widely because substate interconversions are observed in the entire temperature range from below 1 K up to the temperature of denaturation.

This chapter reviews some results of the energy landscape exploration in carbonmonoxy myoglobin (MbCO), with particular emphasis on the characterization of the so-called taxonomic substates. They can be studied individually, and their connection to the ligand-binding function is most obvious. Within each of the taxonomic substates, a large number of statistical substates exists. The connection between the energy landscape of myoglobin and its function has been studied in great detail with ligand-binding experiments after photodissociation of the CO ligand. In recent years, time-resolved spectroscopy and X-ray crystallography over wide temperature ranges have provided exciting new insights into the details of the ligand-binding reaction.

Energy landscape ideas have been especially popular in the field of heme protein research. Application to other proteins will be beneficial for further progress towards the goal of understanding protein function on the basis of the molecular structure and dynamics.

INTRODUCTION

Energy landscape is an overarching concept that has found wide application to systems as diverse as glasses, spin glasses, and synthetic polymers, proteins, evolution,

Simplicity and Complexity in Proteins and Nucleic Acids
Edited by H. Frauenfelder, J. Deisenhofer, and P.G. Wolynes

immunology, and neural networks. Here we will focus on folded proteins. In contrast to the picture frequently conveyed in the field of protein folding (Dobson et al. 1998), proteins do not possess a unique folded state but can assume a large number of slightly different structures, called conformational substates (CS). These can be represented by local minima in a rough energy landscape, separated by energy barriers with widely differing heights. While the concept of the energy landscape has been employed for many years in structure–dynamics–function studies of folded proteins (Ansari et al. 1985), it has become quite popular in protein folding in recent years and has led to what is now called the "new view" of protein folding (Dill and Chan 1997).

Proteins perform a wide variety of tasks in living systems. Here we are interested in proteins that carry out some kind of complex function, such as an enzymatic reaction. In these proteins, the energy landscape has been engineered by evolution such that coordinated transitions between conformational substates occur that modulate the properties of the protein, for example, its interactions with a substrate, in such a way that the desired sequence of reaction steps can happen.

For a deep understanding of the mechanisms involved in protein function, knowledge of the three-dimensional structure with excellent resolution is a prerequisite. Furthermore, a detailed exploration of the energy landscape, which includes characterization and classification of the conformational substates, is necessary. In particular, those conformational substates that play a central role in functional processes need to be investigated. These are often transient species that are difficult to characterize with diffraction methods, as those still lack time resolution and sensitivity. For studying functional processes, spectroscopic methods such as magnetic resonance (NMR, EPR), optical spectroscopy (absorption, fluorescence), and vibrational spectroscopy (infrared, Raman) are very powerful. While these provide excellent time resolution over about twenty orders of magnitude, it is, unfortunately, often difficult to relate the observed spectral features to the underlying structural changes.

Considering the enormous effort that is involved in the exploration of the energy landscape with the presently available techniques, only a few proteins can be studied in sufficient depth to reveal a detailed view of their energy landscapes. It is obvious that different proteins possess different sets of conformational substates. However, since all proteins are built according to the same design principles, their energy landscapes may show universal features, and insights gained from representative examples may be beneficial for a better understanding of other proteins (Frauenfelder and McMahon 1998; Nienhaus 1999). Among the many proteins known to exist, a few have drawn particular attention by the researchers, owing to their availability, interesting physiological properties, and excellent accessibility to certain experimental techniques. Especially for heme proteins, but also for bacteriorhodopsin (bR) and photosynthetic reaction centers, a substantial amount of information about the connections among structure, dynamics, and function has been gathered that should enable one to get a detailed glimpse at the energy landscape.

At present, myoglobin (Mb) may be the protein for which the most detailed picture of the energy landscape has been obtained (Frauenfelder et al. 1991; Nienhaus and

Young 1996). Mb is simple enough that its substates and motions can be studied in detail yet complex enough that the concepts found may be of general validity. Consequently, I will introduce concepts and techniques of landscape exploration for function using this model system. Before discussing functional aspects, aspects of the energy landscape analysis of carbonmonoxy myoglobin (MbCO) will be reviewed. We conclude with some brief remarks on the energy landscape analysis in other protein systems.

ENERGY LANDSCAPE OF CARBONMONOXY MYOGLOBIN

Mb, the oxygen storage protein in muscle, consists of 153 amino acids. It has a molecular weight of 17.8 kDa and the approximate dimensions $2.5 \times 3.5 \times 4.5$ nm^3 (Figure 4.1). The folded polypeptide chain surrounds a heme group with a central iron atom, which reversibly binds small ligands such as O_2 and CO. The structure of Mb, shown in Figure 4.1, encloses an interior volume, the heme pocket, with the ligand-binding site. Some of the most compelling evidence for the complex energy landscape of proteins comes from monitoring ligand rebinding to myoglobin (and other heme proteins) after photodissociation. In particular, MbCO has been studied in great detail. During the rebinding reaction, equilibrium fluctuations and nonequilibrium relaxations affect the binding kinetics in characteristic ways. Consequently, protein motions involved in the reaction can be investigated through the ligand-binding reaction.

Frauenfelder and collaborators introduced a hierarchical arrangement of conformational substates in MbCO (Ansari et al. 1985, 1987), visualized schematically in Figure 4.1, where the conformational potential energy E_c of the protein is plotted as a function of a (conceptual) conformational coordinate. Of course, the energy landscape is actually a hypersurface in the high-dimensional space of the coordinates of all atoms in the protein and hydration shell. The CSi label denotes the substates in the ith tier of the hierarchy. The top row shows MbCO as a conformational state with a unique structure corresponding to a single energy minimum. Within this state, a small number of substates exists, denoted A_0, A_1, and A_3. Each of these substates in tier 0 of the hierarchy contains a large number of statistical substates in tier 1 (Frauenfelder et al. 1991). This arrangement of substates within substates is assumed to continue over several tiers with successively lower barriers. Tiers 0 and 1 are well established by a variety of experiments, some of which will be discussed below. The existence of lower tiers is also confirmed by experiments; however, an unambiguous classification appears much more problematic.

On the basis of molecular dynamics simulations it has been suggested that actually a continuum of substates exists (within the limited range of fluctuations that can be covered by simulations) and that the postulated tier structure of MbCO may arise from the limited experimental information available, especially from the fact that each tier was assigned by a different marker (see Moffat et al., this volume). For MbCO, we cannot address this issue, as currently no data exist that were taken over a sufficiently wide

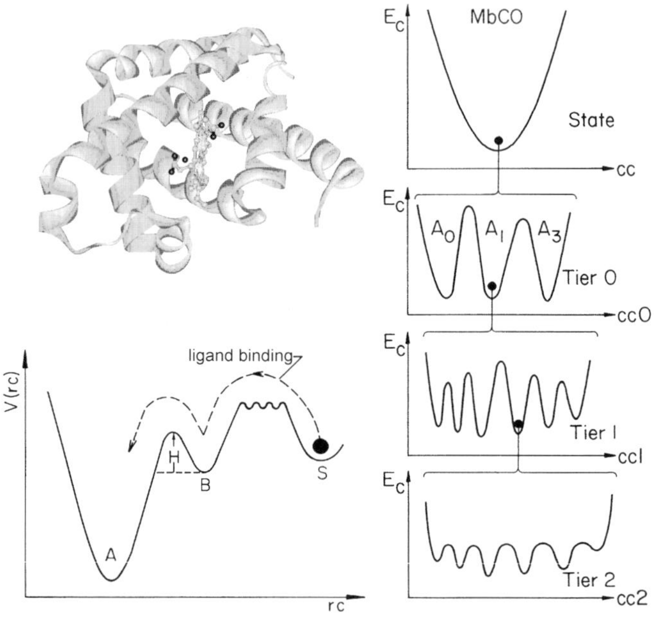

Figure 4.1 Structure, reaction surface, and conformational energy landscape of MbCO. Right column: Simplified scheme of the conformational energy, E_c, in three different tiers of the hierarchical energy landscape, as a function of a conformational coordinate, cc *i*. Left column (top): Ribbon diagram of MbCO including a possible binding path of the CO ligand from the solvent, (bottom) reaction surface for ligand binding, with *A* corresponding to the bound state, *B* to the state where the CO is located in the heme pocket, and *S* to the state where the ligand is outside of the protein. For rebinding, the ligand has to overcome two successive barriers. The enthalpy barrier at the heme iron is denoted with H.

temperature range with a single experimental marker. However, we recently measured the dynamics of photosynthetic reaction centers from purple bacteria after charge separation and recombination, using a single marker to cover the entire temperature range between 5 and 300 K (McMahon et al. 1998). In this work, four different relaxation processes could be distinguished and assigned to four tiers of substates. An Arrhenius parameterization of the temperature dependence of the relaxations revealed well-separated distributions of (apparent) energy barriers in the range between 10 and 110 kJ/mol. On time scales of computer simulations ($t \leq 1$ ns), only motions in the lowest tier of substates were found in this study, consistent with the observation of a continuum of substates in computer simulations. Even within the lower tiers of substates, a hierarchical arrangement of configurations was recently observed in molecular dynamics simulations of crambin (Garcia et al. 1997) and myoglobin (Andrews et al. 1998).

While the energy landscape of MbCO has been studied in considerable detail, our knowledge of the (physiologically more relevant) O_2-ligated form is less complete, mainly because of lack of excellent spectroscopic markers such as the strong infrared (IR) lines of CO. There is, however, evidence from a number of experimental observations that an analogous arrangement of taxonomic and statistical substates exists also in MbO_2.

Taxonomic Substates

The IR spectrum of CO-ligated heme proteins shows a few discrete stretch bands of the bound CO ligand. In MbCO, they have been denoted A_0, A_1, and A_3, with peak wavenumbers $\nu(A_0) \approx 1966\ \mathrm{cm}^{-1}$, $\nu(A_1) \approx 1945\ \mathrm{cm}^{-1}$, $\nu(A_3) \approx 1933\ \mathrm{cm}^{-1}$. The different IR bands arise from the presence of different structural environments that affect the bond order of the heme-bound CO ligand. These bands have been assigned to conformational substates of tier 0 of the substate hierarchy (CS0), and because they can be characterized individually, they are called taxonomic substates (Frauenfelder et al. 1991). Their relative populations depend on environmental variables such as temperature, pressure, pH, and solvent composition. The CO-binding kinetics after photodissociation at cryogenic temperatures are markedly different when monitored in the three A bands (Figure 4.2a).

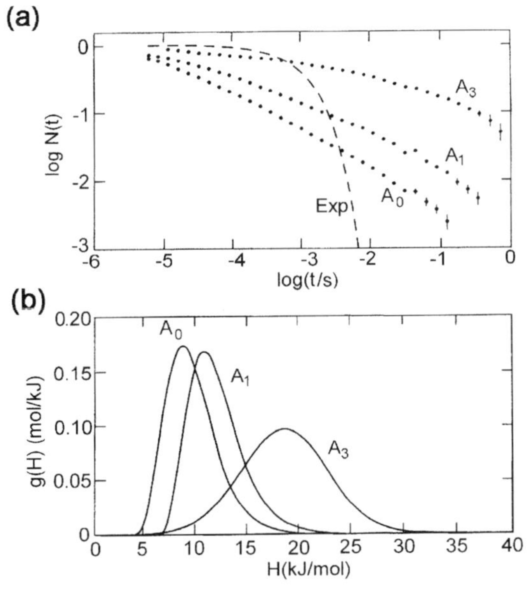

Figure 4.2 Flash photolysis-rebinding kinetics of MbCO (20 mM in 75% glycerol/buffer, pH 5.8). (a) Dots: kinetics at 120 K, monitored in the three CO stretch bands associated with the A_0, A_1, and A_3 taxonomic substates. $N(t)$ denotes the fraction of protein molecules that have not rebound the ligand at time t after photodissociation. Dashed line: exponential. (b) Barrier distributions, $g(H)$, of the individual A substates. They are related to $N(t)$ by Equation 4.5.

Structure

Studies of the pH dependence of the A substate bands have given clues about the structural determinants that cause the deep furcations of the energy landscape into separate A substate basins (Müller et al. 1999). Figure 4.3 shows a pronounced spectral change with pH. At high pH, A_1 and A_3 predominate. With decreasing pH, the population of A_0 grows at the expense of the other two A states. A second, more subtle effect is an upshift of the individual peak wavenumbers by 2–3 cm^{-1} with decreasing pH.

The pH dependence implies that protonations are responsible for the spectral changes. Uptake or release of a proton constitutes a significant energetic perturbation that can also affect the CO stretch vibration in heme proteins through different mechanisms: electrostatic interactions or hydrogen bonding between the CO and the protonated residue, polarization of the extended π-electron system of the heme group, or by alteration of the arrangement of amino acids near the CO. Indeed, protonation changes of amino acid side chains are often key steps in the functional action of enzymes.

Among the amino acids, histidine plays a special role, as its imidazole side-chain titrates near neutral pH. In myoglobin, there are three histidines near the active site, His64, His93 and His97, as shown in Figure 4.4. His93, the proximal histidine, provides the covalent link of the heme group to the polypeptide chain and is known to be unprotonated in the native pH range. His64, the distal histidine, is responsible for most of the dispersion of the stretch frequencies in the A substates, as shown by experiments with site-specific mutants of MbCO where the imidazole side chain of His64 was replaced by other residues (Braunstein et al. 1993; Li et al. 1994). The third histidine, His97, is located on the proximal side of the heme prosthetic group. By comparing the pH dependence of the A substate spectra of the site-specific mutants His64Val and His97Phe with that of the native protein, we were able to show unambiguously that His64 is responsible for the marked changes in the population of the peaks, whereas the subtle shifts of the A bands are due to protonation of His97.

The pH dependence of the IR spectra in Figure 4.3a can be modeled accurately as a linear superposition of IR spectra of four substates, denoted A_{PP}, A_{UP}, A_{PU}, and A_{UU}, where the first and second subscript represent, respectively, the protonation states of His64 and His97. Our model, shown schematically in Figure 4.3b, assumes that the four substates are connected by two energetically uncoupled protonation equilibria following simple Henderson–Hasselbalch relations, with pK values of 4.4 for His64 and 6.0 for His97. At low pH, A_{PP}, with both histidines protonated, predominates, whereas at high pH, the entire population exists in A_{UU}, where both histidines are unprotonated. In the intermediate range, the mixed states A_{UP} and A_{PU} are present but A_{UP}, with protonated His97, is much more abundant. A_{UP} and A_{PU} occur in a fixed ratio at all pH values so that they cannot be obtained independently from pH variation. However, since A_{PU} contributes only 3% to the mixed species, the intermediate spectrum is essentially that of A_{UP}.

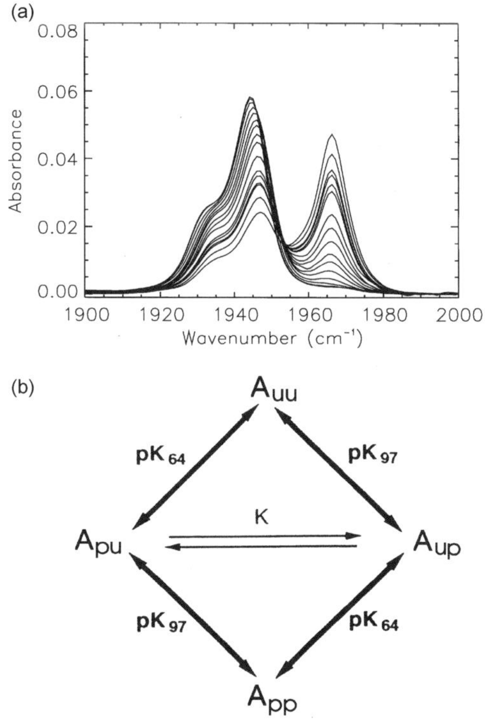

Figure 4.3 Room temperature infrared spectra of CO bound to sperm whale myoglobin in aqueous buffer solution, measured at pH values between 4 and 9. (a) Measured spectra, normalized to 1 OD cm^{-1}; (b) scheme including four substates that are coupled by independent protonations of His64 and His97, which successfully describes the pH dependence.

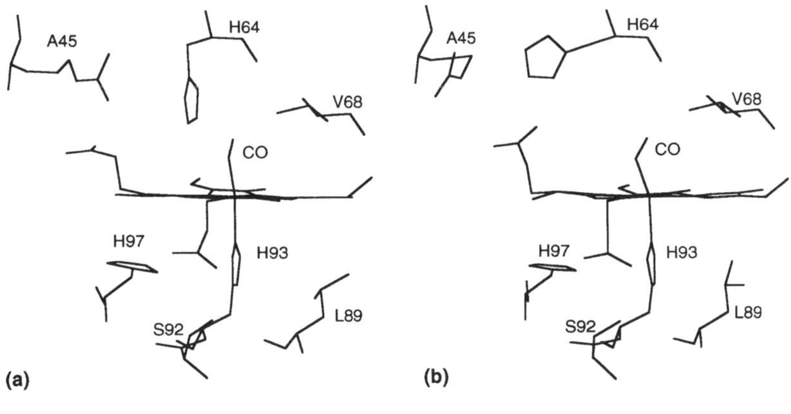

Figure 4.4 Structure of MbCO around the active site at two different pH values: (a) pH 6 and (b) pH 4 (Yang and Phillips 1996). Note especially the change in conformation of the His64 imidazole side chain in response to protonation.

The approach taken here extends earlier work on the A substates (Ansari et al. 1987). Previously, the taxonomic substates were characterized according to the positions of the three main CO stretch bands, A_0, A_1, and A_3. In the present analysis, we used the pH dependence of the IR spectra to separate four substates, A_{PP}, A_{UP}, A_{PU}, and A_{UU}, defined through the protonation states of two histidine residues, 64 and 97. The entire CO stretch spectrum is affected by the protonation, so the different species are characterized by spectra and not individual IR bands. The dominant absorption of the doubly protonated species, A_{PP}, occurs at 1966 cm^{-1}, and so A_0 is a good (though not perfect) marker for this species, as there are also contributions of the A_{PP} species at lower wavenumbers, even at the A_1 position. If His64 is deprotonated, the CO absorbs either in A_1 or A_3. The influence of His97 on the IR spectra creates pairs of adjacent substate bands, A_0 and A_0^+, A_1 and A_1^+, and A_3 and A_3^+. The plus sign indicates that His97 is protonated, and thus the CO absorbs at 2–3 cm^{-1} higher than if it is in the unprotonated form.

Protonation is, however, not the only mechanism creating taxonomic substates, as is evident from the behavior of A_1 and A_3. The population ratio of the two substates is independent of pH, and thus they occur as one species in our four-state protonation scheme and cannot be related by a protonation. A_1 and A_3 could be ring-flip isomers of the His64 imidazole side chain or tautomers, with the imidazole proton either on N^{δ_1} or N^{ε_2}. Moreover, specific interactions of His64 with adjacent residues could also be responsible for the occurrence of A_1 and A_3.

X-ray diffraction experiments on MbCO crystals at different pH values have given direct evidence of these structural alterations that accompany the protonation of His64. As expected from IR spectroscopy on mutant myoglobins, the X-ray structure of MbCO at pH 4, where A_0 dominates, has the His64 imidazole side chain swung out of the distal pocket, as shown in Figure 4.4 (Yang and Phillips 1996).

Dynamics

Interconversion between the taxonomic substates of MbCO has been measured over almost ten orders of magnitude in time by relaxation methods: An external perturbation applied to the sample creates a nonequilibrium situation, and the subsequent relaxation towards equilibrium is followed spectroscopically by monitoring an observable, $M(t,T)$, for example, the area of an A substate band or its peak absorption as a function of time t and temperature T. Two different techniques, pressure relaxation (for times scales longer than 100 s) and flash photolysis kinetics (for faster times scales) were used to measure the dynamics (Johnson et al. 1996).

At temperatures sufficiently high for quick equilibration, hydrostatic pressure was applied to the samples, thus changing the relative A state populations because of their different volumes. Subsequently, the sample was cooled to temperatures where the response was sufficiently slow so that the relaxation after a sudden release of pressure could be measured by FTIR spectroscopy. The change in the area of the A substate bands was parameterized by a relaxation function $\Phi(t,T)$, defined by:

$$\Phi(t,T) = \frac{M(t,T) - M(\infty,T)}{M(0,T) - M(\infty,T)}, \tag{4.1}$$

where $M(0,T)$ is the value of $M(t,T)$ immediately after the perturbation, and $M(\infty,T)$ is the value of the observable when the new equilibrium is reached. Conformational motions in a rough energy landscape generally exhibit dispersive kinetics, and the time dependence of the relaxation function in proteins is often well described by a stretched exponential,

$$\Phi(t,T) = \exp\left[-\left(\kappa(T)t\right)^{\beta}\right], \tag{4.2}$$

where $0 < \beta < 1$. In glassy systems, the exponent β is frequently found in the range 0.3–0.5, which means that the relaxation function extends over several orders of magnitude in time. For the A substate exchange, however, the relaxation functions were close to exponential, as expected for transitions over barriers between basins. Consequently, the apparent rate coefficients were determined from fits of exponentials to the relaxation functions.

A substate interconversions were measured on faster time scales using flash photolysis with IR monitoring at the peak wavenumbers of the A substates. Below ≈ 150 K (in glycerol–water solvents), transitions between the A substates do not occur on any experimental time scale, and nonexponential rebinding is observed, as seen in Figure 4.2a for $T = 120$ K. The significantly different rebinding kinetics in the three A substates create a nonequilibrium population of A substates at times small compared with the exchange rates. At times comparable with the inverse exchange rates, substate interconversions are present that reestablish equilibrium.

In the A_0 substate band, this kinetic generation of a nonequilibrium situation gives rise to quite a spectacular kinetic behavior, as shown in Figure 4.5. After a rebinding phase, where the negative slope in the kinetics represents increase in the bound species, the kinetics turn up again on an intermediate time scale before they finally decay because of loss of A_0 population to A_1 and A_3 (in this process, A_1 and A_3 appear as a single kinetic species because they interconvert three orders of magnitude faster). From the kinetics, average rate coefficients, κ, governing the A substate interconversions, can be evaluated.

Figure 4.6 shows the temperature dependence of κ, plotted logarithmically as a function of $(1000/T)^2$. This plot implies that the Ferry relation,

$$\kappa(T) = A_F \exp\left[-\left(\frac{E_F}{RT}\right)^2\right], \tag{4.3}$$

and not the usual Arrhenius relation,

$$\kappa(T) = A\frac{T}{T_0}\exp\left[-\frac{E_F}{RT}\right], \tag{4.4}$$

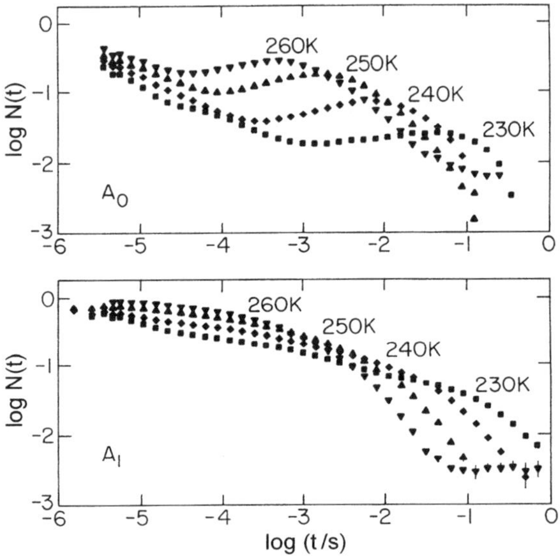

Figure 4.5 Flash photolysis kinetics of MbCO (20 mM in 75% glycerol/buffer, pH 5.8) monitored separately in the A_0 and A_1 substates; squares: 230 K; diamonds: 240 K; triangles (tip up): 250 K; triangles (tip down) 260 K.

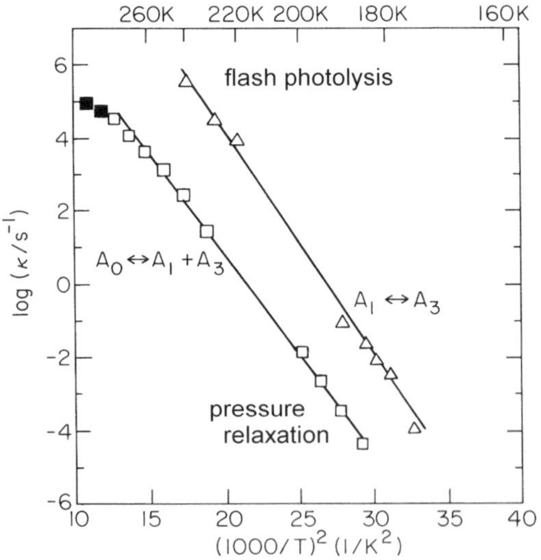

Figure 4.6 Logarithmic plot of the characteristic rate coefficients for the transitions $A_1 \leftrightarrow A_3$ and $A_0 \leftrightarrow A_1 + A_3$ as a function of $(1000/T)^2$, showing data from flash photolysis and pressure release experiments on MbCO (20 mM 75% glycerol/buffer, pH 5.8). The straight lines are best fits of the Ferry law, Equation 4.3, to the data.

is appropriate for describing the temperature activation of these protein motions. Whereas the Ferry relation was introduced originally as an empirical relation to model the temperature dependence of the viscosity of polymer melts, it can be derived theoretically for diffusion in random potentials with a Gaussian distribution of barriers (Zwanzig 1988; Bryngelson and Wolynes 1989). Therefore, application of this relation implies that the interconversion between the A substates involves many nonlocal degrees of freedom that give rise to a rough potential along the reaction coordinate superimposed on the smooth, bistable potential that separates the two energy basins. The temperature dependence of the interconversion rates is governed by the roughness of the energy landscape, and it is not possible to separate the barrier of the smooth part. On the other hand, use of the Arrhenius relation, Equation 4.4, assumes that classical transition state theory can be applied, which describes the reaction as a barrier-crossing event on a smooth reaction surface.

Clearly, the temperature interval covered by the data in Figure 4.6 is insufficient to distinguish between the two alternative temperature dependencies given by Equations 4.3 and 4.4. However, a fit with Equation 4.4 yields an activation enthalpy of 90 kJ/mol and a preexponential of $10^{21.6}$ s^{-1} for the $A_0 \leftrightarrow A_1 + A_3$ transition, and 118 kJ/mol and $10^{30.2}$ s^{-1} for $A_1 \leftrightarrow A_3$. The unrealistically large activation enthalpies and preexponentials (the latter being related to activation entropies) suggest that the Arrhenius relation is inappropriate to model these substate interconversions; they are complex phenomena and not single barrier crossings of the Arrhenius type. With Equation 4.3, we obtain $E_F = 9.3$ kJ/mol and $A_F = 10^{11.8}$ s^{-1} for the $A_0 \leftrightarrow A_1 + A_3$ transition, and $E_F = 9.9$ kJ/mol and $A_F = 10^{16.4}$ s^{-1} for $A_1 \leftrightarrow A_3$. A roughness of about 10 kJ/mol for the energy landscape, as indicated by the parameter E_F, appears quite realistic for the many noncovalent interactions within the protein and with the solvent. If this roughness scale is a universal property of proteins, large-scale motions in other proteins should exhibit a similar behavior. Indeed, in reaction centers, we recently obtained $E_F \approx 8$–10 kJ/mol for protein relaxations observed upon light-induced charge separation (McMahon et al. 1998). Moreover, Halle and collaborators have studied the exchange rate of buried water molecules in bovine pancreatic trypsin inhibitor (BPTI) proteins with nuclear magnetic relaxation dispersion (NMRD) between 4° and 80°C (Denisov et al. 1996). From a Ferry parameterization, they obtained a roughness of about 10 kJ/mol, in agreement with our results. Denisov et al. pointed out that ring flip and disulfide bridge isomerization in BPTI, measured earlier with NMR, also yield $E_F \approx 10$ kJ/mol, whereas an Arrhenius parameterization gave barriers as high as 140 kJ/mol. We conclude that, while apparent Arrhenius energies may allow one to classify different types of motions, or tiers of substates (Ansari et al. 1985), they should not be confused with the (potential) energy barriers.

The very different time scales of the two interconversion processes in Figure 4.6 allow further speculation about the nature of the structural transitions that accompany these motions. The $A_0 \leftrightarrow A_1 + A_3$ transition involves cooperative motions of the protein and the surrounding solvent that enable the imidazole side chain to swing out of the heme pocket. Therefore, it is not surprising that this motion is comparatively slow and strongly dependent on the solvent viscosity (Young et al. 1991). Interconversion of $A_1 \leftrightarrow A_3$ is more than three orders of magnitude faster than $A_0 \leftrightarrow A_1 + A_3$; extrapolation to room temperature yields about 1 ns. The much faster interconversion suggests that less structural rearrangement is necessary than in the $A_0 \leftrightarrow A_1 + A_3$ process.

Statistical Substates

We now turn to the lower-tier substates, sketched crudely in Figure 4.1. At low temperatures, ligand-binding kinetics within each of the taxonomic substates show a smooth, nonexponential time dependence (see Figure 4.2a). Over a wide range of temperatures ($\approx$ 50–150 K), the kinetics can be modeled with a single, approximately Gaussian distribution of enthalpy barriers, $g(H)$,

$$\int g(H) \exp\left[-kt\right] dH \, , \tag{4.5}$$

where the rate coefficient k is given by the Arrhenius equation (Equation 4.4). The temperature-independent barrier distributions, shown for the individual A substates in Figure 4.2b, arise from the presence of a static, heterogeneous ensemble of proteins in different substates within each of the A substates. Their properties must be described by distributions, and we therefore call them statistical substates. In contrast to the interconversion processes discussed in the previous section, at low temperatures each protein molecule is frozen in a particular substate, and consequently ligand binding is likely a simple barrier-crossing event and thus appropriately modeled by the Arrhenius law. The nearly Gaussian shape of the distribution of enthalpy barriers suggests that this heterogeneity arises from a large number of independent degrees of freedom that are mapped linearly to the rebinding enthalpy. Gaussian distributions of rebinding barriers have also been observed in ligand-binding reactions in other proteins (Ehrenstein and Nienhaus 1992).

In heme proteins dissolved in glycerol–water mixtures, the analysis with Equation 4.5 has normally been applied in the temperature range below 160 K (Steinbach et al. 1991). In recent years, kinetic experiments with higher precision and theoretical approaches suggest that a static, temperature-independent barrier distribution below 160 K is only an approximation. The observed deviations from Equation 4.5 are, however, more subtle and demand further exploration.

A number of experimental techniques (hole burning, photon echos, DSC) have given evidence of even lower tiers of substates. Their involvement in the ligand binding function still needs to be explored and will therefore not be discussed here. Despite years of work by many groups, no agreement on the arrangement and classification of the lower tiers in the substate hierarchy has yet been reached.

DYNAMICS AND FUNCTION

To address the problem of function of Mb, we not only need information about the energy landscape of the ligand-bound species, but also about the energy landscape of the unligated species and the transitions between the two states. There has been quite an impressive activity and a number of insightful and technically challenging experimental studies in recent years to explore these issues.

Relevance of CS in Physiological Ligand Binding

Some information on the energy landscape of the deligated species can be obtained from simple flash photolysis experiments on MbCO at room temperature, where rebinding occurs on the millisecond time scale, long after the photodissociating flash, so that the protein molecules have relaxed to their equilibrium deoxy conformation before binding another ligand (Nienhaus et al. 1997). Binding at physiological temperatures is exponential, implying that not only the lower tiers of substates but even the *A* substates are in fast interconversion. The overall rate coefficient is thus an average over all substates, weighted by their average occupancy,

$$\lambda = \sum_i c_i \lambda_i \ , \tag{4.6}$$

where the c_i are fractional weights and λ_i the association rate coefficients of the substates in the ensemble of deoxy Mb molecules in the sample. From the pH dependence of the association rates at room temperature, we were able to show that λ depends on the relative populations of the taxonomic substates, which in turn are governed by the protonation equilibrium of the imidazole side chains of His64 and His97 (Figure 4.7). While the major influence is due to His64, a small effect of His97 on the association rate is clearly observable. The observed change in the population of taxonomic substates with pH exemplifies how proteins can tune their functional properties in response to external influences.

Reaction Energy Landscape

At room temperature, the kinetics of bimolecular ligand binding can be fit to a simple exponential as a consequence of fast fluctuations (compared with the time scale of ligand binding) among all native substates. The reaction is, however, actually very complicated, as shown by detailed kinetic studies over wide ranges in time and temperature. The basic features of the reaction have been clarified, but many details are still not entirely understood and are presently under investigation. A simplistic kinetic scheme of the reaction surface, *V(rc)*, is shown in Figure 4.1 (lower left); it involves three states: *A*, *B*, and *S*. State *A* refers to the ligand-bound state, say MbCO, where the CO molecule is bound to the heme iron. State *B* refers to the unbound state where the CO is in the heme pocket. In state *S*, the ligand has escaped from the protein into the solvent. Two sequential barriers are involved in the binding process. The inner barrier between *A* and *B* is associated with the final binding step at the heme iron, while the outer one between *B* and *S* may represent the gate between the heme pocket and the solvent (Beece et al. 1980). This simplified, one-dimensional model, while pedagogically quite attractive, is only a coarse description of the ligand-binding process and

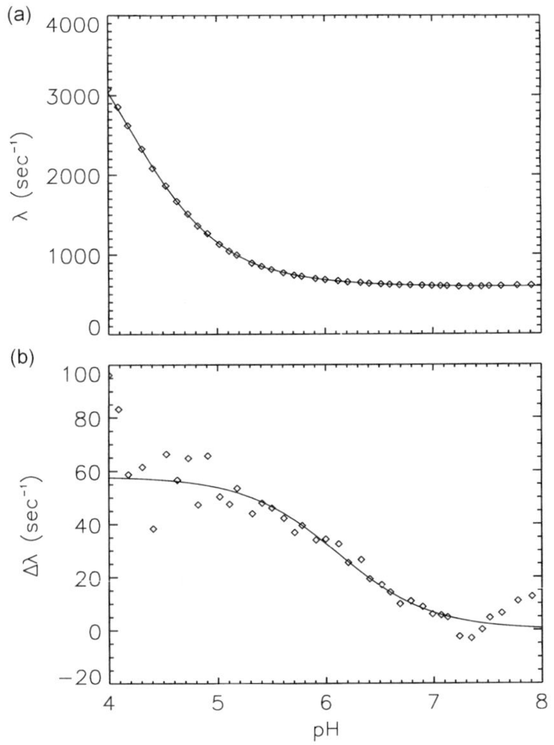

Figure 4.7 Association rates of MbCO after flash photolysis at 296 K; CO pressure = 1 atm. Diamonds: experimental data; solid line: fit. (a) A two-site protonation model (λ / s^{-1} = 592 + 4040 Π(4.2) + 58 Π(6.1), with the Henderson–Hasselbalch function $\Pi(\mathrm{pK}) = (1 + 10^{\mathrm{pK-pH}})^{-1}$. Note that the second term in the sum, with amplitude 4040, originates from protonation of His64, whereas the third term, due to His97, contributes almost negligibly to the overall rate. (b) To show the third term in the equation, in both data and model, the first two terms (592 + 4040 Π(4.2)) were subtracted from the experimental data and the calculated curve in panel (a).

disregards some fundamental properties of heme proteins and, most likely, proteins in general. The crudeness of the model is especially obvious in the physiologically relevant oxygen-binding reaction in heme proteins, as seen in flash photolysis experiments on MbO_2. Even at room temperature, the geminate (internal) rebinding phase is nonexponential out to ≈1 μs and is then terminated by ligand exit to the solvent, as shown, for example, in Figure 16 of Steinbach et al. (1991). The nonexponentiality indicates that the ensemble of proteins is still evolving on the energy landscape ≈ 1 μs after the bond-breaking event occurred. Consequently, species *B* does not correspond to a well-defined chemical intermediate, represented by an ensemble of proteins fluctuating in quasi-equilibrium within a dynamical energy basin. Rather, the reaction occurs as a diffusional process on the energy landscape. This picture, proposed for ligand binding in heme proteins many years ago (Ansari et al. 1985), has become very popular in its application to protein folding (Onuchic et al. 1997; Dobson et al. 1998). On the basis of experimental studies on several protein systems, we believe that simple n-well kinetic schemes will in many instances give only rather coarse descriptions of protein reactions.

Focusing on the Inner Barrier

At sufficiently low temperatures (or high solvent viscosities), protein motions (at least the ones that strongly affect ligand-rebinding barriers) are essentially absent, and the barriers are time and temperature independent. Below 160 K in glycerol–water solvents, the outer barrier, $B \rightarrow S$, is so high that it cannot be overcome on any reasonable time scale, which allows one to focus on the final binding step, $B \rightarrow A$. Under these conditions, where protein motions are strongly suppressed, a small number of discrete CO stretch bands appears, implying different heme pocket structures in the taxonomic substates and a variety of different docking positions of the CO in the heme pocket (Mourant et al. 1993). The main *B* bands are located at 2118, 2131, and 2148 cm^{-1}. Interconversions between *B*2118 and *B*2131 occur even at 10 K via quantum-mechanical tunneling (Alben et al. 1982). The *B*-state lines are very sharp and thus reflect well-defined docking sites and local environments of the CO. Recent room temperature time-resolved IR experiments by Anfinrud and collaborators (Lim et al. 1997) and the first room temperature X-ray structure analysis of the MbCO photoproduct with nanosecond time resolution by Moffat and collaborators (Šrajer et al. 1996) have confirmed the assertion that the *B* states exist transiently at room temperature, providing docking sites for the ligand on the reaction pathway. Moreover, a number of cryo-crystallographic studies have obtained direct structural information on the intermediate states (Schlichting et al. 1994; Teng et al. 1994; Hartmann et al. 1996). Unfortunately, quite pronounced discrepancies exist among the structures, and inconsistencies to spectroscopic experiments still need to be resolved. In X-ray crystallography, detection of small, local changes as well as resolving multiple conformations poses a real challenge, even at highest resolution. Nevertheless, the characterization of suitably prepared photoproduct intermediates with cryo-crystallography has great potential for providing unique information about details of the ligand binding reaction.

The dynamics of ligands in the heme pocket after photodissociation has repeatedly been addressed by molecular dynamics (MD) simulations, but so far, simulations have failed to reveal the sharply defined docking sites evident from IR spectroscopy (Lim et al. 1997). While MD simulations of the MbCO system have been quite popular, to my knowledge, so far only one MD simulation of the barrier-crossing event upon rebinding has been reported (Zheng et al. 1996). An interesting result of this work is the collective nature of the rebinding transition, which involves shifts of several ten atoms near the active site. More work in this direction would be highly desirable.

Fluctuations and Relaxations

At higher temperatures (and especially under physiological conditions), fluctuations among conformational substates occur that render the reaction surface time and temperature dependent. Ligand-binding experiments allow one to separate two dynamic effects: (a) the effective height of the inner barrier increases with time after dissociation of the ligand from the heme iron, and (b) the barrier between the heme pocket and the solvent decreases with increasing temperature.

We consider first the functionally important dynamic process that affects the inner barrier between *B* and *A*. X-ray studies show significant differences in structure between the ligand-bound state *A* (MbCO) and the deoxy state *S* (Mb). For example, the iron in the ligand-bound state, *A*, is nearly in the plane of the heme group while it is significantly out-of-plane (by about 0.3 Å) in deoxy Mb. Agmon and Hopfield (1983) were the first to suggest a protein relaxation to the equilibrium Mb structure following ligand dissociation, and a number of kinetic and spectroscopic studies have given evidence of the relaxation. Indeed, the resultant slowing of the rebinding rate is crucial for allowing efficient escape of the ligand from the protein. While the relaxation has been described by a nonexponential time dependence (Steinbach et al. 1991; Petrich et al. 1991; Lim et al. 1993), more recent results suggest that the relaxation involves a sequence of discrete intermediate steps (Nienhaus et al. 1994). The structural features involved in the relaxation are not yet clear. Both changes of the heme iron geometry due to release of proximal strain and involvement of distal pocket residues have been suggested. Moreover, additional docking sites may become available for the CO at higher temperature, from where binding is slowed (Nienhaus et al. 1998). In any case, the presence of kinetic hole burning effects in the visible, near-IR and mid-IR markers indicates that there are crucial structural coordinates governing the enthalpy barrier that still await being assigned (Ormos et al. 1998).

Motions are also crucial for the ligands for their entry to and exit from the heme pocket because the X-ray structures of myoglobin and hemoglobin do not reveal an open path to the binding site at the heme iron. For the reaction surface in Figure 4.1, this fact implies that the outer barrier between *B* and *S* actually represents a fluctuating gate (Beece et al. 1980). These equilibrium fluctuations are strongly affected by solvent viscosity and composition. The ligand-binding kinetics imply that there is not a unique, specifically designed channel through which the ligands gain access to the heme pocket. Rather, rare fluctuations of the protein matrix transiently lead to locally unfolded structures that allow ligands to pass into the heme pocket.

APPLYING ENERGY LANDSCAPE IDEAS TO OTHER PROTEIN SYSTEMS

To correlate structure and function of proteins, the energetics involved in functional processes need to be well understood. In this regard, energy landscapes have been very popular in the area of heme protein research. Other systems, however, have been studied vigorously by many groups for which an application of these ideas may also be beneficial.

In the bR field, the photocycle intermediates, which we would call taxonomic substates, have been studied extremely well with time-resolved spectroscopic methods. The presence of statistical substates has also been shown for that protein. Although the involvement of protein motions has been pointed out (Warshel 1979), many researchers focused on the retinal chromophore to explain light-induced proton pumping and were reluctant to consider photo-induced tertiary structure changes in the photocycle intermediates. In recent years, however, there have been unambiguous experimental results, including structure determinations of photocycle intermediates, showing that the photocycle is accompanied and possibly even propelled by structural changes in both chromophore and protein moiety.

Also in light-energy conversion in the photosynthetic reaction center evidence has been mounting in recent years for time-dependent changes in the energetic parameters governing electron transfer. These changes arise from protein relaxations following charge separation or recombination. Indeed, by studying the dispersive electron transfer kinetics, a first energy landscape exploration of RC was performed (McMahon et al. 1998).

It is apparent that researchers are becoming increasingly aware of the importance of conformational heterogeneity and interconversions among conformational substates for understanding the function of proteins. It appears quite realistic that further exploration of protein energy landscapes will soon lead to a complete understanding of the function of a protein based on the knowledge of structure and dynamics on the atomic scale.

ACKNOWLEDGMENTS

I would like to thank Hans Frauenfelder and Peter Wolynes for their collaboration and many discussions about energy landscapes. This work was supported by grants GM 18051 from the National Institutes of Health and PHY95–13217 from the National Science Foundation.

REFERENCES

Agmon, N., and J.J. Hopfield. 1983. CO binding to heme proteins: A model for barrier height distributions and slow conformational changes. *J. Chem. Phys.* **79**:2042–2053.

Alben, J.O., D. Beece, S.F. Bowne, W. Doster, L. Eisenstein, H. Frauenfelder, D. Good, J.D. McDonald, M.C. Marden, P.P. Moh, L. Reinisch, A.H. Reynolds, E. Shyamsunder, and K.T. Yue. 1982. Infrared spectroscopy of photodissociated carboxymyoglobin at low temperatures. *Proc. Natl. Acad. Sci. USA* **79**:3744–3748.

Andrews, B.K., T. Romo, J.B. Clarage, B.M. Pettitt, and G.N. Phillips, Jr. 1998. Characterizing global substates of myoglobin. *Structure* **6**:587–594.

Ansari, A., J. Berendzen, S.F. Bowne, H. Frauenfelder, I.E.T. Iben, T.B. Sauke, E. Shyamsunder, and R.D. Young. 1985. Protein states and proteinquakes. *Proc. Natl. Acad. Sci. USA* **82**:5000–5004.

Ansari, A., J. Berendzen, D. Braunstein, B.R. Cowen, H. Frauenfelder, M.K. Hong, I.E.T. Iben, J.B. Johnson, P. Ormos, T. B.Sauke, R. Scholl, A. Schulte, P.J. Steinbach, J. Vittitow, and R.D. Young. 1987. Rebinding and relaxation in the myoglobin pocket. *Biophys. Chem.* **26**:337–355.

Beece, D., L. Eisenstein, H. Frauenfelder, D. Good, M.C. Marden, L. Reinisch, A.H. Reynolds, L.B. Sorensen, and K.T. Yue. 1980. Solvent viscosity and protein dynamics. *Biochemistry* **19**:5147–5157.

Braunstein, D.P., K. Chu, K.D. Egeberg, H. Frauenfelder, J.R. Mourant, G.U. Nienhaus, P. Ormos, S.G. Sligar, B.A. Springer, and R.D. Young. 1993. Ligand binding to heme proteins: III. FTIR studies of His-E7 and Val-E11 mutants of carbonmonoxymyoglobin. *Biophys. J.* **65**:2447–2454.

Bryngelson, J.D., and P.G. Wolynes. 1989. Intermediates and barrier crossing in a random energy model (with applications to protein folding). *J. Phys. Chem.* **93**:6902–6915.

Denisov, V.P., J. Peters, H.D. Hörlein, and B. Halle. 1996. Using buried water molecules to explore the energy landscape of proteins. *Nature Struct. Biol.* **3**:505–509.

Dill, K.A., and H.S. Chan. 1997. From Levinthal to pathways to funnels. *Nature Struct. Biol.* **4**:10–19.

Dobson, C.M., A. Salí, and M. Karplus. 1998. Protein folding: A perspective from theory and experiment. *Angew. Chem. Intl. Ed.* **37**:868–893.

Ehrenstein, D., and G.U. Nienhaus. 1992. Conformational substates in azurin. *Proc. Natl. Acad. Sci. USA* **89**:9681–9685.

Frauenfelder, H., and B. McMahon. 1998. Dynamics and function of proteins: The search for general concepts. *Proc. Natl. Acad. Sci. USA* **95**:4795–4797.

Frauenfelder, H., S.G. Sligar, and P.G. Wolynes. 1991. The energy landscapes and motions of proteins. *Science* **254**:1598–1603.

Garcia, A.E., R. Blumenfeld, G. Hummer, and J.A. Krumhansl. 1997. Multi-basin dynamics of a protein in a crystal environment. *Physica D* **197**:225–239.

Hartmann, H., S. Zinser, P. Komninos, R.T. Schneider, G.U. Nienhaus, and F. Parak. 1996. X-ray structure determination of a metastable state of carbonmonoxy myoglobin after photodissociation. *Proc. Natl. Acad. Sci. USA* **93**:7013–7016.

Johnson, J.B., D.C. Lamb, H. Frauenfelder, J.D. Müller, B. McMahon, G.U. Nienhaus, and R.D. Young. 1996. Ligand binding to heme proteins: VI. Interconversion of taxonomic substates in carbonmonoxy myoglobin. *Biophys. J.* **71**:1563–1573.

Li, T., M.L. Quillin, N. Phillips, Jr., and J.S. Olson. 1994. Structural determinants of the stretching frequency of CO bound to myoglobin. *Biochemistry* **33**:1433–1446.

Lim, M., T.A. Jackson, and P.A. Anfinrud. 1993. Nonexponential protein relaxation: Dynamics of conformational change in myoglobin. *Proc. Natl. Acad. Sci. USA* **90**:5801–5804.

Lim, M., T.A. Jackson, and P.A. Anfinrud. 1997. Ultrafast rotation and trapping of carbon monoxide dissociated from myoglobin. *Nature Struct. Biol.* **4**:209–214.

McMahon, B.H., J.D. Müller, C.A. Wraight, and G.U. Nienhaus. 1998. Electron transfer and protein dynamics in the photosynthetic reaction center. *Biophys. J.* **74**:2567–2587.

Mourant, J.R., D.P. Braunstein, K. Chu, H. Frauenfelder, G.U. Nienhaus, P. Ormos, and R.D. Young. 1993. Ligand binding to heme proteins: II. Transitions in the heme pocket of myoglobin. *Biophys. J.* **65**:1496–1507.

Müller, J.D., B.H. McMahon, E.Y.T. Chien, S.G. Sligar, and G.U. Nienhaus. 1999. Connection between the taxonomic substates and protonation of histidines 64 and 97 in carbonmonoxy myoglobin. *Biophys. J.* **77**:1036–1051.

Nienhaus, G.U. 1999. Conformational motions in the rugged energy landscape of proteins. In: AIP Conf. Proc., Third Intl. Symp. on Biological Physics, ed. H. Frauenfelder, G. Hummer, and R. Garcia. New York: Am. Inst. of Physics, in press.

Nienhaus, G.U., K. Chu, and K. Jesse. 1998. Structural heterogeneity and ligand binding in carbonmonoxy myoglobin crystals at cryogenic temperatures. *Biochemistry* **37**:6819–6823.

Nienhaus, G.U., J.R. Mourant, K. Chu, and H. Frauenfelder. 1994. Ligand binding to heme proteins. The effect of light on ligand binding in myoglobin. *Biochemistry* **33**:13,413–13,430.

Nienhaus, G.U., J.D. Müller, B.H. McMahon, and H. Frauenfelder. 1997. Exploring the conformational energy landscape of proteins. *Physica D* **197**:297–311.

Nienhaus, G.U., and R.D. Young. 1996. Protein dynamics. In: Encyclopedia of Applied Physics, ed. G. Trigg, vol. 15, pp. 163–184. New York: VCH Publishers.

Onuchic, J.N., Z. Luthey-Schulten, and P.G. Wolynes. 1997. Theory of protein folding: The energy landscape perspective. *Ann. Rev. Phys. Chem.* **48**:545–600.

Ormos, P., S. Szaraz, A. Cupane, and G.U. Nienhaus. 1998. Structural factors controlling ligand binding to myoglobin: A kinetic hole burning study. *Proc. Natl. Acad. Sci. USA* **95**:6762–6767.

Petrich, J.W., J.-C. Lambry, K. Kuczera, M. Karplus, C. Poyart, and J.L. Martin. 1991. Ligand binding and protein relaxation in heme proteins: A room temperature analysis of NO geminate recombination. *Biochemistry* **30**:3975–3987.

Schlichting, I., J. Berendzen, G.N. Phillips, Jr., and R.M. Sweet. 1994. Crystal structure of photolysed carbonmonoxy-myoglobin. *Nature* **371**:808–812.

Šrajer, V., T. Teng, T. Ursby, C. Pradervand, Z. Ren, S. Adachi, W. Schildkamp, D. Bourgeois, M. Wulff, and K. Moffat. 1996. Photolysis of the carbon monoxide complex of myoglobin: Nanosecond time-resolved crystallography. *Science* **274**:1726–1729.

Steinbach, P.J., A. Ansari, J. Berendzen, D. Braunstein, K. Chu, B.R. Cowen, D. Ehrenstein, H. Frauenfelder, J.B. Johnson, D.C. Lamb, S. Luck, J.R. Mourant, G.U. Nienhaus, P. Ormos, R. Philipp, A. Xie, and R.D. Young 1991. Ligand binding to heme proteins: The connection between dynamics and function. *Biochemistry* **30**:3988–4001.

Teng, T.-Y., V. Srajer, and K. Moffat. 1994. Photolysis-induced structural changes in single crystals of carbonmonoxy myoglobin at 40 K. *Nature Struct. Biol.* **1**:701–705.

Yang, F., and G.N. Phillips, Jr. 1996. Crystal structures of CO-, deoxy- and met-myoglobins at various pH values. *J. Mol. Biol.* **256**:762–774.

Young, R.D., H. Frauenfelder, J.B. Johnson, D.C. Lamb, G.U. Nienhaus, R. Philipp, and R. Scholl. 1991. Time and temperature dependence of large-scale conformational transitions in myoglobin. *Chem. Phys.* **158**:315–327.

Warshel, A. 1979. Conversion of light energy to electrostatic energy in the proton pump of *Halobacterium halobium*. *Photochem. Photobiol.* **30**:285–290.

Zheng, C., V. Makarov, and P.G. Wolynes. 1996. Statistical survey of transition states and conformational substates of the sperm whale myoglobin-CO reaction system. *J. Am. Chem. Soc.* **118**:2818–2824.

Zwanzig, R. 1988. Diffusion in a rough potential. *Proc. Natl. Acad. Sci. USA* **85**:2029–2030.

5

Information Contained in Protein Structures Determined by NMR in Solution or by X-ray Diffraction in Crystals

M. GEYER[1] and I. SCHLICHTING[2]

[1] Max-Planck-Institut für medizinische Forschung, Abt. Biophysik, 69120 Heidelberg, Germany
[2] Max-Planck-Institut für molekulare Physiologie, Abt. Physikalische Biochemie, Otto-Hahn-Str. 11, 44227 Dortmund, Germany

Keywords: X-ray crystallography, NMR spectroscopy, protein structure determination dynamics, Debye–Waller factor

INTRODUCTION

Protein structures have become part of everyman's life by gracing the covers of the major scientific journals. They have become the modern icons for biochemists and structural biologists, providing a pictorial means for relating to these molecular entities.

Three-dimensional structures can be determined to atomic resolution by crystallography and nuclear magnetic resonance (NMR) spectroscopy (see Table 5.1). For signal-to-noise reasons, both methods study ensembles of molecules (typically 10^{14}–10^{15} molecules in the case of X-ray crystallography and 10^{16}–10^{17} molecules in the case of high-resolution solution NMR), averaging over space and time during data collection. Thus, structures determined by these methods represent the average of the ensemble. Although a great deal of information concerning the workings of enzymes has been deduced from their three-dimensional architectures, dynamics is essential to function: conformational changes, induced fit, and allostery are some widely

Simplicity and Complexity in Proteins and Nucleic Acids
Edited by H. Frauenfelder, J. Deisenhofer, and P.G. Wolynes

Table 5.1 Comparison of protein structure determination by X-ray crystallography and NMR spectroscopy.

X-ray Crystallography	NMR Spectroscopy
Principle:	
Scattering of X-rays by electrons, sampled at reciprocal lattice points (Bragg reflections) or overall in reciprocal space (thermal diffuse scattering).	Dipolar interaction of nuclear spins. Distance information between individual nuclei obtained by nuclear Overhauser effect (NOE) measurements. Assignment of nuclei based on their specific Larmor frequencies.
First Protein Structure:	
Myoglobin, John Kendrew, 1960	Proteinase Inhibitor IIA, Kurt Wüthrich, 1985
Sample condition:	
Protein crystals	Protein in solution
10–50 mM protein (10^{14}–10^{15} molecules)	0.5–2 mM protein, 250 µl (10^{16}–10^{17} molecules)
30%–80% solvent	100% solvent
Restrictions given by crystallization conditions derivatized for phase determination (SIR, MIR, MAD).	Low pH (<pH 8); low salt/ionic strength; isotope labeled (^{15}N, ^{13}C, ^{2}H)
Limitations:	
Crystal formation	Long-time stability (>2d) at room temperature
	High sample purity (no unspecific modification)
Molecular weight: any	Currently limited to ~35 kDa; difficulties due to protein aggregation, protein oligomerization, high hydrophobicity, high solvent viscosity
Data Collection:	
Acquisition time: days to weeks (home source) or hours (synchrotron), the single measurement takes ps–µs (Laue method)	Acquisition time: days to weeks; distinction in assignment, *J*-coupling, NOE, relaxation, and exchange experiments
Spatial and temporal average	Spatial and temporal average over the single experiment time (0.2 h to 80 h)
Low temperature is desirable for reduction of thermal noise and radiation damage	High sample stability is requested
X-rays cause radiation damage	No sample injury

recognized examples. Dynamics can be assessed directly by NMR spectroscopy, which is in this respect superior to X-ray diffraction. However, although most X-ray diffraction experiments are insensitive to time, the data do contain information about conformational fluctuations (Rejto and Freer 1996).

Table 5.1 *continued*

X-ray Crystallography	NMR Spectroscopy
Information Content:	
Spatial structure	Interatomic distances
Information about spatial distribution of protein dynamics but not about time scales	Torsion angles (φ, ψ, χ^1)
	Exchange rates ($^1H \rightarrow {}^2H$) indicating H-bonds
	Individual dynamic properties for each residue including the time scale can be determined. Analysis of backbone dynamics by longitudinal and transverse ^{15}N spin relaxation rates (ps/ns), the overall rotational diffusion tensor and the order parameter S^2 can be determined.
	Conformational exchange (μs/ms) can be observed and the temperature dependence of model-free parameters can be used to characterize the energetics of intramolecular motions.
	Side-chain rotamer analysis by *J*-coupling experiments.
Resolution:	
Given by Bragg spacing	Given by number of restraints/residue
	High resolution: 15 NOEs/res.
Limited by crystal (and wavelength)	Limited by resolution (field strength) and flexibility of the protein.
Refinement:	
Quality marker: free R-factor	Relaxation matrix back calculation approach to improve structure quality.

A structure can only be as good as the data that go into the structure determination process and the theory or models that form the base of the latter. Therefore, it is important to discuss how structures are determined before concentrating on the information contained in the structures. Below we will describe the proceedings of both methods and, by emphasizing the strengths and limitations, show that both methods can provide highly complementary information (Wüthrich 1995; Brünger 1997).

X-RAY CRYSTALLOGRAPHY

X-ray crystallography is currently the dominant technique in structural biology. This is partly due to the fact that it is the oldest of the structure determination methods (the foundations were laid at the beginning of this century; the first X-ray structure of a protein was solved about forty years ago by Kendrew et al. 1960) and partly due to the relative ease and low cost of such a structure determination.

The fundamental process of X-ray diffraction is the coherent elastic scattering of photons by the electrons of the crystalline sample (see Table 5.1). The amplitudes of the scattered X-rays are related by a Fourier transformation to the electron density distribution. To perform the reverse process, a Fourier synthesis, phases are needed in addition to the structure factor amplitudes. The phases cannot be measured directly ("phase problem") but have to be determined indirectly by derivatizing the macromolecule with heavy (i.e., high Z) atoms. In the case where there is no resonance between the incident radiation and the natural frequencies of electrons bound in atoms of the sample, this translates into the traditional multiple isomorphous replacement (MIR) method. The accuracy of the obtainable phase information depends to a large extent on the similarity (isomorphism) of the native and derivatized crystals. This limitation does not apply to the anomalous diffraction method, which makes explicit use of the resonance phenomenon that modifies the scattering of X-rays in an energy-dependent fashion. By collecting data at different wavelengths (multiple anomalous diffraction, MAD) native and derivative data can be obtained from one crystal. The most general and straightforward approach (made possible by recombinant DNA technologies in *Escherichia coli*) is to incorporate seleno-methionine instead of methionine (Hendrickson and Ogata 1997).

The underlying assumption in the description of the scattering process is that the sample consists of point atoms arranged rigidly in an ideal lattice. To account for the effects of thermal motion of the atoms on the scattered intensity, Debye and Waller introduced a factor, the Debye–Waller factor (DWF), that assigns a Gaussian distribution to the probability that an atom fluctuates by x around its equilibrium value x_0 (see, e.g., Frauenfelder 1989). The fluctuation is described by a mean-square displacement $<x^2>$ and taken to be harmonic, isotropic, and independent of the ones from neighboring atoms. The intensity missing from the Bragg reflections due to DW-induced damping is not lost but shows up as thermal diffuse scattering (TDS, see below). The DWF $<e^{-i.\mathbf{kx}}>_{\text{harmonic}} = e^{-½<\mathbf{kx}>^2}$ with the scattering vector $\mathbf{k} = \mathbf{k_i} - \mathbf{k_s}$, $\mathbf{k_i}$ incident, $\mathbf{k_s}$ scattered wavevectors) relates to the crystallographic temperature or B-factor by $e^{-2B(\sin 2\theta/\lambda 2)}$ (θ is the scattering angle); hence $<x^2> \sim (3/8\pi^2)$B.

Structures are refined by minimizing an energy function, E, which is usually given by $E = E_{forcefield} + E_{X\text{-}ray}$, where the X-ray term is given by

$$E_{X-ray} = \tfrac{1}{2} W \sum (|F_o(h)| - k|F_c(h)|)^2 \quad (5.1)$$

with F_o, F_c being the observed and calculated structure factors, respectively, k is a scale factor, and W is the weight of the X-ray term relative to the force field term. Refinement is done by fitting observed and calculated structure factors while maintaining restraints imposed by stereochemistry. The coordinate error of the model is estimated based on a statistical description of the expected errors in the model and the diffraction data (Read 1986) or obtained by inverting the least squares matrix. The quality of the refined structure is reflected in the R-factor, the residual error between computed and observed structure factor amplitudes:

$$R = \frac{\sum(|F_o(h)| - k|F_c(h)|)}{\sum(|F_o(h)|)} . \tag{5.2}$$

To prevent overfitting of the data, a fraction (typically 5%–10%) of randomly selected reflections is not included in the refinement but is used instead to monitor convergence. In analogy to the R-factor, the quality of the cross-validation is measured by R_{free} (Brünger 1992).

The variables to be fitted during refinement are the three coordinates of the (nonhydrogen) atoms, their occupancies, and B-factors. Depending on the resolution of the data — and thus the number of observables — the temperature factors are refined for groups of atoms (i.e., residues), or individually or anisotropically for atoms. Anisotropic analysis of atomic disorder — requiring six times as many variables as isotropic analysis and thus limited to very high-resolution data — allow additional questions about the dynamics of the molecule to be addressed within the framework of the harmonic description by the DW model. Correlations in disorder (and, therefore, presumably rigidity) of parts of small molecules can be rigorously elucidated from the anisotropic displacement parameters by analysis of the Δ-matrix (Dunitz et al. 1988), in which calculations are made of differences in the components of anisotropic displacement parameters of two atoms along the vector joining them. Such analyses are common in small-molecule crystallography but not yet in macromolecular crystallography. As a first attempt at considering the wealth of information represented by the anisotropic disorder information, a translation-libration-screw (TLS) model of the rigid groups is fitted to the experimentally derived DWFs (Schomaker and Trueblood 1968). Agreement of the model with the data gives a measure of the applicability of the rigid-body model and, in cases of good agreement, the three resulting tensors describe the character of the protein disorder.

In the absence of high-resolution data that permit refinement of anisotropic B-factors, correlations in atomic motion among groups of atoms in the protein can be analyzed by normal mode analysis and normal mode refinement (Diamond 1990; Kidera and Go 1992). The basic concept is to approximate the conformational energy surface by multidimensional parabola within the range of thermal fluctuations and to express the dynamics of the protein by a superposition of collective harmonic motions, the "normal modes" (Kidera and Go 1992; Go et al. 1983; Brooks and Karplus 1983; Levitt et al. 1983). Since relatively few normal modes have dominant contributions to the atomic fluctuations, normal mode refinement provides an approach that uses only a few parameters. In addition, external effects — such as static disorder (see below), rigid-body (TLS) motion of the entire molecule (Edwards et al. 1990), and overall anisotropy caused by uneven data correction (Sheriff and Hendrickson 1987) — can be distinguished from the "real" internal fluctuations in the protein molecule.

Protein disorder has been divided into dynamic and static contributions (Frauenfelder et al. 1979; Petsko and Ringe 1984; Ringe and Petsko 1986). The latter originates partly from lattice disorder and can be identified by comparing the displacements obtained by a time scale-insensitive method (e.g., X-ray diffraction) with those determined by a time scale-sensitive technique (e.g., Mössbauer spectroscopy, time scale 100 ns; see, e.g., Parak and Reinisch 1986). Static and dynamic disorder can be distinguished by their temperature dependence. However, static disorder can be assigned erroneously to disorder caused by collective motions that are frozen out at certain temperatures. Depending on the resolution and the spatial separation of the states, these effects are often modeled by alternative conformations, each having a fractional occupancy and a DWF. Another, more recent way of looking at the origins of atomic displacement is to divide the motions in unimodal quasi-harmonic motions that affect the thermal B-factors from refinement, and in anharmonic multimodal valley hopping motions. Although the latter are sometimes difficult to distinguish from static disorder (see above), they are rooted in a dynamic phenomenon (Garcia et al. 1997).

The DWF provides an example of how the remark made at the beginning of this text comes to full effect: the information extracted from the data can be only as good as the underlying model. The assumption of solely harmonic, isotropic displacements made by Debye and Waller is not general enough, and other approaches are needed, such as discussed, for example, by Garcia et al. (1997).

Related to the limitations of the DW model for discussing "disorder" is the fact that the R-factors of macromolecular structures only refine to much higher values as expected from the statistical quality of the data. At low resolution this is caused by lack of a good model for the solvent. At higher resolution, it could be due to errors in the force fields (Lattman 1996); however, it could also be caused by fitting only one structure with ideal geometry to the electron density originating from an ensemble of molecules. Indeed, the R-factor decreases significantly in time-averaged refinement, where an ensemble of structures (obtained by molecular dynamics [MD] from an initial starting structure) is used for refinement (Gros et al. 1990; Clarage and Phillips 1994). At high resolution no individual B-factors have to be assigned, so that the spread of the atoms is represented by the spatial distribution of the atoms in the ensemble. At lower resolution, the addition of temperature factors decreases R_{free} (Schiffer et al. 1995).

The major objection usually raised concerning crystallographically determined structures is the effect of the crystal lattice on the structure of the molecule. This is a valid objection that needs to be analyzed carefully. It can be done most easily when structures have been determined from several crystal forms or when the asymmetric unit contains more than one copy of the molecule. In both cases the molecules have different environments, which allow the effect of the crystal lattice on the structure to be estimated. In addition, structures determined by crystallography and NMR are usually rather similar. The reason for the usually limited effect of the crystal lattice on the structure is that protein crystals contain extended networks of solvent-filled channels that make up 30%–80% of the volume of a crystal. As a result of the high solvent contents and the limited number of (mostly weak) interactions that hold the crystal together, the environment of the molecules in the crystal resembles that of a solution — though a very concentrated (10–50 mM) one — which allows for some motion of the molecules. This explains why most enzyme crystals are enzymatically active and why large ligands can diffuse deep into the interior of the macromolecule.

There are very few limitations on the kind of structures that can be determined crystallographically as long as good crystals are obtained. There is no inherent limitation on the size of the molecule or its character (membrane protein vs. soluble protein). Structures of short-lived species (e.g., enzyme substrate complexes, reaction intermediates) can be determined by time-resolved or kinetic X-ray crystallography (Moffat 1989; Hajdu and Andersen 1993) if the kinetics of the system allow it and conditions for rapid reaction initiation in the crystal can be found (Schlichting and Goody 1997). With the revival of the polychromatic Laue method, time has become a "real" variable (with a resolution up to ~150 ps) in crystallographic experiments (Bourgeois et al. 1996). Alternatively to collecting the data of a short-lived complex on a fast time scale, the reaction may be slowed by, for example, low temperature to experimentally convenient time scales.

The information that can be extracted from crystal structures is mainly about the spatial location of atoms; the resolution obtainable is determined by the quality of the crystal lattice and its vibrations. In the case of X-ray diffraction, hydrogen atoms can only be visualized at atomic resolution. Information on the dynamics of the crystallized macromolecule is limited; in addition, the DWFs contain no information about motion, only about the space taken up by the atoms. Since the thermal motion[1] of the molecules in the lattice breaks the translational symmetry of the lattice, TDS is not restricted to the reciprocal lattice points (Bragg reflections) but is spread continuously through reciprocal space (Clarage and Phillips 1997). TDS can be used to analyze correlated motions within or between unit cells; however, very few studies have been carried out so far (Clarage and Phillips 1997). Such investigations provide experimental data that can be used to test the predictions on atomic displacements and their correlations obtained by molecular dynamics and normal-mode simulation. It seems that liquid-like motions (approximate exponential decay of correlations in atomic positions with pair separation) describe the diffuse scattering of lysozyme crystals better than the lowest frequency modes (Clarage et al. 1992) and that normal modes do better than nanosecond dynamic simulations (Faure et al. 1994).

NMR SPECTROSCOPY

Protein structure determination by NMR spectroscopy became feasible with the introduction of pulsed NMR techniques and two-dimensional experiments at the beginning

---

[1] Thermal vibrations can be divided into those that are similar to the ones observed in infrared or Raman spectra (optic modes) and those that are caused by the propagation of ultrasonic waves in the crystal (accoustic modes). The latter result primarily in halos centered on the Bragg reflections, whereas the optical modes result in TDS that is spread continously but nonuniformly through reciprocal space.

of the 1970s (Ernst et al. 1987; Wüthrich 1986). The subsequent development of the method was closely linked to the progress in NMR instrumentation, in particular the introduction of superconducting high-field magnets, computer technology for data accumulation, processing and evaluation, and also protein expression and purification techniques that allow specific spin labeling. Today, NMR is the most powerful spectroscopic technique, allowing observation of static and dynamic aspects of individual atoms. For a recent textbook, see Cavanagh et al. (1996).

In NMR spectroscopy, spatial structural information is obtained from interatomic distances and dihedral angles (see Table 5.1). The interatomic distances can be determined between specific nuclei (mainly ^1H, but separated for all covalently bound nuclei with spin I = ½, e.g., ^{13}C, ^{15}N, ^{19}F, ^{31}P) by nuclear Overhauser effect spectroscopy (NOESY) measurements. This experiment makes use of the dipolar coupling between two spins through space (NOE); the intensity of the correlation I_{NOE} strongly decreases with the distance r between the two nuclei: $I_{NOE} \sim r^{-6}$. On average, NOEs can be detected up to distances of 5 Å. The identification of the nuclei is based on their characteristic Larmor precession frequency in the static magnetic field, which depends on the specific chemical environment for each nucleus. This so-called "chemical shift" has a dispersion of only 10^{-5} to 10^{-4} of the magnetic resonance frequency and therefore requires high-resolution spectroscopy to avoid signal overlap. The nuclei are assigned to their resonance frequencies by correlation spectroscopy techniques. This enables one to observe spin systems, e.g., single residues, by magnetization transfer along the covalent bonds resulting in typical spin system patterns. Usually, the spin system assignment takes most of the time in the structure determination process, once the protein gets into solution.

The determination of torsion angles is based on J-coupling experiments, which have been developed heavily over the last years and are tailored to measure almost every dihedral angle in the sequence. However, often the result calculated from the spin-spin coupling constant in the Karplus equation is ambiguous, showing up to four solutions for one angle. Hydrogen bonds can be detected indirectly by proton amide exchange experiments. Observation of a slow amide proton exchange rate implies that the H^N atom is involved in a hydrogen bond but does not identify the atom acting as a hydrogen-bond acceptor. Once the sequence assignment is done, a reliable secondary structure can be proposed easily from secondary chemical shift analysis of the C^α carbon and proton shift, $^3J_{H^NH^\alpha}$-couplings and the amide exchange rates. Downfield or upfield shift against the random coil values is closely related to the backbone dihedral angle, indicating helical or stretched secondary structure.

All information on interatomic distances, dihedral angles, and hydrogen bonds is comprised in distance and angle restraints for the structure-building process. Here, the exact estimation of the distance from the NOE intensity is the most critical point, since the NOE is also correlated with the effective rotational correlation times, τ_e, which may vary for different locations in the molecule. The apparent distances manifested in the NOEs correspond to a time average that covers the internal molecular flexibility (motion). In the past, NOEs have usually been classified into strong, medium, and weak according to upper distances of 2.3 Å, 3.5 Å, and 5 Å, respectively, with the lower limits given by the van der Waals radii (1.8 Å). Although today most distance restraints are more precisely calibrated (as a result of increased digital resolution and reduced overlap in multi-dimensional spectra), the deviations and errors can be huge, for example, due to baseline distortions, competitive relaxation processes, or internal conformational exchange. Ambiguous angle constraints derived from the *J*-coupling are usually considered by fitting directly to the Karplus equation in the structure refining process. Hydrogen-bond restraints have a large impact on the nature and precision of the resulting structure and are usually only enforced in well-defined regions of regular secondary structure, in which only one possible hydrogen-bond acceptor is consistent with the NOE data.

The most common approaches for generating NMR structures are distance geometry (DG) and restrained molecular dynamics (rMD). Distance geometry determines ensembles of three-dimensional structures by metric matrix algorithm or variable target function approaches using the table of NOE distance constraints (see, e.g., Havel et al. 1983). Since the number of all interatomic distances in the molecule is limited to constraints up to approximately 5 Å and only few of all distances are experimentally determined with a big error in the NOE determination, DG calculations have a high rate of failure. Restrained molecular dynamics uses force fields for the covalent bond length, bond angles and torsion angles, an electrostatic potential field, a Lennard–Jones potential for attractive and repulsive van der Waals terms, and a supplemented square-well pseudo-energy term for the NMR-derived restraints (see, e.g., van Gunsteren 1988). During a simulated heating and cooling annealing cycle, these potentials drive the structure toward an energetically lowest conformation that minimizes the violations of the restraints. In practice, a hybrid method is often chosen to calculate structures: initial structures are generated with DG that have the correct global fold but poor local geometry. These are refined using rMD by which the annealing processes remove many of the local violations of NMR restraints and covalent inconsistencies in the structure topology (Nilges 1996). The convergence of a structure calculation is judged from the residual constraint violations.

The precision of the structure determination is evaluated by comparing groups of conformers that are calculated from the same experimental data but use different starting conditions. The result of a single restraint molecular dynamic calculation represents one structure that is compatible with the NMR restraints, but it cannot tell whether this solution is unique. Thus the calculation is repeated with different starting structures and results in an ensemble of independently determined solution structures. A statistical theory to connect data error to coordinate error has yet to be developed (Brünger 1997). Thus at present, coordinate error can only be assessed empirically by determining the root mean square (RMS) deviation around the mean of an ensemble. The precision to which a structure can be calculated is directly related to the number of experimental restraints used to generate it. Low-resolution structures may be obtained with as few as four restraints per residue, as long as at least some long-range NOEs are included in the data. High-resolution structures may have more than 15 restraints per residue, corresponding to a mean RMS deviation of backbone atoms of less than 0.5 Å for well-defined regions of secondary structure. Thus, many hundreds of NOE cross-peaks must be assigned unambiguously, using ^{15}N- or ^{13}C-labeled protein. The ability to assign prochiral groups stereospecifically (especially $C^{\beta}H_2$ groups and $(CH_3)_2CH$ groups of valine and leucine residues) is critical to obtaining structures of high precision. The topological quality of a structure is ultimately judged by analysis of its bond lengths, bond angles, and dihedral angle distribution (Ramachandran Plot), using validation tools such as PROCHECK (Laskowski et al. 1996) or What_Check (G. Vriend, EMBL Heidelberg).

An approach to improve NMR protein structures that is becoming increasingly popular is the simulation of NOESY spectra. Once the three-dimensional structures have been determined, interproton distances can be measured from the atomic coordinates and the theoretically expected NOE intensities can be calculated from the dipolar relaxation rate matrix. This method is known as complete relaxation matrix analysis (Keepers and James 1984) or NOE back-calculation and provides one of the few ways to assess independently the accuracy of protein structures derived from NMR data. After suitable scaling, the calculated NOE intensities are compared to the experimentally observed NOE intensities to judge how well the structures reproduce the experimental data. The difference between the NOE intensities can be included as a pseudo-energy term for minimization during the restrained molecular dynamics calculation, which is computationally very demanding in the annealing process. The agreement between observed and calculated intensities can be judged as a figure of merit, like the R-factor in crystallography. However, the limitation to determine NOE intensities accurately from the experimental data (due to peak overlap, spectral artifacts, or baseline distortions) and the absence of a dynamic model capable of reproducing all effects of molecular motion known to be present in the protein prevent the back-calculation approach as an accepted precision criteria. A similar method is the complete cross-validation by Brünger et al. (1993), which is based on the partitioning of NOE data into test sets and the cross-validation of statistical quantities for each of the test sets. The failure of this valuation shows that the accurate determination of a distance restraint from NOE intensities is less important than the fact that the NOE can be detected. This means that quantity overrides quality.

Fortunately, NMR spectroscopy is not limited in application to structure determination. Many NMR-derived parameters depend on both: the time-averaged three-dimensional structures and the intramolecular dynamic properties of proteins. Relaxation measurements provide time-dependent correlation functions that can be interpreted in terms of dynamic models and macromolecular dynamics. NMR experiments are very heterogeneous, in terms of investigations of intramolecular dynamics or intermolecular interactions, and must be matched to the time scale of the phenomenon under investigation. An overview on the recently developed NMR methods for measuring ^{2}H, ^{13}C, and ^{15}N spin relaxation is given by Palmer (1997).

To study intermolecular dynamics, usually, the longitudinal relaxation rates, T_1, the transverse relaxation rates, T_2, and the steady state $\{^1H\}–^{15}N$ nuclear Overhauser effect are usually measured. The principal values and orientation of the overall rotational diffusion tensor for a biological macromolecule can be determined from the dependence of ^{15}N spin relaxation rates on the orientations of the backbone N–H bond vectors in a molecular reference frame. The model-free formalism of Lipari and Szabo (1982) requires a natural separation between isotropic or axially symmetric overall rotational diffusion and internal motions that is particularly useful for globular proteins. The generalized order parameter S^2 and the effective internal correlation time, τ_e, characterize the motions of a bond vector on the pico- to nanosecond time scales. An additional contribution to T_2 from motions on the micro- to millisecond time scale that modulate the magnetic environment of a nucleus is recognized as an apparent increase T_{exp}. For a single two-site conformational exchange, the temperature dependence of T_{exp} can be measured and approximate Arrhenius activation barriers for the exchange process can be determined (Mandel et al. 1996; Geyer et al. 1996). The order parameter S^2 characterizes the distribution of termally accessible orientations of the bond vector defined by the polar and azimuthal angles (ψ, ϕ) in a molecular reference frame. The temperature dependence of S^2 defines the characteristic energy scale for librational motions in proteins.

Fast transverse relaxation, chemical shift anisotropy, signal overlap, and resonance peak ambiguities are the main problems that limit the protein size for and the accuracy of NMR structure determination. During the last ten years, the size of proteins investigated has consequently been extended from ~10 kDa, using ^1H homonuclear two-dimensional NMR spectroscopy, to ~35 kDa using ^1H, ^{13}C, and ^{15}N heteronuclear multidimensional NMR spectroscopy combined with fractional ^2H enrichment. This increase was made possible by the development of heteronuclear multidimensional NMR techniques with pulsed magnetic field gradients. However, this extension also seems to be limited since magnetization transfer efficiency and NOE intensities (signal widths) are dependent on relaxation rates and correlation time. Recent developments and future expectations in NMR are summarized by Kay and Gardner (1997). In the remainder of this section, we will highlight the latest investigations in the solution structure determining field.

Recently, new techniques in structural NMR have been developed based on anisotropic interactions (see, e.g., Prestegard 1998). Traditionally, structure determination of biomolecules has been based on NOE measurements, as described above. The new experiments yield structural constraints that are orientational, rather than distance based (Tolman et al. 1995; Reif et al. 1997; Tjandra and Bax 1997; Pervushin et al. 1997). These experiments rely on the measurement of residual dipolar couplings and chemical shift anisotropy. First, the direct measurement of angles between bond vectors became feasible (Reif et al. 1997), which was used to demonstrate the measurement of the peptide backbone angle, ψ. The angles can be determined directly by using the effects of dipole–dipole cross-correlated relaxation of double-quantum and zero-quantum coherence. Another investigation is the solution NMR in a dilute liquid crystalline medium which enables direct measurements of distances and angles (Tjandra and Bax 1997). By dissolving macromolecules in a dilute aqueous nematic discotic liquid-crystalline medium containing widely spaced magnetically oriented particles, the internuclear dipolar coupling no longer averages to zero as a result of rotational diffusion. Dipolar couplings between ^{1}H–^{1}H, ^{1}H–^{13}C, ^{1}H–^{15}N, and ^{13}C–^{13}C pairs in such an oriented macromolecule are readily measured. This approach promises to improve the accuracy of structures determined by NMR and to extend the size limit. Finally, a new NMR experiment, TROSY (transverse relaxation-optimized spectroscopy), has been developed based on constructive use of interference between dipole–dipole coupling and chemical shift anisotropy (Pervushin et al. 1997). The experiment is an approach for suppression of transverse relaxation in multidimensional NMR experiments. When compared with a conventional two-dimensional ^{1}H–^{15}N correlation experiment, reduction of the line width is about 50%, thus promising a further severalfold increase of the molecular size accessible with solution NMR.

Recent efforts have also been made in the biochemical preparation of NMR samples. In addition to labeling techniques with stable isotopes, including fractionally deuterated samples (Kainosho 1997), the segmental isotope labeling using peptide splicing is most remarkable (Yamazaki et al. 1998). Yamazaki et al. demonstrate a new method for the ligation of labeled and unlabeled peptide fragments using the peptide splicing element intein. Inteins are insertion sequences which are cleaved off after translation, ligating the preceding and succeeding fragments and thus leaving a continous peptide sequence (Perler 1998). The fragments can be separately produced in culture media using different isotopes. Segmental labeling is the most efficient method for dividing a large target into parts of manageable size. This method is expected to facilitate the application of NMR to proteins of almost any size, focusing on separately labeled domains. Another biochemical approach is the dual amino acid-selective and site-directed labeling by cell-free synthesis (Yabuki et al. 1998). Yabuki et al. established a method for producing a milligram quantity of site-specifically labeled protein involving amber suppressor tRNA by *in vitro* transcription with RNA polymerase. In the gene encoding the protein, the codon of a single amino acid was changed to an amber codon. The template DNA and 2 mg of the labeled amino acid were reacted in an *E. coli* cell-free system, yielding 2.2 mg purified selectively labeled protein. This technique enables NMR to focus on specific amino acid residues of particular interest. Local structures of large proteins, dynamics, and protein–protein interactions can be spectroscopically analyzed without preceeding assignment of the total protein.

Finally, NMR has the unique ability to characterize with atomic resolution very weak interactions ($K_D < 10$ mM) between macromolecules and ligands. Once the protein spectrum is assigned, the location of binding sites can immediately be determined by chemical shift changes of resonances near the binding site. This capability has recently been utilized in a method called SAR (structure-activity relationships) by NMR (Shukert et al. 1996), making this technique a major tool in drug discovery. The strategy is to identify two or more ligands that bind to adjacent sites on a protein and to characterize their location. A high-affinity ligand with high specifity to the protein can be constructed from the weaker binding compounds by designing a chemical link, since the free energies of binding of the linked ligands are summed. This technology promises to have a major impact on drug discovery (Wagner 1997).

AN EXAMPLE FOR BOTH TECHNIQUES

A good example to study both the similar and complementary features of X-ray crystallography and NMR spectroscopy structure determination techniques is the oncogene protein Ras in its GDP and GppNHp (a slowly hydrolyzing GTP analog) or GTP-bound forms. The structure was determined by crystallography (Milburn et al. 1990; Pai et al. 1990; Schlichting et al. 1990) in both nucleotide forms and later by NMR spectroscopy in its GDP- and GppNHp-bound forms (Kraulis et al. 1994; Ito et al. 1997, respectively). Crucial to its functioning as a signaling molecule (the GTP bound form represents the "on-state," the GDP-bound form the "off-state") are two regions in the structure that are close to the bound nucleotide: switch 1 (effector loop or loop L2, residues 30–38) and switch 2 (loop L4, residues 60–76). They are named "switch" regions because they have different conformations depending on the bound nucleotide.

The functional flexibility of these loops is in line with their mobility. It is reflected in very high B-factors (resulting in poor electron density) and high RMS deviation values of the structure of the Ras•GDP complex determined by crystallography (Scheidig et al. 1995) and NMR spectroscopy (Kraulis et al. 1994), respectively, as shown in Figures 5.1 and 5.2. The crystal structure of the p21•GDP complex has been determined from several crystal forms, one of which contains four molecules in the asymmetric unit, which allows the influence of the crystal lattice on the conformation on the protein to be probed (Milburn et al. 1990). In the triphosphate complex, the switch 1 region is stabilized by a crystal contact resulting in low B-factors (Pai et al. 1990; Schlichting ct al. 1990). Switch 2 has very high B-factors, the weak electron density (resolution 1.35 Å) was modeled with several conformations, one of which was used to explain the catalytic mechanism (Pai et al. 1990). In addition, the enzymatic reaction was followed by a time-resolved crystallographic experiment using Laue diffraction (Schlichting et al. 1990; Scheidig et al. 1995) showing the structural transition from the substrate to the product complex.

The Ras•GDP NMR structure was derived with 3300 distance restraints and 69 torsion angle restraints from 2D, 3D, and 4D ^{13}C- and ^{15}N-separated NOESY spectra, giving an average of 20 restraints per residue (Kraulis et al. 1994). This results in a RMS deviation between the final 40 solution structures and the mean atomic coordinates of 0.78 Å for the backbone heavy atoms and 1.29 Å for all nonhydrogen atoms if all residues (1–166) are included (Figure 5.1). Analysis of the dynamics of the backbone ^{15}N nuclei in the protein shows that residues within the regions 58–66, 107–109

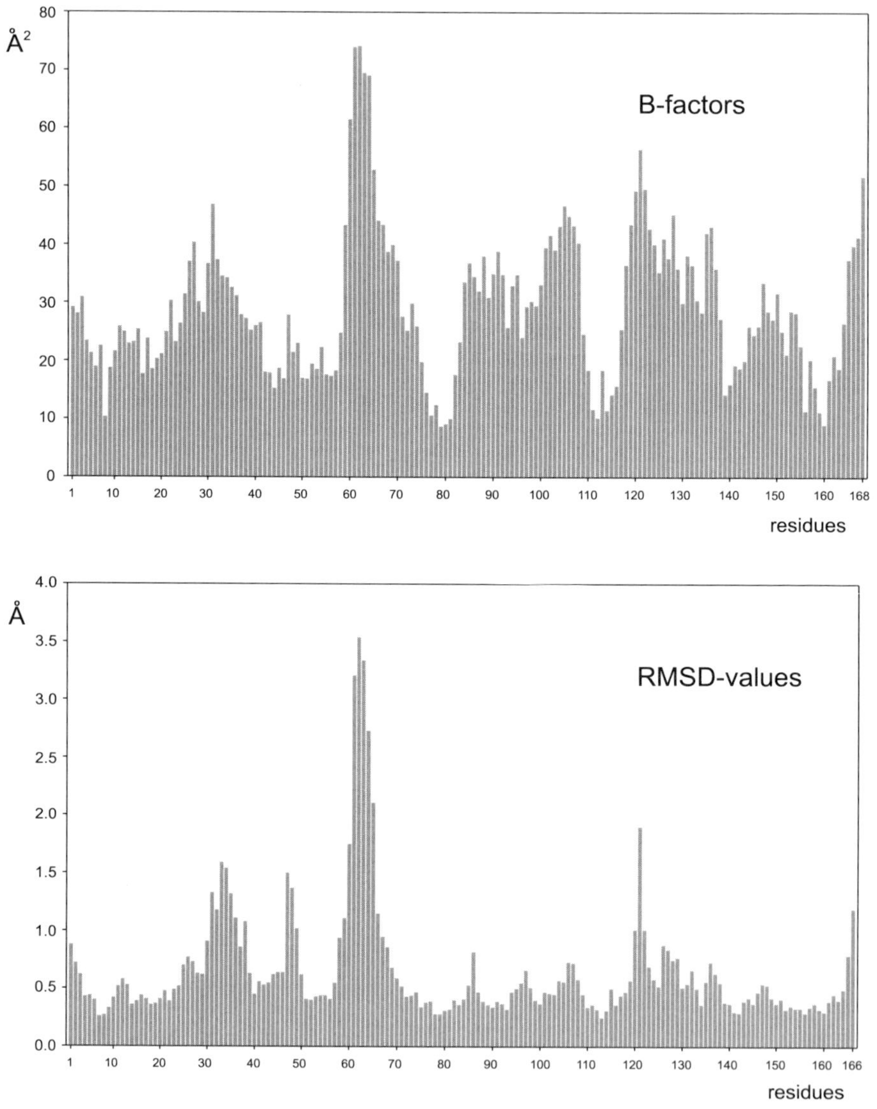

Figure 5.1 B-factors and RMS deviation values of Ras•GDP. The crystallographic B-factors of the C^{α} backbone atoms were taken from the Ras•GDP structure of full-length Ras (1–189) with disordered regions from residue 60–64 and 168–189 (Milburn et al. 1990; PDB accession number: 4q21). The NMR root mean square deviation values of Ras•GDP (1–166) correspond to the C^{α} backbone atoms of the average structure to the final 40 energy minimized structures (Kraulis et al. 1994; PDB accession number: 1crq).

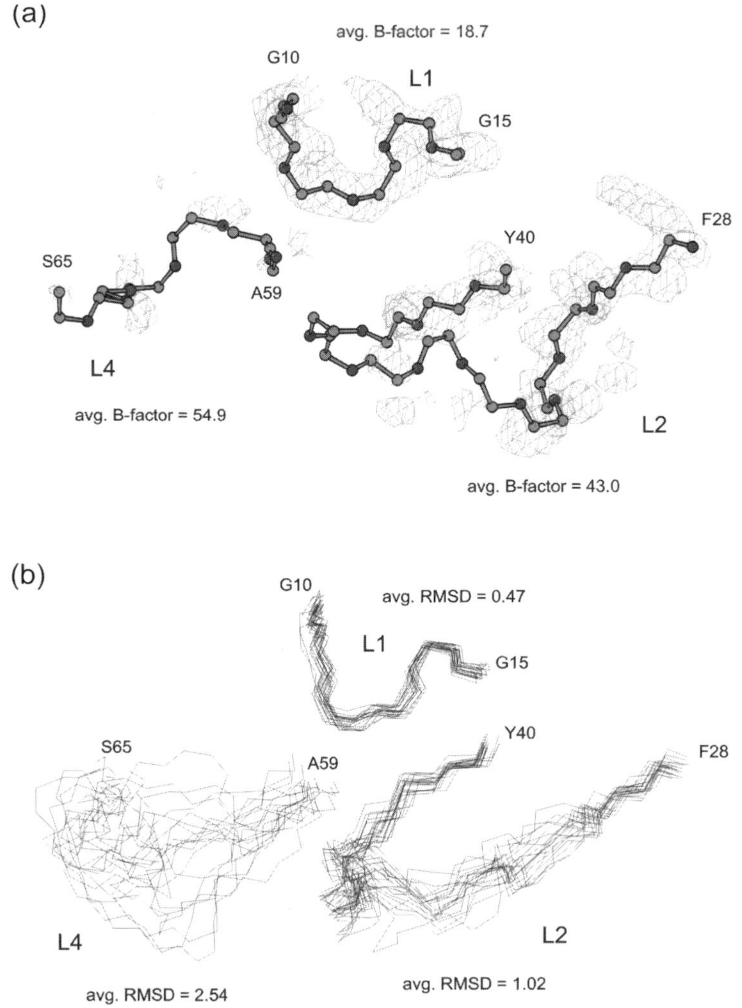

Figure 5.2 Comparison of a structure generated by X-ray diffraction (a), and NMR spectroscopy (b). The functional loop regions L1 (G10–G15), L2 (F28–Y40), and L4 (A59–S65) of Ras•GDP are shown. (a) Electron density map of the crystal structure (Scheidig et al. 1995). The average B-factor for the C^{α} backbone atoms in the respective region is marked. (b) Ensemble of 20 of the final 40 NMR-derived structures (Kraulis et al. 1994). The average RMS deviation value for the C^{α} backbone atoms in the respective region is marked. Structures were plotted using Molscript (Kraulis 1991) and MOLMOL (Koradi et al. 1996).

and, to a lesser degree, 30–38 are dynamically mobile on the nanosecond time scale. This observation corresponds only partially to the RMS deviation analysis of the C^{α} atom that shows the regions 31–38, 47–49, 59–66, and 121 to be less defined than the average (RMS deviation 1 Å) but not 107–109. This shows that RMS deviation values are not necessarily related to highly flexible regions and vice versa since assignment errors can cause undetectable NOE restraint violations, and the force field in the molecular dynamic simulation can pretend stronger fixation than observed, for example, by the electrostatic potential. The structures derived by crystallography and NMR are shown in Figure 5.2.

In the NMR structure of the Ras·GppNHp complex there are some regions that cannot be detected at all, such as residues 10–13, 21, 31–39, 57–64, and 71 (Ito et al. 1997). This phenomenon was named a regional polysterism. These regions are localized in the three functional loops, L1, L2, and L4 of Ras, reflecting the dynamics in the active center of the protein close to the phosphate groups. Complex formation of the effector protein c-Raf-1 with Ras eliminates the regional polysterism nearly completely by reduction of the mobility. The intermolecular dynamics in triphosphate Ras can be described as a two-site conformational exchange process by ^{31}P NMR. The activation enthalpy was calculated from a temperature series to 90 kJ/mol (Geyer et al. 1996). The flexibility of the effector loop can be monitored conveniently by a large shift of a tyrosine side chain covering a trajectory of several Å as observed from crystal structures of uncomplexed Ras and the Ras·Raf complex (Nassar et al. 1995). The stabilization of one conformation of the Ras effector loop upon complexation with effector molecules is reflected by a reduction of the B-factor values (I.R. Vetter, pers. comm.).

The dynamic properties of the flexible loops in the active center of triphosphate-bound Ras still has to be investigated. This information is necessary for the understanding of the intrinsic hydrolysis mechanism, and triggering possibilities of interacting proteins. This data may also complement the crystal structures of Ras·GppNHp and Ras·GDP and the conformational changes in between. The reaction path of the conformational change in Ras was recently simulated with the targeted molecular dynamics method in the presence of a complete solvent shell (Ma and Karplus 1997). The Ras protein was also used for an approach to global fold determination using limited NMR data from larger proteins selectively protonated at specific residue types (Smith et al. 1996). NOE restraints from selectively protonated residues for Ile, Leu, Val, and Phe, Tyr in the completely deuterated 19 kDa protein are already sufficient for the rapid determination of the global fold.

CONCLUSION

In the general excitement of a time when new protein structures are published every week, it is often forgotten that a structure in itself does not tell one how the protein works. For that one needs to know a great deal about mechanism, intermediates, and dynamics. A structure is only a beginning, albeit an important one. Failure to appreciate this distinction has been very expensive for the structure–function field, and many failures of so-called "rationale" drug design are rooted in this outlook. Although the old lock-and-key metaphor of Fischer was overturned by Koshland's discovery of the induced-fit mechanism and by other strong evidence that proteins are flexible objects (and indeed *must* be so to do their jobs), many researchers were surprised at the shortcomings of a design process based on the assumption that active sites of proteins do not respond dynamically to compounds that bind to them. Defining the dynamics-structure-function relationships is one of the most important goals for future research efforts in NMR spectroscopy. It is in the study of dynamic aspects (including the observation of short-lived water molecules, which are often of crucial functional importance (Surridge 1994; Otting 1997)) where NMR is clearly superior to crystallography.

X-ray crystallography has its strengths in the precision of the three-dimensional structures obtainable, but is limited in the information about dynamics. Currently there are about 10 proteins that diffract to about 0.8 Å resolution, allowing application of the sophisticated methods from small molecule crystallography. Multi-harmonic well descriptions (several conformations with anisotropic DWF) seem to provide the best model (T.R. Schneider, pers. comm.). Extraction of the anharmonic contributions of protein dynamics from crystallographic data is difficult since global, local, dynamic, and static disorder is present simultaneously. TDS contains a wealth of information that has hardly been looked at so far. However, mainstream crystallography is most likely to diverge into the three-dimensional analog of the genome project (Pennisi 1998; Shapiro and Lima 1998).

ACKNOWLEDGMENTS

We thank Axel Scheidig for the structure factor amplitudes used for the calculation of the electron density shown in Figure 5.2

REFERENCES

Bourgeois, D., T. Ursby, M. Wulff, C. Pradervand, A. Legrand, W. Schildkamp, and S. Laboure. 1996. Feasibility and realization of single-pulse Laue diffraction on macromolecular crystals at ESRF. *J. Synchrotron Radiat.* **3**:65–74.

Brooks, C.L., III, and M. Karplus. 1983. Harmonic dynamics of proteins: Normal modes and fluctuations in bovine pancreatic trypsin inhibitor. *Proc. Natl. Acad. Sci. USA* **80**:6571–6575.

Brünger, A.T. 1992. The free R value: A novel statistical quantity for assessing the accuracy of crystal structures. *Nature* **355**:472–474.

Brünger, A.T. 1997. X-ray crystallography and NMR reveal complementary views of structure and dynamics. *Nature Struct. Biol.* **4(NMR suppl. I)**:862–865.

Brünger, A.T., G.M. Clore, A.M. Gronenborn, R. Saffrich, and M. Nilges. 1993. Assessing the quality of solution nuclear magnetic resonance structures by complete cross-validation. *Science* **261**:328–331.

Cavanagh, J., W.J. Fairbrother, A.G. Palmer, III, and N.J. Skelton. 1996. Protein NMR Spectroscopy. Principles and Practice. San Diego: Academic.

Clarage, J.B., M.S. Clarage,W.C. Phillips, R.M. Sweet, and D.L. Caspar. 1992. Correlations of atomic movements in lysozyme crystals. *Proteins* **12**:145–157.

Clarage, J.B., and G.N. Phillips, Jr. 1994. Cross-validation tests of time-averaged molecular dynamics refinements for determination of protein structures by X-ray crystallography. *Acta Cryst.* **D50**:24–36.

Clarage, J.B., and G.N. Phillips, Jr. 1997. Analysis of diffuse scattering and relation to molecular motion. *Meth. Enzymol.* **277**:407–432.

Diamond, R. 1990. On the use of normal modes in thermal parameter refinement: Theory and application to the bovine pancreatic trypsin inhibitor. *Acta Cryst.* **A46**:425–435.

Dunnitz, J.D.,V. Shomaker, and K.N. Trueblood. 1988. Interpretation of atomic displacement parameters from diffraction studies of crystals. *J. Phys. Chem.* **92**:856–867.

Edwards, C., S.B. Palmer, P. Emsley, J.R. Helliwell, I.D. Glover, G.W. Harris, and D.S. Moss. 1990. Thermal motion in protein crystals estimated using laser-generated ultrasound and Young's modulus measurements. *Acta Cryst.* **A46**:315–320.

Ernst, R.R., G. Bodenhausen, and A.Wokaun. 1987. Principles of Nuclear Magnetic Resonance in One and Two Dimensions. Oxford: Clarendon.

Faure, P., A. Micu, D. Perahia, J. Doucet, J.C. Smith, and J.P. Benoit. 1994. Correlated intramolecular motions and diffuse X-ray scattering in lysozyme. *Nature Struct. Biol.* **1**:124–128.

Frauenfelder, H. 1989. The Debye–Waller factor: From villain to hero in protein crystallography. *Intl. J. Quant. Chem.* **35**:711–715.

Frauenfelder, H., G.A. Petsko, and D. Tsernoglu. 1979. Temperature-dependent X-ray diffraction as a probe of protein structural dynamics. *Nature* **280**:558–563.

Garcia, A.E., J.A. Krumhansl, and H. Frauenfelder. 1997. Variations on a theme by Debye and Waller: From simple crystals to proteins. *Proteins* **29**:153–160.

Geyer, M., T. Schweins, C. Herrmann, T. Prisner, A.Wittinghofer, and H.R. Kalbitzer. 1996. Conformational transitions in p21^{ras} and in its complexes with the effector protein Raf-RBD and the GTPase activating protein GAP. *Biochemistry* **35**:10,308–10,320.

Go, N., T. Noguti, and T. Nishikawa. 1983. Dynamics of a small globular protein in terms of low-frequency vibrational modes. *Proc. Natl. Acad. Sci. USA* **80**:3693–3700.

Gros, P., W.F. van Gunsteren, and W.G.J. Hol. 1990. Inclusion of thermal motions in crystallographic structures by restrained molecular dynamics. *Science* **249**:1149–1152.

Hajdu, J., and I. Andersson. 1993. Fast crystallography and time-resolved structures. *Ann. Rev. Biophys. Biomol. Struct.* **22**:467–498.

Havel, T.F., I.D. Kuntz, and G.M. Crippen. 1983. Theory and practice of distance geometry. *Bull. Math. Biol.* **45**:665–720.

Hendrickson, W.A., and C.M. Ogata. 1997. Phase determination from multiwavelenth anomalous diffraction measurements. *Meth. Enzymol.* **276**:494–523.

Ito, Y., K. Yamasaki, J. Iwahara, T. Terada, A. Kamiya, M. Shirouzu, Y. Muto, G. Kawai, S. Yokoyama, E.D. Laue, M. Wälchli, T. Shibata, S. Nishimura, and T. Miyazawa. 1997. Regional polysterism in the GTP-bound form of the human c-Ha-Ras protein. *Biochemistry* **36**:9109–9119.

Kainosho, M. 1997. Isotope labelling of macromolecules for structural determinations. *Nature Struct. Biol.* **4(NMR suppl. I)**:858–861.

Kay, L.E., and K.H. Gardner. 1997. Solution NMR spectroscopy beyond 25 kDa. *Curr. Opin. Struct. Biol.* **7**:722–731.

Keepers, J.W., and T. James. 1984. A theoretical study of distance determinations from NMR. Two-dimensional nuclear Overhauser Effect spectra. *J. Magn. Reson.* **57**:404–426.

Kendrew, J.C., R.E. Dickerson, B.E.Strandberg, R.G. Hart, D.R. Davies, D.C. Phillips, and V.C. Shore. 1960. Structure of myoglobin. *Nature* **185**:422–427.

Kidera, A., and N. Go. 1992. Normal mode refinement: Crystallographic refinement of protein dynamic structure I. Theory and test by simulated diffraction data. *J. Mol. Biol.* **225**:457–475.

Koradi, R., M. Billeter, and K. Wüthrich. 1996. MOLMOL: A program for display and analysis of macromolecular structures. *J. Mol. Graphics* **14**:51–55.

Kraulis, P.J. 1991. MOLSCRIPT, a program to produce both detailed and schematic plots of protein structures. *J. Appl. Cryst.* **24**:945–950.

Kraulis, P.J., P.J. Domaille, S.L. Campbell-Burk, T. van Aken, and E.D. Laue. 1994. Solution structure and dynamics of Ras p21·GDP determined by heteronuclear three- and four-dimensional NMR spectroscopy. *Biochemistry* **33**:3515–3531.

Laskowski, R.A., J.A.C. Rullmann, M.W. MacArthur, R. Kaptein, and J.M. Thornton. 1996. AQUA and PROCHECK-NMR: Programs for checking the quality of protein structures solved by NMR. *J. Biomol. NMR* **8**:477–486.

Lattman, E.E. 1996. Why are protein crystallographic R-values so high? *Proteins* **25**:R1–R2.

Levitt, M., C. Sander, and P.S. Stern. 1983. The normal modes of a protein: Native bovine pancreatic trypsin inhibitor. *Intl. J. Quant. Chem. Quant. Biol. Symp.* **10**:181–199.

Lipari, G., and A. Szabo. 1982. Model-free approach to the interpretation of nuclear magnetic resonance relaxation in macromolecules. 1. Theory and range of validity. *J. Am. Chem. Soc.* **104**:4546–4559.

Ma, J., and M. Karplus. 1997. Molecular switch in signal transduction: Reaction paths of the conformational changes in *ras* p21. *Proc. Natl. Acad. Sci. USA* **94**:11,905–11,910.

Mandel, A.M., M. Akke, and A.G. Palmer, III. 1996. Dynamics of ribonuclease H: Temperature dependence of motions on multiple time scales. *Biochemistry* **35**:16,009–16,023.

Milburn, M.V., L. Tong, A.M. de Vos, A. Brünger, Z. Yamaizumi, S. Nishimura, and S.-H. Kim. 1990. Molecular switch for signal transduction: Structural differences between active and inactive forms of protooncogenic Ras proteins. *Science* **247**:939–945.

Moffat, K. 1989. Time-resolved macromolecular crystallography. *Ann. Rev. Biophys. Chem.* **18**:309–332.

Nassar, N., G. Horn, C. Herrmann, A. Scherer, F. McCormick, and A. Wittinghofer. 1995. The 2.2 Å crystal structure of the Ras binding domain of the serine/threonine kinase c-Raf1 in complex with Rap1A and a GTP analogue. *Nature* **375**:554–560.

Nilges, M. 1996. Structure calculation from NMR data. *Curr. Opin. Struct. Biol.* **6**:617–623.

Otting, G. 1997. NMR studies of water bound to biological molecules. *J. Prog. NMR. Spect.* **31**:259–285.

Pai, E.F., U. Krengel, G.A. Petsko, R.S. Goody, W. Kabsch, and A. Wittinghofer. 1990. Refined crystal structure of the triphosphate conformation of H-*ras* p21 at 1.35 Å resolution: Implications for the mechanism of GTP hydrolysis. *EMBO J.* **9**:2351–2359.

Palmer, III, A.G. 1997. Probing molecular motion by NMR. *Curr. Op. Struct. Biol.* **7**:732–737.

Parak, F., and L. Reinisch. 1986. Mössbauer effect in the study of structure dynamics. *Meth. Enzymol.* **131**:568–609.

Pennisi, E. 1998. Taking a structural approach to understanding proteins. *Science* **279**:978–979.

Perler, F.B. 1998. Protein splicing of inteins and hedgehog autoproteolysis: Structure, function, and evolution. *Cell* **92**:1–4.

Pervushin, K., R. Riek, G. Wider, and K. Wüthrich. 1997. Attenuated T_2 relaxation by mutual cancellation of dipole–dipole coupling and chemical shift anisotropy indicates an avenue to NMR structures of very large biological macromolecules. *Proc. Natl. Acad. Sci. USA* **94**:12,366–12,371.

Petsko, G.A., and D. Ringe. 1984. Fluctuations in protein structure from X-ray diffraction. *Ann. Rev. Biophys. Bioeng.* **13**:331–371.

Prestegard, J.H. 1998. New techniques in structural NMR – Anisotropic interactions. *Nature Struct. Biol.* **5(NMR suppl. II)**:517–522.

Read, R.J. 1986. Improved Fourier coefficients for maps using phases from partial structures with errors. *Acta Cryst.* **A42**:140–149.

Reif, B., M. Hennig, and C. Griesinger. 1997. Direct measurement of angles between bond vectors in high-resolution NMR. *Science* **276**:1230–1233.

Rejto, P.A., and S.T. Freer. 1996. Protein conformational substates from X-ray crystallography. *Prog. Biophys. Mol. Biol.* **666**:167–196.

Ringe, D., and G.A. Petsko. 1986. Study of protein dynamics by X-ray diffraction. *Meth. Enzymol.* **131**:389–433.

Scheidig, A.J., S.M. Franken, J.E.T. Corrie, G.P. Reid, A. Wittinghofer, E.F. Pai, and R.S. Goody. 1995. X-ray crystal structure analysis of the catalytic domain of the oncogene product p21$^{H\text{-}ras}$ complexed with caged GTP and Mant dGppNHp. *J. Mol. Biol.* **253**:132–150.

Schiffer, C.A., P. Gros, and W.F. van Gunsteren. 1995. Time-averaging crystallographic refinement: Possibilities and limitations using α-cyclodextrin as a test system. *Acta Cryst.* **D51**:85–92.

Schlichting, I., S.C. Almo, G. Rapp, K. Wilson, K. Petratos, A. Lentfer, A. Wittinghofer, W. Kabsch, E.F. Pai, G.A. Petsko, and R.S. Goody. 1990. Time resolved X-ray crystallographic study of the conformational change in Ha-Ras p21 protein on GTP hydrolysis. *Nature* **345**:309–315.

Schlichting, I., and R.S. Goody. 1997. Triggering methods in crystallographic enzyme kinetics. *Meth. Enzymol.* **277**:467–490.

Schomaker, V., and K.N. Trueblood. 1968. On the rigid-body motion of molecules in crystals. *Acta Cryst.* **B24**:63–76.

Shapiro, L., and C.D. Lima. 1998. The Argonne Structural Genomics Workshop: Lamaze class for the birth of a new science. *Structure* **6**:265–267.

Sheriff, S., and W.A. Hendrickson. 1987. Description of overall anisotropy in diffraction from macromolecular crystals. *Acta Cryst.* **A43**:118–121.

Shukert, S.B., P.J. Hajduk, R.P. Meadows, and S.W. Fesik. 1996. Discovering high-affinity ligands for proteins: SAR by NMR. *Science* **274**:1531–1534.

Smith, B.O., Y. Ito, A. Raine, S. Teichmann, L. Ben-Tovim, D. Nietlispach, R.W. Broadhurst, T. Terada, M. Kelly, H. Oschkinat, T. Shibata, S. Yokoyama, and E.D. Laue. 1996. An approach to global fold determination using limited NMR data from larger proteins selectively protonated at specific residue types. *J. Biomol. NMR* **8**:360–368.

Surridge, C. 1994. Watershed for the structure of cytochrome c. *Nature* **369**:426.

Tjandra, N., and A. Bax. 1997. Direct measurement of distances and angles in biomolecules by NMR in a dilute liquid crystalline medium. *Science* **278**:1111–1114.

Tolman, J.R., J.M. Flanagan, M.A. Kennedy, and J.H. Prestegard. 1995. Nuclear magnetic dipole interactions in field-oriented proteins: Information for structure determination in solution. *Proc. Natl. Acad. Sci. USA* **92**:9279–9283.

van Gunsteren, W.F. 1988. The role of computer techniques in protein engineering. *Protein Eng.* **2**:5–13.

Wagner, G. 1997. An account of NMR in structural biology. *Nature Struct. Biol.* **4(NMR suppl. I)**:841–844.

Wüthrich, K. 1986. NMR of Proteins and Nucleic Acids. New York: Wiley.

Wüthrich, K. 1995. NMR – This other method for protein and nucleic acid structure determination. *Acta Cryst.* **D51**:249–270.

Yabuki, T., T. Kigawa, N. Dohmae, K. Takio, T. Terada, Y. Ito, E.D. Laue, J.A. Cooper, M. Kainosho, and S. Yokoyama. 1998. Dual amino acid-selective and site-directed stable-isotope labeling of the human c-Ha-Ras protein by cell-free synthesis. *J. Biomol. NMR* **11**:295–306.

Yamazaki, T., T. Otomo, N. Oda, Y. Kyogoku, K. Uegaki, N. Ito, Y. Ishino, and H. Nakamura. 1998. Segmental isotope labeling for protein NMR using peptide splicing. *J. Am. Chem. Soc.* **120**:5591–5592.

Standing, left to right:
Andrew McCammon, Ken Dill, Graham Fleming, Klaus Gerwert
Seated, left to right: Hans Frauenfelder, Hans Deisenhofer, Hartmut Michel

6

Group Report: What Do We Need to Know about Proteins and Nucleic Acids?

K.A. DILL, Rapporteur
J. DEISENHOFER, G.R. FLEMING, H. FRAUENFELDER, K. GERWERT, J.A. MCCAMMON, and H. MICHEL

Keywords: experimental and theoretical methods, future needs, structural biology, physical principles

INTRODUCTION

What do we need to know about the structures of proteins and nucleic acids? What do we need to know about their mechanisms of action? What methodological advances in theory/simulation/experiment will lead to progress in our understanding of structure, dynamics, and function? In this chapter we explore these issues.

There are three main reasons for being interested in the structures and properties of biomolecules. First, it is a major intellectual challenge to understand the basic principles of biology, involving some of the most beautiful and complex systems known. Second, this type of research underpins the molecular basis of medicine and biotechnology, and can provide strategies for disease intervention and drug design. Third, it may lead to new materials.

THE UNIQUE PROPERTIES OF BIOPOLYMERS

Living systems are based on two classes of polymers in which information is encoded: proteins and nucleic acids. Nucleic acids are informational polymers; proteins perform chemical and biological functions. The functions and mechanisms of action of proteins are determined by their native structures and fluctuations, which in turn, are

---

Simplicity and Complexity in Proteins and Nucleic Acids
Edited by H. Frauenfelder, J. Deisenhofer, and P.G. Wolynes

encoded in their amino acid sequences. By "sequence" we mean the linear order of monomers covalently linked together in the chain. By "structure" we mean the atomic coordinates of the native molecule. By "function" we mean what the molecule does (hemoglobin carries oxygen); "mechanism" refers to how it does it.

In proteins, conformations are strongly dependent on the monomer sequence. In contrast, it would be disadvantageous for informational polymers, such as DNA, if conformations were highly dependent on the monomer sequence. DNA functions as a repository for information; it must be able to carry any sequence of information with uniform effectiveness, otherwise it will not serve its primary function. The principal structure of DNA is the double helix. The main driving forces are hydrogen bonding and base stacking. One strand complements the other; in this way each strand is protected from sticking to itself.

However, DNA can undergo some sequence-dependent conformational changes when proteins bind DNA and bend it, when proteins open DNA for transcription, and when DNA is packaged with associated proteins into viruses and cells. Sequence dependence results from the mechanical properties of the double helix.

RNA molecules, like DNA, can be single-stranded or double-stranded and can carry information, particularly in the genomes of some classes of viruses. In addition, RNA plays a major role in information processing from DNA to protein and acts as a catalyst in some RNA reactions. Such RNA is usually single-stranded and folds upon itself. Its structure is then determined by its sequence. DNA catalysts are also known. There is relatively little difference between the intrinsic structures and properties of RNA and DNA strands. The folds of RNA, like DNA, are mainly driven by hydrogen bonding and base stacking. Since each base is satisfied by pairing to a single partner base, RNA structures are mainly secondary structures dominated by helices, but sometimes they also have tertiary structures. Functional RNA usually involves tertiary structures.

Proteins can self-assemble with other biomolecules into hierarchical assemblies in which they have coupled functionalities. Specific functions, such as catalysis, require that a biomolecule have specific average structures that are stable over a reasonable time scale (at least milliseconds or seconds), otherwise Brownian processes would dominate at the atomic-size scale. By folding properly, a protein mitigates Brownian randomness. The ability of different amino acid sequences to encode different folds and functions leads to the ability for coordinated and hierarchical assemblies that can perform the sophisticated functions of life. Whereas self-assembly in chemistry refers to micelles, bilayers, and other very simple structures having little or no function, self-assembly in biomolecules refers to higher life forms that can walk and talk and think.

To understand biology, we must ultimately understand these hierarchies, and how individual biomolecules couple to perform high-level functions. "Understanding" means different things to biologists, chemists, and physicists. To achieve quantitative predictive power requires much deeper insights than we currently have. The current state-of-the-art in structural complexity is at the level of viruses and the ribosome. To make the advances we need, all of our theoretical and experimental methodologies must be pushed much further, and new methods are needed.

The future is in two directions. First, it is essential to study at least a few well-characterized systems in greater depth and detail in order to uncover the principles for understanding all systems. Second, it is important to understand systems having larger sizes, greater complexity, and more intricate couplings. Understanding requires moving beyond static structures since dynamics, motions, and distributions of conformations play key roles in function.

STRUCTURE IS THE BASIS FOR PHYSICOCHEMICAL BIOLOGY

The present basis for the current molecular understanding of biology is a set of about 7000 protein, nucleic acid, and membrane protein structures that are known at atomic resolution as a result of mainly two experimental techniques, X-ray crystallography and NMR spectroscopy. Most of these are protein structures that are collected together in the Protein Data Bank (PDB). The PDB is commonly used to draw many types of inferences about the energetic driving forces and the folding code, mechanisms of action, evolutionary relationships, and sometimes folding pathways. How successful are these inferences? If we have more structures, or know structures in better detail, can we improve our understanding of the molecular basis of biology and medicine?

The main organizing principle of the PDB is families or folds, which define how proteins cluster into closely related structures. A "fold" defines a class of conformations; a "family" takes evolutionary relationships into account. The general premise is that sequence similarity implies structural similarity implies functional similarity. Unfortunately, the reverse does not always hold: dissimilarity between two sequences does not necessarily imply dissimilarity between structures. A given globin fold, for example, can be encoded by very different sequences. Moreover, sometimes similar folds can perform different functions.

How many families are there? How are families and folds connected? Identifying and counting families requires a measure of structural similarity between two proteins. There are several such measures, some based on structural criteria alone, and others based also on assumptions that derive from biological and evolutionary relationships. There is some arbitrariness involved in these measures. By some measures, folds or families are relatively tight-knit clustering units, and by others clustering is weak (Yee and Dill 1993). Nevertheless, the process is becoming more automated, for example, by a computer program by Holm and Sander (1993). Increasingly, crystallographers are assigning the fold classes themselves.

Cyrus Chothia estimates that there were about 400 families known in 1995. In contrast, 100 were known in 1989. Families may vary in size. He estimates that most proteins will ultimately come from about 1000 families (see Boxer et al., this volume). The data base may have some biases since (a) crystallography only handles proteins that crystallize, for example, (b) some proteins are easy to purify, and (c) a few proteins are represented in many copies because they are popular targets for mutational studies. Nevertheless, there is some evidence that PDB structures cover "fold space" relatively well (Yee and Dill 1993). Extrapolations based on reasonable assumptions indicate that within the next ten years, most folds will be represented in the PDB. The implication is that a considerable fraction of all protein folds are now known, and a much more complete data base will be available in the foreseeable future. Since the human genome is estimated to encode about 10^5 proteins, about 100 different human proteins will on average be in each family. To "understand" a family will undoubtedly require that at least a few member structures be known in each family.

The atomic structures in the PDB allow informed guesses about a protein's biochemical mechanisms, its structural family, and the nature of its binding sites. The PDB is a powerful tool for threading and multiple sequence alignments as well as homology predictions of a structure based on information about related sequences. These uses of the PDB will benefit from its growth. The PDB currently contains few structures of DNA, RNA, or membrane proteins. New ones would be of considerable value. A bigger data base will teach us rules about conserved residues and evolutionary constraints. Phylogenetic constraints are crucial for predicting RNA structures. When a base change in RNA is accompanied by a change in its base-pairing partner, this is taken as evidence for a nucleic acid helix and is a strong constraint on the possible structure.

For other applications, growth of the PDB appears to be less important. The PDB serves an archival purpose. By analyzing statistical distributions of conformations in the many different protein structures now stored in the PDB, putative energy functions have been constructed that are used in computational protein folding and binding algorithms. The size of the PDB is probably not the main obstacle to improving these energy models. There are two alternative views about what we can learn from the data base of structures. First, it may be similar to the Periodic Table; perhaps uncovering key principles requires a complete set of structures.

An alternative view is that our knowledge becomes asymptotic with the size of the data base, and that there will be diminishing returns. The power of inference from taxonomic information alone may be limited in the absence of a deeper understanding of underlying principles. You cannot understand an organism by anatomy alone; you also need physiology. An example is secondary structure prediction. Chou/Fasman rules (1978) had about 50%–60% success rates in 1972 based on a very small PDB. Current methods have achieved around 70%–75% success with a PDB 100-fold larger. (It has been noted that there is sufficient disagreement over the definitions of helices and strands that 70%–75% may mean that current secondary structure predictions are already adequate [Colloc'h et al. 1993].)

Another 100-fold increase in the PDB (from the order of 10^3 structures today) will probably not advance secondary structure predictions much further. Rather, new insights are necessary. The same applies to statistical contact potentials, such as those of Miyazawa and Jernigan (1985), taken from a small PDB in 1985, and reparameterized from a bigger data base in 1996 (Miyazawa and Jernigan 1996). The energies are not significantly more predictive. (From 7000 proteins 100 amino acids long, we have about $7000 \times 100^2 \times (1/2) = 3.5 \times 10^7$ contacts, from which to fit 210 pairwise contact parameters. It seems unlikely that statistical errors are limiting.)

What are the missing principles? Thermodynamics tells us that the predictor of stable states is the free energy, which is a sum of energy and entropy components. Structural information sometimes reveals energies, but it seldom reveals entropies. In biology and polymer science, entropies are strong driving forces. Visualizing proteins can identify hydrogen bonds, good steric fit, hydrophobic cores, and salt bridges. However, the conformational entropy, a large driving force, is partially hidden from view. Native structures cannot show us all the alternative conformations, and their energies, or the difference in free energy (between native and denatured states) that defines the driving force to fold. To predict driving forces requires more than a native structure alone. Statistical mechanics as well as modeling and simulations are essential. To learn about hydrogen bonding, hydrophobic interactions, and other solvation energies, we also need more classical studies of solution thermodynamics and kinetics. Systematic studies of the mean-square deviations (Debye–Waller factors) may also provide some of the missing information (Frauenfelder et al. 1979; Rejto and Freer 1996; Garcia et al. 1997).

What can we learn from higher-resolution studies? Better structures may help us understand atomic packing within biomolecules. Packing involves subtle detail. Moreover, since small changes in structure can lead to large changes in energies, details may help refine energy functions. An important advantage of higher-resolution structures is that they will allow the development of better models of conformational motion (Debye–Waller) factors. Fluctuations in proteins are usually treated as Gaussian distributions that are uniform in all directions. Small-molecule crystallographers, who work with highly refined structures, model motions as ellipsoids instead. To treat protein motions using more degrees of freedom requires better knowledge of the anisotopies in structures. Debye–Waller factors could lead to a better understanding of the dynamics that are important for mechanisms of action.

For the users of the PDB, it would be of great value if a simple rating of each structure were provided, indicating some measure of the absolute quality of the structure, say with a ranking of 1–5 stars, as hotels are ranked. Various measures, such as resolution, are currently given, but none provides an absolute and overall assessment.

Protein Structure and Folding: The Quest for General Concepts

Some properties of proteins depend on details. Experiments and modeling, however, lead us to believe that there are also universal principles that do not depend on details. This field is young, so the principles are not yet quantitative and predictive, as they are in fields of engineering. Here are a few of the principles that have emerged so far that appear to apply to globular protein structure, dynamics, folding, and function. (The principles for stable globular proteins, which is our main focus here, may be different than for fibrous or multi-chain proteins.)

1. *Energy landscapes can describe folding.* Some proteins fold quickly with simple exponential kinetics, described by funnel-like energy landscapes, and others fold with more complex kinetics, described by bumpy landscapes (Bryngelson et al. 1995; Dill and Chan 1997).
2. *Protein folding is often reversible.* Most simple proteins can be unfolded and will refold to their native states *in vitro*. This implies that the native structure is at the global minimum of free energy, relative to all other conformations that are accessible on the folding time scale (Dill 1990). It implies that chaperones, ribosomes, and other biological helper molecules affect only the speed of folding and not the native structure to which the protein folds.
3. *Collapse and structure formation are mainly driven by hydrophobic forces, steric interactions, and hydrogen bonding.* The dominant component of the folding code appears to be largely the binary pattern of hydrophobic and polar monomer units along the chain (Dill 1990; Dill et al. 1995). Hydrophobic interactions are strong. Within the hydrophobic cores of proteins, hydrogen bonding is a key feature of secondary structures. Ionic interactions are weaker determinants of structure, since charged residues almost invariably are located on protein surfaces, in contact with water, where the dielectric coefficient is high and interactions are weak.
4. *Protein stability appears to be a small difference of large driving forces.* Protein folding involves a sum of many small interactions. The large force driving a protein to fold is opposed by a large conformational entropy that is partially lost when the protein adopts a single native "state," which is itself an ensemble of conformations.
5. *Protein cores are well packed.* The cores in most native folded proteins have tight steric fits (Richards and Lim 1994). Space is well filled inside a protein.
6. *Protein surfaces can be mutated extensively; hydrophobic cores are more sensitive.* Many different sequences will fold to a given native structure. Mutations in the core are more likely to disrupt a native structure than mutations at the surface (Matthews 1993). However, surface mutations can be important for mechanisms of action.

RATIONALIZING MECHANISMS OF ACTION REQUIRES BOTH STRUCTURES AND DYNAMICS

To predict the functions or mechanisms of action of biomolecules from their structures has been a major challenge. The main success has been the DNA Watson–Crick double helix, from which it is immediately clear how information is stored, protected, and duplicated. For proteins, the more modest aim of current practice is to use the known structure to rationalize a known mechanism. Even this goal requires insight, model hypotheses, and additional data. Some proteins have multiple functions, but it is not known whether multiplicity serves a purpose and is widespread or if the additional functions are accidents.

What advances in knowledge of structures would lead to new knowledge of mechanisms? One example is that if protein structures were known at true atomic resolution, we could see protons, and protons are fundamentally important for understanding enzymatic mechanisms.

Yet protein structures alone may not be sufficient to rationalize and understand their mechanisms of action. Many mechanisms, including those involved in chemical catalysis, are dynamic and involve transiently populated conformations that are different from a protein's native structure. The best understood example of the conformational transitions are those associated with the binding of oxygen and carbon monoxide to heme proteins.

How Can We Study Mechanisms?

Biological phenomena normally take place over a very limited range of temperatures and pressures (except for deep-diving animals). To determine crucial parameters that govern biological rate processes, however, experiments over wide ranges in temperature, pressure, and other variables such as viscosity are needed, up to the limits of disruption of the molecular function. Such experiments will provide information about essential barrier properties such as activation enthalpies, entropies, and volumes. Extension of the range to very low temperatures permits detailed observation of the distribution of reaction and relaxation rates by slowing down or freezing the transitions among substates. There have been too few experiments with pressures up to a few kilobars (Frauenfelder et al. 1990). The types of studies sketched here are needed to elucidate reaction mechanisms. Rates and mechanisms can depend on titration state, where protons are located, how they move, fluctuations of the protein, water structure and dynamics at the protein surface, viscosity or pH of the solvent, steric crowding by other proteins in the cell, and other properties, in ways that are not yet well understood.

Are Proteins Homogeneous or Heterogeneous?

Protein folding and functional relaxations are often observed to be nonexponential in time, particularly at low temperatures. This observation can, in principle, be explained by two models. In the *homogeneous* model, every protein molecule in the sample has the same structure and properties as every other molecule. All react or relax with identical nonexponential time dependence. In the *heterogeneous* model, each protein (which has the same primary sequence as every other molecule) can be in a different conformational substate, having a different structure. In the latter model, each protein molecule reacts or relaxes essentially exponentially in time, but with a different rate coefficient. The observed nonexponential time dependence is a property of the ensemble, not the individual protein molecule.

Remarkably, the experimental evidence that is currently available shows that proteins are *heterogeneous* (Austin et al. 1975; Frauenfelder et al. 1991). At ambient temperatures, proteins jump rapidly from substate to substate, often leading to an overall exponential reaction time dependence. Photosynthetic light-harvesting protein is an example of functional heterogeneity. In this case, the relevant property — the electronic energy gap between ground and excited states — is distributed over a wide range for chemically identical chromophores in different molecules. The result of this distribution is that proteins have enhanced spectral coverage and energy funnels toward the reaction center.

Specificity

Specificity refers to how a molecule interacts with one type of substrate differently than with another. Thermodynamic specificity refers to different equilibrium binding affinities. However, specificity can also arise from fluctuations. The following example, which describes the kinetics of the enzyme acetylcholinesterase (AChE), comes from computer simulations from the McCammon lab (Zhou et al. 1998). It illustrates that specificity requires more than knowledge of the native state alone; it also requires knowledge of the internal motions of the native protein.

AChE has a long channel that leads from the surface of the enzyme to the active site. The narrowest region of the channel represents a "gate" that opens and closes due to the thermal fluctuations of the AChE structure. This gate opens less frequently for larger substrates, because such substrates require larger fluctuations to pass through the channel.

At low concentrations of the natural substrate acetylcholine (ACh), the gate is likely to open one or more times during any diffusional encounter with AChE. Thus, ACh has a high probability of entering the active site during an encounter, and it is hydrolyzed at essentially the diffusion-controlled rate that would obtain if the gate were always open. However, because of steric hindrance, the frequency of gate opening decreases very sharply with increasing size of substrates. Larger substrates are more likely to escape by diffusion rather than to enter the active site. As a result, the steady-state reaction rate coefficient in this model decreases by several orders of magnitude with just a modest (10%) increase in the effective radius of the substrate. It seems likely that biology may involve many other examples of dynamic selectivity.

Protein Function: The Quest for General Concepts

Since the number of proteins, and all possible mutations, is extremely large, it is impossible to determine structure, energy landscape, dynamics, and functions for more than a small fraction of the protein universe. As a consequence, it is important to find universal concepts and laws that hold for essentially all proteins and that can be used to predict the properties of newly discovered proteins. At present, only a small number of apparently general concepts are known. Examples are:

1. *Mechanisms can depend on conformational distributions*. Reaction rates can be characterized by distributions. At low temperatures, or at times short compared to the reaction time, the entire distribution is involved. At higher

temperatures and longer times, the transitions among substates collapse the distribution and the reaction rate can be described by a single value. Since function often involves fast and slow steps, distributions can control function effectively.

2. *Proteins are heterogeneous* (Austin et al. 1975; Frauenfelder et al. 1991). A given protein, having a given primary sequence, can assume a very large number of different structures, called conformational substates (CS) with somewhat different properties, for instance, different reaction rates or different volumes.
3. *Mechanisms can be described by energy landscapes* (Frauenfelder 1991). Different CS can be represented as valleys in a very high-dimensional energy (or conformational) landscape (Nienhaus and Young 1996; Nienhaus et al. 1997). The details of the landscape depend on the protein structure, and on factors such as the hydration shell, pH, pressure, and allosteric effector.
4. *Protein substates appear to be hierarchical* (Ansari et al. 1985). CS may be organized in a hierarchy of different tiers, characterized by different average barrier heights. In myoglobin, for instance, barrier heights in different tiers can vary from 1 to 100 kJ/mol. A detailed understanding of the hierarchy is still missing.
5. *Proteins have substate transitions*: *The protein glass transition.* A given protein can jump from CS to CS, changing its structure and properties. The effect of these transitions on dynamics depends strongly on temperature. Well below a certain temperature, known as the protein glass temperature, T_g, transitions among most substates cease and the protein becomes more rigid.

METHODOLOGICAL ADVANCES ARE NEEDED

The ultimate goal of protein science is to predict the structure, energy landscape, dynamics, and function of each biomolecule from its primary sequence. We are far from a solution. The penultimate goal is to determine these properties experimentally; however, even here we are very far away from even a partial solution. For instance, there is, as yet, no general technique that explores the energy landscape and the dynamics. The only problem for which (nearly) general methods of solution are available is protein structure determination. We first discuss structure determination, then other techniques. For the future, it is essential to develop new techniques and close collaboration among the many disciplines involved, including biology, chemistry, and physics.

X-ray Crystallography

One of the main tools of structural biology is X-ray crystallography. Advances in intense X-ray sources, sensitive detectors, multiple anomalous dispersion (MAD) phasing (see below), and cryo-crystallography have led to the high rate at which new crystal structures are appearing. Would stronger X-ray sources (say, 10^3 times stronger than the strongest today) be of use for X-ray crystallography?

The answer depends on the type of crystallographic problem. In diffraction studies with monochromatic X-rays, which require a systematic coverage of reciprocal space, the third-generation synchrotron X-ray sources (ESRF, APS, ALS, Spring-8) have revealed an unexpected problem: the vulnerability of protein crystals to radiation damage even at the temperature of liquid N_2. The reason for this phenomenon is not known at present. It could be due to the heating of the crystal by the intense X-ray beam or to short interruptions of the stream of cold N_2 gas at the crystal position. Whatever the reason, the decay problem sometimes dictates attenuation of the X-ray beam. In such cases, a stronger source would not help. The alternative of using shorter exposure times will require faster beam shutters, different methods to rotate the crystal, and faster methods to record, process, and store the data. When these advances occur, stronger X-ray sources would increase the throughput of synchrotron beamlines. However, there will always be time needed to mount crystals (and to think about the experiment).

The situation is different with the Laue method, where a broad range of X-ray wavelengths is used. Here, a set of diffraction data from a crystal can be obtained with a single exposure without moving the crystal; the completeness of the data depends on the crystal symmetry. The Laue technique has its problems; for example, the inaccessibility of some reflections due to the superposition of harmonics and the need to make simultaneous measurements of intensity from X-rays of different energies. The difficulties do not stem from with the huge amount of data; this is a characteristic of the X-ray method in general, especially in the context of synchrotron data. Rather, problems arise from the use of multiple wavelengths which cause harmonically related reflections to overlap (e.g., the (1,1,0), (2,2,0), (3,3,0), ... reflections). Errors are also introduced through the use of the different wavelengths, for example, different crystal absorption and different detector efficiency. However, the Laue method is the best way to do time-resolved crystallography. At synchrotrons, where X-rays are generated by "bunches" of electrons or positrons accelerated in a storage ring, the time resolution due to the X-ray source is given by the time it takes for a single "bunch" to pass by a window (~100 ps). Data collection could be vastly improved by a much stronger source. The application of time-resolved crystallography would then be limited by other factors.

It is highly desirable that the structures of proteins be determined at true atomic resolution. In the crystallographic methods, the resolution (Bragg spacing) should be 1 Å or lower. This goal has been achieved for very few proteins; crambin and the photoactive yellow protein are among them.

The limits for determining structures at higher resolution are no longer set by the strength of the X-ray beam or the quality of the detector but by the quality of the crystals. An atomic model can be established even if the crystals diffract only to 3.5 Å resolution, since the structures of the building blocks — the amino acids — and the structure of peptide bond are known precisely. However, at 3.5 Å resolution, it is difficult to identify the correct side-chain rotamers and the exact positions of the backbone atoms.

How can we get better crystals? First, we can search for new, better diffracting crystal forms. Much effort should be invested in getting cleaner and more homogeneous protein preparations. Also, there are many parameters to vary, including pH, buffer, precipitation agents, additives, temperature, and speed of reaching supersaturation, just to name the most important ones. If varying these parameters does not work, then the source of the protein, namely the organism, should be varied. It may help to crystallize the protein from a different strain of the same species. The surface residues that are not involved in the function of the protein may have a high degree of variability. It may help to remove them. Proteases like trypsin can be used to remove protruding surface loops and/or termini, and the "shaved" protein might yield well-diffracting crystals. On the other hand, the interaction of surface residues is responsible for the formation of the crystal lattices. It is also a good idea to try enzymes from hyperthermophilic organisms since they seem to have less flexible surface loops, shorter termini, and possibly more rigid structures. Such studies have the added advantage that they can help understand the physical basis for stability and mechanisms. Hyperthermophilic enzymes are often too rigid at room temperature to be enzymatically active.

For the structure determination itself, the use of MAD to obtain the phases has been a major breakthrough. MAD became possible by the availability of tunable X-ray sources at synchrotons. Unfortunately, this method frequently can only be used with frozen crystals, since it is necessary to collect several data sets using the same crystal. Many protein crystals cannot be frozen without an increase in mosaicity and a severe loss of resolution, in which case MAD-phasing cannot be used.

An important technical advance would be the development of X-ray lenses. In combination with coherent sources, these would allow the application of holographic techniques to solve the phase problem and obtain 3D images of macromolecules.

Neutrons for Diffraction Experiments

"Cold" neutrons have de Broglie wavelengths of a few angstroms and are thus suitable for diffraction experiments. However, in contrast to X-rays, neutron sources are weak, expensive, and inconvenient (because they require nuclear reactors or spallation sources). If high-intensity neutron sources were available, macromolecular crystallography would benefit enormously because what neutrons "see" is different than what X-rays see. X-rays interact with electrons and cannot detect hydrogens unless very high-resolution diffraction is available. Neutrons interact with nuclei; in particular they are strongly scattered by hydrogens, which are the most important atoms in proteins in many respects. In addition, neutrons interact very differently with protons and deutrons, making it easy to detect hydrogen–deuterium exchange. In many enzymatic reactions, protons play crucial roles, and the determination of their locations is one of the goals of structural studies. Due to the small number of available neutron sources (approaching zero at times) and their weakness, only very few such studies have been done (McDowell and Kossiakoff 1995; Niimura et al. 1997; Kossiakoff 1985; Korszun 1997). Neutrons would also be suitable for measuring inelastic scattering. In the near future we will have detectors with sufficient time resolution to allow the application of the Laue technique with pulsed neutron sources.

Membrane Proteins

Our knowledge of the structures of membrane proteins is very limited. At the present, the structures of about a dozen sequence- and structure-unrelated membrane proteins are available. Still, based on the numbers of protein sequences that have stretches of hydrophobic amino acids of sufficient lengths to span the membrane as helices, 10%–15% of all prokaryotic proteins and 35% of all eukaryotic "gene products" are membrane proteins. Since β-strand-rich membrane proteins like the porins cannot be identified by inspecting protein sequences, the actual percentage of membrane proteins might be even higher.

There are two reasons for the shortage of membrane protein structures. First, membrane proteins have hydrophobic and hydrophilic surface patches. Detergents must be used to cover the hydrophobic parts of membrane proteins and keep them in solution, a prerequisite for purification and crystallization. However, most membrane proteins are rather unstable in the detergent-solubilized form because detergents replace and mimic the lipid molecules only very poorly. Second, there are no established ways to produce membrane proteins in recombinant form using heterologous expression systems. This limitation is especially important for most receptors, which are only found in tiny amounts in their native membranes.

The current rationale is to try to crystallize membrane proteins within detergent micelles. The choice of the detergent is crucial. It must provide long-term stability of the protein, and it should have a small polar headgroup since a big one would cover much of the protein's polar surface. At first glance, the observation that membrane protein crystallization depends critically on the length of the alkyl chain of the detergent is unexpected. Well-ordered crystals of the two-subunit cytochrome c oxidase from the soil bacterium *Paracoccus denitrificans* have been possible only using undecyl-b-D-maltoside, and not decyl-or dodecyl-b-D-maltoside. The most likely reason is that only an optimal alkyl chain length allows polar interactions between detergents of neighboring membrane proteins.

A very promising — but time-consuming — way to crystallize membrane proteins is to cocrystallize them with conformation-specific Fv-fragments of monoclonal antibodies. The antibodies are thought to fix a certain conformation of the membrane protein and, at the same time, to enlarge the polar surface of the membrane protein.

A novel approach is the use of lipidic bicontinuous cubic phases. In such phases, lipid bilayers form a continuous three-dimensional network. Membrane proteins are thought to be able to diffuse within such a lipid bilayer to form two-dimensional crystals within the plane of the bilayer, and to form three-dimensional crystals through polar interactions between these two-dimensional crystals. It remains to be seen how successful this approach will be. It appears to be the only hope for those membrane proteins that are unstable in the absence of lipid bilayers.

Two membrane protein structures have been determined by electron crystallography using two-dimensional crystals. This promising method is still in an experimental stage and is technically more difficult than X-ray crystallography. Like X-ray crystallography, it requires well-ordered (now two- instead of three-dimensional) crystals, but has the advantage that phases can be obtained directly from electron microscopic images.

In principle, electron microscopy of single molecules can provide high-resolution structural information, at least for large protein complexes. The structures must be averaged using hundreds or thousands of electron microscopic images. Alignment of these images is needed, but precise alignment is only possible with large protein complexes.

For membrane proteins, solution NMR appears to be of little use since the detergent micelles slow down the rotation of the protein molecules, leading to a broadening of the NMR signals. The usefulness of solid state NMR still has to be explored.

NMR Spectroscopy

NMR spectroscopy has emerged as a powerful technique for determining the structures of proteins and nucleic acids. It is highly complementary to diffraction methods because it is most readily applied to molecules in solution, adding to our knowledge base molecules that do not crystallize, and enabling us to study molecules over a wide range of conditions and without the constraints of a crystal lattice. In addition, NMR can probe dynamical events at the level of individual atoms over a very wide range of time scales (from picoseconds upwards) and reveal other properties of macromolecules including the nature of their interactions with solvent.

The determination of macromolecular structure in solution by NMR relies on the resolution of signals from many individual atoms. If this is achieved, information on local conformational properties, particularly internuclear distances (from NOEs) and torsion angles (from coupling coefficients), can be obtained. Structures compatible with such properties can then be computed and refined. It is now possible to determine the structures of small proteins and nucleic acids by NMR at a level of detail approaching that of medium-resolution X-ray crystallography. As the size of the molecule increases, however, it is increasingly difficult to collect enough data to generate well-defined structures. This arises both because of the overlap of the increasing numbers of signals, and because slower molecular tumbling gives rise to altered nuclear relaxation behavior and, notably, to larger intrinsic linewidths.

An important method of enhancing resolution has been to increase the magnetic field strength at which experiments are performed. This has the important additional advantage of increasing signal intensities, improving signal-to-noise ratios, and enabling us to study proteins that are less soluble. Presently, the highest available magnetic fields correspond to proton frequencies of 800 MHz. This represents an increase of a factor of 20 from the earliest spectrometers, but severe practical problems inhibit further gains in this direction. 1000 MHz appears to be the accepted absolute upper limit of present technology. The introduction of new superconducting materials and new approaches to magnet design will be needed for further progress. This is an important long-term objective for future research.

The power of NMR for studying macromolecules has, however, come at least as much from the development of new experimental methods as from increasing magnetic field strengths. This research area has been transformed by the use of ^{13}C and ^{15}N isotopically enriched samples (now readily obtained through the methods of molecular biology), in conjunction with procedures to collect multidimensional spectra. The size limit for complete structure determination of a protein has increased from about 10 kDa to 50 kDa through such procedures. Advances are continuing in the development of methods to increase still further the upper size limit.

Approaches to reducing linewidths, for example, by replacement of protons by deuterons (which have smaller magnetic moments) and by using novel pulse sequences, are in an exciting developmental phase. New methods of extracting structural information, for example, by using magnetic ordering techniques, have tremendous potential for the future. Developments in solid-state NMR are also continuing and provide opportunities for avoiding limitations imposed by molecular tumbling rates. These are likely to be of particular significance in the field of membrane-bound proteins.

One of the exciting features of NMR spectroscopy is that its application is not restricted to highly ordered systems. This allows its application to a wide range of important systems that cannot be studied by diffraction methods, including denatured ensembles of protein molecules and noncrystalline aggregates. This area is ripe for substantial advances in approach and methodology. Particularly interesting subjects for development are the analysis of complex conformational ensembles. NMR spectroscopy is best used in conjunction with a range of complementary biochemical, biophysical, and theoretical approaches. There is tremendous scope for extending structural studies to protein folding and to understanding the relationship between misfolding and disease.

Structure Determination by Infrared (IR) Spectroscopy

X-ray crystallography and multidimensional solution NMR spectroscopy are limited by the need for crystals or high solubility. In principle, IR spectroscopy can provide three-dimensional structures without such limitations. In addition, due to its high sensitivity, much less material is needed. In the future — probably the distant future — two different approaches may provide structural details of parts of proteins, and possibly full three-dimensional structures: linear dichroism measurements, and multidimensional IR spectroscopy. Both have to be combined with molecular biology techniques to mark individual peptide bonds in order to assign vibrational modes.

Linear Dichroism

In principle, it is possible to determine the orientations of individual peptide bonds or other bonds in amino acids, relative to the axis of an oriented sample. To assign the peptide vibration, an individual amino acid would be isotopically labeled. For example, in bacteriorhodopsin, which has no cysteines, site-directed mutagenesis could replace the desired amino acid by biosynthetic isotopically labeled cysteines, during the bacterial growth. If peptide bonds are labeled consecutively, all the angles that are important to yield the three-dimensional structure can be determined in combination with molecular dynamics (MD) calculations. This approach should be applicable to almost any recombinant protein. In addition, by comparison to the IR spectra of known structures, it may be possible to identify the structural family of an unknown protein.

Multidimensional FTIR

Optical spectroscopy typically gives limited structural information. To obtain structures, a method should be sensitive to interactions between spatially separate components. If the distance (and orientation) dependence is known, then structural parameters can be determined, in analogy with multidimensional NMR. Recently a two-dimensional Raman experiment was proposed by Tanimura and Mukamel (1993). Tokmakoff et al. (1997) have shown in simple liquids and liquid mixtures that a 5th-order nonlinear two-dimensional femtosecond Raman experiment directly reveals interactions between different vibrational modes. The liquid mixture experiments show that interactions can be observed between modes on different molecules. A complete understanding of the interactions that includes nonlinear polarizability and anharmonicity remains to be developed. The equivalent experiment with femtosecond IR pulses may be much easier to interpret, and to carry out, on proteins. Such two-dimensional IR experiments hold much promise for studies of protein dynamics, particularly when combined with optical triggering (Hamm et al. 1997)

Time-resolved Crystallography

Protein functions involve structural intermediates that have lifetimes from picoseconds to microseconds, at physiological temperatures. Time-resolved crystallography can follow the sequential appearance and disappearance of intermediates in real time. Contrast this with more conventional approaches, in which intermediates (or structures related to them) are stabilized by chemical trapping, or by physical trapping at low temperatures.

Time-resolved experiments today yield nanosecond time resolution and ~ 1.6 Å structural resolution. This is done with pump-probe experiments in which a significant fraction of the molecules in a single crystal undergo an initial reaction by absorption of a brief pulse of visible laser light (the pump), and are then interrogated by a 150 ps X-ray pulse (the probe) after an adjustable time delay. The X-ray pulse is focused pink radiation (in contrast to white or monochromatic) which falls on a stationary crystal and generates a Laue diffraction pattern. The entire experiment is repeated several times in order to acquire sufficient signal.

From the time-dependent intensities of each spot, the time-dependent structure amplitudes $|F(hkl;t)|$ and difference (or derivative) amplitudes are extracted, for example, from the form $|F(hkl;t)| - |F(hkl;0)|$. This procedure yields $\Delta\rho(t)$, the variation of the difference in the electron density with time, t. In the cases of carboxymyoglobin and photoactive yellow protein, the time-dependent difference maps yield interesting structural information. A challenge is to interpret these data in terms of the time-dependent variation of the populations of time-dependent intermediate structures. Unscrambling the superposition of structures into their component substructures may provide some knowledge of the reaction mechanism. It should be aided by spectroscopic studies that parallel the crystallographic work.

Time-resolved crystallography requires a mechanism for triggering the reaction. Current triggering mechanisms involve light activation. Other schemes for reaction initiation can be envisioned, such as temperature jumps or pressure jumps, to disturb a preexisting structural equilibrium. Ways of conferring light-sensitivity on light-inert systems are being sought. The classic example is the liberation of ATP from a biochemically inert but photochemically sensitive precursor, caged ATP. Ways of speeding up such reactions can be explored to enhance the time resolution to better than a few hundred microseconds, the present limit.

Several interesting systems exist that are naturally light-sensitive including heme proteins, photoactive yellow protein, DNA photolyase, photosynthetic reaction centers, bacteriorhodopsin, green fluorescent protein, and, possibly, flavoproteins.

It would be valuable to extend the present time resolution from a few nanoseconds to around 100 picoseconds, the shortest likely duration of the X-ray pulse emitted by high-energy storage rings. To probe elementary chemical processes, it would be useful to reach even faster times, down to tens of picoseconds. This will require either chopping the storage ring X-ray pulses, or developing X-ray sources based on different and novel physical mechanisms.

NMR Methods for Studying Protein Dynamics

NMR can provide a wealth of information about the conformational dynamics of proteins in solution. Nuclear spin relaxation times and line shapes measured by NMR spectroscopy provide information about the conformational dynamics of proteins and other biopolymers under nearly physiological solution conditions (Palmer et al. 1996; Dayie et al. 1996; Kay 1998). It is now possible to probe conformational flexibility and fluctuations at atomic resolution over more than ten decades of frequency, thanks to hardware improvements, the development of sophisticated pulse sequences, and access to isotopically labeled proteins. The exceptional versatility of NMR for protein dynamics stems from its multinuclear capability (mainly ^{1}H, ^{2}H, ^{13}C, and ^{15}N in proteins), the multitude of types of nuclear couplings (Zeeman, scalar, dipolar, quadrupolar, shielding anisotropy), and the variety of relaxation observables (auto- and cross-relaxation rates of magnetic polarizations and coherences of different ranks and quantum orders).

To date, backbone fluctuations in several dozen proteins (5–20 kDa) have been investigated by ^{15}N and ^{13}C relaxation (with uniform isotopic labeling), yielding accurate orientational order parameters for mainchain N–H and C–H bonds and approximate correlation times (in the range 10–1000 ps) for restricted rotational motions. With specifically deuterated and uniformly ^{13}C labeled proteins, the functionally relevant motions of side chains are now also becoming accessible (Kay 1998). Since ^{15}N, ^{13}C, and ^{2}H spin relaxation rates probe conformational dynamics via the associated fluctuations of anisotropic nuclear couplings, they cannot access conformational motions slower than the fastest isotropic motion, i.e., the overall tumbling of the protein (typically, 5–10 nanoseconds). Slower motions, in the range of microseconds to milliseconds, can be probed if they affect isotropic nuclear couplings, particularly the chemical shift. Still slower motions, in the range of milliseconds to seconds, can be studied by magnetization transfer experiments, where cross-peak intensities give information on exchange between conformations.

Solid-state NMR is less developed than solution NMR but offers certain advantages for studies of protein dynamics: (a) protein tumbling does not truncate the accessible motional spectrum, (b) the relaxation data carry more information, and (c) membrane proteins can be studied under native-like conditions. There have been few solid-state NMR studies of proteins. Most of them have measured the ^{2}H relaxation in proteins with specifically deuterated side chains. There are three experimentally accessible motional time windows. Motions that are fast compared to the anisotropic nuclear coupling (0.1–1 microseconds) can be studied via orientation-dependent spin relaxation rates. Slower motions (1–100 microseconds), such as aromatic ring flips, can be studied by quadrupolar echo line shapes, and still slower motions (milliseconds to seconds) by magnetization exchange and hole-burning experiments.

The information about molecular motions derivable from nuclear spin relaxation rates is encoded in the spectral density function (SDF). In ^{15}N and ^{13}C solution NMR relaxation studies, the SDF is probed at a few discrete frequencies in the high-frequency tail (and at zero frequency, where chemical shift modulation also contributes, however). To extract the full information content of the SDF, the resonance frequency (proportional to the static magnetic field) must be varied over a wide range. With conventional techniques, the field can only be varied by a factor of 2–3 since sensitivity and resolution deteriorate at lower fields. The sensitivity problem can be circumvented by field cycling (FC). In the FC method, polarization and detection take place at a relatively high field, while relaxation proceeds at an arbitrarily low field (Noack 1986). The rapid (millisecond) variation of the magnetic field is accomplished electronically (by switching the current through an air core magnet) or mechanically (by pneumatically shuttling the sample between two magnets). For proteins, the FC technique has so far been applied mainly to studies of water and cosolvents, where resolution is not an issue (Halle et al. 1998).

For NMR studies of protein dynamics, methodological advances are needed. While the quest for ever higher magnetic fields (with high-temperature superconducting magnets on the horizon) is primarily driven by structural studies, dynamic NMR studies would benefit more from access to very low fields. The ultimate development here would be the marrying of the FC technique with high-resolution multidimensional NMR spectroscopy, so that the entire SDF could be mapped on a residue-by-residue basis. Field variation would also benefit solid-state NMR, where the relaxation dispersion extends to lower frequencies (due to the absence of protein tumbling). In solution NMR, there is also a pressing need to explore the 0.01–1 microsecond range. To interpret NMR relaxation data, we also need to develop and test theoretical models of thermally activated conformational motions in the nanosecond-millisecond range in native and nonnative proteins.

Determination of Protein Reaction Mechanisms by Time-Resolved FTIR Difference Spectroscopy

To go beyond the static ground state structures of biomolecules and probe the dynamics of reaction mechanisms, we need methods that can record molecular reactions with high resolution in space and time, and we need ways to trigger the reactions. Time-resolved FTIR difference spectroscopy can satisfy these criteria. Difference spectra between a ground state and a transient intermediate state of a protein can be used to follow selectively absorbance bands of groups that undergo reactions (Gerwert 1993). These bands can be assigned unequivocally using site-directed mutagenesis or isotopic labeling. The difference spectra can be recorded with a time resolution from 30 nanoseconds in several minutes using the step-scan FTIR technique (Rammelsberg et al. 1997).

It may be possible to increase the time resolution to femtoseconds using IR laser pulses. If the IR difference spectrum is known, the time-dependent absorbance change can be used to follow the reactions of individual groups. Time-resolved FTIR can probe protonation states of individual groups (Gerwert et al. 1990), changes in H-bonds (Brudler et al. 1994), isomerization reactions (Gerwert and Siebert 1986), and reactions of individual water molecules (Hatanaka et al. 1997). In addition, using linear dichroism techniques, structural movements of the protein backbone may be spatially resolvable in the future (Ludlam et al. 1996). Due to the high sensitivity of the method, thin proteins films can be used. The reactions can be triggered by light using the chromophores of colored proteins or caged compounds (Cepus, Goody, et al. 1998; Cepus, Ulbrich, et al. 1998).

A promising new approach may be to use a synchrotron to provide an ultra-short, pulsed light source to improve time-resolved IR measurements to extend the spectral range into the far infrared. In the far infrared, global motions and fluctuations of the protein backbone can be determined and directly compared to molecular dynamics simulations.

Single-molecule Studies of Biomolecules

Spectacular advances have been achieved in recent years in probing single molecules (Xie and Trautman 1998). Spectroscopic methods and mechanical methods using the atomic force microscope (AFM) and "optical tweezers" now allow studies of individual molecules.

Spectroscopic Methods

There has been considerable progress in recent years in extending the sensitivity of spectroscopic methods so that it is now possible to observe single molecules. Currently, these efforts are based on fluorescence spectroscopy. To achieve a signal-to-noise ratio sufficient for single-molecule detection, confocal microscopy is used to collect fluorescence quanta from an extremely small volume, of the order of 1 femtoliter. Time-resolved detection allows the extraction of dynamic information from single molecules by analyzing the fluorescence intensity, polarization, lifetime, and spectral shape. These experiments are limited by the finite number of excitation/de-excitation cycles that the fluorophores can endure (typically 10^4–10^6). Given a collection efficiency of a few percent at best, only a small number of photons can be collected before the fluorophore falls victim to photodestruction.

A key advantage of single-molecule methods is the avoidance of the ensemble-averaging that is at the heart of traditional methods. Breaking free of this restriction now allows us to determine whether behavior is homogeneous — all molecules contribute identically to the ensemble-averaged properties — or heterogeneous — some molecules act differently than others. We can address fundamentals, such as ergodicity, the stochastic nature of transitions, and "Schrödinger's cat," i.e., the nature of quantum superposition. For example, if we can follow the trajectories of individual proteins or RNA molecules as they fold, we could map out folding energy landscapes. There is currently no other experimental way to do this.

Owing to their enormous sensitivity and their ability to study biomolecules in their natural environments, single-molecule methods have the potential to monitor protein association, structural changes, and ligand binding in both *in vivo* and *in vitro*. Some examples are given below.

Thermal Fluctuations, Single Molecules, and Time Correlations

It is now possible to watch thermal fluctuations in molecules. With the advent of single-molecule detection in solids (Dickson et al. 1997) and in solution, the analysis of thermal fluctuations and their time dependence has become possible through the calculation of time correlation functions.

Molecular Motions and Correlated Events in Biomolecules

The time frame from picoseconds to seconds has become accessible for the analysis of molecular motions in biostructures. The stochastic fluctuations of selected residues (e.g., aromatic residues) can be observed by excitation with sharply focused laser beams in confocally confined volume elements. The analysis of cross-correlated fluctuations (two color cross-correlations) will allow us to establish relationships between different molecular motions in macromolecular structures. Such correlations are on the wish lists of theorists who are interested in the shapes of folding landscapes. Also, the time interval from nanoseconds to microseconds, during which important classes of motions occur (domain and hinge motions, for example) may become accessible through single-molecule experiments.

Single Molecules, Conformational Transitions, and Ensemble Averages

The development of confocal single-molecule detection and imaging has led to a variety of experiments directed towards understanding the physical–chemical behaviors of individual biomolecules and their relationship to ensemble averages. One example is a 217-base-paired piece of double-stranded DNA, for which it has been possible to monitor conformational transitions related to base stacking. Transitions are detected by a tethered sensor. The single DNA molecules are attached to a streptavidinized glass surface by a biotin group. An analysis of about 100 single molecules shows that each molecule interconverts with its own transition rate and that the distribution of individual rates constitutes the ensemble behavior. These molecules show heterogeneous behavior. The observation that a small fraction (15%) of the analyzed single DNA molecules can operate in two distinctly separate frequencies (rates) suggests the existence of thermal fluctuations of the transition barriers.

Single-molecule Enzyme Reactions

A particularly interesting application to protein dynamics is the observation of enzymatic activity by real-time fluorescence changes as a ligand (substrate or cofactor [e.g., FADH]) binds and releases. If each enzyme samples its configuration space faster than the overall reaction rate, it should lead to simple Poisson statistics. However, proteins may have "memory," i.e., the reaction probabilities of a particular binding event will depend on the previous binding/release events. Systematic variation of external parameters (temperature, pressure, solvent viscosity, electromagnetic fields, etc.) affecting the reaction rates will allow us to examine transitions between conformational substates in detail.

Atomic Force Microscopy and Molecular Tweezers

Another class of single-molecule experiments is based on mechanical pulling and pushing, using optical tweezers (see, e.g., Nishizaka et al. 1995) and single-molecule atomic force microscopy (Binnig et al. 1986). These methods can pull ligands from their receptors (e.g., see Lee et al. 1994; Florin et al. 1994) or unfold proteins (Rief, Gautel et al. 1997). By measuring rupture forces as a function of pulling velocity, force spectroscopy reveals information on (un)binding or (un)folding free-energy landscapes (Rief, Oesterhelt et al. 1997). Theoretical studies and computer simulation (Grubmüller et al. 1996; Evans and Ritchie 1997; Izrailev et al. 1997; Marrink et al. 1998) complement this mesoscopic information with atomistic models that (a) explain how the observed forces and specificity arise from microscopic interactions, (b) extract additional information on unbinding reaction pathways and free-energy landscapes, and (c) characterize structural heterogeneities, e.g., of unbinding reactions.

Theory and Computer Simulation

The goal of science is to be predictive. Experiments lay the foundations for principles, which are captured in theories and models. Theory and simulation, in turn, give direction to the research enterprise as a whole and make it more efficient. In more mature fields such as physics and chemistry, experiments are commonly designed to test and improve theoretical models and principles. Modeling in biochemistry is a relatively new enterprise. Theory in structural biology is often not yet up to the task of making reliable experimental predictions. Nevertheless, the interface between theory and experiment in chemical and physical biology is developing rapidly and merits active development.

A major thrust in computational modeling involves atomic-resolution force fields that are usually sampled by molecular dynamics or Metropolis Monte Carlo methods. Energy functions are semi-empirical and parameterized using small molecule data. Such models are used to predict structural and energetic properties of biomolecules. Two main problems are: the need for large amounts of computer time, and that energy functions are not perfect. A general concern about energy functions is that they do not yet treat polarizabilities of atoms, i.e., atoms are currently considered to have fixed charges and the subtle changes in electron distribution, when an atom is near an electric field, are usually neglected. One difficulty in developing better force fields for proteins is finding benchmarks. Many important problems, such as predicting pKa's and the effect of single-site mutations, remain challenging. Exciting developments are in combining quantum mechanics with classical mechanics for problems that require both, such as bond breakage or light-harvesting. Valence bond, density functional methods, and Car–Parrinello methods are promising. Other good news is that computer speed continues to follow (and even exceed) Moore's law (named after Gordon Moore at Intel Corporation): advancing computer technology leads to computer speeds that double about every 18 months.

A limitation at the present time is that simulations usually follow only a single trajectory of the chain through time, rather than following the ensemble average, which is what most experiments measure. Even more challenging is the prediction of native structures, folding kinetics, the spectrum of internal motions, the temperature dependence or solvent dependence of cooperative transitions, effects of pressure, viscosity, or pH, aggregation processes, and the binding affinities of ligands to biomolecules.

To describe the large-scale conformational changes of proteins, statistical mechanical models have emerged during the past few years. Such modeling has explored the dominant forces of folding, the relationships between sequence, structure, and folding dynamics, evolutionary and mutational change, aggregation, assembly, catalytic rates in multienzyme complexes, structures of cytoskeletal elements, and the nature of cooperative transitions. While these models can treat general principles, long time scales, and large conformational changes, they involve severe simplifications that normally preclude detailed predictions. However, such highly reductionist models may be increasingly needed for exploring more complex and multi-protein processes of interest in cell biology. The more complex the process, the simpler the model that is needed to treat it.

While computer simulations continue to be essential for studying biomolecules, and show increasing success in rationalizing and predicting thermodynamic and kinetic properties of enzymes, there is also an important need for theory. Theory is not the same as simulation. Theory refers to analytical models. The advantage of theory, where it applies, is the breadth of its predictions. The more rigorous the theory, the more convincing is the connection between the premises of the model and its inferences. Both theory and computer simulation are essential tools for modeling biopolymers.

CONCLUSIONS

Proteins, nucleic acids, and membranes are at the center of chemical and physical biology. The fundamental starting point for understanding biology at the deepest level is knowledge of molecular structures, which will continue to come from X-ray crystallography, NMR spectroscopy, and may be possible by IR methods. Advances in these areas will continue to be important. We need more structures of biomolecules that are currently poorly represented in the data base — DNAs, RNAs, and membrane proteins. At the same time, if we could push the limits to get structures at higher resolution, or if we had more intense neutron sources available, we could see hydrogen atoms. Such experiments are central to understanding enzymatic functions. Another frontier is the study of molecular machines: assemblies of multiple biological molecules that function together.

However, higher resolution — and structures alone — will not solve many of our most fundamental problems. To understand biological mechanisms of action better will also require dynamics, fluctuations, ensembles, and distributions of substates, and an understanding of molecular motions. This is made challenging by the extraordinarily broad range of time scales: from femtoseconds to days. There are big gaps from nanoseconds to milliseconds where new methods are needed. Important techniques include time-resolved FTIR and X-ray crystallography. A most exciting recent large-scale advance is the ability to study single molecules, rather than their ensemble averages. To meet these challenges, it will be crucial to bring together different methods and disciplines to bear on all aspects of the structures and functions of biomolecules.

ACKNOWLEDGMENTS

We gratefully acknowledge the assistance of C.M. Dobson, C.H. Chothia, D.M. Crothers, H. Grubmüller, B. Halle, B.W. Matthews, K. Moffat, G.U. Nienhaus, F.M. Richards, and R. Rigler.

REFERENCES

Ansari, A., J. Berendzen, S.F. Bowne, H. Frauenfelder, I.E.T. Iben, T.B. Sauke, E. Shyamsunder, and R.D. Young. 1985. Protein states and proteinquakes. *Proc. Natl. Acad. Sci. USA* **82**:5000–5004.

Austin, R.H., K.W. Beeson, L. Eisenstein, H. Frauenfelder, and I.C. Gunsallus. 1975. Dynamics of ligand binding to myoglobin. *Biochemistry* **14**:5355–5373.

Binnig, G., C.F. Quate, and C. Gerber. 1986. Atomic force microscope. *Phys. Rev. Lett.* **56**:930–933.

Brundler, R., H.J.M. de Groot, W.B.S. van Liemt, W.F. Steggerda, R. Esmeijer, P. Gast, A.J. Hoff, J. Lugtenburg, and K. Gerwert. 1986. Asymmetric binding of the 1- and 4-C=O groups of Q_A in rhodobacter sphaeroides R26 reaction centres monitored by Fourier transform infra-red spectroscopy using site-specific isotopically labelled ubiquinone-10. *EMBO J.* **5**:805–811.

Bryngelson, J.D., J.N. Onuchic, N.D. Socci, and P.G. Wolynes. 1995. Funnels, pathways, and the energy landscape of protein folding. *Proteins* **21**:167–195.

Cepus, V., R.S. Goody, A.J. Scheidig, and K. Gerwert. 1998. Time-resolved FTIR studies of the GTPase reaction of H-ras p21 reveal a key role for the b-phosphophate. *Biochemistry* **37**:10,263–10,271.

Cepus, V., C. Ulbrich, Chr. Allin, A. Troullier, and K. Gerwert. 1998. Fourier transform infrared photolysis studies of caged compounds. *Meth. Enzymol.* **291**:223–245.

Chou, P.Y., and G.D. Fasman. 1978. Prediction of the secondary structure of proteins from their amino acid sequences. *Adv. Enzymol.* **47**:45.

Colloc'h, N., C. Etchebest, E. Thoreau, B. Henrissat, and J.-P. Mornon. 1993. Comparison of three algorithms for the assignment of secondary structure in proteins: The advantages of a consensus assignment. *Protein Eng.* **6**:377–382.

Dayie, K.T., G. Wagner, and J.-F. Lefèvre. 1996. Theory and practice of nuclear spin relaxation in proteins. *Ann. Rev. Phys. Chem.* **47**:243–282.

Dickson, R.M., A.B. Cubitt, R.Y. Tsien, and W.E. Moerner. 1997. On/off blinking and switching behavior of single molecules of green fluorescent protein. *Nature* **388**:355–358.

Dill, K.A. 1990. Dominant forces in protein folding. *Biochemistry* **29**:7133–7155.

Dill, K.A., S. Bromberg, K. Yue, K.M. Fiebig, D.P. Yee, P.D. Thomas, and H.S. Chan. 1995. Principles of protein folding — A perspective from simple exact models. *Protein Sci.* **4**:561–602.

Dill, K.A., and H.S. Chan. 1997. From Levinthal to pathways to funnels. *Nature Struct. Biol.* **4**:10–19.

Evans, E., and K. Ritchie. 1997. Dynamic strength of molecular adhesion bonds. *Biophys. J.* **72**:1541–1555.

Florin, E.-L., V.T. Moy, and H.E. Gaub. 1994. Adhesion forces between individual ligand-receptor pairs. *Science* **264**:415–417.

Frauenfelder, H., N.A. Alberding, A. Ansari, D. Braunstein, B.R. Cowen, M.K. Hong, I.E.T. Iben, J.B. Johnson, S. Luck, M.C. Marden, J.R. Mourant, P. Ormos, L. Reinisch, R. Scholl, A. Schulte, E. Shyamsunder, L.B. Sorensen, P.J. Steinbach, A. Xie, R.D. Young, and K.T. Yue. 1990. Proteins and pressure. *J. Phys. Chem.* **94**:1024–1037.

Frauenfelder, H., and B. McMahon. 1998. Dynamics and function of proteins: The search for general concepts. *Proc. Natl. Acad. Sci. USA* **95**:4795–4797.

Frauenfelder, H., G.A. Petsko, and D. Tsernoglou. 1979. Temperature-dependent X-ray diffraction as a probe of protein structural dynamics. *Nature* **280**:558–563.

Frauenfelder, H., S.G. Sligar, and P.G. Wolynes. 1991. The energy landscapes and motions of proteins. *Science* **254**:1598–1603.

Frauenfelder, H., and P.G. Wolynes. 1994. Biomolecules: Where the physics of complexity and simplicity meet. *Phys. Today* **47**:58–64.

Garcia, A.E., J.A. Krumhansl, and H. Frauenfelder. 1997. Variations on a theme by Debye and Waller: From simple crystals to proteins. *Proteins* **29**:153–160.

Gerwert, K. 1993. Molecular reaction mechanisms of proteins as monitored by time-resolved FTIR spectroscopy. *Curr. Op. Struct. Biol.* **3**:769–773.

Gerwert, K., and F. Siebert. 1994. Evidence for light-induced 13-cis, 14-cis, isomerization in bacteriohodopsin obtained by FTIR difference spectroscopy using isotopically labelled retinals. *EMBO J.* **5**:5523–5530.

Gerwert, K., G. Souvignier, and B. Hess. 1990. Simultaneous monitoring of light-induced changes in protein side-group protonation, chromophore, isomerization, and backbone motion of bacteriorhodopsin by time-resolved Fourier-transform infrared spectroscopy. *Proc. Natl. Acad. Sci. USA* **87**:9774–9778.

Grubmüller, H., B. Heymann, and P. Tavan. 1996. Ligand binding: Molecular mechanics calculation of the streptavidin-biotin rupture force. *Science* **271**:997–999.

Halle, B., V.P. Denisov, and K. Venu. 1998. Multinuclear relaxation dispersion studies of protein hydration. In: Modern Techniques in Protein NMR, vol. 17, ed. L.J. Berliner and N.R. Krishna. New York: Plenum.

Hamm, P., M. Lim, and R.M. Hochstrasser. 1997. Vibrational energy relaxation of the cyanide ion in water. *J. Chem. Phys.* **107**:10,523–10,531.

Hatanaka, M., H. Kandori, and A. Maeda. 1997. Localization and orientation of functional water molecules in bacteriorhodopsin as revealed by polarized Fourier transform infrared spectroscopy. *Biophys. J.* **73**:1001–1006.

Holm, L., and C. Sander. 1993. Protein structure comparison by alignment of distance matrices. *J. Mol. Biol.* **233**:123–138.

Izrailev, S., S. Stepaniants, M. Balsera, Y. Oono, and K. Schulten. 1997. Molecular dynamics study of unbinding of the avidin-biotin complex. *Biophys. J.* **72**:1568–1581.

Kay, L.E. 1998. Protein dynamics from NMR. *Nature Struct. Biol.* **NMR Suppl.**:513–517.

Korszun, Z.R. 1997. Neutron macromolecular crystallography. *Meth. Enzymol.* **276**:218–232.

Kossiakoff, A.A. 1985. The application of neutron crystallography to the study of dynamic and hydration properties of proteins. *Ann. Rev. Biochem.* **54**:1195–1227.

Lee, G.U., D.A. Kidwell, and R.J. Colton. 1994. Sensing discrete streptavidin-biotin interactions with atomic force microscopy. *Langmuir* **10**:354–357.

Ludlam, C.F., I.T. Arkin, X.M. Liu, M.S. Rothman, P. Rath, S. Aimoto, S.O. Smith, D.M. Engelmann, and K.J. Rothschild. 1996. Fourier transform infrared spectroscopy and site-directed isotope labeling as a probe of local secondary structure in the transmembrane domain of phospholamban. *Biophys. J.* **70**:1728–1736.

Marrink, S.-J., O. Berger, P. Tieleman, and F. Jähnig. 1998. Adhesion forces of lipids in a phospholipid membrane studies by molecular dynamics simulations. *Biophys. J.* **74**:931–943.

Matthews, B.W. 1993. Structural and genetic analysis of protein stability. *Ann. Rev. Biochem.* **62**:139–160.

McDowell, R.S., and A.A. Kossiakoff. 1995. A comparison of neutron diffraction and molecular dynamics structures: Hydroxyl group and water molecule orientations in trypsin. *J. Mol. Biol.* **250**:553–570.

Miyazawa, S., and R.L. Jernigan. 1985. Estimation of effective interresidue contact energies from protein crystal structures: Quasi-chemical approximation. *Macromolecules* **18**:534.

Miyazawa, S., and R.L. Jernigan. 1996. Residue-residue potentials with a favorable contact pair term and an unfavorable high packing density term, for simulation and threading. *J. Mol. Biol.* **256**:623–644.

Nienhaus, G.U., J.D. Mueller, B. McMahon, and H. Frauenfelder. 1997. Exploring the energy landscape of proteins. *Physica D* **197**:297–311.

Nienhaus, G.U., and R.D. Young. 1996. Protein dynamics. In: Encyclopedia of Applied Physics, ed. G.L. Trigg, vol. 15, pp. 163–185. New York: VCH Publishers.

Niimura, N., Y. Minezaki, T. Nonaka, J.-C. Castagna, F. Cipriani, P. Hoghoj, M.S. Lehmann, and C. Wilkinson. 1997. Neutron Laue diffractometry with an imaging plate provides an effective data collection regime for neutron protein crystallography. *Nature Struct. Biol.* **4**:909–914.

Nishizaka, T., H. Miyata, H. Yoshikawa, S. Ishiwata, and K. Kinosita, Jr. 1995. Unbinding force of a single motor molecule of muscle measured using optical tweezers. *Nature* **377**:251–254.

Noack, F. 1986. NMR field-cycling spectroscopy: Principles and applications. *Prog. NMR Spect.* **18**:171–276.

Ostermeier, C., A. Harrenga, U. Ermler, and H. Michel. 1997. Structure at 2.7 Angstrom resolution of the *Paracoccus denitrificans* two-subunit cytochrome-c oxidase complexed with an antibody F-V fragment. *Proc. Natl. Acad. Sci. USA* **94**:10,547–10,553.

Palmer, A.G., J. Williams, and A. McDermott. 1996. Nuclear magnetic resonance studies of biopolymer dynamics. *J. Phys. Chem.* **100**:13,293–13,310.

Rammelsberg, R., B. Heßling, H. Chorongiewski, and K. Gerwert. 1997. Molecular reaction mechanisms of proteins monitored by nanosecond step-scan FT-IR difference spectroscopy. *Appl. Spectrosc.* **51**:558–562.

Rejto, P.A., and S.T. Freer. 1996. Protein conformational substates from X-ray crystallography. *Prog. Biophys. Mol. Biol.* **66**:167–296.

Richards, F.M., and W.A. Lim. 1994. An analysis of packing in the protein folding problem. *Qtly. Rev. Biophys.* **26**:423–498.

Rief, M., M. Gautel, F. Oesterhelt, J.M. Fernandez, and H.E. Gaub. 1997. Reversible unfolding of individual titin immunoglobulin domains by AFM. *Science* **267**:1109–1112.

Rief, M., F. Oesterhelt, B. Heymann, and H.E. Gaub. 1997. Single molecule force spectroscopy reveals conformational change in polysaccharides. *Science* **275**:1295–1297.

Tanimura, Y., and S. Muklamel. 1993. Two-dimensional femtosecond vibrational spectroscopy of liquids. *J. Chem. Phys.* **99**:9496–9511.

Tokmakoff, A., M.J. Lang, D.S. Larsen, G.R. Fleming, et al. 1997. Two-dimensional Raman spectroscopy of vibrational interactions in liquids. *Phys. Rev. Lett.* **79**:2702–2705.

Xie, X.S., and J.K. Trautman. 1998. Optical studies of single molecules at room temperature. *Ann. Rev. Phys. Chem.* **49**: 441–480.

Yee, D.P., and K.A. Dill. 1993. Families and the structural relatedness among globular proteins. *Protein Sci.* **2**:884–899.

Zhou, H.X., S. Wlodek, and J.A. McCammon. 1998. Conformation gating as a mechanism for enzyme specificity. *Proc. Natl Acad. Sci. USA* **95**: 9280–9283.

7

Is Complexity of Biological Macromolecules a Result of Randomness or Not?

T. GAREL and H. ORLAND

Service de Physique Théorique, CE-Saclay, 91191 Gif-sur-Yvette Cedex, France

Keywords: complexity, heteropolymers, collapse transitions, chiral dipolar chain, domain structure

ABSTRACT

A protein can be described as a homopolymeric main chain with specific side chains. The respective role of the backbone and side chains is discussed in the framework of several simplified models. It is shown that the chemistry of the backbone selects universal domain-like structures (secondary structures) and that the fine-tuning of the native structure is due to a delicate interplay between backbone defects (glycine and proline) and side chains.

INTRODUCTION

From a physicist's point of view, proteins are small heteropolymers that have the property of folding reversibly into a single geometrical shape with biological activity (Creighton 1992). The active site is generally located on the surface of the protein. Although the folding process depends on the experimental conditions (*in vivo* or *in vitro*, pH, salt concentration, etc.), the stable folded structure is in general unique. Some typical numbers of interest in proteins are:

1. the number of monomers $N \sim 150$;
2. the linear size of the folded molecule $R \sim 40$ Å;
3. the range of folding times, from the initial ($\tau_i \sim 10^{-6}$ s) to the final stage ($\tau_f \sim 10^{-3}$–1 s); the latter estimate suggests that the phase space of a protein may have many trapping local minima;

Simplicity and Complexity in Proteins and Nucleic Acids
Edited by H. Frauenfelder, J. Deisenhofer, and P.G. Wolynes

4. the first-order character of the folding transition, with an entropy loss of order k_B per monomer.

A closer look at a proteic chain reveals that this heteropolymer consists of an almost periodic backbone and different side chains (residues). For globular proteins (as opposed to fibrous or membrane proteins), the presence of water around the protein complicates the problem even further: it is commonly believed, and sometimes proven, that the folding transition is driven by some very hydrophobic parts of the chain. On the other hand, the final folded structure of proteins contains universal features such as helices and sheets, resembling the familiar domain patterns of crystalline materials.

Therefore, the question arises as to whether proteins should be approached as small heteropolymers (stressing the diversity of the protein residues) or small homopolymers (where the stress is on the (almost) periodic protein backbone). In any case, these remarkably shaped memory alloys have a finite size, pointing towards an important role of the surface and, therefore, of the solvent.

From a rigorous point of view, one may define a complex system as one that possesses an infinite number of thermodynamic phases in the thermodynamic limit (Mézard et al. 1993). To tackle as many situations as possible, we loosely define the physical complexity in the following way: a system is said to be complex if it has an exponential number of local (symmetry-unrelated) free-energy minima, separated by high-energy barriers. This definition is in broad agreement with that given by Palmer (1982), but is more specific than the simple counting argument of Levinthal (1968). Indeed, this argument neglects the role of the energy bias and is purely based on combinatorics. Nevertheless, the Levinthal Paradox partially survives, since the kinetic accessibility of the native state in such a complex phase space remains to be explained.

To specify the topology of phase space, an energy function (which assigns an energy to a chain conformation) is required. Three theoretical paths have so far been trodden (Elber 1996); each of them considers the folded structure as thermodynamically stable (as opposed to kinetically accessible).

The first one starts from a phenomenological (as opposed to microscopic) form of the interactions (both bonded and nonbonded), where one attributes some adjustable residue-dependent value to the elastic constants, bond lengths or angles, and electrostatic charges present in the Hamiltonian of the protein. One may then perform molecular dynamics or Monte Carlo calculations. So far, no one has succeeded in folding a protein in this way, partly because of Pt. 3 above, not to mention the reliability (or lack thereof) of the phenomenological Hamiltonian.

The second approach identifies the protein with a heteropolymer of some sort and follows, within the framework of simple random Hamiltonians, the (spin) glass appeal; in particular, the existence of a unique ground state structure is due to subtle nonextensive disorder-dependent terms, such as the ones present in Derrida's random energy model (Derrida 1981). Apart from irrealistic mean field treatments, these random Hamiltonians have been studied numerically and seem to point towards a kind of (magic) cluster-like criterion: stable structures have the largest gap between the ground state and the first excited states (Salí et al. 1994).

Interestingly enough, the link between protein folding and cluster physics is gaining ground quite rapidly (Berry et al. 1997; Zhou et al. 1997). This closing gap between the two fields shows up also in the third approach, where one models a protein by a homopolymer with simple specific interactions between the monomers. The folding transition may then be identified with a crystallization phase transition, where helices and sheets appear as slight variants of the same order parameter (Pitard et al. 1997; Doniach et al. 1996).

One weak point common to the above theories is their (lack of) treatment of the solvent. The balance between unfolded and folded free energies (enthalpies) is extremely delicate; for instance, the entropy loss per monomer is of order k_B at the folding transition, whereas the differences in free enthalpies are only a fraction of k_BT. Water is a complicated liquid and to disentangle the hydrogen-bond effects from the more classical electrostatic interactions is difficult. One must, therefore, take all of the above models with a grain of salt and consider the division of the Hamiltonian between hydrophilic or hydrophobic terms, on the one hand, and various electrostatic terms (charges, dipoles, etc.), on the other, as a rather uncontrolled procedure.

In this chapter, we will try to adopt a pragmatic point of view, by starting from a physical assessment of the relevant interactions in the protein + water system (see section on **INTERACTIONS**). This will lead us to consider first a lattice model of a long homopolymer chain in a bad solvent, i.e., the familiar (bulk) θ collapse transition (see section on **THE θ POINT**). This preliminary study shows that the entropy of the collapsed phase is extensive, and thus such a simplified model cannot account for the uniqueness of the native state. One must therefore add specific energy contributions to eliminate this extensive entropy.

We will remain first in a homopolymeric framework, by adding short-range (hydrogen bond-like) or long-range (dipolar-like) contributions (see section on **HOMOPOLYMERIC ASPECTS: BACKBONE AND SECONDARY STRUCTURES**). The model can be further complicated by the introduction of chirality. In their ground state, all of these models possess a very small (possibly zero) entropy per monomer. The dipolar chiral chain is indeed a very attractive model for the protein backbone and provides an explanation for the ubiquitous secondary structures of proteins; similarly, glycine and proline residues play a role in defects. It must be noted, however, that there may exist above the ground state many metastable states, which can be viewed as originating from the extensive entropy of the collapsed homopolymer.

Another possibility to eliminate the extensive entropy of a θ collapsed homopolymer is to introduce a specific primary sequence (technically modeled by disorder) in the Hamiltonian (see section on **HETEROPOLYMERIC ASPECTS: SIDE CHAINS**). In this case, a monomer can be thought of as representing a residue, and the level of description is that of the tertiary structure. Most prominent among the "random" characteristics of the residues are their electric charges and their affinities with water. These model heteropolymers are difficult to treat, even in mean field, and a simpler model emerges: the random energy model (REM), which is exactly soluble and has given interesting insights into the physics of heteropolymers (including proteins). A noteworthy feature of these rough models is that they provide a typical size for a single domain protein (possible quaternary structures are not considered in this chapter).

Finally, one may draw attention to the phase diagrams of the above models, and argue that the kinetic accessibility of their ground state suggests that proteins have been evolution-selected so that they may reach a definite shape in a reasonable time (see **CONCLUSION**). By definite shape, we mean a shape that would allow for a surface site with specific biochemical activity.

INTERACTIONS

As mentioned above, it is necessary to work with a Hamiltonian. The first Hamiltonian to be considered in this field is the CHARMM potential or its OPLS, GROMOS, ECEPP, etc., cousins. These phenomenological force fields, as they are often called, contain bonded and nonbonded interaction terms. The bonded terms contain the geometrical information of the protein (bond lengths, valence angles, dihedral angles, etc.), including chirality; the nonbonded terms account for the van der Waals and Coulomb interactions. The van der Waals term between atoms i and j reads:

$$V_{ij}^{vdW} = 4\varepsilon_{ij}\left(\left(\frac{\sigma_{ij}}{r_{ij}}\right)^{12} - \left(\frac{\sigma_{ij}}{r_{ij}}\right)^{6}\right). \tag{7.1}$$

The Coulomb interactions between "ions" i and j are given by

$$V_{ij}^{C} = \frac{332}{\varepsilon_r}\frac{q_i q_j}{r_{ij}} \tag{7.2}$$

in kCal/mol, where ε_r is an effective dielectric constant. The physical challenge is to go from the microscopic Coulomb interactions to a trustworthy mesoscopic description; this is essentially the role of Equations 7.1 and 7.2. In particular, there is no specific term for the hydrogen bonds that the protein makes with the water molecules, or with itself. The protein backbone is, in particular, represented by charges q_i and its dominating part is the CONH peptide bond, which is roughly equivalent to a dipole (its experimental value being $\sim 3.7\ D$). It is quite difficult to find a reliable value for the effective dielectric constant of Equation 7.2 since bulk water is characterized by $\varepsilon_r \simeq 80$, and the interior of the protein is a rather low ε_r medium. The CHARMM Hamiltonian is probably a first try that may be valid in the folded state but most likely not in the unfolded state. In any case, a dielectric constant is appropriate at the level of the Maxwell (averaged) fields and not at the microscopic level. Since these force fields claim to work in a nonmicroscopic effective way, it is perfectly legitimate to use a dielectric constant — to put an intelligent guess on its value is another story.

Furthermore, we notice that the same problem arises for water molecules (hydrogen bonds, dipoles, etc.). The only proper way is to work with a microscopic Hamiltonian for both the protein and the solvent. Since this is not possible (at least analytically), the CHARMM Hamiltonian and its different approximations are to be trusted as long as they yield results that are not too far from experiments. In this type of approach, the affinity of atom i for water can possibly be interpreted in terms of the energy scale ε_{ii} in Eq. 7.1, a very large value of ε_{ii} denoting a strong hydrophobic character of atom i.

Some energy scales (in degrees Kelvin) present in the system are: (a) 10^5 for a covalent bond; (b) 300 for a single hydrogen bond, or for a typical ε_{ij}; (c) $1300/\varepsilon_r$ for the CONH–CONH nearest neighbor dipole energy.

As far as theoretical models are concerned, a final word of caution is in order: many theoreticians favor a Landau continuum approach to phase transitions (including the idea of domain, domain walls, etc.). In a 40 Å-long system, this is certainly not obvious (it is part of the folding problem). It is indeed hard to eliminate the short-range interactions to get a "universal" behavior in such a short heteropolymer chain.

We will start this complicated program by considering a very simplified version of the CHARMM potential, namely, a homopolymer chain on a (cubic) lattice in a bad (but simple) solvent.

THE θ POINT

Well-known Flory arguments show that there is a transition between an entropy-dominated swollen coil and a low-temperature energy-dominated collapsed coil. If ε measures the attraction energy between two monomers, the θ transition temperature is given by $T_\theta \sim 3.7\,|\varepsilon|$ (see, e.g., Doniach et al. 1996 and references therein). An important point is that the dimensionless entropy per monomer, s, is generally finite below the transition. Because of the chain constraint, this entropy is very dependent on the boundary conditions (Gordon et al. 1976). For periodic boundary conditions it is possible to estimate this entropy for a Hamiltonian path (fully packed chain) as $s = 6/e$. Of course, periodic boundary conditions are not appropriate to the one-chain problem: however, the above estimate of s suggests that the modeling of the folding transition (and therefore of a unique ground state or shape) should at least eliminate the extensive part of the entropy. Then, one may focus on the role of the surface in a collapsed globule. If N is the number of monomers, the surface of an (isotropic) globule contains a number of monomers of order $N^{2/3}$. For values of N corresponding to proteins, this shows that approximately half of the atoms are on the surface.

HOMOPOLYMERIC ASPECTS: BACKBONE AND SECONDARY STRUCTURES

Staying within homopolymeric models, one may first add forces which mimic the electrostatic term of Equation 7.2: we mentioned previously that the electrostatics of the backbone was dominated by the peptide bond CONH, which has a large dipole moment. Dipolar interactions are difficult to treat since they are both long ranged and anisotropic. In a finite system, they clearly "feel" the surface. Since dipoles have preferred (and subtle) configurations, one expects that the low-temperature phase of these models have less (possibly zero) bulk entropy.

To address the issues of bulk and surface contributions separately, we have first considered short-ranged dipolar-like interactions, both in mean field and (numerically) in three dimensions (Doniach et al. 1996). These models can be defined as self-avoiding walks on a lattice, with nearest-neighbor attraction ε, and a bending rigidity ε_h to mimic the rigidity induced by (dipolar-like) hydrogen bonds. These models turn out to have a low-temperature frozen phase which can be viewed as quasi-crystalline; this phase can be reached either from the liquid globule ($T < T_\theta$) or directly from the coil (unfolded) state. Note that the dynamics of the transition are very different in the two cases: in the first-order direct transition, one expects metastable states to be present in the coil phase. In the ground state, the bulk entropy is negligibly small, leading to a natural identification of this crystal to the "crystal of secondary structures" description of a protein. However, a surface entropy remains.

When one takes into account the full long-ranged dipolar interaction to describe the protein backbone, the surface of the protein is to be optimized since surface and bulk contributions now compete (hence a nonlocal coding of the shape by the interactions). It is well known from solid state physics that (dipole-induced) domain structures frequently differ by a nonextensive amount. It is thus meaningless to separate dipole and surface contributions. An interesting feature of the dipolar interaction is that one may include the very relevant issue of chirality in the homopolymer Hamiltonian. The low-temperature phase of the dipolar chiral chain possesses remarkable analogies with the liquid crystal blue phases; it has been shown that α-helices and β-sheets can be obtained naturally from a unique tensorial-order parameter (Pitard et al. 1997). This tensorial-order parameter is spatially modulated but has the very unusual property of possessing no harmonics. Therefore, it describes a kind of frozen phase with very little entropy.

In summary, the universal features of secondary structures in folded proteins can be explained by the almost homopolymeric character of the protein backbone, modeled through a chiral dipolar chain. The low-temperature phase of this model is frozen, except for the natural defects between the ordered dipolar domains (i.e., the helices and sheets). In real proteins, these defects are likely to be the glycine and proline residues, which indeed are most of the time found in surface loops. The theoretical unity of secondary structures is supported by the experimental finding of a chameleon-like sequence, which can adopt both helix or sheet conformations (Minor and Kim 1996). This is also in agreement with the findings of Head-Gordon et al. (1992): in this paper, the authors test the "polyalanine" hypothesis, namely that the native structure of a protein is a locally stable structure for the corresponding polyalanine.

HETEROPOLYMERIC ASPECTS: SIDE CHAINS

To achieve a unique native state with a specific biologically active site, it is necessary to allow for primary sequence diversity. The physicist's naive approach is to replace this diversity by randomness and to represent proteins as random heteropolymers.

The approach of the folding transition through the heteropolymer path is detailed in Garel et al. (1997). Basically, the models rely on a rewriting of Equations 7.1 and 7.2 as tractable disorder models. These equations contain atom-dependent charges q_i as well as pair-dependent parameters ε_{ij} and σ_{ij}. As noted previously, these parameters can be partially interpreted in terms of water–protein interactions. Once again, the goal is to eliminate the extensive part of the entropy by introducing disorder, until some kind of freezing transition occurs. Denoting by $f(r)$ a short-range function (e.g., $f(r) = \frac{e^{-kr}}{r}$), some random models of interest are:

(a) the random charged (RC) chain, where the Hamiltonian reads

$$H_{RC} = \sum_{ij} q_i q_j f(r_{ij}), \tag{7.3}$$

(b) the random hydrophilic–hydrophobic (RHH) chain, where one has

$$H_{RHH} = \sum_{ij} (h_i + h_j) f(r_{ij}), \tag{7.4}$$

(c) the REM chain, where one has

$$H_{REM} = \sum_{i<j} v_{ij} f(r_{ij}), \tag{7.5}$$

and uncorrelated values of the interactions v_{ij} and v_{kl}.

The phase diagram of these models has been studied in mean-field for large N, in the case where the random variables (a) q_i and (b) h_i are independent random variables. Although the physics of these three models is very different, they share in common the possible existence of a liquid globular phase, between the coil phase and the frozen low temperature ("native-like") phase. Below we comment on the results and refer the reader to the literature.

The RC Chain: Existence of a Critical Size

Proteins have essentially four charged residues out of twenty, histidine notwithstanding. Assuming equipartition of the amino acids in proteins implies that there is of order $N/5$ charged residues in a N-monomer chain. It is an experimental fact that charges are (almost) always on the surface of the folded protein. If we take this experimental fact as a necessary condition for the folded protein, we get $N/5 \sim N^{2/3}$ (since $N^{2/3}$ is the typical size of the surface) and therefore a critical size $N_c \sim 125$, in good agreement with the typical (monodomain) protein size. For larger N, the folded protein will make several domains. For $N \sim N_c$, the entropy of the chain is basically of order unity. The shape of the frozen globule is then connected to the distribution of charges along the chain. This naive argument neglects the interactions between charges (which may induce an additional Rayleigh droplet instability), their possible delocalization on the surface, not to mention the role of water; nevertheless, it illustrates the enhanced importance of the surface due to the chain constraint.

The RHH Chain

The degree of hydrophobicity of the amino acids is encoded in the h_i parameters. The random heteropolymer case may exhibit, at the mean field level, a direct transition to the native state, through an unusual first-order transition (driven by an attractive third virial coefficient). In this approach, there is no obvious way to calculate the size of dominantly hydrophobic clusters.

Since proteins are not random as far as hydrophobicity is concerned (Irbäck et al. 1996), one may turn to the periodic chain. Let us consider the periodic chain, with equal (absolute) values of the h_i (Orlandini and Garel 1998). If the period of the h_i is p monomer long and a is a geometric constant, one will get a single hydrophobic core if $N < ap^3$, and multiple cores in the opposite case. The value of p appropriate for proteins is not an easy parameter to determine since it depends on the secondary structures ($p \sim 3.6$ for α-helices, and $p \sim 2.3$ for β-sheets). This may suggest that the existence of a single core is partially due to the hydrophobic interaction for an all α-helix protein, and less so for an all β-sheet protein. In the general case of a RHH, where both the signs and magnitudes of the h_i are random, one may conservatively conclude that there is a maximum size N'_c for the existence of a single hydrophobic core.

The REM Chain

This simple case is a fixed point in the mean field physics of disordered systems. In the context of protein folding, it has become very popular, because the low-temperature properties are governed by a few dominant folds. This dominance stems from nonextensive effects in the free energy of these folds (typically of order one in mean field, as compared to the extensive part of the free energy). Several remarks are in order:

1. In some generalized random energy models (GREM), where there are correlations between the energy levels (e.g., between the v_{ij}'s of equation [Elber 1996]), these dominant folds do not exist anymore.
2. It is not obvious how the (disorder-dependent) nonextensive part of the free energy compares with the nonextensive surface contribution.

To conclude, the physics of heteropolymers displays features of both disordered and polymer systems. Surface properties are not easily included in this scheme. As in the homopolymeric approach, depending on physical parameters such as denaturant concentration and pH, one may reach the native-like phase either from a liquid globule or directly from a coil state. This implies that there is in general several coil states, with or without metastable states.

It can be argued that an efficient (i.e., quick) folding transition towards a definite shape must avoid the slow dynamics of the liquid globule and the trapping of the metastable glassy states. In other words, evolution-selected proteins should lie around a particular line in the phase diagram (Shakhnovich 1994; Klimov and Thirumalai 1996).

CONCLUSION

Proteins have both homo- and heteropolymeric characters. The mesoscopic size of proteins, as well as the peculiar nature of water, confers a decisive role to the surface of the protein. This is clearly apparent in our approach. If one considers that proteins are made to have a biologically active site on the surface, then the variety of presently observed biological functions supports the heteropolymeric point of view, since different functions require different side chains.

We nevertheless think that the homopolymeric view of proteins has been underestimated in the literature, with some notable exceptions (Head-Gordon et al. 1992). For instance, dipolar (backbone) effects have universal features such as secondary structures for which a unified theory is now available (Pitard et al. 1997). A related issue concerns the description of secondary structures as minimal surfaces (Hyde et al. 1997): there is still a great deal to be learned from the Bonnet transformation, which transforms a helicoid into a catenoid (β-barrel) (Louie and Somorjai 1982).

The remarkable features of proteins lie in the subtle interplay between the homo- and heteropolymeric aspects. For instance, dipolar effects from both the substrate and ligand should affect the active site. The dynamics of folding also reflects this duality. One may, for instance, consider the disorder as necessary to restrict the dipolar search through a large number of domain structures. It is perfectly conceivable that a hydrophobic collapse occurs first (due to the most hydrophobic residues) and that the hydrophobic core acts as a kind of boundary conditions for the dipolar domain structure, yielding a single solution after some time.

The answer to the question raised in the title of this paper is clearly "no." In our opinion, sequence diversity (or its poor man's version, randomness) is necessary for the appearance or selection of biological functions through the design of a proper active site.

Finally, the great flexibility of proteins, required to achieve biological activity, involves excitation energies which are fractions of k_{BT}. It is not clear that classical statistical physics is the proper tool to handle these complex molecules.

REFERENCES

Berry, R.S., N. Elmaci, J.P. Rose, and B. Vekhter. 1997. Linking topography of its potential surface with the dynamics of folding of a protein model. *Proc. Natl. Acad. Sci. USA* **94**:9520–9524 (and references therein).

Creighton, T.E., ed. 1992. Protein Folding. New York: W.H. Freeman.

Derrida, B. 1981. Random Energy Model: An exactly solvable model of disordered systems. *Phys. Rev. B* **24**:2613.

Doniach, S., T. Garel, and H. Orland. 1996. Phase diagram of a semiflexible polymer chain in a Θ solvent: Application to protein folding. *J. Chem. Phys.* **105**:1601.

Elber, R., ed. 1996. Recent Developments in Theoretical Studies of Proteins. Singapore: World Scientific.

Garel, T., H. Orland, and E. Pitard. 1997. Protein folding and heteropolymers. In: Spin Glasses and Random Fields, ed. A.P. Young, pp. 387–443. Singapore: World Scientific.

Gordon, M., P. Kapadia, and A. Malakis. 1976. The graph-like state of matter VII. The glass transition of polymers and Hamiltonian walks. *J. Phys.* **A9**:751.

Head-Gordon, T., F.H. Stillinger, M.H. Wright, and D.M. Gay. 1992. Poly (L-alanine) as a universal reference material for understanding protein energies and structures. *Proc. Natl. Acad. Sci. USA* **89**:11,513.

Hyde, S., S. Andersson, K. Larsson, Z. Blum, T. Landh, S. Lidin, and B.W. Ninham. 1997. The Language of Shape. Amsterdam: Elsevier (and references therein).

Irbäck, A., C. Peterson, and F. Rotthast. 1996. Evidence for non-random hydrophobicity in protein chains. *Proc. Natl. Acad. Sci. USA* **93**:9533.

Klimov, D.K., and D. Thirumalai. 1996. A criterion that determines foldability of proteins. *Phys. Rev. Lett.* **76**:4070.

Levinthal, C.J. 1968. Are there pathways for protein folding? *Chim. Phys.* **65**:44.

Louie, A.H., and R.L. Somorjai. 1982. Differential geometry of proteins: A structural and dynamical representation. *J. Theor. Biol.* **98**:189.

Mézard, M., G. Parisi, and M.A. Virasoro. 1993. Spin Glass Theory and Beyond. Singapore: World Scientific.

Minor, D.L., and P.S. Kim. 1996. Context dependent secondary structure formation of a designed protein sequence. *Nature* **380**:730.

Orlandini, E., and T. Garel. 1998. Collapse transitins of a periodic hydrophilic hydrophobic chain. *Eur. Phys. J.* **B6**:101.

Palmer, R.G. 1982. Broken ergodicity. *Adv. Phys.* **31**:669.

Pitard, E., T. Garel, and H. Orland. 1997. Protein folding, anisotropic collapse, and blue phases. *J. Phys.* **7**:1201.

Šalí, A., E.I. Shakhnovich, and M. Karplus. 1994. How does a protein fold? *Nature* **336**:451.

Shakhnovich, E.I. 1994. Proteins with selected sequences fold into their unique native conformation. *Phys. Rev. Lett.* **72**:3907.

Zhou, Y., M. Karplus, J.M. Wichert, and C.K. Hall. 1997. Equilibrium thermodynamics of homopolymers and clusters: Molecular dynamics and Monte Carlo simulations of systems with square well interactions. *J. Chem. Phys.* **107**:10,691 (and references therein).

8

Hemoglobin and Allostery

Is it Understood?

W.A. EATON

Laboratory of Chemical Physics, Bldg. 5, National Institute of Diabetes and Digestive and Kidney Disease, National Institutes of Health, Bethesda, MD 20892–520, U.S.A.

Keywords: structural biology, protein function, hemoglobin, allostery

ABSTRACT

The explosion in structural biology has created a challenge to the physical scientist. Can quantitative studies provide a truly deeper understanding of protein function than is obtained from static structures and qualitative analysis? This question is addressed by presenting the key concepts and experimental results that have led to our current understanding of how hemoglobin works, the paradigm of allostery. It is concluded that the underlying simplicity in the mechanism of cooperative oxygen binding by hemoglobin could not have been exposed without novel physical experiments and a rigorous quantitative analysis.

INTRODUCTION

A scientific question can be answered at many levels. An answer that may be quite satisfying for scientists from one discipline may be totally unsatisfactory to another, and vice versa. This is an increasingly important issue as scientists from different disciplines focus more and more on what are, at least superficially, very similar questions concerning biological problems. Several such questions were addressed at this workshop. Perhaps foremost among these was: how do proteins function? "Protein function" has come to take on a variety of meanings. To a genomist, faced with trying to understand the role of tens of thousands of proteins, "function" is a brief descriptor of what the protein does in a cell, such as "bind ligand *x*," or "oxidize protein *y*," or "phosphorylate protein *z*," etc. To many biologists the problem of understanding, say, the ligand-binding function may be considered solved once the three-dimensional atomic structure is known, and the binding site identified as the constellation of amino

---

Simplicity and Complexity in Proteins and Nucleic Acids
Edited by H. Frauenfelder, J. Deisenhofer, and P.G. Wolynes. Published 1999 Dahlem University Press

acid residues that influences the reaction in experiments on site-directed mutants. To more physically oriented scientists, solution of the structure marks just the first, albeit absolutely essential, step in understanding how this protein functions. However, an interesting challenge for the physical scientist is to demonstrate that a truly deeper understanding of protein function can be obtained from quantitative studies which is not already apparent from the beautiful, static color pictures of the structural biologist and qualitative analysis. This is the subject that I will address in this article. I will confine my discussion to a single protein molecule, hemoglobin (Figure 8.1). Hemoglobin is the most intensively studied of all proteins, having consumed more human time and effort than any other. It is widely considered as the paradigm for understanding the relation between structure and function in proteins at the most basic physical level. There is, moreover, scarcely a single fundamental issue concerning the biology, chemistry, or physics of proteins that has not arisen in research on hemoglobin.

My selection of topics and citations from the vast hemoglobin literature reflects a personal view of what I have found interesting and important to understand. It is my hope, however, that the issues raised will coincide with those of other biophysical scientists trying to understand how proteins work. For the sake of both clarity and brevity, the structure of hemoglobin is not considered in any detail. This is necessary to expose the fundamental issues, as well as the simplicity that underlies hemoglobin's complex

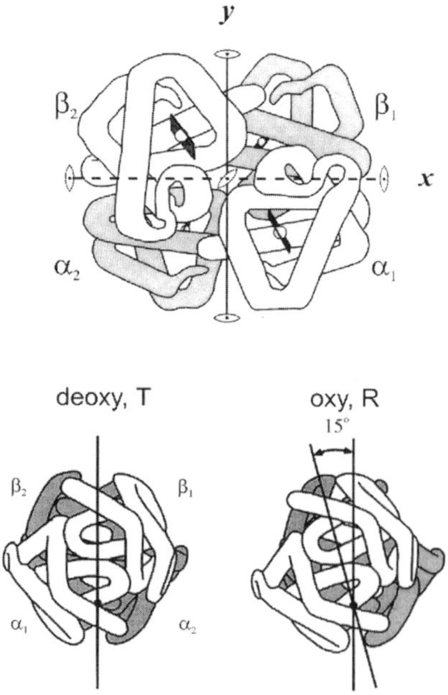

Figure 8.1 Schematic structures of hemoglobin (adapted from Dickerson and Geis 1983). The views at the bottom are looking down the pseudo twofold x axis. The quaternary conformational change consists of a rotation of the symmetrically related $\alpha\beta$ dimers by about 15 degrees relative to each other and a translation of about 0.1 nm along the rotation axis.

structure and behavior. Although the discussion will be confined to concepts and key experimental results, I should emphasize that almost every breakthrough in understanding followed major technological advances in experimentation.

HISTORICAL BACKGROUND: FROM BOHR TO PERUTZ

To make the issues more transparent, it is instructive to examine the evolution of our understanding of hemoglobin in historical sequence. Even though this history spans almost a century (Edsall 1980), it is quite relevant to the current study of all types of proteins. With today's technology, comparable results on a newly discovered protein might be obtained in a much shorter time, but the questions would be very similar.

Hemoglobin belongs to a class of multisubunit proteins, now called allosteric proteins. The biologically important and physically interesting property of these proteins is that they exhibit cooperative interactions between ligand-binding sites distant from each other in the structure. The study of cooperative interactions in proteins began with the work of the physiologist Christian Bohr, father of the atomic physicist, Niels Bohr. Bohr made two seminal observations (Bohr 1904; Edsall 1980). First, his careful measurements of hemoglobin (Hb) oxygenation showed that the binding curve is sigmoidal instead of hyperbolic, as would occur for a simple $Hb + O_2 \leftrightarrow HbO_2$ equilibrium. As the number of oxygen molecules bound increases in the association reaction, the apparent binding affinity increases; and conversely, in dissociation, the apparent affinity decreases as the number of bound oxygen molecules decreases (Figure 8.2). Second, Bohr discovered that carbon dioxide lowers the oxygen affinity (the necessary reciprocal effect of increased affinity for carbon dioxide and protons as the oxygen pressure is lowered was measured much later). These cooperative properties make hemoglobin an efficient transporter of oxygen from the lungs to the tissues and of carbon dioxide from the tissues to the lungs (Figure 8.2). This was all known by the year 1904. Both the protein and its biological function had been clearly identified, and the functional significance of cooperative binding had been explained. Would a genomist want to know more? This of course is only the very beginning of the subject, and a fundamental question remained: *how* does hemoglobin exhibit this cooperative behavior?

The next major step came with the work of Gilbert Adair. Adair made extremely precise osmotic pressure measurements over a very wide range of hemoglobin concentrations (Adair 1925; Edsall 1980). From these he determined the molecular weight of hemoglobin to be 68,000. It was known by then that about 17,000 grams of protein contained 1 mole of iron atoms, so Adair had discovered that hemoglobin contained four binding sites and formulated binding in terms of four equilibrium constants. In this phenomenological description, cooperativity was quantitatively described by the increase in the four apparent binding constants with the addition of each oxygen molecule to hemoglobin. In Adair's scheme, the fourth oxygen molecule binds with an affinity ~100-fold greater than the first.

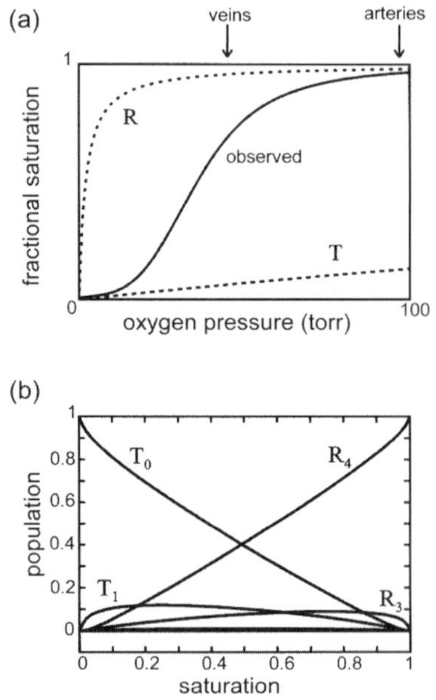

Figure 8.2 Monod–Wyman–Changeux (MWC) description of hemoglobin oxygenation. (a) The observed binding curve and the corresponding hyperbolic (noncooperative) binding curves for the R- and T-quaternary structures. Binding begins at low oxygen pressure along the T state binding curve and ends at high pressure along the R state binding curve. As successive oxygen molecules bind, the equilibrium between the two quaternary structures shifts towards R. The sigmoidal shape of the overall curve and its shift to the right by acid and carbon dioxide facilitates loading of oxygen in the lungs and unloading in the tissues. (b) Population of 10 tetrameric species as a function of ligand saturation for a two-state MWC allosteric model with four identical binding sites (parameters of Henry et al. 1997). The populations of T_2, T_3, T_4, R_0, R_1, and R_2 are also shown, but are too small to be distinguished on this scale.

A few years later, the legendary chemist Linus Pauling suggested the first explanation of cooperative binding (Pauling 1935). Pauling found that he could adequately describe the oxygen binding curve with a two-parameter model in which binding oxygen to one heme caused an interaction with neighboring hemes that increased their affinity. One parameter was defined as the intrinsic binding constant and the other as an interaction parameter. He reasoned that a single interaction parameter required hemoglobin to possess either a tetrahedral or square symmetric arrangement of the hemes. Pauling preferred the latter, because it allowed him to explain the change in affinity as somehow resulting from a resonance electronic interaction between the hemes (he assumed that the hemes were on the surface of the molecule and that tetrahedral symmetry would place them too far away for the direct interaction he thought necessary). The year was 1935. At this point hemoglobin is a "P68" protein with four hemes which serve as binding sites for oxygen; its cooperative behavior could be explained by a sequential model in which binding to one heme raises the affinity of its neighbors through a direct heme–heme interaction. Would a modern-day biologist really want more of an explanation? The answer is almost certainly yes, but *not* necessarily to find out if Pauling was right, or to inquire more deeply into the mechanism of cooperativity. Instead the biologist would like to have a structural basis for understanding the alteration in function produced by mutations, particularly to understand how mutations cause diseases such as sickle cell anemia (the first example of a "molecular disease," also discovered by Pauling [Pauling et al. 1948]). To answer these questions the biologist would quite reasonably wish for a three-dimensional structure at atomic resolution.

Enter Max Perutz. After almost three decades of pioneering work in the development of the X-ray crystallographic method for determining the three-dimensional structure of proteins, Perutz solved the structure of hemoglobin. His first major result was a comparison of the crystal structures of fully deoxygenated hemoglobin and fully oxygenated (actually, considerably oxidized) hemoglobin at low resolution (Perutz et al. 1964). Perutz expected to see a change in molecular structure because of early observations by Felix Haurowitz that hemoglobin crystals shattered upon changing oxygenation state. The electron density map indeed showed that the β subunits move closer together upon oxygenation. This was a key experimental fact that helped motivate a completely different and powerful theory on the origin of cooperative interactions, not only for hemoglobin oxygenation but for proteins in general.

Jacques Monod and Jean-Pierre Changeux wanted to understand how metabolic pathways are regulated. They observed that many enzymes are activated and inhibited by substrates and ligands in a cooperative fashion, and that these enzymes contained more than one protein subunit. They also saw the analogy to hemoglobin and enlisted the help of the biophysical chemist and hemoglobin expert, Jeffries Wyman, to develop a mathematical formulation of a theory that would apply to all types of multisubunit proteins (Monod et al. 1965). Their theoretical model was very different from Pauling's model, later elaborated by Daniel Koshland and coworkers (Koshland et al. 1966). Koshland suggested that the change in affinity resulted from some kind of conformational change of subunits upon ligand binding that is transmitted directly to adjacent subunits (Koshland et al. 1966). In contrast, in the theory of Monod, Wyman, and Changeux (MWC), cooperativity arises from an equilibrium between two structures having different arrangements of the subunits, so-called quaternary structures (Figure 8.1). The "tense" or T-quaternary structure has a low affinity for ligand, while the "relaxed" or R-quaternary structure has a high affinity for ligands. In the case of hemoglobin, the ligand is oxygen while for enzymes the ligand is the substrate. According to the MWC model, cooperativity arises from a shift in the population from the low affinity T-quaternary structure to the high-affinity R-quaternary structure with increasing oxygen pressure, as required by Le Chatelier's principle (Figure 8.2a). Binding to each quaternary structure, however, is purely statistical, and the intrinsic affinity of each heme for oxygen is independent of the number of oxygen (or substrate) molecules already bound (Figure 8.2). Cooperative interaction between identical ligands was called "homotropic," while the cooperative interaction between unlike ligands, such as oxygen and protons (or substrate and inhibitor in the case of allosteric enzymes) was called "heterotropic." The model is extremely simple. It is, however, subtle, with important implications for protein structural changes in general, and also for the evolution of protein function.

Perutz took steps far beyond "merely" solving the structures of oxy- and deoxyhemoglobin, now identified with the R- and T-quaternary structures (Figure 8.1). He wanted to explain as much of the existing biochemical and mutational data as possible, and did so with his "stereochemical mechanism" (Perutz 1970). In a tour-de-force Perutz saw through the complexity of his atomic models and proposed a detailed structural explanation for exactly how hemoglobin worked. In comparing the two quaternary structures he discovered a set of salt bridges at the subunit interfaces that were present in the T-quaternary structure but absent in R. Perutz described in great detail how oxygen binding to the heme in the T-quaternary structure with its associated iron displacement moves a helix, breaks a salt bridge, releases a proton, and destabilizes the structure at the subunit interface between the component $\alpha\beta$ dimers of the tetramer (Figures 8.1 and 8.3). In this way, oxygen binding to the T-quaternary structure biases the quaternary equilibrium toward the R state. Perutz attributed three roles to the salt bridges: stabilization of the T-quaternary structure relative to R, lowering of the oxygen affinity in T because of the energy required to break them upon oxygen binding, and release of protons upon breakage explaining why the overall affinity is lowered when the pH decreases (as occurs in the tissues). He viewed the mechanism as a combination of the MWC model and the Koshland, Nemethy, and Filmer (KNF) models, because a key component of the KNF model is that ligand binding induces conformational changes in the protein. Oxygen binding results in conformational changes, but there appeared to be no mechanism for transmitting these changes to the heme of the neighboring subunits other than via a change in quaternary structure. For homotropic effects Perutz's mechanism is therefore pure MWC in that the intrinsic affinity of a subunit is solely determined by the quaternary structure. This issue caused considerable confusion to many hemoglobin researchers and created much controversy, which has persisted until quite recently (Perutz 1989) (*vide infra*).

Perutz's mechanism was also successful in qualitatively rationalizing the altered behavior of dozens of naturally occurring hemoglobin variants containing single amino acid substitutions, many of them producing human pathology (Perutz and Lehman 1968). For example, his mechanism explained how substitutions at the subunit interface, distant from the heme, produced changes in oxygen affinity by their effect on the quaternary equilibrium. He found that the β6glu→val mutation in hemoglobin S is on the molecular surface, creating a sticky hydrophobic patch which promotes polymerization, the cause of sickle cell disease (Eaton and Hofrichter 1990). An analysis of residues that have a specific role in his mechanism, moreover, showed an interesting relation between exons and function and led to a proposal for the evolution of cooperative binding by exon shuffling (Eaton 1980).

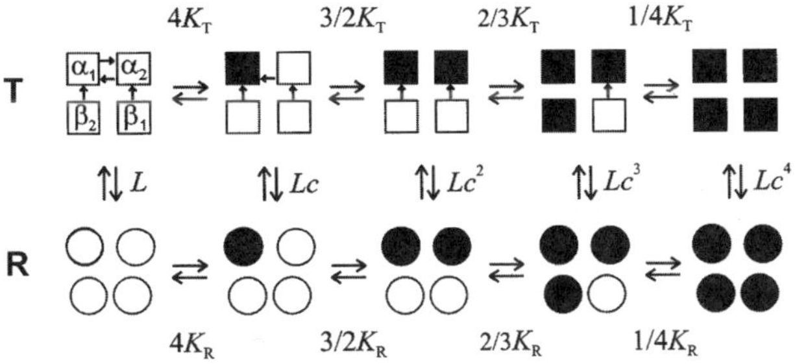

Figure 8.3 Simplified schematic of MWC–PSK model. Open symbols designate unliganded subunits and filled symbols liganded subunits. Arrows connecting subunits represent salt bridges — the quaternary bonds of MWC theory that constrain and stabilize the T-quaternary structure. There are actually six intersubunit salt bridges in the T-quaternary structure (each arrow between α subunits represents a pair of salt bridges) that are not present in the R-quaternary structure and two *intra*subunit salt bridges in the β subunits. Four of the eight salt bridges contain ionizable groups, and their breakage produces a change in pK of the participating residues that leads to a net release of protons. Proton release upon binding to the T-quaternary structure is called the "tertiary Bohr effect," while proton release upon the change of quaternary structure is called the "quaternary Bohr effect." Ligand binding to a subunit in the T-quaternary structure breaks the salt bridge originating from that subunit, while the change from the T- to R-quaternary structure at any ligation state breaks all four salt bridges. Ligand binding to the T-quaternary structure occurs with an association constant that is simply the R-state association constant reduced by a factor which is proportional to the strength of the salt bridge.

Perutz's stereochemical mechanism was a turning point in the history of proteins, for it represented the beginning of an era in which structural changes at the atomic level were being used to explain how a protein functioned. The year was 1970. To many structural biologists, the "hemoglobin problem was solved"; its function was understood in terms of its three-dimensional structure. The subject was closed, and it was time to move on to the next protein. Can further inquiry by physical scientists lead to a truly deeper understanding of how hemoglobin functions?

UNDERSTANDING HEMOGLOBIN ALLOSTERY BY PHYSICAL SCIENTISTS

Hemoglobin at Equilibrium

Perutz's bold approach to understanding hemoglobin function in detailed structural terms aroused the interest of many physical scientists, both theorists and experimentalists, and stimulated an enormous amount of research. Perutz's mechanism *qualitatively* explained a vast array of experimental facts. Could it, however, also provide a *quantitative* explanation? Monod, Wyman, and Changeux had invented a very general model with little structural detail, while Perutz had proposed a detailed structural mechanism but had no prescription on how to make it quantitative. What was the connection between the two?

The important first step to help settle this issue was taken by Attila Szabo and Martin Karplus (Szabo and Karplus 1972). They developed a detailed statistical mechanical formulation of Perutz's structural model (Figure 8.3). Szabo and Karplus clearly showed that the Perutz mechanism is in fact a two-state allosteric model for homotropic effects (the interaction between oxygen binding sites), and that his mechanism could provide a quantitative explanation of equilibrium experimental data. Ignoring heterotropic effects and assuming equal intrinsic affinity for the α and β hemes, the MWC partition function is simply:

$$\Theta = L(1+K_T p)^4 + (1+K_R p)^4 \tag{8.1}$$

where L is the ratio of the T to R populations in the absence of oxygen, K_R is the affinity of subunits in the R-quaternary structure, and K_T is the affinity of subunits in the T-quaternary structure. In a simplified version of the Szabo and Karplus formulation, the ratio of affinities $(c \equiv K_T/K_R)$ is simply the strength of an intersubunit salt bridge, the structural origin of the lower T-state affinity (Figure 8.3). The model therefore made the interesting prediction that once the salt bridges are completely broken, as in the fully oxygenated molecule, the quaternary equilibrium constant (Lc^4) would be relatively insensitive to solution conditions. This prediction of $Lc^4 \approx$ constant was dramatically confirmed a decade later (Imai 1983). Szabo and Karplus also recognized that the MWC model had to be modified to account for heterotropic effects in the absence of a quaternary change, particularly the influence of pH. This is, however, not a serious criticism of the MWC model. Monod, Wyman, and Changeux clearly recognized that it was unrealistic to expect their model to work exactly for heterotropic effects (Monod et al. 1965).

The Szabo (S) and Karplus (K) model is appropriately called the MWC–PSK model, since it is based on both the structural mechanism of Perutz (P) as well as the two-state allosteric model of MWC. For the past 25 years the MWC–PSK model has provided a powerful framework for synthesizing the results from spectroscopic and equilibrium experiments. The issue that has caused the most controversy during this period is whether the homotropic effect in hemoglobin is adequately described by a two-state MWC model. The applicability of the MWC model was supported by a series of novel spectroscopic studies and insightful analyses by Robert Shulman, John Hopfield, and Seiji Ogawa (Shulman et al. 1975). This group demonstrated in a variety of ways that under a fixed set of solution conditions there are only two populations of molecules: high-affinity molecules with spectroscopic signatures of the R-quaternary structure, and low-affinity molecules with spectroscopic signatures of the T-quaternary structure. They also emphasized that the observation of a spectroscopic change in an unliganded subunit produced by ligand binding to a neighboring subunit without a change in quaternary structure is no evidence for a breakdown of the MWC model (as has often been claimed), unless it is also demonstrated that the affinity is altered.

The question of the quantitative applicability of a two-state model has always really centered on the binding properties of the low-affinity T-quaternary structure. Hemoglobin in the R-quaternary structure binds oxygen with the same high affinity as the $\alpha\beta$ dimer (i.e., the half-molecule, Figure 8.1). The increase in affinity with increasing oxygen saturation, therefore, arises from a release of the quaternary constraints in the T-state tetramer that are responsible for the low affinity. Cooperativity must therefore arise from cooperative binding to T prior to a quaternary structural change to R, from the increase in affinity resulting from the quaternary transition from T to R, or from both. That is, does cooperativity arise solely from the T to R transition as in the MWC–PSK model, or is there significant cooperativity in binding to T, as in a sequential model? To address this question directly, Andrea Mozzarelli and coworkers (Rivetti et al. 1993) measured oxygen binding to single crystals of hemoglobin known to remain in the T-quaternary structure by X-ray crystallography (Liddington et al. 1988). Possible differences between crystal and solution were now not an issue and there was no ambiguity about the quaternary structure. These studies confirmed the essential point of the MWC model that the T-quaternary structure binds oxygen non-cooperatively (Figure 8.4). The crystal binding curve is noncooperative with a Hill *n* of almost exactly 1.0. This is, however, the result of a small amount of cooperativity, amounting to less than 10% of the difference in binding free energy for the first and fourth ligands, which is exactly compensated by a small difference in affinity of the α and β subunits.[1]

Because X-ray crystallography of these crystals showed that oxygenation does not break the salt bridges (Liddington et al. 1988), the crystal-binding studies supported two other major aspects of Perutz's mechanism. According to the Perutz mechanism, there should be no pH-dependence to the T-state affinity (tertiary Bohr effect) unless the salt bridges break. As predicted, none was observed in the crystal. His mechanism also implies that there would be strain in the oxygenated T state if the salt bridges do not break, predicting a lower oxygen affinity in the crystal compared to solution. This lowered affinity in the crystal was also observed (Rivetti et al. 1993). More recently,

[1] The original MWC model is easily extended to include inequivalent binding of α and β subunits, with the affinity of each depending only on quaternary structure (i.e., no cooperative binding within T or R). There is in principle, however, a position dependence on the quaternary equilibrium because of the lack of symmetry. Hemoglobin does not have pseudo-tetrahedral symmetry with only one kind of intersubunit interface, but has pseudo 222 (D_2) symmetry with four kinds (Figure 8.1). For doubly liganded molecules the quaternary equilibrium constant with both ligands on the same $\alpha\beta$ dimer [$\alpha1(x)\beta1(x)\alpha2\beta2$] must therefore differ from that with one ligand on each type of subunit, but on separate dimers [$\alpha1(x)\beta1\alpha2\beta2(x)$]. A consequence of the inequivalence of these two doubly-liganded structures is that the T-quaternary structure could bind ligands cooperatively. The crystal results show that this effect, which may be exaggerated in chemically modified hemoglobins used as analogs for oxygenation intermediates (Ackers 1990; Edelstein 1996), is quite small.

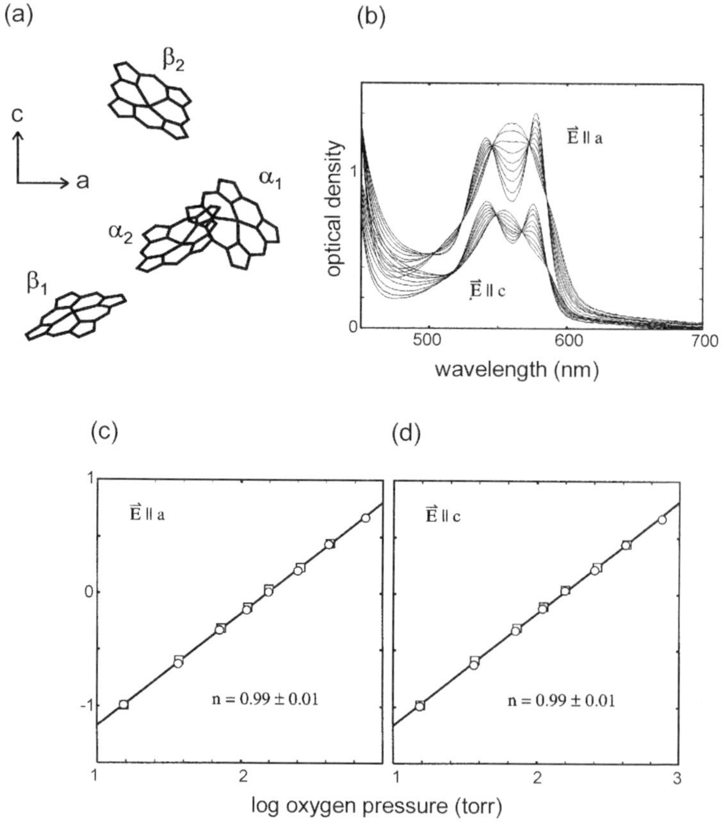

Figure 8.4 Oxygen binding to a single crystal of hemoglobin in the T-quaternary structure (from Rivetti et al. 1993). (a) Projection of four hemes onto the (*ac*) crystal face of the optical measurements. (b) Absorption spectra of a crystal at 9 oxygen pressures between 0 and 760 torr with light linearly polarized parallel to either the *a* or *c* crystal axes. (c) Hill plot of fractional saturation with oxygen (y) versus oxygen pressure from spectra measured with linearly-polarized light parallel to the *a* crystal axis. The circles are increasing oxygen pressure, while the squares are for decreasing oxygen pressure. Binding is perfectly reversible and noncooperative. (d) Same as (c) except that the light is polarized parallel to the *c* crystal axis. Using the polarized absorption spectra and the heme orientations derived from the X-ray structures, the individual binding curves of the α and β subunits can be obtained (Makinen and Eaton 1974; Rivetti et al. 1993). This analysis shows that the subunit has a two- to fivefold higher affinity than the β subunit, and that this subunit inequivalence is exactly compensated by a small amount of cooperativity, resulting in a noncooperative binding curve for the tetramer (Rivetti et al. 1993; Mozzarelli et al. 1997).

Mozzarelli has found that deoxyhemoglobin encapsulated in a silica gel also binds oxygen noncooperatively (observed previously by Shibayama and Saigo 1995), has a higher oxygen affinity than the crystal, and exhibits most of the tertiary Bohr effect (Bettati and Mozzarelli 1997). It would appear that in the silica gel hemoglobin is trapped in the T-quaternary structure, as in the crystal, but in the gel the salt bridges break upon oxygen binding, as envisaged by Perutz for hemoglobin in solution. It is ironic that oxygen binding by hemoglobin is so different in crystal and solution, yet crystal binding studies strongly support Perutz's mechanism for solution binding, a mechanism based on the structure of hemoglobin in the crystal. The MWC–PSK model has no problem explaining this otherwise perplexing result.

At this point every major element of the equilibrium formulation of the MWC–PSK model had been subjected to critical tests, with no major discrepancies between theory and experiments. For biochemists, cooperative oxygen binding by hemoglobin is not only well understood, but possibly better understood than the function of any other protein. There was, however, a large body of work on the complex kinetics of hemoglobin to be explained. In general, models for complex systems are more critically tested by kinetics. In kinetic experiments, thermodynamically unstable intermediates can be generated and examined spectroscopically. Kinetic experiments are particularly important for the hemoglobin mechanism, where there are a large number of postulated intermediate species, most of them at extremely low population at equilibrium (Figure 8.2b).

Hemoglobin Kinetics

The critical observation for understanding kinetic cooperativity was made by Quentin Gibson (Gibson 1959). Gibson discovered that the bimolecular rate of carbon monoxide rebinding following almost complete photodissociation by an intense light pulse is more than 20 times faster than the initial rate obtained by mixing deoxyhemoglobin with carbon monoxide. This result suggested that the rate of ligand binding, and therefore the ligand affinity, is not determined by the number of ligands already bound, as required by the Pauling/KNF sequential models. Gibson also observed that the deoxyhemoglobin optical spectrum produced immediately after the flash is different from that of deoxyhemoglobin at equilibrium. He called this the spectrum of "fast-reacting" hemoglobin, Hb*, and suggested that Hb* has a structure different from equilibrium deoxyhemoglobin. This work was followed by a variety of innovative kinetic experiments using rapid mixing and flash photolysis methods by Gibson, Eraldo Antonini, and Maurizio Brunori. This work is summarized in the now classic book by Antonini and Brunori (1971).

The important connection to the MWC model was made by Hopfield, Shulman, and Ogawa (Hopfield et al. 1971). They identified Hb* as deoxyhemoglobin in the R-quaternary structure. These investigators also showed that almost all of the kinetic experiments on both ligand association and dissociation could be explained with a simple two-state allosteric model. In their kinetic version of the MWC model there are only two binding rates, one for T and one for R, and correspondingly two dissociation rates. There were no direct experimental measurements of any of the 10 quaternary conformational rates (2 for each ligation state), but existing data suggested that they are rapid. Hopfield, Shulman, and Ogawa therefore made the simplifying assumption that all quaternary rates are instantaneous.

Although this kinetic analysis provided additional strong support for the MWC model, the fits to the data were not quantitative. There also remained a major puzzle. Equilibrium studies had suggested that oxygen and carbon monoxide bind with the same cooperative mechanism. Cooperativity is, however, manifested very differently in the kinetics for the two ligands. The oxygen dissociation rate is highly cooperative, speeding up as the fractional saturation with ligand decreases, while the association rate is only weakly cooperative. The reverse is true for carbon monoxide, with greater cooperativity in the association rate than in the dissociation rate. In terms of the MWC model, the dissociation rate of oxygen is much higher in T than in R, while the association rate is only slightly higher in R than in T. Again, the reverse is true for carbon monoxide, with a higher association rate to R compared to T, and a smaller difference in the dissociation rate between R and T. If hemoglobin had a single mechanism why do the two ligands behave so differently? Szabo resolved this apparent paradox by showing how it is a direct result of Perutz's mechanism. He developed a clever argument based on transition state theory, the stereochemistry of carbon monoxide and oxygen binding in the heme pocket, and on the suggestion from Perutz's mechanism that it would be harder to move the iron into the heme plane to reach the transition state in T than in R.[2] It provided the much-needed justification for drawing conclusions on the mechanism of cooperativity from kinetics using carbon monoxide (more frequently employed in kinetic experiments than oxygen because the carbon monoxide complex is more stable to oxidation and has a higher photolysis quantum yield). His analysis also explained the slower binding of carbon monoxide compared to oxygen, even though carbon monoxide has a much higher affinity.

---

2 Szabo postulated that the transition state is close to the reactant (ligand next to the heme, but not bound to the iron) in the case of oxygen, but closer to the product (ligand covalently bonded to the iron) in the case of carbon monoxide. In the reaction with carbon monoxide covalent bonding energy is used to overcome steric obstacles, requiring the iron to move into the plane of the porphyrin ring to form a partial Fe·····C–O bond in the transition state. Because of the additional constraints in T, it is harder to move the iron into the plane in T than in R, resulting in a larger barrier, and, therefore, a slower binding rate. In the dissociation reaction the iron moves less to reach the transition state, so the contribution of the protein conformation to the barrier heights is more similar in R and T. The same reasoning holds for oxygen, except in this case the bent Fe–O–O bond does not encounter steric obstacles at the heme, and the transition state is reached with little Fe·····O–O partial bond formation and therefore little motion of the iron toward the porphyrin plane. An alternative explanation for the slower binding of carbon monoxide compared to oxygen is the larger energy barrier created by crossing of potential surfaces of spin different multiplicity, and the larger net spin change in the reaction of quintet iron of the deoxyheme with singlet carbon monoxide ($\Delta S=2$) versus triplet oxygen ($\Delta S = 1$) (Szabo 1978; Greene et al. 1978; Frauenfelder and Wolynes 1985).

There was still a crucial missing piece to the story. In spite of the importance of protein conformational changes in both the MWC and KNF models, their kinetics had not yet been observed. This prompted Gibson to look carefully at the spectra of deoxyhemoglobin in photolysis experiments. Using a technique with microsecond time-resolution, he observed deoxyheme spectral changes following carbon monoxide photodissociation in the microsecond-millisecond regime which he attributed to R- → T-quaternary conformational changes (Sawicki and Gibson 1976). Gibson was, however, unable to fit the ligand-rebinding kinetics and spectral changes simultaneously with an MWC model (except at high pH), and concluded that the experimental results are inconsistent with this model. This, of course, created a serious problem for the MWC model, a problem that was not solved for 20 years. Its solution required a combination of applications of new laser technologies to proteins pioneered by Robin Hochstrasser (Greene et al. 1978; Hochstrasser 1998), a deeper understanding of protein physics that grew out of the work of Hans Frauenfelder and Peter Wolynes (Austin et al. 1975; Frauenfelder and Wolynes 1985; Frauenfelder et al. 1991), and an enormous increase in computer power. Below I will briefly describe the four key ingredients that went into solving this problem.

Nanosecond-resolved Spectroscopy

The most revealing experiments were carried out by James Hofrichter and coworkers, who developed high-precision, nanosecond-resolved spectroscopy to measure hemoglobin optical spectra following carbon monoxide photodissociation in the time range from a few nanoseconds to ~50 ms (Figure 8.5) (Hofrichter et al. 1983; Jones et al. 1992). This work showed that the spectral changes of the deoxyhemes observed by Gibson begin much earlier than a microsecond, and appear to be a single continuous process extending from less than a nanosecond to milliseconds. No less than 7 exponential relaxation times were required to describe empirically the kinetics of spectral changes and of geminate and bimolecular ligand rebinding (Figure 8.5). Explaining such complex kinetics in a simple way with a physically meaningful set of parameters was a daunting task. There were, however, two observations in these nanosecond studies that simplified matters considerably. First, the ~50-fold slower overall rate of bimolecular binding to the T-quaternary structure could be completely explained by a slower geminate rebinding rate, indicating that there is no significant difference between the R- and T-quaternary structures in the rate of ligand motion into and out of the heme pocket (Murray et al. 1988). Second, both the kinetics of geminate rebinding and the kinetics of deoxyheme spectral changes, attributed to conformational relaxation, are independent of the degree of photolysis (and therefore the number of ligands still bound to the tetramer) for times less than ~1 μs, indicating that the subunits have not yet communicated their ligation state and that the quaternary structure remains R. Conformational changes prior to ~1 μs are therefore purely tertiary, while those occurring later include quaternary as well.

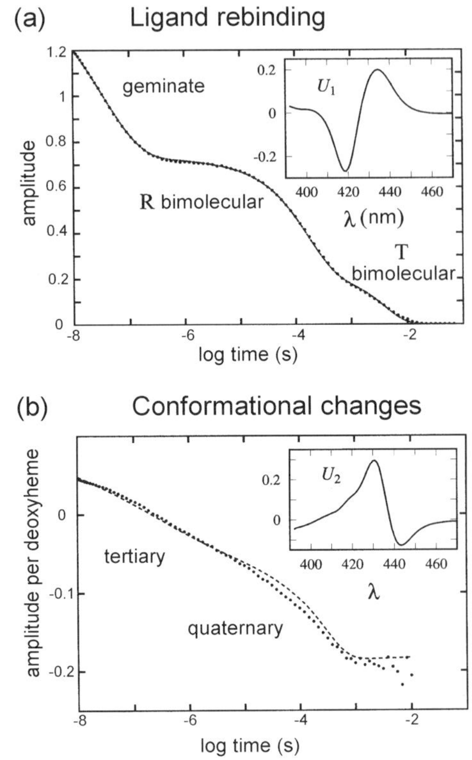

Figure 8.5 Kinetics of hemoglobin following nanosecond photodissociation of carbon monoxide complex (from Henry et al. 1997). (a) Ligand rebinding as determined by the time dependence of the amplitude of the first basis spectrum (inset) in a singular-value-decomposition (SVD) analysis of time resolved spectra. Geminate rebinding to R is followed by bimolecular rebinding to R and then to molecules which have switched from T to R. (b) Protein conformational changes as determined by the time dependence of the amplitude of the second (the only other significant) basis spectrum (inset) in the SVD analysis. In both (a) and (b) the points are experimental, and the dashed curves are calculated from the extended MWC model of Henry et al. (1997). The deviations between the calculated and observed curves in (b) may not be significant. The singular value of the second SVD component is only 7% of the first, so that deviation represents only about a 1% difference between the observed and calculated spectral amplitudes.

The Tertiary Relaxation Is a Single Process

A second key ingredient was the realization that the continuous tertiary relaxation is a single process, in which the protein structure changes from the liganded conformation, that exists prior to photodissociation to the conformation of the relaxed unliganded subunit, still in the R-quaternary structure; its multiexponential shape did not require consideration of new intermediate kinetic species of undetermined structure. This view was motivated by the results on myoglobin. Frauenfelder and Wolynes suggested, by analogy to relaxation in glasses, that conformational relaxation in myoglobin might exhibit an extended time course, often described by a stretched exponential function (Frauenfelder et al. 1991):

$$f(t) \propto \exp\left\{-(kt)^{\beta}\right\} 0 < \beta < 1 . \tag{8.2}$$

The initial observations on this relaxation in myoglobin following nanosecond photodissociation indicated nonexponential behavior (Ansari et al. 1992), but it was not until the work of Philip Anfinrud that there was compelling evidence for a continuous relaxation. Anfinrud observed spectral changes to begin at hundreds of femtoseconds and extend all the way to microseconds ($\beta \approx 0.1$) (Jackson et al. 1994). The extended time course could be explained by a model in which the interconversion among conformational substates is occurring on the same time scale as the overall change from the average liganded conformation to the average unliganded conformation (Hagen and Eaton 1996). These authors assumed a "kinetic rule" for the rate constants in which all transition states have the same energy ($E^{\ddagger}$). The transition rate from conformational substates with energy between E_i and E_i+dE to conformational substates with energy between E_j and E_j+dE is given by:

$$k_{ij} = k_0 g\left(E_j\right) dE \exp\left(-\frac{E^{\ddagger} - E_i}{RT}\right) , \tag{8.3}$$

where $g(E_j)$ is the density of states at E_j. In this model the activation energy becomes progressively larger and the apparent rate for substate interconversion slows dramatically as the system moves from the substates of the unrelaxed (liganded) conformation to those of the relaxed (unliganded) conformation at lower energy ($E^{\ddagger} - E_i$ continuously increases). The numerical solution of the master equation describing the overall conformational change from a Gaussian distribution of substate energies of the liganded conformation to a Gaussian distribution with the unliganded conformation closely approximates a stretched exponential function; more importantly, it fits the experimental kinetic curve perfectly (Hagen and Eaton 1996).

The Tertiary Relaxation Slows Geminate Rebinding

A third ingredient was the suggestion by Noam Agmon and John Hopfield that conformational relaxation raises the barrier to carbon monoxide rebinding (Agmon and Hopfield 1983). The most direct test of this idea was made by embedding myoglobin in a glass of the sugar trehalose at room temperature (Hagen et al. 1996). This experiment was suggested by the relation for the viscosity dependence of the relaxation rate (k_{conf}) in myoglobin (Ansari et al. 1992):

$$k_{conf} \propto (\sigma + \eta)^{-1} \exp\left(-\frac{E_0}{RT}\right) , \tag{8.4}$$

where η is the viscosity of the solvent and σ is the contribution to the overall friction from interaction between protein atoms. The high viscosity of the glass traps the substates, preventing their interconversion, just as in Frauenfelder's cryogenic temperature experiments in a glycerol/water glass ($\eta > 10^{15}$ cP) (Austin et al. 1975). As in the low-temperature glass, myoglobin in a room-temperature glass exhibits distributed geminate rebinding rates and kinetic hole burning. The substate-averaged rate $\langle k_{BA} \rangle$ for geminate rebinding (B→A) in the glass,

$$\langle k_{BA} \rangle = \int_0^\infty g(H_{BA}) k_{BA}(H_{BA}) dH_{BA} \ , \tag{8.5}$$

calculated from the observed energy barrier distribution ($g(H_{BA})$), is ~2,500 times faster than the rate observed for the relaxed protein (in which there is near-averaging and therefore near exponential rebinding). This result is the most striking confirmation of the Agmon/Hopfield idea that conformational relaxation slows the rate of geminate rebinding, and demonstrates again the importance of understanding the properties of conformational substates in proteins (Hagen et al. 1996).

Linear Free-energy Relation between Quaternary Rates and Equilibria

The final key ingredient was the simplification of the description of the quaternary rates. The MWC model does not define any relation among quaternary rates. There are ten such rates, defining the R↔T equilibrium constants at the five ligation states. The observation of a linear free-energy relation, however, between quaternary rates and equilibrium constants,

$$k(R_i \rightarrow T_i) = \gamma\left(Lc^i\right)^{\alpha} \text{ and } k(T_i \rightarrow R_i) = \gamma\left(Lc^i\right)^{\alpha-1}, i = 0,1,2,3,4, \tag{8.6}$$

reduced the description of the quaternary rates to the two MWC parameters and two new kinetic parameters: α which is a measure of the distance along a reaction coordinate between R and T, and γ which sets the time (Eaton et al. 1991). Since $\alpha \approx 0.2$, it was suggested that the transition state for the quaternary change has thermodynamic properties and, therefore, presumably a structure which is much closer to R than T. The same conclusion was reached from a comparison of the activation and equilibrium enthalpies and entropies for the quaternary change of a hemoglobin that has no complicating heterotropic effects (Hofrichter et al. 1991).

The Extended MWC Kinetic Model

The new findings described above motivated a straightforward extension by Eric Henry and coworkers of Gibson's kinetic version of the MWC model (Sawicki and Gibson 1976; Henry et al. 1997). Their extended MWC model adds geminate rebinding, the change in tertiary conformation prior to the quaternary change, and a linear free-energy relation between quaternary rates and equilibria (assumed by Gibson)

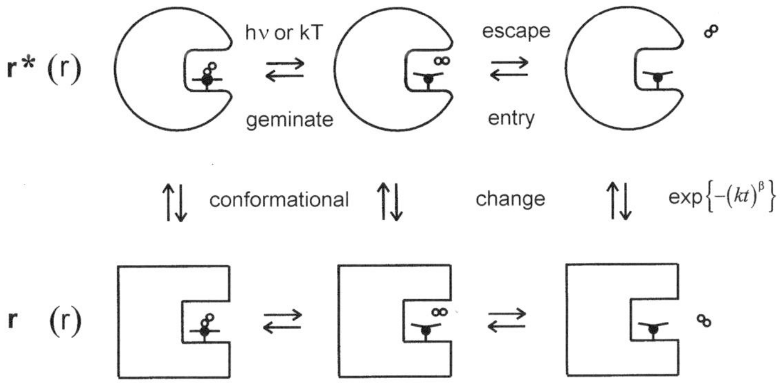

Figure 8.6 Schematic of extended MWC kinetic model for a single subunit in the R-quaternary structure (Henry et al. 1997). A ligand dissociated from the heme or entering the heme pocket from the solvent may either bind or escape into the solvent. The rate of ligand entry and escape is independent of quaternary or tertiary structure, while the bond-making and -breaking steps at the heme depend on both the tertiary and quaternary structure. Following dissociation, the subunit conformation relaxes from r*, the liganded conformation existing prior to dissociation, to r, the liganded conformation in the relaxed equilibrium R-quaternary structure. The rate of the R to T-quaternary transition now depends both on the number of ligands bound and the tertiary conformation of the four subunits. An analogous scheme applies to subunits in the T-quaternary structure, where the subunits would be labeled t* and t. In the modification described by Equation 8.7, where there are only two tertiary conformations, the natural labels would be r and t for both the R- and T-quaternary structures. The liganded subunit in the R-quaternary structure and the unliganded subunit in the T-quaternary structure would almost exclusively populate the r and t conformations, respectively. Unliganded subunits in R and liganded subunits in T, on the other hand, would populate both r and t conformations.

(Figure 8.6). The tertiary conformational change is treated as a stretched exponential process that slows the geminate rebinding rate. Using Monte Carlo methods to search a large parameter space, Henry and coworkers showed that this model does extremely well in simultaneously fitting the ligand-rebinding kinetics and the kinetics of tertiary and quaternary conformational changes (Figure 8.5), as well as the equilibrium data on the population of the five ligation states determined in the low-temperature electrophoresis experiments of Michele Perrella (Perrella et al. 1990). The early attempt by Gibson (Sawicki and Gibson 1976) was unsuccessful because he did not have the necessary time resolution and therefore did not consider the possibility of tertiary conformational changes prior to the quaternary change contributing to the spectral kinetics.

Theory is now also in agreement with kinetic experiments. The study of Henry and coworkers appears to have closed a very long chapter on the problem of reconciling the complex kinetics of homotropic interactions in hemoglobin with an MWC description. What remains to be done is a comparably rigorous analysis of the kinetic data on heterotropic effects, i.e., the kinetic equivalent of the MWC–PSK treatment of the dependence of equilibrium behavior on solution conditions (pH, chloride, 2,3-diphosphoglycerate, etc.) (Szabo and Karplus 1972; Lee et al. 1988). This analysis might be considerably simplified by an interesting modification of the MWC–PSK model (Henry, Hofrichter, and Eaton, unpublished results), suggested by the crystal-binding studies (Rivetti et al. 1993; Mozzarelli et al. 1997), and the indication of only two spectrally distinct conformations in the kinetic studies (Jones et al. 1992). Instead of four subunit conformations, liganded and unliganded in each of the two quaternary structures, there are only two subunit conformations, r and t, with high and low affinities, respectively. The equilibrium between these tertiary conformations is determined both by the ligation state and the quaternary structure (Figure 8.6). Ligation and the R quaternary structure favor r, while deligation and the T-quaternary structure favor t. The partition function for this model is:

$$\Theta = L'\{1 + K_r p + l_T (1 + K_t p)\}^4 + \{1 + K_r p + l_R (1 + K_t p)\}^4 , \qquad (8.7)$$

where L' is the ratio of the T to R populations with all subunits in the r conformation, l_T is the ratio of t to r populations of the unliganded subunits in the T-quaternary structure, l_R is the corresponding equilibrium constant in the R-quaternary structure, and K_t and K_r are the affinities of the subunits in the t and r conformations which are functionally the same in T and R. In this model, heterotropic effectors would influence L', l_R, and l_T, but not K_r or K_t. In the limits where l_T is large and l_R small, equation 7 is identical to the MWC partition function (equation 1) with $L = L' \, l_T{}^4$. Elements of the equilibrium formulation of this model have been previously suggested in the work of others (Szabo and Karplus 1972; Herzfeld and Stanley 1972; Lee et al. 1988). The simplicity of this description makes it very appealing, but it is yet to be tested in detail.

REMAINING QUESTIONS

At this point in time, many physical scientists would argue that at least the homotropic part of cooperative oxygen binding to hemoglobin is sufficiently well understood to treat it as a closed subject. But is it? Many fundamental questions remain. Is the free energy of breaking salt bridges the origin of the low affinity of the T-quaternary structure (or t tertiary conformation in the extended model, Equation 8.7), or are they simply signatures of an energetically more significant conformational change? The difference in binding free energy of R and T (or r and t) is never more than ~6 $k_B T$. One may ask whether it is possible in such a complex machine to determine the structural origin of such a small free-energy difference? This raises the more general question of the free-energy threshold for which it will be impossible to give detailed structural explanations in proteins.

There are many other issues related to the detailed dynamics of hemoglobin that I have not even mentioned, such as: the interesting thermal and photophysics associated with laser excitation, the path of the ligand motion into and out of the protein, and the sequence of structural events in the tertiary and quaternary conformational change. These are fundamental problems in protein physics that are being addressed by powerful experimental methodologies, such as time-resolved electronic and vibrational spectroscopic methods, site-directed mutagenesis, and most recently time-resolved X-ray crystallography (Šrajer et al. 1996; Eaton et al. 1996). All-atom molecular dynamics simulations, which began with the simulation of the ligand dissociation process by Henry et al. (1985), have and will play an important role. There is no doubt that the results of all of these studies will improve our understanding of how hemoglobin works. Should we expect, however, to discover additional underlying simplicity to hemoglobin's function from such studies?

Finally, I should return to the original question. Have the quantitative studies of the physical scientists produced a truly deeper understanding of hemoglobin function than Perutz had already obtained from static structural models and qualitative analysis? The physical scientists would argue yes. First, the theoretical analysis of Szabo and Karplus clarified the relation between the Perutz mechanism and the two-state allosteric model of Monod, Wyman, and Changeux, and showed how to make quantitative predictions from a detailed structural model. Second, the single crystal studies of oxygen binding provided the convincing experimental evidence in support of this MWC–PSK model for describing cooperative effects in equilibrium experiments. Finally, time-resolved spectroscopy, from the femtosecond to the millisecond regime, and an improved understanding of protein physics from studies on myoglobin, led to a long-sought interpretation of hemoglobin's complex kinetics. All together the physical studies have made it possible to provide a quantitative understanding of how an elegant protein machine works. More importantly, the underlying simplicity in the apparently complex behavior of hemoglobin could not have been exposed without novel physical experiments and a rigorous quantitative analysis. The wide applicability of the model of Monod, Wyman, and Changeux, suggests, moreover, that the experimental and theoretical approaches used in investigating hemoglobin will be helpful in uncovering the underlying simplicity in how other proteins function.

ACKNOWLEDGMENTS

The ideas presented here arose from discussions over many years with my collaborators, James Hofrichter, Eric Henry, and Andrea Mozzarelli. I am also indebted to Attila Szabo, Peter Wolynes, and Robert Zwanzig for helpful discussions on theoretical issues, and to Maurizio Brunori for helpful discussions on hemoglobin.

REFERENCES

Ackers, G., M.K. Doyle, D. Myers, and M.A. Daugherty. 1992. Molecular code for cooperativity in hemoglobin. *Science* **255**:54–63.

Adair, G.S. 1925. A critical study of the direct method of measuring osmotic pressure of haemoglobin. *Proc. Roy. Soc. Lond. A* **108A**:627–637.

Agmon, N., and J.J. Hopfield. 1983. CO binding to heme proteins; A model for barrier height distributions and slow conformational changes. *J. Chem. Phys.* **79**:2042–2053.

Ansari, A., C.M. Jones, E.R. Henry, J. Hofrichter, and W.A. Eaton. 1992. The role of solvent viscosity in the dynamics of protein conformational changes. *Science* **256**:1796–1798.

Antonini, E., and M. Brunori. 1971. Hemoglobin and myoglobin in their reactions with ligands. Amsterdam: North-Holland Publishing Co.

Austin, R.H., K.W. Beeson, L. Eisenstein, H. Frauenfelder, and I.C. Gunsalus. 1975. Dynamics of ligand binding to myoglobin. *Biochemistry* **14**:5355–5373.

Bettati, S., and A. Mozzarelli. 1997. T state hemoglobin binds oxygen noncooperatively with allosteric effects of protons, inositol hexaphosphate, and chloride. *J. Biol. Chem.* **272**:32,050–32,055.

Bohr, C., K.A. Hasselbach, and A. Krogh. 1904. Über einen in biologischer Beziehung wichtigen Einfluß, den die Kohlensäurespannung des Blutes auf dessen Sauerstoffbindung übt. *Skand. Arch. Physiol.* **15**:401–412.

Dickerson, R.E., and I. Geis. 1983. Hemoglobin: Structure, Function, Evolution, and Pathology. Benjamin/Cummings: Menlo Park, CA.

Eaton, W.A. 1980. The relation between coding sequences and function in haemoglobin. *Nature* **284**:183–185.

Eaton, W.A., E.R. Henry, and J. Hofrichter. 1991. Application of linear free-energy relations to protein conformational changes: the quaternary structural change of hemoglobin. *Proc. Natl. Acad. Sci. USA* **88**:4472–4475.

Eaton, W.A., E.R. Henry, and J. Hofrichter. 1996. Perspective: Nanosecond crystallographic snapshots of protein structural changes. *Science* **274**:1631–1632.

Eaton, W.A., and J. Hofrichter. 1990. Sickle cell hemoglobin polymerization. *Adv. Prot. Chem.* **40**:63–279.

Edelstein, S.J. 1996. An allosteric theory for hemoglobin incorporating asymmetric states to test the putative molecular code for cooperativity. *J. Mol. Biol.* **257**:737–744.

Edsall, J.T. 1980. Hemoglobin and the origins of the concept of allosterism. *Federation Proc.* **39**:226–235.

Frauenfelder, H., S.G. Sligar, and P.G. Wolynes. 1991. The energy landscapes and motions of proteins. *Science* **254**:1598–1603.

Frauenfelder, H., and P.G. Wolynes, 1985. Rate theories and puzzles. *Science* **229**:337–345.

Gibson, Q.H. 1959. The photochemical formation of a quickly reacting form of haemoglobin. *Biochem. J.* **71**:293

Greene, B.I., R.M. Hochstrasser, R.B. Weisman, and W.A. Eaton. 1978. Spectroscopic studies of oxy- and carbonmonoxyhemoglobin after pulsed optical excitation. *Proc. Natl. Acad. Sci. USA* **75**:5255–5259.

Hagen, S. J., and W.A. Eaton. 1996. Nonexponential structural relaxations in proteins. *J. Chem. Phys*. **104**:3395–3398.

Hagen, S.J., J. Hofrichter, and W.A. Eaton. 1996. Protein reaction kinetics in a room-temperature glass. *Science* **269**:959–962.

Henry, E.R., C.M. Jones, J. Hofrichter, and W.A. Eaton. 1997. Can a two-state MWC allosteric model explain hemoglobin kinetics? *Biochemistry* **36**: 6511–6528.

Henry, E.R., M. Levitt, and W.A. Eaton. 1985. Molecular dynamics of photodissociation of carbon monoxide from hemoglobin. *Proc. Natl. Acad. Sci. USA* **82**:2034–2038.

Herzfeld, J., and H.E. Stanley. 1974. A general approach to cooperativity and its application to the oxygen equilibrium of hemoglobin and its effectors. *J. Mol. Biol.* **82**: 231–265.

Hochstrasser, R. 1997. Ultrafast spectroscopy of protein dynamics. In: Femtochemistry and Femtobiology: Ultrafast Reaction Dynamics at Atomic-scale Resolution. Nobel Symposium 101, ed. V. Sundström, pp. 678–700. Imperial College Press.

Hofrichter, J., E.R. Henry, A. Szabo, L.P. Murray, A. Ansari, C.M. Jones, M. Coletta, G. Falcioni, M. Brunori, and W.A. Eaton. 1991. Dynamics of the quaternary conformational change in trout hemoglobin. *Biochemistry* **30**:6583–6598.

Hofrichter, J., J.H. Sommer, E.R. Henry, and W.A. Eaton. 1983. Nanosecond absorption spectroscopy of hemoglobin, elementary processes in kinetic cooperativity. *Proc. Natl. Acad. Sci. USA* **80**:2235–2239.

Hopfield, J.J., R.G. Shulman, and S. Ogawa. 1971. An allosteric model of hemoglobin: I. Kinetics. *J. Mol. Biol.* **61**:425–443.

Imai, K. 1983. The Monod-Wyman-Changeux allosteric model describes haemoglobin oxygenation with only one adjustable parameter. *J. Mol. Biol.* **167**:741–749.

Jackson, T.A., M. Lim, and P.A. Anfinrud. 1994. Complex nonexponential relaxation in myoglobin after photodissociation of MbCO: Measurement and analysis from 2 ps to 56 ms. *Chem. Phys.* **180**:131–140.

Jones, C.M., A. Ansari, E.R. Henry, G.W. Christoph, J. Hofrichter, and W.A. Eaton. 1992. The speed of intersubunit communication in proteins. *Biochemistry* **31**, 6692–6702.

Koshland, D.E., G. Nemethy, and D. Filmer. 1966. Comparison of experimental binding data and theoretical models in proteins containing subunits. *Biochemistry* **5**: 365–385.

Lee, A., M. Karplus, C. Poyart, and E. Bursaux. 1988. Analysis of proton release in oxygen binding by hemoglobin: implications for the cooperative mechanism. *Biochemistry* **27**:1285–1301.

Liddington, R., Z. Derewenda, G. Dodson, and D. Harris. 1988. Structure of the liganded T state of hemoglobin identifies the origin of cooperative oxygen binding. *Nature* **331**: 725–728.

Makinen, M.W., and W.A. Eaton. 1974. Optically detected conformational changes in haemoglobin single crystals. *Nature* **247**: 62–64.

Monod, J., J. Wyman, and J.-P. Changeux. 1965. On the nature of allosteric transitions: A plausible model. *J. Mol. Biol.* **12**:88–118.

Mozzarelli, A., C. Rivetti, G.L. Rossi, W.A. Eaton, and E.R. Henry. 1997. Allosteric effectors do not alter the oxygen affinity of hemoglobin crystals. *Protein Sci.* **6**:484–489.

Murray, L.P., J. Hofrichter, E.R. Henry, M. Ikeda-Saito, K. Kitagishi, T. Yonetani, and W.A. Eaton. 1988. The effect of quaternary structure on the kinetics of conformational changes and nanosecond geminate rebinding of carbon monoxide to hemoglobin. *Proc. Natl. Acad. Sci. USA* **85**: 2151–2155.

Pauling, L. 1935. The oxygen equilibrium of hemoglobin and its structural interpretation. *Proc. Natl. Acad. Sci. USA* **21**:186–191.

Pauling, L., H.A. Itano, S.J. Singer, and I.C. Wells. 1949. Sickle cell anemia: A molecular disease. *Science* **110**:543–548.

Perrella, M., A. Colosimo, L. Benazzi, M. Ripamonti, and L. Rossi-Bernardi. 1990. What the intermediate compounds in ligand binding to hemoglobin tell about the mechanism of cooperativity. *Biophys. Chem.* **37**:211–223.

Perutz, M.F. 1970. Stereochemistry of cooperative effects in haemoglobin. *Nature* **228**:726–739.

Perutz, M.F. 1989. Mechanisms of cooperativity and allosteric regulation in proteins. *Qtly. Rev. Biophys.* **22**:139–236.

Perutz, M.F., W. Bolton, R. Diamond, H. Muirhead, and H.C. Watson. 1964. Structure of haemoglobin. An X-ray examination of reduced horse haemoglobin. *Nature* **203**:687–690.

Perutz, M.F., and H. Lehmann. 1968. Molecular pathology of human hemoglobin. *Nature* **219**:902–909.

Rivetti, C., A. Mozzarelli, G.L. Rossi, E.R. Henry, and W.A. Eaton. 1993. Oxygen binding by single crystals of hemoglobin. *Biochemistry* **32**:2888–2906.

Sawicki, C.A., and Q.H. Gibson. 1976. Quaternary conformational changes in human hemoglobin studied by laser photolysis of carboxyhemoglobin. *J. Biol. Chem.* **251**:1533–1542.

Shibayama, N., and S. Saigo. 1995. Fixation of the quaternary structures of human adult haemoglobin by encapsulation in transparent porous silica gels. *J. Mol. Biol.* **251**:203–209.

Shulman, R.G., J.J. Hopfield, and S. Ogawa. 1975. Allosteric interpretation of haemoglobin properties. *Qtly. Rev. Biophys.* **8**:325–420.

Šrajer, V., T.-Y. Teng, T. Ursby, C. Pradervand, Z. Ren, S.-i. Adachi, W. Schildkamp, D. Bourgeois, M. Wulff, and K. Moffat. 1996. Photolysis of the carbon monoxide complex of myoglobin: Nanosecond time-resolved crystallography. *Science* **274**: 1726–1729.

Szabo, A. 1978. Kinetics of hemoglobin and transition state theory. *Proc. Natl. Acad. Sci. USA* **75**:2108–2111.

Szabo, A., and M. Karplus. 1972. A mathematical model for structure–function relations in hemoglobin. *J. Mol. Biol.* **72**:163–197.

9

Proteins Reactions and Conformational Change

Simple, Complex, or Both?

M. KARPLUS

Laboratoire de Chimie Biophysique, ISIS, Université Louis Pasteur, 67000 Strasbourg, France, and
Dept. of Chemistry and Chemical Biology, Harvard University, Cambridge, MA 02138, U.S.A.

Keywords: enzyme catalysis, ligand binding , myoglobin, conformational change, quaternary structural changes, protein folding reaction

ABSTRACT

Proteins are complex systems so that their reactions may be expected to have characteristics not often observed in small molecule reactions. Criteria for a simple reaction (exponential time dependence, Arrhenius-like temperature dependence) are used to discuss a variety of reactions in proteins. They concern bond-making and bond-breaking reactions (enzyme catalysis by triosephosphate isomerase and ligand binding by myoglobin) and conformational changes that include aromatic side-chain rotations, loop/lid opening and closing motions, quaternary structural changes, and protein folding. In some of these cases complex behavior has been observed, but it is evident that more information is needed for a complete understanding. The protein folding reaction is of particular interest in this regard because of the large amount of theoretical and experimental effort that is being devoted to it at the present time.

INTRODUCTION

Macromolecules of biological interest include proteins, nucleic acids, lipids and lipid membranes, and carbohydrates. They are complex mesoscopic systems with many degrees of freedom and multidimensional potential energy surfaces. Their motions cover

Simplicity and Complexity in Proteins and Nucleic Acids
Edited by H. Frauenfelder, J. Deisenhofer, and P.G. Wolynes

a wide range of time scales (from femtoseconds to seconds or even longer) and length scales (from 0.01 Å to 10 Å or even larger). A fundamental question concerns the relationship between the microscopic interactions, which govern the potential energy surface ("energy landscape"), and the observed macroscopic behavior. Of particular interest is whether and, if so how, the microscopic complexity manifests itself in the simplicity or complexity of experimental observables.

This chapter focuses on proteins with the expectation that a corresponding discussion is applicable to other macromolecular biological systems. Proteins participate in a wide range of reactions, all of which involve internal motions. Some, as in the binding and release of oxygen in myoglobin and the catalysis by enzymes, concern the making and breaking of chemical bonds. Others involve conformational changes, such as hinge-bending motions, subunit reorientations and the folding and unfolding reactions of the polypeptide chain, which are governed by relatively weak nonbonded interactions arising from the van der Waals and electrostatic terms in the potential energy. The possibility that any or all of these reactions will have complex time and temperature dependence has to be considered because of the range in the time and length scales of the underlying protein motions.

In the native state, it is now well documented that proteins have a large number of substates with similar but not identical configurations (Austin et al. 1975; Elber and Karplus 1987). These substates are likely to have different properties. In considering protein reactions, the role of the individual substates has to be considered. More generally, there is the question of whether the distributions of protein configurations play a role in the kinetics of the biologically important processes at room temperature (or at other temperatures). What is the observed effect on reactions of the rate at which substates interconvert? Are the configurations of the protein that govern the rate of a given phenomenon localized or do they involve the protein as a whole? Does nonexponential behavior arise from relaxation phenomena in a nonequilibrium situation? Questions of this type, which are important for localized events (e.g., ligand binding) become even more important for global reactions like protein folding, for which much less is known about the nature and complexity of the energy surface.

To set the stage, I first define what I mean by "simple" versus "complex" in a reaction. Then I give some examples of what has been found both theoretically and experimentally in reactions involving proteins. The examples include the catalysis of the conversion of dihydroxyacetone phosphate (DHAP) to glyceraldehyde-3-phosphate (GAP) by the enzyme triosephosphate isomerase, ligand rebinding after photolysis of myoglobin, certain cases of conformational change in native proteins and the protein folding reaction which involves the entire polypeptide chain. Since the purpose of this position paper was to aid in the discussion, the descriptions are qualitative and I have avoided equations where possible. The concluding section summarizes the present situation and the outlook for a more complete understanding of reactions involving proteins.

DEFINITION OF SIMPLE VERSUS COMPLEX BEHAVIOR

Simple versus complex behavior for a reaction is defined in terms of two not unrelated criteria. A reaction is considered to have simple behavior if a phenomenological expression involving rate constants describes the time dependence of the reaction and if the temperature dependence of the rate is Arrhenius-like.

We consider the unimolecular reaction ($A \leftrightarrow B$), which is appropriate for the cases of interest here, including geminate rebinding in myoglobin, an enzymatic reaction after the Michaelis complex has formed, conformational change, and protein folding and unfolding. The differential expressions for the rate of the reaction can be written:

$$\frac{dB}{dt} = k_A A(t) - k_B B(t) \tag{9.1a}$$

$$\frac{dA}{dt} = -k_A A(t) + k_B B(t), \tag{9.1b}$$

where k_A, k_B are the rate coefficients for the forward, backward reaction and $A(t)$,$B(t)$ are the concentrations of species A, B, respectively, at time t; low concentrations (limit of infinite dilution) are assumed to avoid the need to consider interactions among molecules. If the rate coefficients k_A and k_B are essentially independent of time (i.e., $k_A(t) = k_A$, $k_B(t) = k_B$), they are referred to as the rate constants for the reaction. The integrated rate expressions are then:

$$\Delta A(t) = A(t) - A_{eq} = \Delta A(0)\, e^{-kt} \tag{9.2a}$$

$$\Delta B(t) = B(t) - B_{eq} = \Delta B(0)\, e^{-kt}, \tag{9.2b}$$

with A_{eq}, B_{eq} the concentrations of A, B at equilibrium and $\Delta A(t)$, $\Delta B(t)$ and $\Delta A(0)$, $\Delta B(0)$ the deviations from equilibrium at time t and $t = 0$, respectively. The constant k in Equation 9.2 is the sum of the rate constants ($k = k_A + k_B$).

Equations 9.2a and 9.2b are the standard rate expressions for a unimolecular reaction and their validity corresponds to the first criterion for simplicity. If k is a function of time, the reaction is not simple by our definition. For higher-order reactions, the corresponding expression involving a rate constant can be written for the differential rate law, though the integrated expression is more complex, except under certain limiting conditions.

The second criterion for a simplicity of a reaction is that the temperature dependence of the rate follows an Arrhenius-like equation. This requires that the rate constant can be written to a good approximation in the form:

$$k = A(T)e^{-\Delta E^{\ddagger}/RT}, \tag{9.3a}$$

where $\Delta E^{\ddagger}$, the activation energy, is approximately independent of temperature, and $A(T)$ is the pre-exponential factor with only a weak temperature dependence. Given Equation 9.3a, we have

$$\frac{d \ln k}{d(1/T)} \cong -\frac{\Delta E^{\ddagger}}{R}, \tag{9.3b}$$

i.e., a plot of ln k versus $(1/T)$ approximates a straight line. More generally we can write:

$$k = A'(T)e^{-\Delta G^{\ddagger}/RT}, \tag{9.4a}$$

and, by the Gibbs–Helmholtz relation,

$$\frac{d \ln k}{d(1/T)} \cong -\frac{\Delta H^{\ddagger}}{R}, \tag{9.4b}$$

where $\Delta G^{\ddagger}$ and $\Delta H^{\ddagger}$ are the free energy and enthalpy of activation and $A'(T)$ is the appropriate pre-exponential factor, which is significantly different from $A(T)$ in many cases.

For elementary gas-phase reactions involving small molecules, both conditions for simplicity are usually satisfied over a wide range of temperatures (Weston and Schwarz 1972). However, even in these systems, where the entropic contribution to the reaction is small, it is important for quantitative interpretations to consider the temperature dependence of the pre-exponential factor. An example is provided by the $H + H_2$ exchange reaction. The temperature dependence of the rate constant calculated by the quasiclassical trajectory method (Karplus et al. 1965) can be fitted by the Arrhenius expression,

$$k(T) = 4.3 \times 10^{13} \exp[-7.435/\mathrm{RT}]\ \mathrm{cm}^3\ \mathrm{mol}^{-1}\ \mathrm{sec}^{-1}, \tag{9.5a}$$

over the temperature range 300 to 1000 K with RT in kcal/mol. A slightly better fit is obtained by taking the temperature dependence of the pre-exponential factor into account. The result is:

$$k(T) = 7.9 \times 10^{9}\ T^{1.18} \exp[-6.23/RT]\ \mathrm{cm}^3\ \mathrm{mol}^{-1}\ \mathrm{sec}^{-1}. \tag{9.5b}$$

Equations 9.5a and 9.5b make clear that even for this simple reaction with a high activation barrier, some care has to be used in interpreting the numerical values obtained by fitting Arrhenius-like equations.

By contrast, for systems like disordered crystals and glasses (Bai and Fayer 1989; Greer et al. 1994; Angell 1995), nonexponential relaxation after a perturbation and non-Arrhenius temperature dependence for motional phenomena are commonly observed. Although phenomenological models exist that ascribe such behavior to the multiplicity of time scales, the details of the origins of the complexity are generally not fully understood at the atomic level. Observing corresponding complexities in protein reactions would not be surprising, given the multiplicity of time scales for their internal motions.

CONDITIONS FOR SIMPLE BEHAVIOR

Criterion for Existence of a Rate Constant

There are three essential aspects of a reaction that are involved in the existence of a well-defined rate constant (e.g., Chandler 1978). We consider a single elementary step in a reaction; if the overall reaction has several steps, as in many enzymes (Fersht 1985), the discussion would apply separately to each one. The first requirement is that it is possible to define a coordinate, the reaction coordinate (or other progress variable), for the transition from reactants to products; the second requirement is that there exists a well-defined barrier with an energy several times kT separating the reactant and product states along the reaction coordinate; the third, intimately related with the other two, is that the relaxation time of the degrees of freedom other than the reaction coordinate, is fast relative to motion along the reaction coordinate. Given that these three requirements are satisfied, a rate constant can be defined and simple exponential behavior is expected. It can be true that a rate constant (simple rate expression) exists at one temperature but not at another because the relaxation times of different modes of the system depend differently on the temperature.

To make clearer what is involved in the three requirements, I shall describe the activated dynamics method that is commonly used for calculating the rate of chemical reactions. The method was originally proposed for gas-phase reactions (Keck 1967; Anderson 1973; see also Wigner 1938) and has been developed and applied in its present form to reactions in solution (Grote and Hynes 1980; Montgomery et al. 1979) and in proteins (e.g., Northrup et al. 1982).

Following the formulation of Chandler (1978), we can write the rate coefficient in the form:

$$k = \frac{1}{2}\kappa\left\langle|\dot{\zeta}|\right\rangle^{\ddagger}\frac{\rho\left(\zeta^{\ddagger}\right)}{\int\limits_{i}\rho(\zeta)\delta\zeta}, \tag{9.6}$$

where ζ is the reaction coordinate and the double dagger (‡) defines the value of the coordinate at the transition state. The symbol $\rho(\zeta)$ represents the probability density for the system as a function of ζ, which corresponds to the free energy (potential of mean force), $W(\zeta)$, defined by:

$$W(\zeta) = -RT\ln\rho(\zeta)/C, \tag{9.7}$$

with appropriate normalization constant, C. The activation free energy in Equation 9.4a can now be written:

$$\Delta G^{\ddagger} = -RT1n\left[\frac{\rho\left(\zeta^{\ddagger}\right)}{\int_{\iota}\rho(\zeta)d\zeta}\right], \tag{9.8}$$

where the integral $\int_1$ is over the values of the reaction coordinate associated with the reactant valley. Such a potential of mean force is meaningful if and only if for each value of the reaction coordinate ζ, the system can be assumed to be in equilibrium with respect to all of the other degrees of freedom, i.e., the latter must relax rapidly relative to motion along the reaction coordinate. That the barrier be sufficiently high and well defined means that the essential step determining the rate of the reaction is the passage over the barrier, which is a rare event. If the barrier were very low or nonexistent, one might expect to observe a diffusive rate for a reaction in solution that would not generally satisfy the conditions for a simple reaction. The quantity $\langle|\dot{\zeta}|\rangle^{\ddagger}$ is an equilibrium average value of the time derivative of $\zeta(\dot{\zeta} = d\zeta / dt)$ evaluated with ζ restrained to the transition state ‡. This can be determined computationally by restricting the system to $\zeta = \zeta^{\ddagger}$ and doing simulations that are long enough to obtain the average velocity; again this procedure is meaningful only if there is a separation of time scales. Finally, κ is the transmission coefficient, which in gas-phase reactions usually has a value close to unity unless surface crossing or very high-collision energies are involved. However, in solution or in proteins, κ often is significantly less than unity. The time-dependent transmission coefficient $\kappa(t)$ can be calculated from the reaction flux expression (Chandler 1978; Grote and Hynes 1980):

$$\kappa(t) = N\left\langle \dot{\zeta}(0)\theta\left[\zeta(t) - \zeta^{\ddagger}\right]\right\rangle_{\ddagger}, \tag{9.9}$$

where $\theta[\zeta(t) - \zeta^{\ddagger}]$ is a step function that equals one when the reaction coordinate is in the product state and is zero otherwise; N is a normalization constant such that $\kappa(0+)$ is equal to unity. Equation 9.7 corresponds to the transition-state theory (TST) expression if $\kappa(t)$ equals unity, i.e., it is assumed that there are no recrossings, which would reduce the rate constant, so that the TST value is an upper limit for the reaction rate. The function $\kappa(t)$ is, in fact, equal to unity at very short times, corresponding to the fact that no recrossings have occurred. For a well-defined rate constant, there is then a short relaxation period in which systems may recross the transition state one or more times until they settle into the reactant well (no contribution to reaction rate) or into the product well (a contribution to reaction rate). The requirement for a high activation barrier (relative to kT) is evident here because it means that in a sufficiently dense system like a macromolecule or a liquid (Skinner and Wolynes 1980), collisions remove kinetic energy from the reaction coordinate as soon as the system has left the transition-state region.

If a separation of time scales were not present, $\kappa(t)$ would not reach a plateau value and the rate coefficient would have a significant time dependence, i.e., the reaction rate would not show the simple exponential behavior corresponding to Equation 9.2. Concomitant with this is the possibility that the definition of a potential of mean force barrier as in Equations 9.6–9.8 is not appropriate for this system, i.e., the coupling of different degrees of freedom may be such that a reaction coordinate, which varies slowly relative to other degrees of freedom, does not exist. Thus, in proteins, where there are many different types of degrees of freedom with different time scales of relaxation, deviations from simple behavior is not unexpected.

There are two additional points. The first is that we have used the description of the activation dynamics method primarily as a way of indicating the factors involved in the existence of a rate constant. It is, of course, not necessary for the existence of simple exponential behavior that we can determine the reaction coordinate or appropriate progress variable. It is required only that there exists such a coordinate with the properties we have described. This is clearly relevant to protein folding reactions where, as we discuss below, the definition of a reaction coordinate is a major problem. More generally, a difficulty in applying the activated dynamics method to complicated many-body systems is the determination of an optimal reaction coordinate and the transition state along that coordinate. This problem is rooted in the conceptual and computational complexity associated with finding the minimal number of atomic coordinates which adequately specify a dividing surface between a "reactant" configuration and a "product" configuration. The second point is that the use of Equation 9.6 does not require knowing the exact transition state, which is essential only for the application of transition-state theory. With Equation 9.6, a poor choice of reaction coordinate or transition state (dividing surface) does not invalidate the method, in principle. The result is that the transmission coefficient will have a reduced value. However, the efficiency of the activated dynamics-transition state sampling rapidly decreases as the choice of reaction coordinate and transition state become less than optimal.

Validity of Arrhenius Equation

The second criterion for simplicity is concerned with the variation of the rate constant, if it exists, with temperature. If the reaction has an Arrhenius-like temperature, as defined in Equations 9.3 and 9.4, it is regarded as simple from this point of view. Gas-phase reactions of small molecules with well-defined barriers at least several times kT tend to have rate constants that obey the Arrhenius law; a plot of $\ln k$ versus 1/T approximates a straight line with an increase in $\ln k$ as T increases (Equation 9.4b with $\Delta H^{\ddagger}$ constant).

Deviations from such a simple Arrhenius temperature dependence in complex systems can be of several types. At high temperature, significant curvature in the Arrhenius plot may appear with the possibility of a decrease in the rate with increasing temperature. This has been observed in the protein folding reactions and is discussed below. At low temperatures, there are several types of deviation from Arrhenius behavior. One of these corresponds to a decrease in the temperature dependence (i.e., the rate constant falls off more slowly than predicted by the Arrhenius law) that arises from the contribution of tunnelling to the reaction rate. The other deviation corresponds to a higher-order temperature dependence, i.e., the reaction rate decreases more rapidly at low temperatures than expected from the Arrhenius law. Such behavior is often referred to as a "super-Arrhenius" temperature dependence. We consider both types of deviations from Arrhenius behavior in the relevant sections.

EXAMPLES OF PROTEIN REACTIONS

Dynamics of Enzymatic Reactions

The enzyme triosephosphate isomerase (TIM) catalyzes the interconversion of DHAP and GAP in a central step of the glycolytic cycle (Fersht 1985). TIM is of particular interest because it has been characterized as a "perfect enzyme," i.e., the chemical steps in the reaction are accelerated such that the overall rate is limited by the diffusion-controlled binding of the substrate and release of the product (Knowles 1991b). The reaction (DHAP $\leftrightarrow$ GAP) has several steps. The first of these, which we consider in what follows, consists of the transfer of a proton from DHAP to the residue Glu 165 of TIM, which acts as a base to form an enediolate intermediate (see Figure 9.1). The rate enhancement of this step is about 5×10^5 relative to the reaction catalyzed by a general base in solution (Richard 1984). As has been shown (Bash et al. 1991), the reduction of the activation barrier is due to interactions of the reactants and the transition state with specific, primarily charged, residues of the enzyme (for a related study, see Åqvist and Fothergill 1996). Displacements of atoms in the enzyme by 1 Å or so can have a large effect on the activation barrier (Knowles 1991a; Joseph-McCarthy et al. 1995). Such a tight coupling of structure and reactivity raises the question of its possible effect on the dynamics of the reaction, i.e., do the structural requirements for transition-state stabilization result in a significant reduction in the effective rate of crossing the barrier? Is the reaction simple in terms of the present definition? To begin to answer these questions, the activated dynamics method summarized above (see section on **Criterion for Existence of a Rate Constant**) was applied to the first step of the TIM reaction (Neria and Karplus 1997); that is, a reaction coordinate was selected, the transition state was determined by calculating the potential of mean force along the reaction coordinate, and the transmission coefficient was evaluated by use of the reaction flux method. This is the first time the transmission coefficient has been determined for an enzymatic reaction. Warshel and coworkers have made related studies of the dynamics of reactions in enzymes and in solution (see Warshel, this volume, and **Comments on Warshel**, p. 150).

Although the reaction has been studied by the semi-empirical molecular orbital QM/MM method (Bash et al. 1991), a simpler representation of the potential energy surface was needed to calculate the large number of trajectories that are required for evaluating the transmission coefficient. The representation used describes the surface for the reaction in terms of the coupling of two states, one corresponding to the reactants and the other to the products. In each state the potential surface is represented by a standard molecular mechanics (MM) force field. The possibility of reaction is introduced by coupling the two electronic states (Brumer and Karplus 1973; Åqvist and

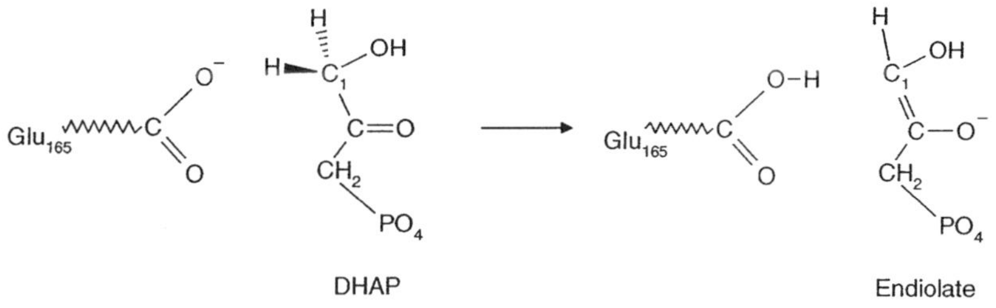

Figure 9.1 The initial proton transfer step in the catalytic mechanism of triosephosphate isomerase. A proton is transferred from the C1 carbon of DHAP to one of the carboxyl oxygens of the GLU165 yields an endiolate intermediate.

Warshel 1993). The potential function $V(\mathbf{R})$ is chosen to have the EVB form (Åqvist and Warshel 1993; Åqvist and Fothergill 1996):

$$V(\mathbf{R}) = \frac{1}{2}(V_R(\mathbf{R}) + V_P(\mathbf{R})) - \frac{1}{2}\sqrt{(V_R(\mathbf{R}) - V_P(\mathbf{R}))^2 + 4V_{RP}^2}\,, \tag{9.10}$$

where $\mathbf{R}$ is the set of all coordinates, $V_R(\mathbf{R})$ and $V_P(\mathbf{R})$ are the MM potential energies in the reactants and products states, respectively, and V_{RP} is approximated by a constant coupling, assumed to be independent of $\mathbf{R}$ for simplicity.

An appropriate reaction coordinate for the proton transfer from carbon C_1 of DHAP to the O of Glu 165 is the asymmetric stretch defined by:

$$q_{as} = \frac{1}{m_C + m_O}(m_C r_{CH} - m_O r_{OH}), \tag{9.11}$$

where r_{CH} and r_{OH} are the distances of the proton from the donor carbon and the acceptor oxygen, respectively, and m_C and m_O are their masses; for a collinear arrangement q_{as} has a reduced mass of 0.97amu. In the reaction, the proton moves from the vicinity of the donor carbon atom ($q_{as} = -1.1$ Å) to the vicinity of the accepting oxygen ($q_{as} =$ 0.3 Å). To determine the location of the transition state, the potential-of-mean force (PMF) (McQuarrie 1976) as a function of q_{as} was determined by use of umbrella sampling (Valleau and Torrie 1977). As already described, the PMF is an effective potential for the reaction coordinate that is based on an equilibrium average over all other degrees of freedom. The resulting free-energy profile for the reaction is shown in Figure 9.2.

As a starting point for the transition-state trajectory calculations, equilibrium configurations in the transition state were sampled by running a 400 ps trajectory; $q_{as}(t)$ was constrained to $q_{as}^{\#}$ with a SHAKE-like algorithm (Ryckaert et al. 1977), and 40 configurations at 10 ps intervals were saved. For each of these configurations, 100 trajectories were initiated with different values of sampled from a Maxwell distribution. The function $\kappa(t)$ calculated from these trajectories is shown in Figure 9.3. It decays rapidly and has almost reached the plateau value by 10 fs. The plateau value is $\kappa = 0.43$

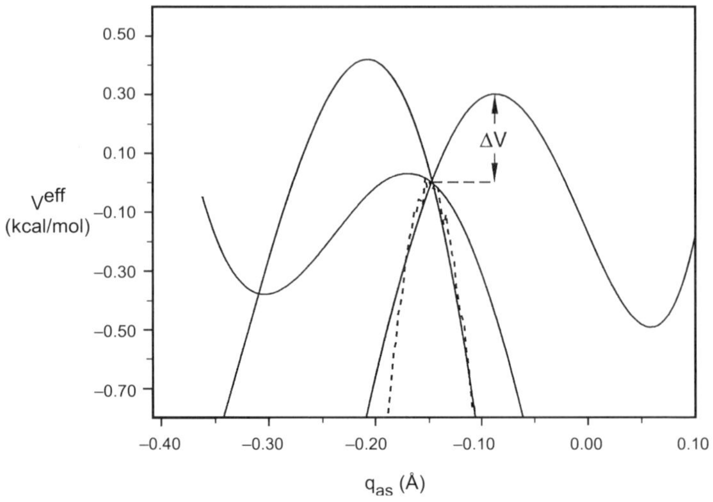

Figure 9.2 The potential of mean force near the transition state (- - -) and the potential energy for three initial configurations used in the calculation of the transmission coefficient. The potential energy was calculated with all degrees of freedom, other than the reaction coordinate, fixed. The zero of energy is at q and ΔV is the height of the instantaneous barrier that trajectories need to surmount.

± 0.08, a significant reduction from the TST value of unity. However, the reduction by a factor of two is a small effect relative to the specific interactions in the protein that lower the potential of mean force barrier for the reaction so as to increase the rate by five orders of magnitude. It will be interesting to compare this result for the enzyme catalyzed reaction with that for the reaction in solution (Cui and Karplus, work in progress).

The rapid decay of $\kappa(t)$ to the plateau value makes it likely that most degrees of freedom coupled to the reaction coordinate do not change significantly during the time required to trap the trajectory in the reactant or product wells. Defining the substrate and the GLU165 atoms as the "intramolecular" part, and the rest of the system as the "environment," the calculation of the transmission coefficient was repeated from the same initial configurations, but with only the intramolecular atoms allowed to move and the environment atoms fixed in their initial ($t = 0$) positions. The values of $\kappa(t)$ obtained from these rigid environment simulations are also shown in Figure 9.3. The behavior of $\kappa(t)$ is very similar to that obtained when all atoms are allowed to move. This demonstrates that the dynamics of the environment atoms is unimportant for the barrier crossing per se and suggests that the structure at $t = 0$ determines whether or not a reaction takes place.

If all degrees of freedom other then the reaction coordinate are frozen ("frozen bath" assumption) and there is a barrier ΔV, in addition to the PMF barrier, for a particular set of bath coordinates (a configuration chosen from the 400 ps trajectory), the

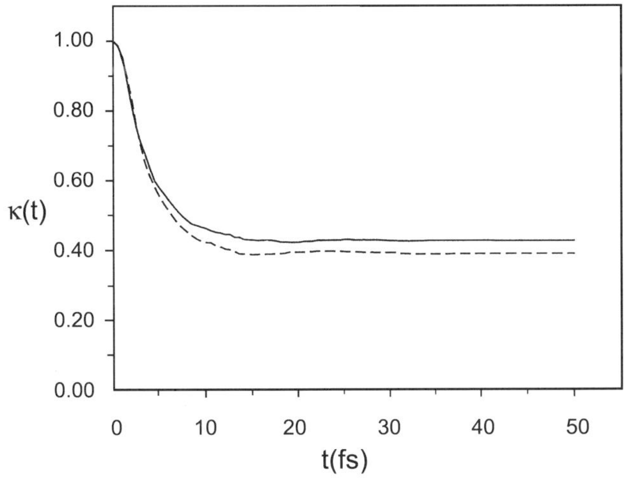

Figure 9.3 The time-dependent transmission coefficient, $\kappa(t)$, when all atoms are allowed to move (—), and in a rigid environment (– – –) (see text).

transition-state trajectory has to surmount that barrier to complete the reaction; the range ΔV observed in the calculations is about 0.5 kcal/mol. This reduces the rate from the TST value, which is based on the PMF barrier. With such a "frozen bath" assumption, the transmission coefficient is given by (Bergsma et al. 1987):

$$\kappa^{froz} = \left\langle e^{-\Delta V / k_B T} \right\rangle, \tag{9.12}$$

where k_B is the Boltzmann constant, T is the is temperature, and the average is over all initial transition-state configurations (Figure 9.2). The average over the 40 transition-state configurations gives $k^{froz} \cong 0.40$, in good agreement with the simulations. This confirms that the frozen bath approximation is valid, i.e., the surrounding system is fixed within the short time scale of the barrier crossing.

In Warshel (this volume), a discussion of the analysis of solvation effects given in Neria and Karplus (1997) is presented; however, the point of that paper appears to have been misinterpreted (see **Comment on Warshel**, p. 150). Two mechanisms were described for the influence of the environment on the instantaneous barrier. The first, found in a study of a symmetric SN_2 reaction in water (Bergsma et al. 1987), corresponds to a nonequilibrium solvation mechanism. In that reaction, the instantaneous barrier arises from different solvation states of the intramolecular subsystem, e.g., a fluctuation in the environment that results in better solvation of the reactant relative to the products introduces an instantaneous barrier located between the transition state and the products state, and vice versa. In an alternative mechanism for the environmental influence, the instantaneous barrier is determined by the configuration of the intramolecular subsystem, and the intramolecular dynamics is coupled to low-frequency fluctuations of the environment, e.g., intramolecular motions that are required

Comment on Warshel (this volume)

At the meeting, questions were addressed to Warshel concerning comments in his paper on our study of the triosephosphate isomerase (TIM) reaction. Since the comments are not discussed elsewhere in this volume, it is useful to address them here.

One point concerns the transmission coefficient and the relation of the dynamics in the enzyme and in solution. Warshel and coworkers have used an approximate correlation function analysis to argue that dynamical effects in solution and in an enzyme are similar. Warshel states (Chapter 12, this volume) that their approximate equation (Equation 12.5) on which the correlation function analysis is based was verified "by averaging many downhill trajectories (Hwang et al. 1988)." However, examination of Hwang et al. (1988), which deals with an S_N2 reaction in solution, does not give any data on this point nor does it give a value for the transmission coefficient. Warshel also states (this volume) that "F is the so-called 'transmission factor' that tells us how many times a trajectory that passes the transition state from the reactant state moves back and forth before falling to the product state." Actually, the transmission "factor" or transmission coefficient determines the fraction of trajectories that, starting in the transition state, end up quenched in the product state; the number of times a trajectory "moves back and forth" does not matter (see also section, **Criterion for Existence of a Rate Constant,** p. 143).

In referring to Neria and Karplus (1997), Warshel states (this volume) that "Unfortunately, it was implied in this work that the rate acceleration of TIM is still associated with dynamical effects." It is difficult for me to understand how Warshel could have misinterpreted our paper in this way. As stated in the Conclusions section of Neria and Karplus (1997): "Of particular interest for enzymes is the question whether dynamical effects are significant; i.e., does the specific coupling to the enzyme degrees of freedom involved in lowering the barrier lead to strong constraints on the motion. Evolution of a 'perfect enzyme' like TIM could involve both energetic and dynamical effects. The simulations demonstrate that there is a *reduction* of the transmission coeffficient (by only a factor of 2.5) from the transiton state theory value of unity, while there is an overall rate enhancement of 10^6, relative to the solution reaction." Thus it is clear that the dominant effect in the rate acceleration is the reduction in the free-energy (potential-of-mean force) barrier, which was analyzed by Bash et al. (1991). This is clearly not a dynamical effect. Perhaps Warshel is suggesting that the transmission coefficient in solution is much smaller than that in the enzyme, but that is unlikely.

Warshel also states (this volume) in relation to the Neria and Karplus (1997) calculation: "these conclusions were reached without attempting to calculate the actual activation barrier..." However, as pointed out above, Bash et al. (1991) did estimate the activation barrier and found that its lowering was dominated by electrostatic effects, as confirmed by the subsequent calculations of Åqvist and Fothergill (1996). I did not discuss these results in detail in the section on TIM in the chapter I wrote for this volume because it is concerned with the question of whether a rate constant description (i.e., a rate coefficient independent of time) is valid for the reaction.

There is no disagreement with Warshel and the more general viewpoint that the major factor in the rate acceleration by most enzymes is the reduction in the activation barrier. What I and my collaborators have done for a series of reactions in proteins and peptide channels is to calculate the transmission coefficient to show its role in such reactions. The values we have found are approximately 0.1 for Na^+ barrier crossing in the interior of the gramicidin channel (Roux and Karplus 1991), 0.22 for tyrosine ring flips in the bovine trypsin inhibitor (Northrup et al. 1982; see also this chapter), and 0.43 for the TIM reaction (Neria and Karplus 1997; see also this chapter).

for the reaction are hindered by the environment, which appears rigid on the time scale of the barrier crossing (~10 fs) but fluctuates on a picosecond time scale. The two mechanisms were tested for the proton transfer in the TIM reaction, and it was demonstrated that the second mechanism is dominant in introducing the effect of the environmental fluctuations on the barrier modulation and in the reduction of the rate from the transition-state limit. Which mechanism operates for the proton transfer in solution from DHAP to acetate, for example, and the magnitude of the effect, i.e., the value of $\kappa(t)$, has not been determined (Cui and Karplus, work in progress).

From the above analysis, there is a well-defined rate constant for the reaction, i.e., the reaction is simple according to the first criterion and Equations 9.2a and 9.2b are applicable. As the calculations have shown, there are three time scales for the reaction. There is the very fast time scale of relaxation of the high-energy transition-state trajectories to a thermal distribution (~10 fs); there is the time scale of the motions associated with the enzyme environment that determines the height of the transition-state barrier (~1 ps). The dominant modulations of the barrier are of a low-frequency character ($\omega < \mathrm{cm}^{-1}$) and originate from the interaction between the intramolecular subsystem and the rest of the enzyme. Finally, there is the time scale associated with overall reaction rate (~1 ms). Since the third is much longer than the other two, simple behavior, as exemplified by the well-defined plateau value for the transmission coefficient, is expected and was obtained in the calculations. However, it should be noted that the trajectories covered a time scale (400 ps) many orders of magnitude shorter than that of the overall reaction rate (ms). Consequently, it is not excluded that there exist very slow structural fluctuations (on the same time scale or longer than the reaction) that alter the barrier. This would lead to nonexponential behavior, but it appears very unlikely at room temperature for such a "slow" reaction; at low temperatures, nonexponential kinetics might occur (see next section). Interestingly, single-molecule studies of alkaline phosphatase have shown significantly different activities for individual molecules (Craig et al. 1996). However, in this case , it is likely that a difference in composition (e.g., glycosylation) was the source of the variation in reaction rates. In the future, single-molecule techniques are likely to provide important new information in this area.

Rebinding of Ligands after Photodissociation of Myoglobin

A well-documented case of complex behavior in a protein reaction is the geminate rebinding of ligands to myoglobin after flash photolysis (Austin et al. 1975; Henry et al. 1983; Ansari et al. 1985; Frauenfelder and Wolynes 1985; Frauenfelder 1995). This reaction has been studied experimentally over a wide range of temperatures and time scales, and numerous theoretical analyses of the results have been made. Moreover, the high-resolution X-ray structures of liganded and unliganded myoglobin do not reveal any path by which ligands can move between the heme-binding site and the outside of the protein (Perutz and Matthews 1965; Takano 1977). Since fluctuations must therefore be involved in the overall ligand-binding reaction (Case and Karplus 1979;

Austin et al. 1975; Frauenfelder et al. 1979; Debrunner and Frauenfelder 1982; Elber and Karplus 1987), myoglobin has become a model system for studying the relation of motion to function in proteins.

The phenomenological description of the kinetics of photodissociation and rebinding of a ligand X to myoglobin (Mb) can be written as:

$$\mathrm{MbX} \leftrightarrow (\mathrm{Mb}+\mathrm{X})_{\mathrm{pocket}} \leftrightarrow (\mathrm{Mb}+\mathrm{X})_{\mathrm{protein}} \leftrightarrow (\mathrm{Mb}+\mathrm{X})_{\mathrm{solution}}, \tag{9.13}$$

where the subscripts refer to the location of the ligand X. Each of the designated species may involve several different states or substates on a microscopic level. At low temperatures (below 160 K in mixed glycerol/water), it is found that the geminate rebinding of the CO ligand $[(\mathrm{Mb+X})_{\mathrm{pocket}} \to \mathrm{MbX}]$ is nonexponential. As the temperature is raised, the geminate recombination of CO becomes essentially exponential. Thus, the first criterion for a simple reaction (Equations 9.1 and 9.2) is satisfied at room temperature but violated at low temperature. For NO, in contrast to CO, the geminate rebinding at room temperature is nonexponential (Petrich et al. 1991); the temperature dependence of NO rebinding has not been studied. Thus, the difference between the CO and NO behavior can aid in elucidating the source of complexity in this reaction.

The very long times over which the CO rebinding is nonexponential and time-delay flash experiments at low temperatures (Austin et al. 1975) suggest that an inhomogeneous model is involved. Thus, a description corresponding to the frozen bath model (i.e., protein molecules are in different substates with different activation barriers; see previous section), with the equilibration between bath configurations slow compared to the overall reaction appears to be appropriate. This is in accord with equilibrium molecular dynamics simulations at 80 K (Kuczera et al. 1990) and incoherent neutron scattering experiments (Doster et al. 1989; Smith et al. 1990). They indicate that protein fluctuations at such temperatures are relatively harmonic and correspond to oscillations in a single well or substate. This means that a distribution of barriers for a population of photolyzed myoglobin molecules is required for describing the observed reaction, rather than a single potential of mean force surface. The number of unliganded Mb hemes, $N(t)$, as a function of time after photolysis can then be written as:

$$N(t) = \int_0^\infty dE\, g(E) \exp[-k(E)t], \tag{9.14}$$

where $g(E)$ is the distribution of barrier heights and $k(E)$ is the rate constant for a barrier of energy E. This results in nonexponential relaxation, i.e., the rate coefficient is time-dependent and its form is determined by the choice of $g(E)$ and $k(E)$. Both power law or stretched exponential expressions for $N(t)$ can be fitted to the experimental data. The temperature dependence of the rebinding was obtained by assuming that $k(E)$ in Equation 9.14 can be expressed in Arrhenius form (Equation 9.3) with a temperature independent expression for $g(E)$, i.e., the distribution of substates with different barrier heights was taken to be frozen in below 160 K within the precision of the experiments. Independent measurements of other markers of internal motion have demonstrated that certain relaxation phenomena in myoglobin (Iben et al. 1989; Steinbach et al. 1991) are nonexponential in time and have a super-Arrhenius temperature dependence that can be expressed in the form:

$$k = A \exp [-(E/RT)^2] , \tag{9.15}$$

over a wide temperature range. These results suggest that most of the rebinding below 160 K is a relatively local phenomenon in which protein relaxation does not play an important role. This is in accord with a dissociation simulation at 10 K (Kuczera and Karplus, unpublished), which demonstrated that the initial motion of the iron out of the heme plane still occurs in 40 fs (as it does at room temperature) and that at least some of the undoming required for rebinding can take place without protein relaxation.

For CO at room temperature, because of the large electronic barrier to ligand binding, one does not even begin to observe recombination until about 100 ns subsequent to photodissociation (Henry et al. 1983). Thus, the transitions between the substates are fast relative to the overall reaction and the protein is expected to have relaxed nearly completely to the unliganded configuration by the time significant CO rebinding occurs. A description of the type outlined above (see section, **Criterion for Existence of a Rate Constant**) is expected to be applicable with $\kappa(t)$ having a well-defined plateau value. This implies that a rate constant should exist and exponential rebinding should occur, in agreement with experiment. At some intermediate temperatures (near 160 K), the rate of protein relaxation and rebinding will be on the same time scale and complex behavior is expected. It has in fact been observed that the rate decreases with increasing temperature in this range (Agmon and Hopfield 1983; Steinbach et al. 1991). This is a clear manifestation of non-Arrhenius temperature dependence.

Due to the difference in the electronic structure of the two ligands, there is a much lower barrier for rebinding of NO than CO. This is exemplified by the fact that in myoglobin at room temperature, 80% of the NO molecules have rebound after 300 ps (Petrich et al. 1991). An analysis of the Mb potential energy surface for the native state (Elber and Karplus 1987), based on a room temperature 300 ps molecular dynamics trajectory (Levy et al. 1985), demonstrated that the interconversion rates between substates range from 0.1 ps for those that are very similar, to at least the length of the simulations (300 ps), for those with larger structural differences; experiments (Lim et al. 1993; Jackson et al. 1994) indicate that there are relaxation processes that extend from femtoseconds into the microsecond range. These results show that the substate transitions are on the same time scale as protein relaxation after photolysis in myoglobin, as would be expected since they cover the same conformational space. In fact, the rms difference between MbCO and Mb in crystal structures is 0.3 Å for the main chain and 0.7 Å for all nonhydrogen atoms (Takano 1977), somewhat smaller than the conformational space (about 2 Å) explored by the single Mb simulation at room temperature (Elber and Karplus 1987).

From these results, it is clear that the rigid bath model used to explain nonexponential rebinding of CO at low temperatures is not applicable to NO at room temperature. Instead, the similarity of the time scales of the protein motions and the rebinding rate suggests that there is a coupling between the two. This is likely to arise from the difference of the position of the heme iron relative to the heme plane in the liganded and unliganded structures, i.e., the heme is essentially planar with the six-coordinate iron in the heme plane in the former and the heme is "domed" with the iron out of the heme plane in the latter. The barrier to rebinding is expected to be lowest when the five-coordinate iron is in the heme plane and to increase as the heme moves out of the plane and the protein relaxes to the unliganded equilibrium structure. To examine the possibility that the nonexponential rebinding of NO can be produced by a time-dependent evolution of the barrier, we describe a simple model; molecular dynamics simulations that support this model have been made (Becker and Karplus, to be published). The disappearance of unliganded protein, $N(t)$, is assumed to follow the modified first-order decay law,

$$-dN/dt = k(t)N \,, \tag{9.16}$$

where $k(t)$ is a time-dependent rate coefficient that is assumed to obey the modified Arrhenius equation,

$$k(t) = A \exp[-E(t)/RT] \,, \tag{9.17}$$

with

$$E(t) = (E_0 - E_{eq}) \exp(-k_{bar}\, t) + E_{eq} \,. \tag{9.18}$$

Here, k_{bar} is the rate constant that determines the decay of the initial barrier height, E_0, to its equilibrium value, E_{eq}. Reasonable parameters (i.e., $E_0 = 0.027$ kcal/mol, $E_{eq} =$ 1.2 kcal/mol, $k_{bar} = 1.5 \times 10^{10}\ \mathrm{s}^{-1}$) give a good fit to NO rebinding for myoglobin at room temperature; the limiting time constants are 42 ps at $t = 0$ and 300 ps at $t = \infty$ (Petrich et al. 1991).

The model just described requires that there be a slow component to the iron motion out of the heme plane. The existence of such a component has been demonstrated in simulations by Kuczera et al. (1993); interestingly, simulations with somewhat different protein models (Schaad et al. 1993; Li et al. 1993) did not find a nonexponential component. The simulations of Kuczera et al. (1993) showed that after photodissociation there is an ultrafast relaxation process with an average time constant of about 40 fs; the range of relaxation times varies from 30 to 70 fs in different trajectories. It consists of an iron out-of-plane displacement coupled to local adjustments of the heme. The ultrafast behavior is very similar to that obtained following the dissociation of a heme-histidine-CO complex in the absence of the protein. The remaining (slower) processes correspond to structural and energetic relaxation involving the protein, as well as the heme. This is due to the fact that the doming of the heme group exerts a force on the surrounding protein atoms. The iron out-of-plane displacement averaged over several trajectories, $\langle z(t)\rangle$, which was used as the reaction coordinate for the structural transition leading from the liganded to the unliganded state, shows nonexponential behavior. Over the time scale of the simulations (100 ps), the results can be fitted to a power law or a stretched exponential. Figure 9.4 shows the calculated

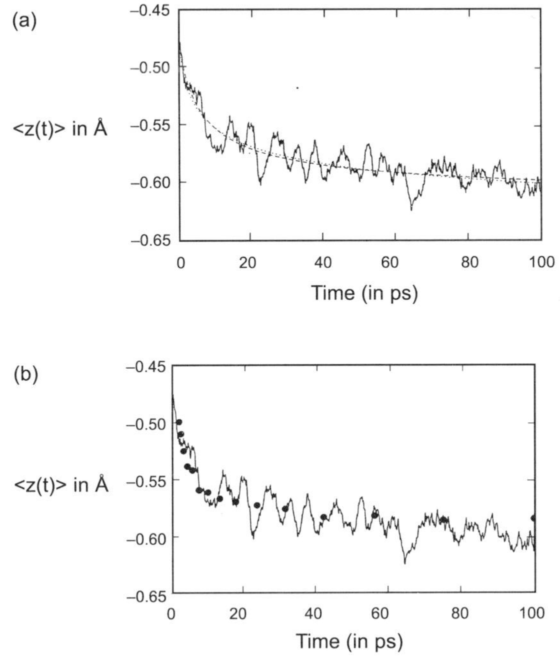

Figure 9.4 Myoglobin photodissociation showing the out-of-plane motion of iron, (t) in Å, as a function of time in ps. The solid line is the average of four simulations smoothed with 2 ps windows. (a) Fit with a power law; two slightly different fits are shown. (b) Superposition of the simulated (t) for the out-of-plane motion of the iron and experimental Dg(t) for the spectral shift of band III (measurements by Lim et al. 1993). The experimental data (filled circles) were subjected to an overall linear scaling (see text). Adapted from Kuczera et al. (1993).

relaxation behavior; the fluctuations in $z(t)$ correspond to high-frequency local oscillations superposed on the global relaxation. Also shown is the measured frequency shift, $\Delta\nu(t)$, of band III (Lim et al. 1993), scaled in magnitude to correspond to the amplitude of $\langle z(t)\rangle$. Band III is a heme charge-transfer band whose frequency is related to the iron out-of-plane distance. The nonexponential time dependence of $\Delta\nu(t)$ and $\langle z(t)\rangle$ are in excellent agreement over the range of the simulation. This supports the simulation results and indicates as well that band III shift is a linear function of the iron out-of-plane displacement. Steinbach et al. (1991) have suggested that there is also a linear relationship between the barrier height for rebinding and the position of band III. Further, the temporal evolution of $\langle z(t)\rangle$ from the simulations is in agreement with time scales found in the experimental studies of NO ligand rebinding (Petrich et al. 1991).

The position of band III is a marker for conformational relaxation of myoglobin after photodissociation. It has been shown, as already mentioned, that the shift in band III is nonexponential and takes place over a time scale from femtoseconds to microseconds. Hagen and Eaton (1996) have developed an interesting model for the nonexponential decay to equilibrium of a protein after a perturbation, such as the photodissociation reaction. The model is based on the assumption of two relaxation time scales, one (faster) being the transition from the liganded to a certain portion of the unliganded manifold of states and the other (slower) the redistribution of states to the equilibrium population in the unliganded manifold. To obtain a description that can be used for simple calculations, Hagen and Eaton assumed a Gaussian (random energy) model for the distribution of energies in the unliganded and liganded manifold of states and introduced a single transition state (Koper and Hilhorst 1987) that governs the rates of all transitions (within each manifold and between manifolds). With reasonable parameters, they are able to fit the model to the stretched exponential type behavior of the measured shift of band III. The details of the relaxation behavior depend on the assumption of the connectivity between states, i.e., whether any state can make a transition to any other state, as in the protein folding model of Bryngelson and Wolynes (1989; see also Savens et al. 1994) or whether the allowed transitions are restricted to states with similar energies. A major question in this type of model concerns the relation between energy and structure, i.e., what is the structure of the postulated single transition state, if one exists, and is it possible to assume that states of similar energy have similar structures since any physical transition will most likely involve two similar structures. Hints concerning the complex nature of the behavior that can be expected are given by the topological and kinetic analysis (Becker and Karplus 1997) of the potential surface calculated for an alanine tetrapeptide by Czerminski and Elber (1989; 1990), as described below (see section on **PROTEIN FOLDING**).

The present analysis suggests that there are two limiting possibilities for protein reactions, in general. In one limit, many protein substates with different barrier heights exist at equilibrium, and it is the substate interconversion rate versus the rebinding rate that determines whether or not exponential behavior is observed. In the other limit the barrier and binding rate are altered by protein relaxation from the liganded to the unliganded structure after dissociation so that the relative rate of relaxation and rebinding determines whether the latter shows nonexponential kinetics. The two possible sources of nonexponential behavior are not mutually exclusive and are likely to play a greater or lesser role, depending on the nature of the ligand, the protein and the temperature, as well as other conditions specifying the system. Single-molecule measurements would be useful to distinguish between these possibilities (Wang and Wolynes 1994; Hagen and Eaton 1996).

I have gone into considerable detail in reviewing myoglobin rebinding after photolysis because it is the protein reaction for which the most detailed experimental and theoretical studies exist. Nevertheless, many questions concerning this well-studied reaction remain unanswered (see also Nienhaus, this volume).

Specific Studies of Conformational Change

There are many reactions in proteins involving conformational change. It is likely that one of the reasons for having molecules as large as proteins carrying out important functions in living systems is that this makes it possible to have the well-defined conformational transitions needed for control and signalling. A conformational change in the context of this section is a reaction in which the initial and final states have well-defined structures, unlike the protein folding or unfolding reactions, where one of the "states" is a superposition of a very large number of different conformations; this very interesting case is considered below in the section, **PROTEIN FOLDING**. The most common examples of conformational changes, of which there are many, involve an alteration in the relative stability of two structures due to the binding of a ligand or a substrate. Such conformational changes include lid opening and closing motions (e.g.,TIM [Joseph et al. 1990], loop reorientation [as in rasp21, Ma and Karplus 1997], hinge-bending motions [as in lysozyme, McCammon et al. 1976, and in GroEL subunits, Xu et al. 1997], and the relative motion of subunits [e.g., as in the quaternary transition of hemoglobin, Perutz 1971]). For many of these cases, a reaction coordinate would be relatively easy to define. This contrasts such transitions with those concerned with globally distributed conformational changes that involve motions of all or most of the protein, as in the transition of myoglobin from the liganded to the unliganded state discussed above and the protein folding transition considered below.

Ring Flips in Bovine Pancreatic Trypsin Inhibitor

One conformational change whose dynamics has been simulated in detail is a "degenerate" one, in that the initial and final states are the same. It consists of aromatic ring isomerization (ring "flips") in the bovine pancreatic trypsin inhibitor (BPTI) (Snyder et al. 1976; Wagner et al. 1976; Northrup et al. 1982). Room temperature activated dynamics simulations very similar to those described for TIM (see section on **Dynamics of Enzymatic Reactions**) were performed for the tyrosine ring (Tyr 35), which has the slowest flipping rate of the four tyrosines in BPTI. Even for this apparently simple case, the obvious reaction coordinate, corresponding to the ring dihedral angle, is not satisfactory; instead, a more complex reaction coordinate that takes account of essential coupling between the ring and the local protein backbone was introduced. Based on this reaction coordinate, the transmission coefficient was calculated (Equation 9.9) and a well-defined plateau value equal to 0.22 was reached in 0.1 ps. Thus, the calculations indicate that the conformational change shows simple behavior, at least at room temperature. This is as expected because the calculated potential of mean force barrier for the ring flip is on the order of 10 kcal/mol and the barrier crossing rate is much slower than the collisional relaxation of the activated ring kinetic energy (McCammon and Karplus 1980). It is not unlikely that at a lower temperature nonexponential behavior would be observed, in analogy to that found for ligand rebinding in myoglobin. However, such low-temperature behavior has not been studied theoretically or experimentally. The experimental data on the ring flip from nuclear magnetic resonance spectroscopy have been used with the assumption of simple kinetic behavior of the first and second type to obtain the reaction rate (Wagner et al. 1976). The measured value of the activation enthalpy and entropy above 300 K are 37 kcal/mol and 68 eu, respectively. This contrasts with calculated values of 10 kcal/mol for the enthalpy and approximately zero for the entropy. Such a large disagreement suggested that the experimental separation into enthalpy and entropy of activation, based on the assumption that both are temperature independent, is in error (Brooks et al. 1988; see also section below on **Temperature Dependence of Folding Rate**). In fact, the measured rates as a function of temperature showed significant curvature suggesting that complexity of the second type may be present. Otting et al. (1993) have remeasured the Tyr 35 ring flips as part of a study of the equilibration of two conformers of the Cys 14–Cys 38 disulfide bond of BPTI. In the range 299 to 323 K, similar to that used by Wagner et al. (1976), the measurements (33 kcal/mol and 53 eu) confirmed the activation parameters found in the earlier work; as before, a kT/h pre-exponential factor was used. For the same temperature range, the disulfide conformational exchange had "similar" activation parameters (values were not given in the paper). What is interesting is that at lower temperatures (277 to 293 K), the disulfide activation parameters were found to be very different; the activation energy was on the order of 10 kcal/mol and the activation entropy was small and negative (on the order of –15 eu). Thus, there is a striking difference between the low- and high-temperature disulfide conformational transition with the former having activation parameters much closer to those calculated for the ring flip; unfortunately, the Tyr 35 ring flip rate could not be measured at the lower temperature. Otting et al. (1993) suggested that at high temperatures the disulfide transition and ring flip are concerted to explain the increased activation parameters. This could indicate that the discrepancy between the earlier experiments and the calculations for the ring flip arises from the fact that the measurements represent the high-temperature process, while the calculations correspond to the low-temperature process; indeed no disulfide flip was observed in the ring flip calculation. It is likely, therefore, that the ring flip transition and the disulfide reorientation are not simple by the second criterion, i.e., in this case, the reaction appears to follow a different path at low and high temperatures and the activation parameters have a complex temperature dependence, even though the native structure of the protein is essentially independent of temperature.

Triosephosphate Isomerase (TIM) Lid/Loop Transition

For most other cases, where the occurrence of a conformational change is demonstrated by the observation of two different crystal structures (for a list and classification of many of the known conformational changes in proteins see, e.g., Gerstein et al. 1994), there are no measurements of the dynamics. One case, which has been studied experimentally and theoretically, is the conformational change of a loop region that protects the active site of TIM from solvent during the reaction. I briefly describe the results because it demonstrates what is known and what is not known about a rather localized conformational change in a protein. An analysis based on the open and closed crystal structures of the enzyme suggested that the peptide loop motion, in going from one structure (unliganded, open) to the other (liganded, closed), is best described as a rigid "lid" motion, i.e., the internal conformation of the loop/lid does not change in the transition and certain residues at the two ends of the lid can be identified as hinges by use of pseudo-dihedral angles based on the Ca atom of each residue (Joseph et al. 1990; Gerstein and Chothia 1991). High-temperature molecular dynamics simulations were required to go from a closed to an open structure (analogous to protein unfolding; see section on **PROTEIN FOLDING**) in a reasonable simulation time. It was found that the loop opens and closes in a "jump-like fashion" within approximately 20 ps at 1000 K. This suggests that there is a barrier to opening even in the absence of substrate or product. Wade et al. (1993) simulated the loop motion by Brownian dynamics in the presence of the electrostatic field due to the rest of the (rigid) protein. They used a simplified representation of the loop residues, analogous to that employed by McCammon et al. (1980) in Brownian dynamics simulations of α-helix unwinding. Use of such a model permitted simulations as long as 100 ns, which would have been impossible with a more detailed description. The simulations showed that the opening and closing motion occurred on a 1 ns time scale. Moreover, many dihedral transitions within the loop/lid were observed, in contrast to the analysis of Joseph et al. (1990). A recent simulation of the lid motion (Derreumaux and Schlick 1998) with an all-atom model using stochastic dynamics supports the rigid-lid description and suggests a sizeable activation barrier for the transition.

There have been two experimental studies aimed at the dynamics of the loop/lid motion. One of these uses line shape analysis of the quadrupole powder pattern determined by solid state deuterium NMR (Williams and McDermott 1995) to probe the motion of the perdeuterated indole side chain of Trp 168. This residue is near the end of the loop (166–176); in fact, one of the hinges involves motion about the pseudo-dihedral angles 166–167 and 167–168, though the displacement of the former is larger than the latter. The experimental data were interpreted by assuming displacements of the Trp 168 side chain that correspond to the magnitude observed in the different crystal structures; with this assumption, a transition rate of about 10^4 s^{-1} was obtained. This is much slower than that observed in the Brownian dynamics simulation. Using a pre-exponential factor of 10^{12} s^{-1}, which may not be appropriate for what is likely to be a diffusive transition, an activation energy of 12 kcal/mol was estimated; no temperature-dependence measurements were made to confirm this. Somewhat surprisingly, the same rate was observed for the free enzyme and that with the bound inhibitors G3P and PGA. Also, the same ratio of 10 to 1 between the probabilities of the two forms (assumed open to closed for the free enzyme and closed to open for the inhibited enzyme) was obtained by fitting the spectra. An obvious complication in the interpretation is that there is no direct evidence that the two conformations used to fit the data are actually the open and closed form of the loop, rather than two unrelated slowly interconverting conformations of Trp 168, such as have been found to exist in studies of tryptophane fluorescence in other proteins. With a barrier of 12 kcal/mol, the 20 ps time for the transition found in the molecular dynamics simulation at 1000 K yields a rate of about 10^4 s^{-1} at room temperature, close to that estimated from the NMR experiment. One other study examined the temperature dependence of a TIM transition by using ^{31}P NMR of Glu 165 modified covalently with the substrate analogue 3-chloroacetol phosphate (Yüksel et al. 1994). Signals corresponding to two conformations were observed (again there is no direct evidence that they represent the open and closed lid) and the temperature dependence of the transition rate between them was fitted to an Arrhenius plot; it yielded a barrier of 34 kcal/mol. The actual data indicate significant curvature, though it is difficult to determine whether this is meaningful since only three temperature points were measured.

In summary, it is evident from the differences in the results obtained from the various experiments and simulations that the present understanding of this transition is incomplete and that more work is needed to determine the details of its behavior, including whether the reaction can be described as simple or complex.

Quaternary Transition in Hemoglobin

A conformational transition that has been studied at various levels of detail over many years is the quaternary transition in hemoglobin. Many measurements of the kinetics of the transition have been made because of its important role in hemoglobin function, as exemplified by the Monod–Wyman–Changeux model (1965). From the structural data of Perutz and coworkers (Perutz 1971) it is known that there are two quaternary structures, often referred to as *T* and *R*, with the former more stable in the unliganded state (no O_2 bound) and the latter more stable in the fully liganded state (one O_2 bound to each heme). Since both tertiary and quaternary transitions play a role in the different states of hemoglobin, a full analysis of the kinetics is rather complicated (Hofrichter et al. 1991; Henry et al. 1997). However, for the present report we are not concerned with these aspects. To a good approximation, the quaternary transition of the hemoglobin tetramer can be described in terms of two $\alpha\beta$ dimers ($\alpha_1\beta_2$ and $\alpha_2\beta_2$) that undergo a relative rotation of 15° with respect to each other; there is also a small relative translation coupled to the rotation (Baldwin and Chothia 1979). The most complete kinetic data are available for component I of trout hemoglobin, whose quaternary transition is similar to that of human hemoglobin (Hofrichter et al. 1991). The experiments were done by photolysis of hemoglobin fully liganded with CO, which provides a trigger analogous to that used in the myoglobin experiments described above. It was possible to isolate spectrally the *R* to *T* transition of the fully photodissociated tetramers (i.e., those with no CO bound). A simple rate expression was assumed in the analysis of the data; there is not enough information to determine a nonexponential time dependence for the reaction, even if it were to occur (W. Eaton, pers. comm.). Measurements of the temperature dependence over the range 275 to 338 K showed simple Arrhenius behavior. Thus, within the limits of the available experimental data, the quaternary transition is simple by both criteria described earlier (see section on **Definition of Simple versus Complex Behavior**). The measured activation energy was 8 kcal/mol. Use of this value, a gas-phase pre-exponential factor of (kT/h), and a transmission coefficient of unity yielded an activation entropy equal to -11.7 cal mol^{-1} K^{-1}. However, a simple diffusive model is probably better for obtaining the pre-exponential factor. The Kramers equation in the high friction limit (Kramers 1940; McCammon and Karplus 1977) and a simplified sphere representation of the $\alpha\beta$ dimer with a 20 Å radius yield a time of ~10^9 s^{-1} for the 15° motion involved in the *R* to *T* transition, in the absence of an activation barrier. This suggests that the factor $\kappa(kT/h)$ is significantly smaller than the value used by Hofrichter et al. (1991), in agreement with a suggestion in their paper. Use of Equation 9.4a with this pre-exponential factor yields a positive entropy of activation. This seems more reasonable since simulations suggest that there is a loosening of the tetramer in the transition (Fischer et al., unpublished).

PROTEIN FOLDING

Protein folding is the protein reaction of greatest complexity. The entire molecule is directly involved in the reaction, unlike some of the cases considered above. Moreover, the nonbonded van der Waals and electrostatic interactions that determine the potential energy surface for protein folding (often referred to pictorially as the "energy landscape" [Frauenfelder 1988]) and lead to the stability of the native state are individually weak (between 0.1 and 2 to 3 kcal/mol). If one is considering effective energy surfaces or potentials of mean force surfaces, these energy terms correspond to the effective interaction energy between pairs of atoms in the presence of a canonically averaged solvent and so depend on temperature (Karplus and Shakhnovich 1992). There are a very large number of these contributions; even for a small protein, such as the well-studied C terminal fragment of CI2 with 64 residues, there are on the order of 10^5 potential energy terms if they are represented as atom–atom interactions. Nevertheless, the free-energy difference between the native and denatured state is only on the order of tens of kcal/mol at ambient temperatures (e.g., 27 kcal/mol for a rather stable protein like CI2 at 298 K), which corresponds to about 0.4 kcal per residue and 0.04 kcal/atom (the enthalpy of unfolding is larger by an order of magnitude or so; it is about 135 kcal/mol for CI2 at 298 K). Clearly, a delicate balance of interactions is involved in the effective energy and free-energy surface. Another element in the complexity of the protein folding reaction is the large number of conformations accessible to the polypeptide chain. This has two aspects. One, of course, is the search problem, embodied in the so-called Levinthal Paradox (Levinthal 1969; Karplus 1997); the other, closely related, is the possibility that the configurational entropy has an important role in determining the activation barrier, if one exists.

In spite of the microscopic complexity, most measurements of folding kinetics (see, for example, Jackson and Fersht 1991; Schindler and Schmid 1996; Scalley and Baker 1997) indicate that the rate of formation of the native state obeys the simple unimolecular rate law given in Equations 9.1 and 9.2. More specifically, the observed rates of folding (and unfolding) can be represented either by a simple rate law or by a kinetic scheme consisting of a small number of intermediates, in addition to the denatured and native state, and each step of the overall reaction can be described by a simple rate law (Fersht et al. 1994). Thus, the kinetics of the protein folding reaction appears to be simple in terms of the first criterion, i.e., an exponential time dependence for the folding rate is observed within the accuracy of available experiments. However, the second criterion, that temperature dependence obeys an Arrhenius law, is not satisfied by the protein folding reactions whose temperature dependence has been studied (see, for example, Oliveberg et al. 1995). It appears to be Arrhenius-like at physiological temperatures, but very clearly deviates from Arrhenius behavior at higher temperatures. Thus, essential questions concerning protein folding are why the reaction is simple in terms of the first criterion, in spite of the complexity of the energy surface and the many degrees of freedom involved, and why the protein folding reaction is complex in terms of the second criterion.

Exponential Time Dependence of Protein Folding

From the discussion of reaction rate theory (see section, **Criterion for Existence of a Rate Constant**) and its applications to reactions involving proteins, a number of points are clear. A simple kinetic rate law is applicable if and only if there is a separation of time scales, i.e., the rate of the reaction (measured as flux passing over a barrier) is slow compared to the elementary collisional events that lead to equilibration of the other degrees of freedom of the system. This requirement is expected to be satisfied if there exists a significant free-energy barrier (several times kT) for the reaction. Experiments show that at room temperature proteins have activation free-energy barriers for the folding reaction that are large by this criterion (for CI2, the value of $\Delta G^{\ddagger}$ estimated from experiments is 15 kcal/mol at 298 K when a preexponential factor of 6×10^{12} s^{-1} is used; the latter is almost certainly too large by several orders of magnitude, as indicated below) so that simple exponential behavior is in accord with this qualitative criterion. It raises the question why there is *a* dominant free-energy barrier, if there is one: a question which does not have a simple answer. There is a biologically relevant reason for a large barrier for unfolding so as to satisfy the thermodynamic criterion for a unique native structure that is stable under physiological conditions. However, this does not provide reason for having a large (several kT) activation barrier for folding.

In previous sections I have briefly described the multiminimum surface of the native state and its effects on reactions in native proteins. It is very likely that the polypeptide energy surface for nonnative states is also a complex multiminimum surface. However, no detailed calculations of effective potential energy surfaces for proteins, including a satisfactory sampling of the nonnative conformations, are available. One of the few studies of a complete potential surface considers the alanine "tetrapeptide" (Czerminski and Elber 1989, 1990). Even this ultrasmall system, with only seven soft dihedral angles, has a number of points of interest for the protein folding problem. The surface has 139 minima and 502 transition states. A diagram of a part of the surface is shown in Figure 9.5 (Becker and Karplus 1997; Becker 1997). The minima can be organized into local basins by an energy criterion. Each basin contains several minima but the barriers between them are such that there is generally rapid

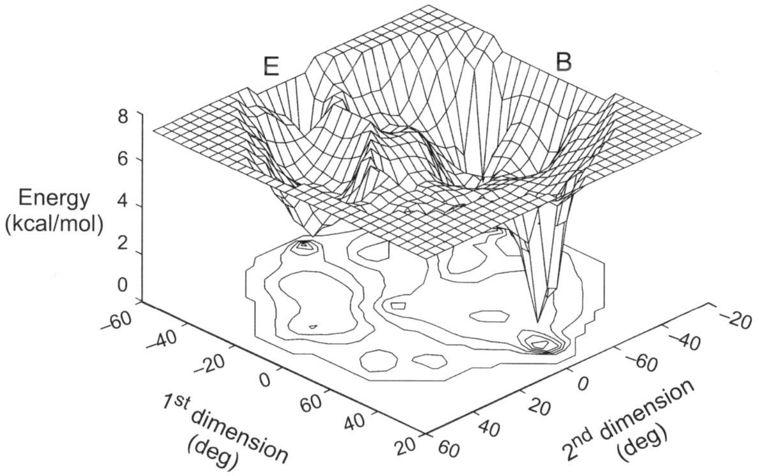

Figure 9.5 A simplified diagram of the energy surface of the main basin, which includes the minimum energy conformer (basin D), projected on to two principal components (a reduced representation relative to the seven soft dihedral angles specifying the conformation). It is interesting to note that the accessible region is rather broad down to an energy of about 4 kcal mol^{-1}, when the different basins separate. Adapted from Becker (1997).

equilibration within a basin, relative to the passage from one basin to another. A kinetic analysis of the approach of the system to the native state from an initial distribution over the denatured states shows approximately exponential behavior at 300 K. If such results can be extrapolated to the effective energy surface for protein folding, they suggest that, in spite of the existence of many minima, there is a dominant barrier ("the" barrier to folding) and that transition rates connecting different basins are fast relative to the passage along the reaction coordinate (progress variable) over the "rate-determining" barrier. The overall folding kinetics should then obey a simple rate law, possibly complicated by intermediates, and the global description in terms of simple kinetic schemes is a useful one, even if the species involved may represent populations with a wide range of structures (Dobson et al. 1998).

The most direct support for this view of protein folding is provided by simulations of a random heteropolymer bead model on a cubic lattices (Šali et al. 1994a, b); related studies have been made by Socci et al. (1996) for a very similar model and by Chan and Dill (1994) on square lattices. Although oversimplified relative to actual proteins, such models have important protein-like properties, i.e., they have a large number of conformations (e.g., 10^{16} for a 27-mer) and a unique ground state; the "unique" ground state in this model encompasses the large number of substates with very similar structures (within 2Å rms) that in reality contribute to the native state of the proteins and have a role in the complex kinetic behavior exhibited in the ligand rebinding to myoglobin, for example (see section on **Validity of Arrhenius Equation**). Unlike more detailed protein models, calculations of the folding kinetics and

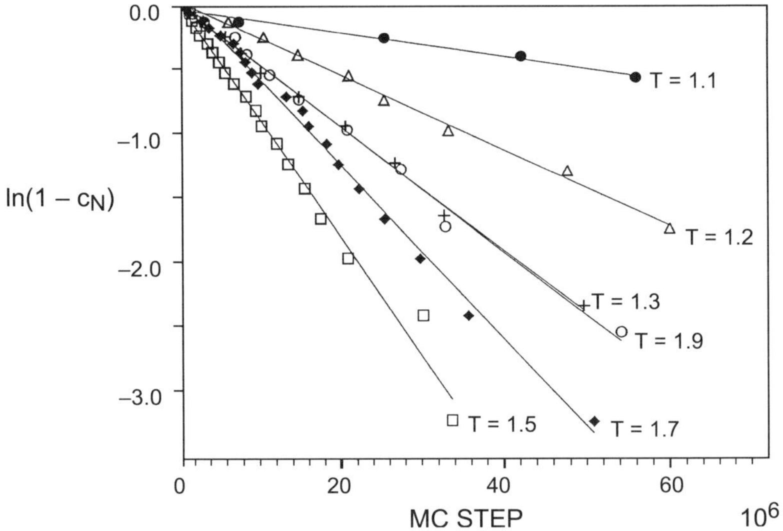

Figure 9.6 Distribution of the mean first passage times for a 27-mer folding sequence. The distribution of the mean first passage times is shown at the temperatures indicated on the plot. The lines are linear least-squares fits to the points. One hundred independent folding trials were done to obtain each of the points. Adapted from Karplus et al. (1995).

thermodynamics are sufficiently fast that they can be explored in detail for a large number of sequences at different temperatures. Figure 9.6 shows the observed kinetics for a 27-mer sequence (Karplus et al. 1995; Karplus 1997; Dobson et al. 1998). For all temperatures studied, it is found that the rate of folding is exponential, i.e., the heteropolymer folding reaction is simple by the first criterion. This is true even though the time dependence of the decay of the correlation function for the structural overlap between coordinate sets is best fitted by a stretched exponential with the exponent of the time near 0.5, corresponding to an approximately diffusive time dependence (Šali and Karplus, unpublished). For some designed sequences (i.e., sequences that are optimized by selecting the interactions to speed up folding) more complex folding behavior has been found (Abkevich et al. 1994). At high temperatures, folding can be described by a single exponential; however, at lower temperatures there is both a fast and slow exponential phase. This was explained, in analogy to interpretations of experiments on protein folding, by the existence of an (off-pathway) intermediate.

The effective free-energy surfaces at two different temperatures drawn as a function of the progress variable Q, which is the fractional overlap (number of native contacts divided by the total number of native contacts) with the native state, are in accord with the calculated simple rate behavior (see Figures 9.7b and 9.7d). For both cases, the free-energy surface has a significant barrier near the native state that dominates the folding kinetics. In the (high- and low-temperature) limits shown (for the more general temperature dependence of the surfaces, see Šali et al. 1994b), the free-energy surface

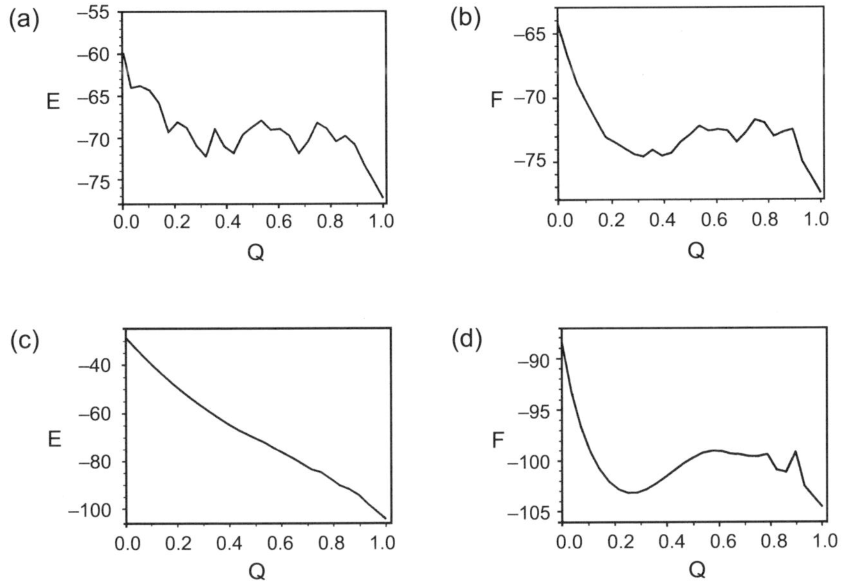

Figure 9.7 Folding effective energy and free-energy surfaces of a 27-bead heteropolymer on a cubic lattice. (a) The average effective energy, *E*, as a function of *Q*, the fraction of native contacts at a low temperature (T = 0.7); there are 28 in the native state, which is a $3 \times 3 \times 3$ cube. (b) The average effective free energy, *F*, as a function of *Q* for *T* = 0.7. (c) and (d) same as (a) and (b) at a high temperature (*T* = 2). Results adapted from Figure 4 of Šali et al. (1994b). Details of the definition and methods are given in that paper.

appears relatively smooth and has a single important barrier, as required for the validity of a phenomenological rate equation. This is true even at the lower temperature for which the effective energy surface is quite "rough" (see Figure 9.7d) due to a partial cancellation between the "roughness" in the energy and entropy (Šali et al. 1994b; Dobson et al. 1998).

In Figure 9.7 we have used *Q*, the fractional overlap with the native state, as the progress variable to describe the free energy of the reaction. Because of the complexity of the protein folding reaction and the many degrees of freedom involved, it is much less simple to define a reaction coordinate than for small molecule reactions. In fact, it is likely that a reaction coordinate corresponding to that for small molecule reactions does not exist, i.e., there is no well-defined linear combination of coordinates that describes the motion from the reactant to the product state. This follows if the folding trajectories of individual polypeptide chains are very different and cover a wide range of possibilities. Clearly, if the fractional number of native contacts, *Q*, is used to specify the progress of the reaction, many different sets of contacts correspond to the same value of the progress variable, except for the unique native state. (For use of *Q* as the progress variable in a protein unfolding simulation with an atomic model, see Lazaridis and Karplus 1997.) It is not self-evident that motion along *Q* is always an appropriate description of the protein folding reaction, i.e., there is the question of whether motion along *Q* corresponds to the slow step so that the effective free energy as a function of its value can be used to analyze the reaction, e.g., to show that there is a single dominant free-energy barrier. For the random heteropolymers studied by Šali et al. (1994a, b), consistent results are obtained with *Q* as the progress variable. Moreover, the transition states as defined by the free-energy barriers (see Figures 9.7b and 9.7d) correspond to the values of *Q*, where the probability of reaching the native state increases rapidly from essentially zero to about 0.5 (Šali and Karplus, unpublished). Use of *Q* for short chains, as in the 27-mer, has been supported by the work of Pande et al. (1997). For longer chains, more than one progress variables may be required to describe the folding process (Dinner and Karplus, submitted; Dobson et al. 1998). Alternative progress variables for protein folding have been considered by Du et al. (1998) and by Chan and Dill (1998). As pointed out above, for the 27-mers, *Q* satisfies the conditions for a useful progress variable or "transition coordinate" suggested in Du et al. (1998). Chan and Dill (1998) found cases in lattice folding simulations where there is no separation in time scale for the reaction and the other coordinates of the system; it would be interesting to know whether the folding rate is nonexponential for these systems. In this regard, it is important to point out again that for the validity of the simple phenomenological rate expression, it is not necessary to "know" the appropriate progress variable; it is necessary only that such a coordinate exists and that motion along it is slow relative to the relaxation of other degrees of freedom of the system.

Given the simulations, which demonstrate simple kinetic behavior of the first type (Equations 9.1 and 9.2), a number of models have been introduced to justify the experimental and theoretical results. The essential idea (Chan and Dill 1994; Zwanzig 1997; Pande et al. 1997; Gutin et al. 1998; Neria and Karplus, unpublished) is that the native state is coupled to the very large number of configurations making up the denatured state by a much smaller number of "doorway" states. The analysis of the 27-mer lattice simulations at low temperature suggests on the order of 1000 such doorway states (the "transition-state ensemble"). If the doorway states have similar free energies and the equilibration among the denatured states is fast relative to the motion that leads to the doorway states, a transition-state like expression for the flux through the doorway states can be derived and an expression for the rate constant can be obtained. Several of the models (Chan and Dill 1994; Zwanzig 1997; Neria and Karplus, unpublished) have used a master equation approach to calculate the kinetics. A somewhat different model (Gutin et al. 1998) has elements in common with that of Hagen and Eaton (1996) for the conformational relaxation in myoglobin. The two essential assumptions of the model of Gutin et al. (1998) are that the mean energy of the system relaxes rapidly (fast time scale) to the equilibrium value given by the random energy model; below a critical temperature the equilibrium energy is essentially that of the native state. On a slower time scale, the system finds one of the small number of transition states from which it folds rapidly to the native state. It is assumed that the transition states all have essentially the same energy in line with the kinetic models of Koper and Hilhorst (1982), Hagen and Eaton (1996), and Saven et al. (1994). The model differs from that of Hagen and Eaton (1996) in that the equilibration of degrees of freedom other than the reaction coordinate is rapid, so that exponential kinetics is obtained.

It is important to consider briefly the pre-exponential factor in Equation 9.4a since given the activation enthalpy of the reaction from a measurement of the rate as a function of temperature (Equation 9.4b), the activation entropy depends on its value. Although the gas-phase (Eyring) value, (kT)/h, which equals ~6×10^{12} at 300 K, is often used, it is not generally valid for solution reactions, as has already been mentioned (see section, **Rebinding of Ligands after Photodissociation of Myoglobin**). For the motions involved in the folding reaction, the system is likely to be in the diffusive regime because of the size of the particles (residues or larger groups). However, the microscopic complexity is such that a range of different structural entities may be involved during various portions of the folding reaction. This is clearly one of the problems that still needs to be resolved. In fact, it may be very difficult to determine the appropriate "average" value of the pre-exponential factor for use in Equation 9.4a . Values such as 10^{10} s^{-1} (Scalley and Baker 1997) and 6×10^{12} s^{-1} (Oliveberg et al. 1995) have been used, but there is little justification for these or any other values. Since the elementary step in residue-based treatments of protein folding (e.g., Monte Carlo lattice simulations) is likely to be a dihedral or pseudo-dihedral angle transition, a value in the neighborhood of 10^9 s^{-1} may be appropriate (McCammon et al. 1980). However, if larger entities are involved (e.g., as in the diffusion-collision model), values closer to 10^7 to 10^8 s^{-1} may be needed (Karplus and Weaver 1976; 1994). Henry et al. (1996) have suggested a value in the neighborhood of 10^6 s^{-1} for the fastest forming loops in proteins based on measurements of cytochrome c (Jones et al. 1993) and a theoretical model for the formation of a contact by collapse of a random chain. This value seems somewhat too small, not for the process specifically considered but for use generally in the protein folding reaction, particularly because some proteins fold very rapidly; for example, an Arc repressor mutant has a bimolecular rate constant for folding of 3×10^8 M^{-1} s^{-1} (Waldburger et al. 1996), although this number cannot be compared directly with unimolecular folding rates. There is also the question of whether the diffusive rate slows down significantly as the transition state for the folding reaction is approached (for a diffusive protein folding model, see Socci et al. 1996). It is not clear whether this is a major factor since both the effective diffusion constant for the entities involved and the range over which diffusion is required are expected to decrease simultaneously. With the large uncertainty concerning this often essential factor, as well as the questions concerning the appropriate progress variable for the folding reaction, a discussion would be very useful (see Eaton et al., this volume).

Temperature Dependence of Folding Rate

One of the striking differences between small molecule kinetics and protein folding is the temperature dependence of the rate constant. Although the unfolding rate is often approximately Arrhenius-like (i.e., a plot of ln k vs $1/T$ is essentially a straight line)

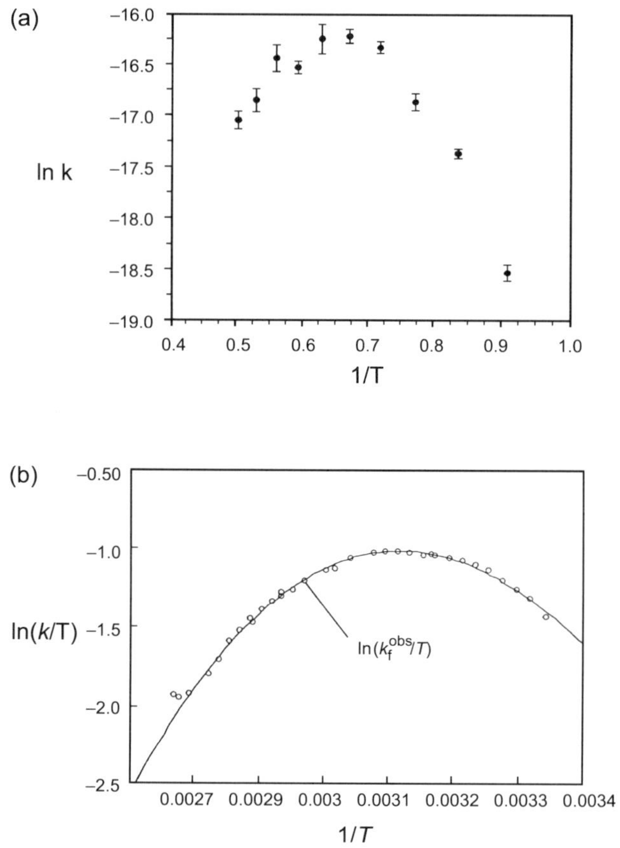

Figure 9.8 (a) Calculated Arrhenius plot of the folding reaction rate constant as a function of temperature for the sequence used in Figure 9.6; results adapted from Karplus et al. (1993). (b) Arrhenius temperature dependence of the measured refolding rate constant for CI2; adapted from Oliveberg et al. (1995).

(e.g., Segawa and Sugihara 1984), the folding rate is not. Figures 9.8a and 9.8b show the results obtained, respectively, for folding with the 27-mer cube lattice model that we have already discussed (Šali et al. 1994a, b; Karplus et al. 1995) and for folding of CI2 from the experiments of Oliveberg et al. (1995). The interpretation of the lattice model results is straightforward and provides an increased understanding of the folding dynamics. At low temperatures ($1/T$ large in Figure 9.8a), there is Arrhenius-like behavior associated with a positive activation energy (or enthalpy), as in Equations 9.3 and 9.4. This means the barrier to folding is dominated by the activation energy in correspondence with the effective energy surface in Figure 9.7a. As the temperature increases ($1/T$ small in Figure 9.8a), the Arrhenius plot turns over, and the activation energy becomes negative. In the present model, this arises from temperature independent interactions and an effective energy surface (funnel-like) in which there is a monotonic decrease in the energy. The corresponding surface is shown in Figure 9.7c, which yields a negative energy of activation, as confirmed by analyzing the results in Figure 9.8a with Equation 9.4b. However, the free energy of activation, $\Delta G^{\ddagger}$, is still positive due to the activation entropy ($-T\Delta S^{\ddagger}$), whose effect increases with increasing temperature. The activation entropy is associated with the decrease in the number of configurations accessible with increasing Q, i.e., there is an entropic "bottleneck" associated with the transition state and this bottleneck becomes more important at higher temperatures because of the larger number of states accessible in the denatured region. At very low temperatures, there is no indication in the lattice calculations (Figure 9.8b and Gutin et al. 1998) or in the available experiments (Figure 9.8b) that there is a deviation from the Arrhenius law, i.e., no "super-Arrhenius" behavior, analogous to that described by Equation 9.15, for example, has been observed in protein folding, although it was predicted by Bryngelson and Wolynes (1989).

The experimental Arrhenius plot shown in Figure 9.8b, is very similar in form to that found in the lattice simulations (Figure 9.8a). Consequently, the formal interpretation in terms of activation enthalpy, $\Delta H^{\ddagger}$, and activation entropy $\Delta S^{\ddagger}$, is exactly the same. As the temperature increases, the activation enthalpy becomes negative, while the activation free energy remains positive. One explanation of these results is that the effective enthalpy surface of protein CI2 (and other proteins) corresponds, in fact, to those shown in Figures 9.7a and 9.7c at low and high temperatures, respectively, and that the positive activation free energy at high temperatures arises from the configurational entropy term, as in the lattice models. If $T = 1.6$, the maximum in the rate of the lattice simulations (Figure 9.8a) is set equal to 325 K, a factor of 10 decrease in the rate at higher temperatures corresponds to about 400 K, somewhat larger than the experimental value of 360 K for CI2; this is in accord with the fact that the entropic contribution in CI2 is expected to be larger than for the 27-mer. The effective energy as a function of the number of native contacts obtained in unfolding simulations of CI2 at high temperature (Lazaridis and Karplus 1997) does show a monotonically decreasing energy with increasing Q, in accord with this model. This is a very interesting result (Lazaridis and Karplus 1997; and to be published) because it suggests that the experimental temperature dependence of the protein folding reaction is in accord with funnel-like effective energy surfaces at high temperatures but not at room temperature.

It has also been suggested (Segawa and Sugihara 1984; Scalley and Baker 1997; Chan and Dill 1998) that the temperature dependence of the interactions themselves, due to the hydrophobic effect, is involved in the non-Arrhenius behavior. The hydrophobic effect increases with increasing temperature over the range of interest (275 to 330 K), as based on small molecule studies and their application to protein thermodynamics (Makhatadze and Privalov 1995). The dominant contribution of the hydrophobic effect is on the denatured state, which is enthalpically less stabilized as the temperature increases. If the transition state is less unfolded than in the denatured state (for an analysis of CI2, see Lazaridis and Karplus 1997), it would have a proportionally smaller contribution from the hydrophobic effect and the magnitude of $\Delta H^{\ddagger}$ would be expected to decrease with increasing temperature. This contribution to the curvature in the Arrhenius plots contrasts with the lattice results in which the interaction are independent of temperature and the change in activation enthalpy is due to the difference in the energies of the configurations sampled as a function of temperature.

In a recent study, Scalley and Baker (1997) showed that by simultaneously varying the temperature and denaturant concentration to keep $\Delta G/T$ fixed, where ΔG is the free energy of unfolding, much of the temperature dependence of $\Delta H^{\ddagger}$ (i.e., the curvature of the Arrhenius plot) could be eliminated. They argued that this meant that the temperature dependence of the interactions played a dominant role in the curved Arrhenius plot. Although the results of Scalley and Baker are interesting, their conclusion that it "convincingly demonstrates that the anomalous temperature of folding derives from the temperature dependence of the interactions that stabilize proteins" does not seem warranted. Chan and Dill (1998) have recently published an analysis of a modified HP folding simulation on a square lattice and interpreted their results as supporting the interpretation of Scalley and Baker (1997).

SUMMARIZING DISCUSSION

In this chapter, I have tried to provide a brief background in chemical kinetics with specific emphasis on the conditions for simple behavior in reactions. Simple behavior was defined by two (related) aspects. The first is that a simple phenomenological rate law with an exponential time dependence for the rate applies and the second is that the temperature dependence of the rate follows the Arrhenius law. We have seen that although simple behavior is found in some protein reactions, significant deviations from both types of simplicity have been documented and interpreted theoretically. Ligand rebinding after photolysis in myoglobin was discussed in considerable detail because this "simple" reaction has been studied more than any other and has been shown to deviate from both requirements for simplicity under certain conditions.

For protein folding the situation is somewhat different. Although deviations from simplicity might have been expected, there is considerable evidence that a simple rate law with an exponential overall folding rate applies to most proteins that have been studied so far; there are exceptions (intermediates) but they are "trivial" from the present viewpoint (Roder and Colon 1997). By contrast, the second criterion for simplicity is not satisfied. The temperature dependence of the folding rate of all proteins that have been studied show significant deviations from Arrhenius behavior. I have given an interpretation of the results based on lattice simulations, but the isolation of the actual mechanism for the non-Arrhenius type behavior in proteins will require additional studies.

From the history of the experimental and theoretical investigations of ligand rebinding after photolysis in myoglobin, it seems likely that the applicability of a simple phenomenological rate law to the overall protein folding reaction hides an underlying complexity. One set of experiments that will help to reveal this complexity involves studies with a short time resolution for systems under nonequilibrium conditions. A range of methods for triggering the folding (and unfolding) reaction have been developed, and these processes can be studied, in principle, on time scales from nanoseconds to seconds or longer (Eaton et al. 1996; Dobson et al. 1998). A recent protein folding study (Sabelko et al. 1999), which makes use of ns beating of cold denatured proteins, has observed nonexponential folding, although the origin of this "complex" behavior is not clear. However, techniques for monitoring the structural details as a function of time need to be improved. Techniques that are likely to make important contributions are time-resolved X-ray crystallography (Šrajer et al. 1996), real time NMR (Dobson et al. 1998), as well as infrared spectroscopy (Phillips et al. 1995) and vibrational Raman optical activity (Wilson et al. 1996). More generally for protein dynamics, infrared photon echo measurements (Tokmakoff and Fayer 1995; Rector et al. 1997) and optical line broadening studies (Frauenfelder 1995; Thorn Leeson and Wiersma 1995; Thorn Leeson et al. 1997) are expected to yield significant new results.

Another type of approach that is likely to be of great importance for many reactions involving proteins, including enzyme catalysis, ligand binding, and protein folding, is the monitoring of single molecules (e.g., Craig et al. 1996; Ashkin 1997; Lu and Xie 1997; Xie and Trautman 1998; Zocchi 1997). This provides a direct way of determining the diversity that exists for the various reactions and will make it possible to distinguish, in principle, whether the macroscopic behavior is evident in the individual molecules or whether it is a consequence of averaging over a population of molecules. In this regard it is interesting to mention that single-molecule behavior has been studied for a long time for the case of membrane channel conductance, in which the molecules involved undergo a significant structural change between open and closed states. For most channel systems, the structural data required for atom-based simulations are not available. However, the high-resolution structure of a potassium channel determined very recently (Doyle et al. 1998) opens the way for detailed theoretical studies on this very interesting system; many simulations will certainly be made in the near future. Finally, the more global variation of experimental conditions (e.g., folding at low temperatures) will aid in determining the difference in the temperature (or other) dependence of different relaxation processes.

The future holds great promise for this exciting field. It is likely that, as measurements and simulations become more precise and cover a wider range of conditions, complex behavior of the type discussed in this chapter will be found for an increasing number of reactions in proteins.

ACKNOWLEDGMENTS

Partial support for the research reported here was provided by the National Science Foundation and the National Institutes of Health.

REFERENCES

Abkevich, V.I., A.M. Gutin, and E.I. Shakhnovich. 1994. Free energy landscape for protein folding kinetics. *J. Chem. Phys.* **101**:6052–6062.

Agmon, N., W. Doster, and F. Post. 1994. The transition from inhomogeneous to homogeneous kinetics in CO binding of myoglobin. *Biophys. J.* **66**:1612–1622.

Agmon, M., and J.J. Hopfield. 1983. CO binding to heme proteins: A model for barrier height distributions and slow conformational changes. *J. Chem. Phys.* **79**:2042–2053.

Anderson, J.B. 1973. Statistical theories of chemical reactions. Distributions in the transition region. *J. Chem. Phys.* **58**:4684–4692.

Angell, C.A. 1995. Formation of glasses from liquids and biopolymers. *Science* **267**:1924–1927.

Ansari, A., J. Berendzen, S.F. Bowne, H. Frauenfelder, I.E.T. Iben, T.B. Sauke, E. Shyamsunder, and R.D. Young. 1985. Protein states and proteinquakes. *Proc. Natl. Acad. Sci. USA* **82**:5000–5004.

Åqvist, J., and M. Fothergill. 1996. Computer simulation of the triophosphate isomerase catalyzed reaction. *J. Biol. Chem.* **271**:10,010–10,016.

Åqvist, J., and A. Warshel. 1993. Simulation of enzyme reactions using valence bond force fields and other hybrid quantum/classical approaches. *Chem. Rev.* **93**:2523–2544.

Ashkin, A. 1997. Optical trapping and manipulation of neutral particles using lasers. *Proc. Natl. Acad. Sci. USA* **94**:4853–4860.

Austin, R.H., K.W. Beeson, L. Eisenstein, H. Frauenfelder, and I.C. Gunsalus. 1975. Dynamics of ligand binding to myoglobin. *Biochemistry* **14**:5355–5373.

Bai, Y.S., and M.D. Fayer. 1989. Time scales and optical dephasing measurements: Investigation of dynamics in complex systems. *Phys. Rev. B* **39**:11,066–11,084.

Baldwin, J., and C. Chothia. 1979. Haemoglobin: The structural changes related to ligand binding and its allosteric mechanism. *J. Mol. Biol.* **129**:175–200.

Bash, P.A., M.J. Field, R.C. Davenport, G.A. Petsko, D. Ringe, and M. Karplus. 1991. Computer simulation and analysis of the reaction pathway of triosephosphate isomerase. *Biochemistry* **30**:5826–5832.

Becker, O.M. 1997. Quantitative visualization of a macromolecular potential energy 'funnel'. *J. Mol. Struct.* (*THEOCHEM*) **398–399**:507–516.

Becker, O.M., and M. Karplus. 1997. The topology of multidimensional potential energy surfaces: Theory and application to peptide structure and kinetics, *J. Chem. Phys.* **106**:1495–1517.

Bergsma, J.P., B.J. Gartner, K.R. Wilson, and J.T. Hynes. 1987. Molecular dynamics of a model SN_2 reaction in water. *J. Chem. Phys.* **86**:1356–1376.

Brooks, C.L., III, M. Karplus, and B.M. Pettitt. 1988. Proteins: A Theoretical Perspective of Dynamics, Structure, and Thermodynamics. Adv. Chem. Phys. **LXXI**. New York: Wiley.

Brumer, P., and M. Karplus. 1973. Collision complex dynamics in alkali halide exchange reactions. *Faraday Soc. Disc.* **55**:80–91.

Bryngelson, J.D., and P.G. Wolynes. 1989. Intermediates and barrier crossing in a random energy model (with applications to protein folding). *J. Phys. Chem.* **93**:6902–6915.

Case, D.A., and M. Karplus. 1979. Dynamics of ligand binding to heme proteins. *J. Mol. Biol.* **132**:343–368.

Chan, H.S., and K.A. Dill. 1994. Transition states and folding dynamics of proteins and heteropolymers. *J. Chem. Phys.* **100**:9238–9257.

Chan, H.S., and K.A. Dill. 1998. Protein folding in the landscape perspective: Chevron plots and non-Arrhenius kinetics. *Proteins* **30**:2–33.

Chandler, D. 1978. Statistical mechanics of isomerization dynamics in liquids and the transition-state approximation. *J. Chem. Phys.* **68**:2959–2970.

Craig, D.B., E.A. Arriaga, J.C.Y. Wong, H. Lu, and N.J. Douichi. 1996. Studies on single alkaline phosphatase molecules. *J. Am. Chem. Soc.* **118**:5245–5283.

Czerminski, R., and R. Elber. 1989. Reaction path study of conformational transitions and helix formation in a tetrapeptide. *Proc. Nat. Acad. Sci. USA* **86**:6963–6967.

Czerminski, R., and R. Elber. 1990. Reaction path study of conformational transitions in flexible systems: Applications to peptides. *J. Chem. Phys.* **92**:5580–5601.

Debrunner, P.G., and H. Frauenfelder. 1982. Dynamics of proteins. *Ann. Rev. Phys. Chem.* **33**:283–299.

Derreumaux, P., and T. Schlick. 1998. The loop opening/closing motion of the enzyme triosephosphate isomerase. *Biophys. J.* **74**:72–81.

Dobson, C. M., A. Šalí , and M. Karplus. 1998. Protein folding: A perspective from theory and experiment. *Angew. Chem. Intl. Ed.* **37**:869–893.

Doster, W., S. Cusack, and W. Petry. 1989. Dynamic transition of myoglobin revealed by inelastic neutron scattering. *Nature* **337**:754–756.

Doyle, D.A., J.M. Cabral, R.A. Pfuetzner, A. Kuo, J.M. Gulbis, S.L. Cohen, B.T. Chait, and R. MacKinnon. 1998. The structure of the potassium channel: Molecular basis of K^+ conduction and selectivity. *Science* **280**:69–77.

Du, R., V.J. Pande, A.Y. Grasberg, T. Tanaka, and E. Shakhnovich. 1998. On the transition coordinate for protein folding. *J. Chem. Phys.* **108**:334–350.

Eaton, W. A., D.A. Thomson, C.-K. Chan, S.J. Hagen, and J. Hofrichter. 1996. Fast events in protein folding. *Structure* **4**:113–119.

Elber, R., and M. Karplus. 1987. Multiple conformational states of proteins: A molecular dynamics analysis of myoglobin. *Science* **235**:318–321.

Fersht, A. 1985. Enzyme Structure and Mechanism, 2nd ed. New York: Freeman.

Fersht, A.R., L.S. Itzhaki, N.F. El Masry, J.M. Matthews, and D.E. Otzen. 1994. Single versus parallel pathways of protein folding and fractional formation of structure in the transition state. *Proc. Natl. Acad. Sci. USA* **91**:10,426–10,429.

Frauenfelder, H. 1988. Physics and biology. In: Physics in a Technological World, ed. A.P. French, pp. 255–267. New York: AIP Press.

Frauenfelder, H. 1995. Complexity in proteins. *Nature Struct. Biol.* **2**:821–823.

Frauenfelder, H., G.A. Petsko, and D. Tsernoglu. 1979. Temperature-dependent X-ray diffraction as a probe of protein structural dynamics. *Nature* **280**:558–563.

Frauenfelder, H., and P.G. Wolynes. 1985. Rate theories and puzzles of hemeprotein kinetics. *Science* **229**:337–345.

Gerstein, M., and C. Chothia. 1991. Analysis of protein loop closure. Two types of hinges produce one motion in lactate dehydrogenase. *J. Mol. Biol.* **220**:133–149.

Gerstein, M., A.M. Lesk, and C. Chothia. 1994. Structural mechanisms for domain movements in proteins. *Biochemistry* **33**:6739–6749.

Greer, J.-L., J. Fan, and C.A. Angell. 1994. The protein–glass analogy: Some insights from homopeptide comparisons. *J. Phys. Chem.* **98**:13,780–13,790.

Grote R.F., and J.T. Hynes. 1980. The stable state picture of chemical reactions. *J. Chem. Phys.* **73**:2715–2732.

Gutin, A., A. Šalí, V. Abkevich, M. Karplus, and E. Shakhnovich. 1998. Temperature dependence of the folding rate in a simple protein model: Search for a "glass" transition. *J. Chem. Phys.* **108**:6466–6483.

Hagen S. J., and W. A. Eaton. 1996. Nonexponential structural relaxations in proteins. *J. Chem. Phys.* **104**:3395–3398.

Henry, E.R., C.M. Jones, J. Hofrichter, and W.A. Eaton. 1996. Diffusion limited contact formation in unfolded cytochrome c: Estimating the maximum rate of protein folding. *Proc. Natl. Acad. Sci. USA* **93**:11,615–11,617.

Henry, E.R., C.M. Jones, J. Hofrichter, and W.A. Eaton. 1997. Can a two-state MWC allosteric model explain hemoglobin kinetics? *Biochemistry* **36**:6511–6528.

Henry, E.R., J.H. Sommer, J. Hofrichter, and W.A. Eaton. 1983. Geminate recombination of carbon monoxide to myoglobin. *J. Mol. Biol.* **166**:443–451.

Hofrichter, J., E.R. Henry, A. Szabo, L.P. Murray, A. Ansari, C.M. Jones, M. Coletta, G. Falcioni, M. Brunori, and W.A. Eaton. 1991. Dynamics of the quaternary conformational change in trout hemoglobin. *Biochemistry* **30**:6583–6598.

Iben, I., D. Braunstein, W. Doster, H. Frauenfelder, M.K. Hong, J.B. Johnson, S. Luck, P. Ormos, A. Schulte, P.J. Steinbach, A.H. Xie, and R.D. Young. 1989. Glassy behavior of a protein. *Phys. Rev. Lett.* **62**:1916–1919.

Jackson, S.E., and A.R. Fersht. 1991. Folding of chymotrypsin inhibitor 2. 1. Evidence for a two-state transition. *Biochemistry* **30**:10,428–10,435.

Jackson, T.A., M. Lim, and P.A. Anfinrud. 1994. Complex nonexponential relaxation in myoglobin after photodissociation of MbCO. *Chem. Phys.* **180**:131–140.

Jones, C.M., et al. 1993. Fast events in protein folding initiated by nanosecond laser photolysis. *Proc. Natl. Acad. Sci. USA* **90**:11,860–11,864.

Joseph, D., G.A. Petsko, and M. Karplus. 1990. Anatomy of a protein conformational change: Hinged "lid" motion of the triosephosphate isomerase loop. *Science* **249**:1425–1428.

Joseph-McCarthy, D., G.A. Petsko, and M. Karplus. 1995. Use of a minimum perturbation approach to predict TIM mutant structures. *Protein Eng.* **8**:1103–1115.

Karplus, M. 1997. The Levinthal paradox: Yesterday and today. *Folding and Design* **2**:569–576.

Karplus, M., A. Caflisch, A. Šalí, and E. Shakhnovich. 1995. Protein dynamics: From the native to the unfolded state and back again. In: Modelling of Biomolecular Structures and Mechanisms, ed. A. Pullman et al., pp. 69–84. Kluwer Academic.

Karplus, M., R.N. Porter, and R.D. Sharma. 1965. Exchange reactions with activation energy. I. Simple barrier potential for (H, H_2). *J. Chem. Phys.* **43**:3259–3287.

Karplus, M., and E. Shakhnovich. 1992. Protein folding: Theoretical studies of thermodynamics and dynamics. In: Protein Folding, ed. T. Creighton, pp. 127–195. W.H. Freeman.

Karplus, M., and D.L. Weaver. 1976. Protein-folding dynamics. *Nature* **260**:404–406.

Karplus, M., and D.L. Weaver. 1994. Folding dynamics: The diffusion-collision model and experimental data. *Protein Sci.* **3**:650–668.

Keck, J.C. 1967. Variational theory of reaction rates. *Adv. Chem. Phys.* **13**:85–121.

Knowles, J.R. 1991a. Enzyme catalysis: Not different, just better. *Nature* **350**:121–124.

Knowles, J.R. 1991b. To build an enzyme... *Phil. Trans. Roy. Soc. Lond. B* **332**:115–121.

Koper, G., and H. Hilhorst. 1987. Power law relaxation in the random energy model. *Europhys. Lett.* **3**:1213–1217.

Kramers, H.A. 1940. Brownian motion in a field of force and the difference model of chemical kinetics. *Physica* **7**:284–293.

Kuczera, K., J.-C. Lambry, J.-L. Martin, and M. Karplus. 1993. Nonexponential relaxation after ligand dissociation from myoglobin: A molecular dynamics simulation. *Proc. Natl. Acad. Sci. USA* **90**:5805–5807.

Kuczera, K., J. Kuriyan, and M. Karplus. 1990. Temperature dependence of the structure and dynamics of myoglobin: A simulation approach. *J. Mol. Biol.* **213**:351–373.

Lazaridis, T., and M. Karplus. 1997. "New view" of protein folding reconciled with the old through multiple unfolding simulations. *Science* **278**:1928–1931.

Levinthal, C. 1969. How to fold graciously. In: Mossbauer Spectroscopy in Biological Systems, Proc. of a Meeting held at Allerton House, Monticello, Illinois, ed. P. Debrunner, J.C.M. Tsibris, and E. Münck. Urbana: Univ. of Illinois Press.

Levy, R.M., R.P. Sheridan, J.W. Keepers, G.S. Dubey, S. Swaminathan, and M. Karplus. 1985. Molecular dynamics of myoglobin at 298 K: Results from a 300 ps computer simulation. *Biophys. J.* **48**:509–518.

Li, H., R. Elber, and J.E. Straub. 1993. Molecular dynamics simulation of NO recombination to myoglobin mutants. *J. Biol. Chem.* **268**:17,908–17,916.

Lim, M., T.A. Jackson, and P.A. Anfinrud. 1993. Nonexponential protein relaxation: Dynamics of conformational change in myoglobin. *Proc. Natl. Acad. Sci. USA* **90**:5801–5804.

Lu, H.P., and S. Xie. 1997. Single molecular spectral fluctuations at room temperature. *Nature* **385**:143–146.

Ma, J., and M. Karplus. 1997. Molecular switch in signal transduction: Reaction paths of the conformational changes in *ras* p21. *Proc. Natl. Acad. Sci. USA* **94**:11,905–11,910.

Makhatadze, G.I., and P.L. Privalov. 1995. Energetics of protein structure. *Adv. Prot. Chem.* **47**:307–425.

McCammon, J.A., B.R. Gelin, M. Karplus, and P.G. Wolynes. 1976. The hinge-bending mode in lysozyme. *Nature* **262**:325–326.

McCammon, J.A., and M. Karplus. 1977. Internal motions of antibody molecules. *Nature* **268**:765–766.

McCammon, J.A., and M. Karplus. 1980. Dynamics of tyrosine ring rotations in a globular protein. *Biopolymers* **19**:1375–1405.

McCammon, J.A., S.H. Northrup, M. Karplus, and R.M. Levy. 1980. Helix-coil transitions in a simple polypeptide model. *Biopolymers* **19**:2033–2045.

McQuarrie, D.A. 1976. Statistical Mechanics. New York: Harper and Row.

Monod, J., J. Wyman, and J.-P. Changeux. 1965. On the nature of allosteric transitions: A plausible model. *J. Mol. Biol.* **12**:88–118.

Montgomery, J. A., D. Chandler, and B.J. Berne. 1979. Trajectory analysis of a kinetic theory for isomerization dynamics in condensed phases. *J. Chem. Phys.* **70**:4056–4066.

Neria, E., and M. Karplus. 1997. Molecular dynamics of an enzyme reaction: Proton transfer in TIM. *Chem. Phys. Lett.* **267**:23–30.

Northrup, S.H., M.R. Pear, C.-Y. Lee, J.A. McCammon, and M. Karplus. 1982. Dynamical theory of activated processes in globular proteins. *Proc. Natl. Acad. Sci. USA* **79**:4035–4039.

Oliveberg, M., Y.-J. Tan, and A.R. Fersht. 1995. Negative activation enthalpies in the kinetics of protein folding. *Proc. Natl. Acad. Sci. USA* **92**:8926–8929.

Otting, G., E. Liepinsh, and K. Wüthrich. 1993. Disulfide bond isomerization in BPTI and BPTI(G36S): An NMR study of correlated mobility in proteins. *Biochemistry* **32**:3571–3582.

Pande, V.S., A.Y. Grasberg, and T. Tanaka. 1997. The theory of folding kinetics for short proteins. *Folding and Design* **2**:109–114.

Perutz, M. 1971. Stereochemistry of cooperative effects in haemoglobin. *Nature* **232**:408–413.
Perutz, M., and B.T. Matthews. 1965. An X-ray study of azide methaemoglobin. *J. Mol. Biol.* **21**:199–202.
Petrich, J.W., J.-C. Lambry, K. Kuczera, M. Karplus, C. Poyart, and J.-L. Martin. 1991. Ligand binding and protein relaxation in heme proteins: A room temperature analysis of NO geminate recombination. *Biochemistry* **30**:3975–3987.
Phillips, C.M., Y. Mizutami, and R.M. Hochstrasser. 1995. Ultrafast thermal-induced unfolding of RNase A. *Proc. Natl. Acad. Sci. USA* **92**:7292–7296.
Rector, K.D., C.W. Rella, J.R. Hill, A.S. Kwok, S.G. Sligar, E.Y.T. Chien, D.D. Dlott, and M.D. Fayer. 1997. Mutant and wild-type myoglobin–CO protein dynamics: Vibrational echo experiments. *J. Phys. Chem. B* **101**:1468–1475.
Richard, J.P. 1984. Acid-based catalysis of the elimation and isomerization reactions of triosephosphates. *J. Am. Chem. Soc.* **106**:4926–4936.
Roder, H., and W. Colon. 1997. Kinetic role of early intermediates in protein folding. *Curr. Opin. Struct. Biol.* **7**:15–28.
Roux, B., and M. Karplus. 1991. Ion transport in a gramicidin-like channel: Dynamics and mobility. *J. Phys. Chem.* **95**:4856–4868.
Ryckaert, J.-P., G. Ciccotti, and H.J.-C. Berendsen. 1977. Numerical integration of the Cartesian equations of motion of a system with constraints: Molecular dynamics of n-alkanes. *J. Comp. Phys.* **23**:327–341.
Sabelko, J., J. Ervin, and M. Gruebele. 1999. Observation of strange kinetics in protein folding. *Proc. Natl. Acad. Sci. USA* **96**:6031–6036.
Šalí, A., E. Shakhnovich, and M. Karplus. 1994a. How does a protein fold? *Nature* **369**:248–251.
Šalí, A., E. Shakhnovich, and M. Karplus. 1994b. Kinetics of protein folding: A lattice model study of the requirements for folding to the native state. *J. Mol. Biol.* **235**:1614–1636.
Saven, J.G., J. Wang, and P.G. Wolynes. 1994. Kinetics of protein folding. *J. Chem. Phys.* **101**:11,037–11,043.
Scalley, M.L., and D. Baker. 1997. Protein folding kinetics exhibits an Arrhenius temperature dependence when corrected for the temperature dependence of protein stability. *Proc. Natl. Acad. Sci. USA* **94**:10,636–10,640.
Schaad, O., H.X. Zhan, A. Szabo, W.A. Eaton, and E.R. Handy. 1993. Simulations of the kinetics of ligand binding to a protein by molecular dynamics: Geminate rebinding of nitric oxide to myoglobin. *Proc. Natl. Acad. Sci. USA* **90**:9547–9551.
Schindler, T., and F.X. Schmid. 1996. Thermodynamic properties of an extremely rapid protein folding reaction. *Biochemistry* **35**:16,833–16,842.
Segawa, J.J., and M. Sugihara. 1984. Characterizing the transition state of lysozyme unfolding. 1. Effect of protein–solvent interactions on the transition state. *Biopolymers* **23**:2473–2488.
Skinner, J.L., and P.G. Wolynes. 1980. General kinetic models of activated processes in condensed phases. *J. Chem. Phys.* **72**:4913–4927.
Smith, J., K. Kuczera, and M. Karplus. 1990. Dynamics of myoglobin: Comparison of simulation results with neutron scattering spectra. *Proc. Natl. Acad. Sci. USA* **87**:1601–1605.
Snyder, G.H., R. Rowan, and B.D. Sykes. 1976. Complete tyrosine assignments in the high-field ^{1}H nuclear magnetic resonance spectrum of bovine pancreatic trypsin inhibitor selectively reduced and carboxamidomethylated at cystine 14-38. *Biochemistry* **15**:2275–2283.

Socci, N.D., J.N. Onuchic, and P.G. Wolynes. 1996. Diffusion dynamics of the reaction coordinate in protein folding. *J. Chem. Phys.* **104**:5860–5868.

Šrajer, V., T. Teng, T. Ursby, C. Pradervand, Z. Ren, S. Adachi, W. Schildkamp, D. Bourgeois, M. Wulff, and K. Moffat. 1996. Photolysis of the carbon monoxide complex of myoglobin: Nanosecond time-resolved crystallography. *Science* **274**:1726–1729.

Steinbach, P., A. Ansari, J. Berendzen, D. Braunstein, K. Chu, B. Cowen, D. Ehrenstein, H. Frauenfelder, J.B. Johnson, D. Lamb, S. Luck, J. Mourant, G. Nienhaus, P. Ormos, R. Philipp, A. Xie, and R.D. Young. 1991. Ligand binding to heme proteins: Connection between dynamics and function. *Biochemistry* **30**:3988–4001.

Takano, T. 1977. Structure of myoglobin refined at 2.0A Resolution I. Crystallographic refinement of a metmyoglobin from sperm whale. *J. Mol. Biol.* **110**:537–568.

Thorn Leeson, D., and D.A. Wiersma. 1995. Looking into the energy landscape of myoglobin. *Nature Struct. Biol.* **2**:848–851.

Thorn Leeson, D., D.A. Wiersma, K. Fritch, and J. Friedrich. 1997. The energy landscape of myoglobin. *J. Phys. Chem. B* **101**:6331–6340.

Tokmakoff, A., and M.D. Fayer. 1995. Infrared photon echo experiments. Exploring vibrational dynamics and liquids and gases. *Accts. Chem. Res.* **28**:437–445.

Valleau, J.P., and G.M. Torrie. 1977. Byways. In: Statistical Mechanics Part A, ed. B.J. Berne, pp. 169–194. New York: Plenum.

Wade, R.C., M.E. Davis, B.A. Luty, J.D. Madura, and J.A. McCammon. 1993. Gating of the active site of triose phosphate isomerase: Brownian dynamics simulations of flexible peptide loops in the enzyme. *Biophys. Soc.* **64**:9–15.

Wagner, G., A. DeMarco, and K. Wüthrich. 1976. Dynamics of the aromatic amino acid sidechains in the globular conformation of the basic pancreatic trypsin inhibitor. I. IHNMR Studies. *Biochem. Struct. Mech.* **2**:139–149.

Waldburger, C.D., T. Jonsson, and R.T. Sauer. 1996. Barriers to protein folding. *Proc. Natl. Acad. Sci. USA* **93**:2629–2634.

Wang, J., and P.G. Wolynes. 1994. Intermittency of single molecule reaction dynamics in fluctuating environments. *Phys. Rev. Lett.* **74**:4317–4320.

Weston, R.E., and H.A. Schwarz. 1972. Chemical Kinetics. New Jersey: Prentice Hall.

Wigner, E.P. 1938. The transition state method. *Trans. Faraday Soc.* **34**:29–41.

Williams, J.C., and A.E. McDermott. 1995. Dynamics of the flexible loop of triosephosphate isomerase: The loop motion is not ligand gates. *Biochemistry* **34**:8309–8319.

Xie, X.S., and J.K. Trautman. 1998. Optical studies of single molecules at room temperature. *Ann. Rev. Phys. Chem.* **49**:441–490.

Xu, Z., A.L. Horwich, and R.B. Sigler. 1997. The crystal structure of the asymmetric GroEL-GroES-CADP$_7$ chaperonin coupling. *Nature* **388**:741–750.

Yüksel, K.Ü., A.-Q. Sun, R.W. Gracy, and K.D. Schnackerz. 1994. The hinged lid of yeast triosephosphate isomerase. Determination of the energy barrier between the two conformations. *J. Biol. Chem.* **269**:5005–5008.

Zocchi, G. 1997. Proteins unfold in steps. *Proc. Natl. Acad. Sci. USA* **94**:10,647–10,651.

Zwanzig, R. 1997. Two-state models of protein folding kinetics. *Proc. Natl. Acad. Sci. USA* **94**:148–150.

Standing, left to right:
Bob Austin, Bertil Halle, Henri Orland, Bill Eaton, Martin Karplus
Seated, left to right:
Peter Wolynes, Fred Richards, Ilme Schlichting, Chris Dobson

10

Group Report: Are There Common Themes in the Function, Energy Landscape, and Dynamics of Proteins?

W.A. EATON, Rapporteur

R.H. AUSTIN, C.M. DOBSON, B. HALLE, M. KARPLUS, H. ORLAND, F.M. RICHARDS, I. SCHLICHTING, and P.G. WOLYNES

INTRODUCTION

Many biomolecular processes involve conformational changes in which the shape of the molecule is modified significantly. These include tertiary structural changes in folded functioning proteins and quaternary structural changes in multisubunit proteins like hemoglobin. An intensively studied conformational process is protein folding, in which very large structural rearrangements occur (Ptitsyn 1995; Orland et al. 1996; Dill and Chan 1997; Eaton et al. 1997; Fersht 1997; Onuchic et al. 1997; Roder and Colon 1997; Dobson et al. 1998). The study of protein folding has been motivated by the intellectual challenge of understanding how the myriad of conformations that make up the denatured state of a protein rearrange to an essentially unique structure. It has also been stimulated by the worldwide genome project, in which tens of thousands of new protein sequences are being discovered each year. These problems have attracted theorists and experimentalists from many different areas of the biophysical sciences. Understanding conformational changes of folded proteins and the folding process are the two principal subjects of the overview papers by Wolynes, Dobson, and Nienhaus, as well as the background papers by Garel and Orland, Eaton, and Karplus (all this volume). As is evident from the background and overview papers, there are a variety of theoretical and experimental approaches. The theoretical approaches include approximate analytical theories, simulations of simplified representations of proteins, and all-atom molecular dynamics simulations. The framework provided by

Simplicity and Complexity in Proteins and Nucleic Acids
Edited by H. Frauenfelder, J. Deisenhofer, and P.G. Wolynes. Published by 1999 Dahlem University Press

the theory has not only suggested new types of experiments but has also aided the interpretation of many existing experiments and design of computer algorithms aimed at protein structure prediction.

Our group's discussion, described below, centered on the role of energy landscapes and analytical theory in understanding conformational processes in proteins, with the major emphasis on protein folding. Because of the complex nature of the protein folding problem, there was considerable divergence in the views expressed during the meeting, and on several points, some of them mentioned below, disagreements were narrowed but not completely resolved. In contrast to other groups at this conference, the discussion was limited in scope. Rather a few fundamental conceptual issues were probed in some depth. In particular, the problem of finding useful ways for reducing the description of the complex protein energy landscape to a few relevant variables was considered at length. There was also discussion of the insights to be gained from homopolymeric aspects of protein folding, and of issues concerning the role of solvent in protein folding, function, and dynamics. One outcome of this discussion was the uniform agreement that there is a pressing need for systematic data from laboratory experiments that can be used to test and refine the theoretical concepts. Indeed, apart from protein engineering experiments on the nature of the transition state ensemble in folding and NMR experiments measuring protection factors for proton exchange, theory has been primarily compared with computer simulations of simplified representations of proteins as the "experimental system."

ENERGY LANDSCAPES IN PROTEIN DYNAMICS

The microscopic description of protein conformational processes involves many different configurations and individual isomerization events. This becomes obvious when one realizes that the unfolded state of a protein may have more thermally accessible microscopic configurations than there are individual molecules in an experimental sample. Although much of this conformational diversity disappears in the folding reaction, even a native protein has a complex energy landscape. For an understanding of the dynamics, it is necessary to begin at the microscopic level and simplify the description of the process in a way that captures the most important elements of the underlying physics and chemistry, while being useful for explaining experiments.

Potential energy surfaces play an essential role in understanding any process involving a structural change. The term "energy landscape" is introduced to emphasize the complexity of the potential energy surface of a protein, in contrast to that for simple molecule reactions. In many cases, the energy landscape of interest is a potential of mean force based on a canonical average over the solvent degrees of freedom. A useful and sometimes critical step for our understanding of a protein conformational process involves simplifying the multidimensional surface, which for a reasonably sized protein at the outset deals with several thousand degrees of freedom, by selecting the essential coordinates. Describing the ensembles of protein conformations in terms of these coordinates enormously reduces the complexity, and thereby aids in obtaining an intuitive grasp of the underlying physics and chemistry.

One result of this contraction of information is a free-energy profile for a conformational process. A useful step is to discover a minimal set of appropriate simplifying coordinates for these profiles. These may be specific geometrical coordinates (such as the native dihedral angles for specific residues) or collective coordinates for ensembles such as the native contact density or fraction correct dihedral angles. If done in a sufficiently complete manner, the resulting free-energy profile would allow the calculation of experimentally measured rates. This puts a great premium on how to find these degrees of freedom — a major conceptual problem for theorists and for experimentalists, as well. The compression to such a minimal set may often miss a crucial feature of the kinetic process, namely that some configurations specified by the missing degree of freedom are local energy minima and can be very long lived, leading to important modifications of the kinetics.

A confusing feature of some of the discussions of energy surfaces and landscapes arises from the fact that it is not always made clear which degrees of freedom are being referred to in these drawings and which degrees of freedom are being omitted. Solvent degrees of freedom are generally in the omitted category because the essential information concerning the large-scale motions in protein folding resides in the complicated rearrangements of the biopolymer itself. Thus, drawings of the energy landscapes for a biomolecule generally refer to an effective energy or partial free energy. This assumes solvent motions are faster than isomerizations of the polypeptide chain, a problematic assumption for experiments in cryogenic solvents where evidence exists for a "slaved" glass transition. Fast degrees of freedom, in addition to those of the solvent, may also be projected out as done in nonequilibrium statistical mechanics, including, for example, the local vibrations of the atoms in their close-packed environment. Less clear is which other small-scale motions of the polypeptide chain can be integrated out. Thus energy landscape diagrams can include information about the configurational entropy of the polymer chain. In small systems, such as a tetrapeptide, it is possible to introduce the specific relevant additional collective coordinates. For computer simulations of minimalist models of small protein folding, good approximations to the essential coordinates have been found (Onuchic et al. 1997; Dobson et al. 1998; Guo and Thirumalai 1997; Plotkin and Wolynes 1998; Socci et al. 1998). More work is needed to develop ideas along these lines for larger systems and more realistic representations (Lazaridis and Karplus 1997; Guo et al. 1997). It is not yet clear what the minimal size polypeptide that captures the essence of folding would be; however, it will certainly be larger than a single α-helix, and likely to be larger than a β-hairpin.

Under some conditions where protein folding is slow, the free-energy profile alone makes clear why the process is activated. On the other hand, for the simplest choice of folding reaction coordinate under stabilizing conditions the profile may have an apparent barrier that is small. The statistical landscape theory (*vide infra*) gives a semiquantitative explanation for the slowness of folding observed in low-temperature simulations where the free-energy profile alone yields a small barrier and therefore a relatively fast process. The slowness must then arise from a small configurational diffusion constant. The magnitude of this additional slowing quantifies the ruggedness of the landscape. An alternative explanation for the slowness is that the true reaction coordinate is almost orthogonal to the ones that have been used to plot profiles, but such coordinates have not yet been described in the literature.

STATISTICAL ENERGY LANDSCAPE THEORY

There now exists a partially developed statistical formulation of the commonly used concepts and drawings of energy landscapes, first introduced for liquids and glasses in the 1960s, and widely used to explain low-temperature experiments. This approach was first illustrated using the random energy model but now makes use of further advances in spin glass theory. The current theory, incorporating some details about the correlations in the energy surface, provides expressions for the rates of conformational motions in terms of energy distributions and barrier heights.

A crucial prediction of the statistical energy landscape theory is that a characteristic temperature related to the roughness emerges, a so-called ideal glass transition temperature, T_g. While the predictions of the thermodynamics at T_g and the statistics of structures that are populated below that temperature have been amply confirmed in lattice simulations, there is controversy about the detailed kinetic effects above and below T_g. Some hallmarks of glassy dynamics have been seen, such as nonexponential rearrangement processes in partially optimized sequences (Abkevich et al. 1994; Nymeyer et al. 1998; Socci et al. 1998). Others have fit simulation data using only exponential or bi-exponential kinetic progress curves. Also, a recent paper reports no deviation from an Arrhenius temperature dependence at or below T_g (Gutin et al. 1998). The existing statistical theory focuses on kinetic behavior above T_g. The accessible temperature range above T_g is sufficiently small that both empirical Arrhenius expressions, and the microscopic expressions of correlated landscape theory appear to fit the data. Reliable simulation studies under the conditions envisioned by the theory and further refinements of the approximations to deal with varying conditions of collapse, etc., are needed before this can be fully resolved. Precise experiments that determine whether there are deviations from Arrhenius temperature dependence in protein folding in the accessible range would be valuable; however, the temperature dependence of the hydrophobic forces makes this difficult to interpret. An essential question is whether it is appropriate to emphasize a glass transition in the statistical landscape theory if it does not occur for natural protein folding in the accessible temperature range. Studies on the more poorly designed artificial sequences using reduced numbers of amino acids may help to highlight the glassy features (*vide infra*).

The original statistical landscape theory predicted that the speed of folding scales with the ratio of the folding temperature to the glass transition temperature (T_f/T_g). Simulation studies have emphasized an analogous parameter controlling the rate —

the energy gap between the native state and some average of the energy over the compact configurations. It has also been suggested that the rate scales inversely as the quantity $(T_\theta - T_f)/T_\theta$ where T_θ is the collapse temperature of the protein, but this has recently been questioned. Several recent studies show that these quantities are all correlated to some extent (Klimov and Thirumalai 1998; Dinner et al. 1998; Melim et al. 1998).

Energy landscapes provide a framework for interpreting the results of protein folding experiments. Studies of minimalist models have indicated that although the underlying folding mechanism is the same, there can be considerable variation in the observed kinetics. These have been described in some detail and compared with experiments (Onuchic et al. 1997; Dobson et al. 1998). The lattice results suggest that folding at high temperatures leads to a smooth decrease in the effective energy with an entropy bottleneck. By contrast, at low temperatures, the slow step is found to be the search through a rough energy landscape. The statistical energy landscape theory has been used for classifying folding scenarios (Socci et al. 1998). The theory, through the calculation of T_g and its dependence on the degree of collapse and foldedness of a protein, gives some scenarios in which there are specific intermediates. In these cases the statistical Ansatz can only discuss the distribution of folding times over different sequences: mutation effects on kinetics are large. In another scenario, which arises when the glass transition temperature is low compared to the ambient conditions, so many states are involved in the bottleneck for folding that smooth linear free-energy relations are expected. In addition, there is a regime of downhill nonexponential folding which has been seen in computer simulations, but there are no published reports of this in the laboratory for natural proteins. Small designed peptides have shown nonexponential folding behavior (Volk et al. 1997).

Linear free-energy relations have been used in analyzing the protein folding reaction. In general, the linear free-energy relations determined by site-directed mutagenesis experiments (Fersht 1997) suggest that only part of the stabilization of the native state is realized in the ensemble of structures that represent the kinetic bottleneck (transition state). This was originally described by saying there are weakened bonds (ϕ values less than unity) in the transition state. Lattice simulations, all-atom simulations (Guo et al. 1997; Sheinerman and Brooks 1998; Lazaridis and Karplus 1997), and more general considerations indicate that it is more appropriate to interpret the ϕ values, as energy-weighted populations of particular interactions. The fact that essentially all ϕ's are fractional supports some kind of transition state ensemble. As in other areas of chemical kinetics, such as electron transfer, if the ϕ values can be related to quasi-equilibrium quantities it may be possible to determine them by using experiments for an equilibrium partially denatured state under different thermodynamic conditions. These extrathermodynamic free-energy relations only arise on a funnel-like energy landscape and detailed calculations for ideal funnel landscapes qualitatively agree with experiment (Shoemaker et al. 1997; Portman et al. 1998).

Direct experimental determination of the parameters, on which theoretical models can be based, is sorely needed. One of these is the configurational relaxation time required for determining the prefactor in the rate expression for cases where the transition state search is over collapsed states. Such a collapse has been suggested from the *m*-values in kinetic experiments (rate dependence on denaturant concentration); however, this does not take account of the fact that the denatured state may also be collapsed under some conditions. A knowledge of the prefactor is needed to assess models for thermodynamic barrier heights. Another type of important information concerns the amount of structure to be found in the collapsed state. It is crucial to determine the fraction of time that specific native contacts are made and, equally important, the nonspecific contacts incongruent with the native structure. This might be done by high-resolution NMR on slowly folding proteins. Much information of this type being sought from experimentalists will be of general interest for a better understanding of protein folding independent of the theoretical model being used.

THE PROBLEM OF FINDING A REACTION COORDINATE

In many complex systems, there is a clear separation between microscopic time scales (fast degrees of freedom) and macroscopic time scales (slow degrees of freedom). These slow modes are the so-called collective modes present in many areas of physics, such as nuclear physics (fission), condensed matter physics, etc., and allow for a "hydrodynamical" description of processes. The important coordinates obtained from the reduced description of an energy landscape, which are often referred to as reaction coordinates or progress variables, are usually chosen to be these slow modes. Once these reaction coordinates $\{q\}$ are clearly identified, it is possible to define an effective free energy $F\{q\}$ as a function of these reaction coordinates, by integrating out all fast modes. As discussed earlier, the main achievement in doing so is of course a great reduction in the number of degrees of freedom, and the possibility of using a local equilibrium description of the slow dynamics of the reactions. In particular, in this approach, slow dynamics might result from a rugged landscape, or from a very small effective diffusion constant.

For chemical reactions, the reaction coordinate globally describes the transition from reactant to product. For simple chemical reactions, of which the $H + H_2$ reaction is the best studied example, a reaction coordinate is well defined and easy to determine. Moreover, the reactive trajectories at most temperatures remain close to the minimum energy path. The question is how these ideas from the field of reaction dynamics of small molecules can be transferred to macromolecules of biological interest. For chemical reactions in proteins (e.g., enzyme-catalyzed reactions, binding and release of CO in myoglobin), the ideas from small molecule reactions carry over as a good zeroth-order approximation. In an aqueous solution of myoglobin at room temperature, for example, a straightforward reaction coordinate is a linear combination of the CO distance from the iron and certain of the heme atoms. However, at lower temperatures (but above the glass transition of the solvent) or higher viscosities, it is clear that the conformation of the rest of the protein plays a role in the reaction rate. Rebinding after photolysis is nonexponential, and it appears likely that this is due in part to relaxation of the protein conformation (also nonexponential) on the same time scale as the rebinding process (see chapters by Eaton, Karplus, and Nienhaus, all this volume). In the case of NO binding, this is already evident at room temperature and normal aqueous viscosities in picosecond kinetic experiments (although complex ligand dynamics may also produce nonexponential behavior). A single reaction coordinate no longer describes the rebinding process, and at least one other coordinate representing the protein conformational change is required. At low temperatures when equilibration between conformations with different barriers is slow relative to the rebinding reaction, an inhomogeneous model is appropriate.

One way of obtaining more information on the protein coordinates involved in the reaction is to use time-resolved X-ray crystallography. This provides the change in atomic coordinates as a function of time during the reaction. Such data provides very useful constraints on any simulations of the reaction, although by themselves they cannot determine the reaction coordinates or the energy surface involved. One problem with such experimental studies is that, unless the reaction barrier is low, the data describe mainly the reactants and products. If the barrier is low enough, however, the populations of intermediate structures may be sufficiently large to be directly observed in the time-resolved electron density maps. This would give the most direct structural information on the mechanism. Even for the high-barrier case, X-ray data are likely to be very useful in increasing our understanding of reactions in proteins.

In contrast to chemical reactions in proteins, protein folding poses fundamental problems in defining the reaction coordinate. This is one of the main outstanding issues that is under intense experimental and theoretical investigation. Simulations of lattice proteins suggest that there are average progress variables that can be defined, but that much larger fluctuations will occur than in chemical reactions. It is not clear, however, whether a reaction coordinate for protein folding (in the sense defined above) can always be determined. One hopeful sign is the observation for some proteins of two-state (exponential) kinetic and thermodynamic behavior, which suggests that such a coordinate may not be too difficult to find.

Statistical energy landscape theory and other approaches that use reaction coordinates based on Monte Carlo models and molecular dynamics (MD) simulations agree on this aspect of a description for the dynamics. In the statistical energy landscape theory, slow dynamics arise from trapping of the system in the numerous local minima (which describe more or less long-lived metastable intermediates). In other approaches, the integration over the fast modes renormalizes the diffusion constant, and describes the system as evolving on a smooth effective free-energy landscape with an anomalously small diffusion constant. Since a small configurational diffusion constant may arise from mechanisms other than "traps" by for example excluded volume effects or topological constraints, it would be useful to relate the energy landscape picture to detailed underlying microscopic models.

SOLVENT IN ENERGY LANDSCAPES AND DYNAMICS

Since a large fraction of the amino acid residues of a typical protein are at the surface, the energetics, structure, and dynamics of the protein–water interface are clearly important for the behavior of native proteins as well as for protein folding and association. It would seem, therefore, that this interface deserves more attention than it has so far received, particularly on the experimental side. The last few years have seen major advances in our understanding of protein hydration, largely as a result of NMR relaxation studies using the complementary techniques of nuclear magnetic relaxation dispersion (NMRD) (Halle et al. 1998) and high-resolution NOE spectroscopy (Otting 1997). In addition, data on the solvent has become available from simulations of proteins in water (Phillips and Pettitt 1995).

Nearly all native globular proteins contain water molecules buried within the polypeptide matrix (internal water) (Baker 1995). It appears that such internal water molecules heal packing defects, extend the hydrogen-bond framework, and are sometimes directly involved in function. They are conserved among homologous proteins to the same extent as amino acid residues and should clearly be regarded as an integral part of the protein. According to NMRD studies, most internal water molecules have residence times in the range 10^{-8} to 10^{-6} s at ambient temperature (Denisov and Halle 1996), but deeply buried waters can be more long-lived (10^{-4} s) (Denisov et al. 1996). These results indicate that native proteins undergo conformational fluctuations large enough for exit and entrance of water molecules on the microsecond time scale. Such behavior is not different from that demonstrated in MD simulations of the exit of CO and NO ligands from the heme pocket of myoglobin, which is on the hundreds of nanosecond time scale at room temperature.

The protein-induced perturbations of the structure and dynamics of the external solvent are short-ranged, being essentially confined to the first layer of water molecules in direct contact with surface atoms. This has been established by NMR studies of a variety of model systems where the water content can be varied without altering the surface (Halle 1998). NMR relaxation and MD simulations are consistent with a similarly short-ranged perturbation at protein surfaces. These findings are *not* incompatible with long-range effects of hydration on the interaction between surfaces or macromolecules, which involves the cumulative effect of a very large number of very weakly perturbed water molecules ($< k_BT$ per water molecule).

Rotational and translational dynamics of water molecules in the first hydration layer of proteins is slowed down *on average* by 5-fold relative to bulk water (Halle 1998). In view of the heterogeneous nature of the protein surface, however, the distribution of local motional correlation times is likely to be broad. For a majority of the hydration sites at protein surfaces, the dynamic retardation is probably similar to that of small solutes (a factor of two). At ambient temperature, water molecules at surface sites invariably have sub-nanosecond residence times (with an average of about 30 ps) unless they reside in narrow pockets or crevices. MD simulations are in qualitative agreement with the NMR picture but tend to give somewhat too fast dynamics (possibly due to force-field imperfections) (Halle 1998).

The energetics of the protein–solvent interface is not so easily characterized experimentally (for proteins in solution). It is an experimental fact that proteins can undergo substantial conformational changes, and even unfold, at a free-energy cost that is only a tiny fraction of the hydration free energies involved. Enthalpy–entropy compensation is presumably responsible for much of this invariance; the essential compensating factor arises from a configurational entropy gain in many cases. While hydration clearly is important for protein stability, the few direct studies of nonnative protein hydration suggest that the water occluded in denatured proteins may be rather different from bulk water (Denisov et al. 1999). If this is the case, we cannot expect to rationalize protein stability quantitatively in terms of measured free energies of small molecule transfer into bulk water. MD simulations show that the local topography of the protein surface is important for the energetics and structure of hydrophobic hydration (Cheng and Rossky 1998). This is in line with the well-known dependence of interfacial tension on interfacial curvature and suggests that interfacial free energies of proteins cannot be reliably estimated from the solvent accessible surface area using a global measure of interfacial tension, although empirical studies give satisfactory correlations (Makhatadze and Privalov 1995).

As pointed out earlier, one of the difficult problems in investigations of protein conformational kinetics is to disentangle the effects of dynamical effects of the solvent and (free)-energy barriers intrinsic to the protein–solvent system. This requires studying the kinetics as a function of solvent viscosity without altering the energy landscape. Such experimental studies have suggested that the dramatic slowing in the conformational kinetics in myoglobin as the solvent glass transition is approached is primarily the result of the very large solvent viscosity. This view has recently received support from MD simulations of myoglobin solvated with explicit water (D. Vitkup and M. Karplus, submitted). By use of a Nose–Hoover thermostat, the temperature of the protein and the water could be controlled separately, something that is not possible to do experimentally. In this strongly nonequilibrium situation, the protein fluctuations in these simulations were shown to be governed by the temperature of the surrounding water. For example, a simulation with the protein at 180 K and the solvent at 300 K yielded the same (large) fluctuation magnitudes as a simulation with both the protein and solvent at 300 K. Conversely, if the solvent was at 180 K, the same (small) protein fluctuations were obtained for the protein at 300 or 180 K. When the temperature of the system was further lowered to 140 K, the intrinsic protein potential and temperature become important. The behavior at 180 K may thus be described as a "slaved" glass transition.

ANALOGIES BETWEEN PROTEINS AND HOMOPOLYMERS

It is well known from polymer theory that a globular homopolymer can exist in an extremely large number of shapes (exponential in the number of monomers) with quasi-degenerate energies (Garel and Orland, this volume). For a small system, the size of a protein, this quasi-degeneracy is not an exact degeneracy; however, usually the energy gap between the ground state conformation and the state of next-lowest energy is very small and would not fully account for the stability of natural proteins. The precise structure of the ground state is determined by a delicate balance between the bulk and the surface energy of the polymer. This phenomenon is similar to what is observed in the study of small atomic or molecular clusters, for which the determination of the ground state is a very difficult problem (Ball et al. 1996). The shape of these clusters depends in a very subtle way on the number of their constituents. Although their ground state is nondegenerate, they are very easily destabilized by thermal fluctuations (note, however, the existence of "magic" cluster sizes with high symmetry, for which the stability gap is unusually large [Lindgard and Bohr 1996; Nelson et al. 1997]).

This is in striking contrast with proteins for which the native state is unique (ignoring conformational substates for the present discussion), nondegenerate, and stable. A major characteristic of native protein structure is the close-packing of secondary structures. Ramachandran plots (based on steric considerations alone) together with the full compactness of the structure are not sufficient to explain the omnipresence of secondary structures. Indeed, structures satisfying Ramachandran statistics may still have very little secondary structure content. To account for the locally ordered structures, one has to take into account the possibility of hydrogen-bond formation, either through a specific model for the H-bond (Luthey-Schulten et al. 1995), or through forces acting between the strong dipole moments carried by the peptide bond. A simplified theoretical model of a homopolymer where each monomer carries a dipole moment, with steric repulsion and chiral carbons, possesses liquid crystalline-like phases, in which the local ordering can be shown to consist of α-helices or β-sheets. The native phase in these models is filled with secondary structures. However, the low-temperature phase still has substantial conformational entropy and thus has important thermal fluctuations. Although the homopolymeric dipolar backbone has considerably more structure and less degeneracy than an ordinary globular polymer, the location and nature of the secondary structure is not uniquely determined. The freezing-in of the secondary structure and the uniqueness of the native state ultimately result from the sequence diversity. Indeed, it is clear that a specific sequence of residues will select one of the possible arrangements of secondary structures as its ground state, and thus open a gap from the continuum of quasi-degenerate states of the protein. Realistic (though still schematic) statistical mechanical models for proteins should thus take into account both the backbone specificity (e.g., through dipole moments) and the sequence diversity. Only then can a unique native shape with a specific active site appear.

Nevertheless, the description of a folded protein as containing many characteristics of a homopolymer is supported by structural observations on proteins and provides a valuable starting point. One of these is that regular elements of structure satisfy simultaneously the criteria for minimizing different interactions. For example an α-helix arises from the repeat of one of the lowest energy ϕ/ψ values and is optimally close packed for van der Waals interactions. An antiparallel β-sheet has similar optimization. The close-packed nature of a protein with secondary structure can therefore satisfy the steric constraints very well but avoid uncompensated electrostatic interactions in an environment that excludes water. Interestingly, the averaged propensities for the different amino acids in folded proteins correlates closely with the average populations of the different conformers accessible to the protein in its unfolded "random coil" state.

There are many states of a protein that are potentially accessible. Which of these is formed under given conditions depends on the interplay of multiple rates and equilibria (Dobson, this volume). For an isolated polypeptide chain, the native state interconverts with denatured configurations. Under some conditions, partially folded states with common characteristics may be highly populated. In the presence of other polypeptides, intermolecular interactions can take place and compete with intramolecular ones. As the concentration of molecules is increased under conditions where the native fold is preserved, aggregates and crystals can form. Under conditions where nonnative states interact, disordered aggregates can form but so can ordered structures such as amyloid fibrils. These fibrils appear to be closely similar, regardless of the structure of the soluble native state of the protein, and to be composed of a cross-β structure with β-sheets parallel to the fiber axis (Sunde et al. 1997). Amyloid seems therefore to be a generic state of proteins, whose basic structure is independent of the specific polypeptide sequence (Dobson, this volume). This may be the thermodynamically stable state of a protein under some conditions (e.g., where the native state is destabilized by factors such as mutations or exposure to low pH) provided the concentration is sufficiently high. It may form in biological systems as a result of being exposed to such conditions for a sufficiently long period of time, and then even if returned to conditions where it is thermodynamically unstable relative to monomeric states may be extremely stable kinetically. It may form as a result of the failure of biological control mechanisms, e.g., in aging. It is not clear if there are any examples of aging of isolated proteins (other than chemical degradation) on biologically relevant time scales.

An intriguing possibility is that the β-structure of amyloid might bear some resemblance to the structure of crystalline organic homopolymers such as polyethylene. The same kinds of theoretical approaches described above for homopolymers may therefore give new insights into the structure and mechanism of amyloid formation. Reports of nucleated self-assembly of amyloid suggest that the extensive studies of the physics and physical chemistry of nucleated self-assembly of other biological structures, notably hemoglobin S fibers in sickle cell disease (Eaton and Hofrichter 1990), will provide new directions for further studies of the amyloid state.

REFERENCES

Abkevich, V.I., A.M. Gutin, and E.I. Shakhnovich. 1994. Free-energy landscape for protein-folding kinetics — Intermediates, traps, and multiple pathways in theory and lattice model simulations. *J. Chem. Phys.* **101**:6052–6062.

Baker, E.N. 1995. Solvent interactions with proteins as revealed by X-ray crystallographic studies. In: Protein Solvent Interactions, ed. R.B. Gregory. New York: M. Dekker.

Ball, K.D., R.S. Berry, R.E. Kunz, F.Y. Li, A. Proykova, and D.J. Wales. 1996. From topographies to dynamics on multidimensional potential surfaces of atomic clusters. *Science* **271**:963–966.

Cheng, Y.K., and P.J. Rossky. 1998. Surface topography dependence of biomolecular hydrophobic hydration. *Nature* **392**:696–699.

Denisov, V.P., and B. Halle. 1996. Protein hydration dynamics in aqueous solution. *Faraday Soc. Disc.* **103**:227–244.

Denisov, V.P., B.-H. Jonsson, and B. Halle. 1999. Hydration of denatured and molten globule proteins. *Nature Struct. Biol.* **6**:253–260.

Denisov, V.P., J. Peters, H.D. Horlein, and B. Halle. 1996. Using buried water molecules to explore the energy landscapes of proteins. *Nature Struct. Biol.* **3**:505–509.

Dill, K.A., and H.S. Chan. 1997. From Levinthal to pathways to funnels. *Nature Struct. Biol.* **4**:10–19.

Dinner, A.R., V.I. Abkevich, E.I. Shakhnovich, and M. Karplus. 1998. Factors that affect the folding ability of proteins. Los Alamos Condensed Matter Preprint 9805034.

Dobson, C.M., A. Šalí, and M. Karplus. 1998. Protein folding: A perspective from theory and experiments. *Angew. Chem. Intl. Ed.* **37**:869–893.

Eaton, W.A., and J. Hofrichter. 1990. Sickle cell hemoglobin polymerization. *Adv. Prot. Chem.* **40**:63–69.

Eaton, W.A., V. Munoz, P.A. Thompson, C.-K. Chan, and J. Hofrichter. 1997. Submillisecond kinetics of protein folding. *Curr. Opin. Struct. Biol.* **7**:10–14.

Fersht, A.R. 1997. Nucleation mechanisms in protein folding. *Curr. Opin. Struct. Biol.* **7**:3–9.

Guo, Z., C.L. Brooks, III, and E.M. Boczko. 1997. Exploring the folding free-energy surface of a three helix bundle protein. *Proc. Natl. Acad. Sci. USA* **94**:10,161–10,166.

Guo, Z., and D. Thirumalai. 1997. The nucleation collapse mechanism in protein folding: Evidence for the nonuniqueness of the folding nucleus. *Folding and Design* **2**:377–391.

Gutin, A., A. Šalí, V. Abkevich, M. Karplus, and E.I. Shakhnovich. 1998. Temperature dependence of the folding rate in a simple protein model: Search for a "glass" transition. *J. Chem. Phys.* **108**:6466–6483.

Halle, B. 1998. Water in biological systems: The NMR picture. In: Hydration Processes in Biology: Theoretical and Experimental Approaches, ed. M.-C. Bellissent-Funel. Amsterdam: The NATO Science Series, IOS Press.

Halle, B., V.P. Denisov, and K. Venu. 1998. Multinuclear relaxation dispersion studies of protein hydration. In: Modern Techniques in Protein NMR, Vol. 16BB, ed. L.J. Berliner and N.R. Krishna. New York: Plenum.

Klimov, D.K., and D. Thirumalai. 1998. Linking rates of folding in lattice models of proteins with underlying thermodynamic characteristics. Los Alamos Condensed Matter Preprint 9805061.

Lazaridis, T., and M. Karplus. 1997. "New view" of protein folding reconciled with the old through multiple unfolding simulations. *Science* **278**:1928–1931.

Lindgaard, P.A., and H. Bohr. 1996. Magic numbers in protein structures. *Phys. Rev. Lett.* **27**:779–782.

Luthey-Schulten, Z., B.E. Ramirez, and P.G. Wolynes. 1995. Helix-coil, liquid crystal, and spin glass transitions of a collapsed heteropolymer. *J. Phys. Chem.* **99**:2177–2185.

Makhatadze, G.I., and P.L. Privalov. 1995. Energetics of protein structure. *Adv. Prot. Chem.* **47**:307–425.

Melim, R., H. Li, N.S. Wingreen, and C. Tang. 1998. Designability, thermodynamic stability, and dynamics in protein folding: A lattice model study. Los Alamos Condensed Matter Preprint 9806197.

Nelson, E.D., L.F. Teneyck, and J.N. Onuchic. 1997. Symmetry and kinetic optimization of protein-like heteropolymers. *Phys. Rev. Lett.* **79**:3534–3537.

Nymeyer, H., A.E. Garcia, and J.N. Onuchic. 1998. Folding funnels and frustration in off-lattice minimalist protein folding landscapes. *Proc. Natl. Acad. Sci. USA* **95**:5921–5928.

Onuchic, J.N., Z. Luthey-Schulten, and P.G. Wolynes. 1997. Theory of protein folding: The energy landscape perspective. *Ann. Rev. Phys. Chem.* **48**:545–600.

Orland, H., T. Garel, and D.T. Thirumalai. 1996. Analytical theories of protein folding. In: Recent Developments in Theoretical Studies of Proteins, ed. R. Elber, pp. 197–268. Singapore: World Scientific.

Otting, G. 1997. NMR studies of water bound to biological molecules. *Prog. NMR Spect.* **31**:259–285.

Phillips, G.N., and B.M. Pettitt. 1995. Structure and dynamics of water around myoglobin. *Protein Sci.* **4**:149–158.

Plotkin, S., and P.G. Wolynes. 1998. Non-Markovian configurational diffusion and reaction coordinates for protein folding. *Phys. Rev. Lett.* **80**:5015–5019.

Ptitsyn, O. 1995. Molten globule and protein folding. *Adv. Prot. Chem.* **47**:83–229.

Portman, J.J., S. Takada, and P.G. Wolynes. 1998. Variational theory for site resolved protein folding free energy surfaces. *Phys. Rev. Lett.* **81**:5237–5240.

Roder, H., and W. Colon. 1997. The role of early intermediates in protein folding. *Curr. Opin. Struct. Biol.* **7**:15–28.

Sheinerman, F.B., and C.L. Brooks, III. 1998. Molecular picture of folding of a small α/β protein. *Proc. Natl. Acad. Sci. USA* **95**:1562–1567.

Shoemaker, B.A., J. Wang, and P.G. Wolynes. 1997. Structural correlations in protein folding funnels. *Proc. Natl. Acad. Sci. USA* **94**:2318–2322.

Socci, N.D., J.N. Onuchic, and P.G. Wolynes. 1998. Protein folding mechanisms and the multidimensional folding funnel. *Proteins* **32**:136–158.

Sunde, M., L.C. Serpell, M. Bartlam, P.E. Fraser, M.B. Pepys, and C.C.F. Blake. 1997. Common core structure of amyloid fibrils by synchrotron X-ray diffraction. *J. Mol. Biol.* **273**:729–739

Volk, M., Y. Kholodenko, H.S.M. Lu, E.A. Gooding, W.F. DeGrado, and R.M. Hochstrasser. 1997. Peptide conformational dynamics and vibrational stark effects following photoinitiated disulfide cleavage. *J. Phys. Chem.* **101**:8607–8616.

11

Are Molecular Motions Important for Function?

J.A. McCammon

Dept. of Chemistry and Biochemistry, Dept. of Pharmacology, University of California at San Diego, La Jolla, CA 92093–0365, U.S.A.

Keywords: correlation, diffusion, dynamic selectivity, gating

ABSTRACT

Many of the functions of biopolymers are governed by the time dependence of their internal motions. Several examples are described. It is concluded that evolution has selected biological molecules for dynamical as well as structural properties.

INTRODUCTION

Crystallographic, spectroscopic, and theoretical studies have made clear that internal motion is necessary for the function of most biopolymers. Examples of such motion include the displacement of amino acid residues in enzymes to allow substrate binding and product release, the separation and displacement of nucleic acids during DNA replication, etc. But how important is the actual time dependence of such motion? Most of this brief chapter addresses this question and points to a number of cases where certain details of biomolecular dynamics have functional importance.

MOLECULAR MOTIONS

Some internal motions of biopolymers can be described relatively simply. These range from localized vibrations within covalently bonded groups to elastic vibrations that involve coherent small-amplitude displacements of larger portions of the molecule. The local vibrations may be rather weakly coupled to other motions in the molecule and its

Simplicity and Complexity in Proteins and Nucleic Acids
Edited by H. Frauenfelder, J. Deisenhofer, and P.G. Wolynes

surroundings, while the large-scale vibrations may be insensitive to local conformational or other variations from one molecule to another.

Generally, motions in biopolymers are more complex, and more interesting. The ease with which dihedral angles can be varied in biopolymers, together with the relatively soft nature of their nonbonded interactions other than the short-range interatomic repulsions, and the dense packing of groups within many biopolymers, combine to yield the rugged energy landscape that is now familiar from much experimental and theoretical work (Frauenfelder et al. 1991). Variations in the protonation states of titratable groups and in the binding of water molecules and ions to sites in the biopolymer also contribute to the structure of the energy landscape.

Motions in biopolymers correspond to excursions on this energy landscape and may be correspondingly complex. Even the "simple" motions mentioned at the outset of this section will be perturbed by transitions over barriers in the biopolymer's energy landscape, e.g., the localized vibrations of a covalently bonded group will differ to some extent, depending on in which energy well the biopolymer resides in the landscape.

Spectroscopic studies suggest that myoglobin has a hierarchical energy landscape and that it has three conformational substates in its "taxonomic" tier with the largest barriers (Frauenfelder et al. 1991). One current picture from such studies shows barriers on the order of 100 kJ/mol between these taxonomic conformational substates, and three lower tiers with about four conformational substates in each tier (Thorn Leeson et al. 1997). A recent NMR study of the most slowly exchanging buried water molecule in the bovine pancreatic trypsin inhibitor indicates that its exchange can be modeled as a diffusion process on an energy landscape with the crossing of barriers on the order of 10 kJ/mol (Denisov et al. 1996). The exchange of this water molecule occurs with a characteristic time of about 170 μs at 300 K. Examination of the structure of the protein shows that not only side-chain motions, but also significant backbone motions must occur during the exchange of this particular water molecule, which is consistent with many conformational substates being involved with the exchange process. Denisov et al. (1996) suggest that any local process in a protein that occurs on the nanosecond-to-millisecond time scale, and that requires substantial displacements of groups in the protein, may be rate-limited by interconversions of conformational substates and display similar features to those observed in their study.

Most experimental results to date concerning motions in biopolymers reflect the behavior of ensembles of molecules, which has complicated the interpretation of the results. The advent of single-molecule spectroscopic methods is thus of very great interest (Dickson et al. 1997; Jia et al. 1997; Kojima et al. 1997; Wennmalm et al. 1997).

BINDING AND SELECTIVITY

In a number of cases, particularly where ligand-receptor binding is fast, it appears that certain features of the internal motion of one or both partners have evolved to be rapid enough to avoid kinetic bottlenecks. The enzyme acetylcholinesterase represents one such case. In the crystallographic structures of forms of the enzyme from two different species, a gorge or channel extending approximately 2 nm from the surface of the enzyme to the active site is apparent (Sussman et al. 1991; Bourne et al. 1995). This is the most likely route for binding the substrate acetylcholine. However, in both structures, a constriction exists midway down the channel that, if static, would preclude passage of substrate. Despite this, the enzyme binds substrate at or near the diffusion-controlled limit. It has been known for some time that, if such obstacles can be removed frequently enough by the fluctuations in an enzyme or other receptor, the obstacles will not slow the overall rate of binding (McCammon and Northrup 1981). Recent molecular dynamics simulations suggest that this is the case for acetylcholinesterase (Wlodek et al. 1997). Fluctuations in the enzyme open the channel every few ps, which is often enough to allow capture of the substrate before it can diffuse away over times on the order of a few ns. A recent analysis by Zhou (1998) presents the most complete current theory of such "gated" diffusional binding processes, and suggests that a similar picture describes the classic example of myoglobin. Another group of "gated" enzymes comprises those that have a peptide loop that opens and closes over the active site. Wade et al. (1994) suggest that somewhat slower but still rapid gating (times on the order of one ns) allows one such enzyme, triosephosphate isomerase, to operate in the diffusion-controlled regime. As discussed by Karplus (this volume), further study will be required to clarify the detailed nature of the loop motion in this enzyme.

It must be noted that the molecular dynamics simulations of acetylcholinesterase mentioned above are far too short to sample transitions over barriers that separate many conformational substates of the protein. However, for acetylcholinesterase, similar behavior is observed for the two subunits of the homodimeric enzyme (Wlodek et al. 1998). Because only small displacements in the wall of the channel are required to open the gate, it may be that a relatively simple "elastic" picture is sufficient here. For myoglobin, where ligand binding is thought to involve more complex motions of the protein, the gate dynamics may still be sufficiently rapid at 300 K to allow for "simple" binding.

What are the possible functional implications of such gating motions? Insuring the maximum possible speed of binding is clearly one function. For example, it is necessary to create a special environment around a substrate for enzymatic catalysis, but evolutionary pressure has forced the creation of this environment to happen very rapidly for certain enzymes.

Another function may well be the contribution of gating to the selectivity of binding (Zhou et al. 1998). For ligands that are only slightly larger than the natural ones, the gate may not open frequently enough to allow unhindered binding, so that the larger ligands are less likely to be bound before diffusing away. The overall rate of binding can decrease very rapidly with the increasing size of the ligand, and this will be reflected in the probability of binding one ligand compared to another in the nonequilibrium regime typical of living systems (Zhou et al. 1998).

Although the above discussion focused on the binding of small molecules to biopolymers, similar issues arise in connection with the binding of biopolymers to one another. In particular, rapid motion (times of a few ns) of surface loops of proteins may facilitate the assembly of chaperonins (Landry et al. 1997) and allow for the binding of multiple receptors in the case of certain fibronectin domains (Carr et al. 1997).

Because only small displacements are required to open the gates in some of the systems mentioned above, biopolymer motion on short time scales (ps to ns) can influence function. In other cases, larger displacements and longer time scales are important, as discussed below.

CONCERTED BINDING AND RELEASE

In the case of biopolymers or assemblies of biopolymers that bind more than one ligand, it appears that the binding of one ligand sometimes drives the release or relocation of another ligand. One example is the enzyme dihydrofolate reductase from *Escherichia coli*, in which the binding of the cofactor NADPH leads to structural changes that tend to expel the cofactor THF from a different site, as part of the cyclic activity of the enzyme (Sawaya and Kraut 1997). In some ATP synthases, the proton-driven rotation of an asymmetric axle centered in an enzymatic cluster causes changes in the conformations of these enzymes that in turn enable substrate binding, and drive catalysis and product release, all at different sites in the cluster; a recent review is given by Zhou et al. (1997). The actual dynamics of the transitions involved remains to be determined, but is undoubtedly complex. To have useful rates of turnover (time of about 100 ms for ATP synthase), there must be upper bounds on the roughness of the energy landscape.

MOLECULAR CLOCKS

The preceding discussion has considered processes that are fast, or at least closely correlated in time. Other functional processes in biopolymers may require delay times, which in some cases may imply lower bounds on the roughness of the energy landscape. Enzymes that bind two or more substrates or cofactors that interact in the active site may in some cases require that one of these species be held for some time in a particular conformation. This has been suggested to be the case in lactate dehydrogenase, where the cofactor NADH may have to retain the conformation found in its binary complex with the enzyme during the binding and reaction of substrate (van Beek et al. 1997). The nicotinamide ring is sterically hindered from rotating during the reaction, which occurs on the ms time scale, and this is thought to contribute to the stereospecificity of the reaction. Necessary delay times may also occur in biopolymer conformational changes. This appears to occur for example in certain proteins that effect the fusion of viral and host membranes. In the case of the influenza protein hemagglutinin HA2, post-translational cleavage is thought to leave the protein in a long-lived metastable conformation, which is only induced to change in response to a reduction in pH within the host cell (Steinhauer et al. 1996). In another context, it has been suggested on the basis of molecular dynamics simulations that the relaxation of the bacterial photosynthetic reaction center is slow on the time scale of the initial electron transfer steps following photoexcitation, and that this slow relaxation leads to a smaller reorganization energy and faster electron transfer than would obtain in the case of fast relaxation (Parson et al. 1998).

The delay of product release is crucial for the time-dependent organization of the cell cytoskeleton. Actin filaments are dynamic polymers whose assembly and disassembly in the cytoplasm drives cell shape changes, cell locomotion, and chemotactic migration. The ATP hydrolysis that accompanies actin polymerization and the subsequent release of the cleaved phosphate destabilizes the filaments, and, therefore, must be slow compared to their elongation (Pollard et al. 1992). The results of molecular dynamics simulations suggest that the phosphate is stabilized by a tightly bound divalent cation and by a salt bridge formed with His73 (Wriggers and Schulten 1999). Consistent with this model, certain His73 mutants exhibit rapid depolymerization or decreased stability (Solomon and Rubenstein 1987). Actin's phosphate release appears to act as a clock, altering in a time-dependent manner the mechanical properties of the filament and its propensity to depolymerize.

ACKNOWLEDGMENTS

Dr. Volkhard Helms and Dr. Willy Wriggers provided helpful suggestions. Work in the author's laboratory is supported in part by grants from the National Institutes of Health, the National Science Foundation, and the NSF Partnerships for Advanced Computational Infrastructure.

REFERENCES

Bourne, Y., P. Taylor, and P. Marchot. 1995. Acetylcholinesterase inhibition by fasciculin: Crystal structure of the complex. *Cell* **83**:502–512.

Carr, P.A., H.P. Erickson, and A.G. Palmer. 1997. Backbone dynamics of homologous fibronectin type III cell adhesion domains from fibronectin and tenascin. *Structure* **5**:949–959.

Denisov, V.P., J. Peters, H.D. Horlein, and B. Halle. 1996. Using buried water molecules to explore the energy landscape of proteins. *Nature Struct. Biol.* **3**:505–509.

Dickson, R.M., A.B. Cubitt, R.Y. Tsien, and W.E. Moerner. 1997. On/off blinking and switching behaviour of single molecules of green fluorescent protein. *Nature* **388**:355–358.

Frauenfelder, H., S.G. Sligar, and P.G. Wolynes. 1991. The energy landscapes and motions of proteins. *Science* **254**:1598–1603.

Jia, Y.W., A. Sytnik, L.Q. Li, S. Vladimirov, B.S.Cooperman, and R.M. Hochstrasser. 1997. Nonexponential kinetics of a single tRNA(Phe) molecule under physiological conditions. *Proc. Natl. Acad. Sci. USA* **94**:7932–7936.

Kojima, H., E. Muto, H. Higuchi, and T. Yanagida. 1997. Mechanics of single kinesin molecules measured by optical trapping nanometry. *Biophys. J.* **73**:2012–2022.

Landry, S.J., N.K. Steede, and K. Maskos. 1997. Temperature dependence of backbone dynamics in loops of human mitochondrial heat shock protein 10. *Biochemistry* **36**:10,975–10,986.

McCammon, J.A., and S.H. Northrup. 1981. Gated binding of ligands to proteins. *Nature* **293**:316–317.

Parson, W.W., Z.T. Chu, and A. Warshel. 1998. Reorganization energy of the initial electron-transfer step in photosynthetic bacterial reaction centers. *Biophys. J.* **74**:182–191.

Pollard, T.D., I. Goldberg, and W.H. Schwarz. 1992. Nucleotide exchange, structure, and mechanical properties of filaments assembled from ATP-actin and ADP-actin. *J. Biol. Chem.* **267**:20,339–20,345.

Sawaya, M.R., and J. Kraut. 1997. Loop and subdomain movements in the mechanism of *Escherichia coli* dihydrofolate reductase: Crystallographic evidence. *Biochemistry* **36**:586–603.

Solomon L.R., and P.A. Rubenstein. 1987. Studies on the role of actin's N-methylhistidine using oligodeoxynucleotide-directed site-specific mutagenesis. *J. Biol. Chem.* **262**:11,382–11,388.

Steinhauer, D.A., J. Martin, Y.P. Lin, S.A. Wharton, M.B.A. Oldstone, J.J. Skehel, and D.C. Wiley. 1996. Studies using double mutants of the conformational transitions in influenza hemagglutinin required for its membrane fusion activity. *Proc. Natl. Acad. Sci. USA* **93**:12,873–12,878.

Sussman, J.L., M. Harel, F. Frolow, C. Oefner, A. Goldman, L. Toker, and I. Silman. 1991. Atomic structure of acetylcholinesterase from *Torpedo californica* — A prototypic acetylcholine-binding protein. *Science* **253**:872–879.

Thorn Leeson, D., D.A. Wiersma, K. Fritsch, and J. Friedrich. 1997. The energy landscape of myoglobin: An optical study. *J. Phys. Chem. B* **101**:6331–6340.

van Beek, J., R. Callender, and M.R. Gunner. 1997. The contribution of electrostatic and van der Waals interactions to the stereospecificity of the reaction catalyzed by lactate dehydrogenase. *Biophys. J.* **72**:619–626.

Wade, R.C., B.A. Luty, E. Demchuk, J.D. Madura, M.E. Davis, J.M. Briggs, and J.A. McCammon. 1994. Simulation of enzyme-substrate encounter with gated active sites. *Nature Struct. Biol.* **1**:65–69.

Wennmalm, S., L. Edman, and R. Rigler. 1997. Conformational fluctuations in single DNA molecules. *Proc. Natl. Acad. Sci. USA* **94**:10,641–10,646.

Wlodek, S.T., T.W. Clark, L.R. Scott, and J.A. McCammon. 1997. Molecular dynamics of acetylcholinesterase dimer complexed with tacrine. *J. Am. Chem. Soc.* **119**:9513–9522.

Wriggers, W., and K. Schulten. 1999. Investigating a back door mechanism of actin phosphate release by steered molecular dynamics. *Proteins* **35**:262–273.

Zhou, H.X. 1998. Theory of the diffusion-influenced substrate binding rate to a buried and gated active site. *J. Chem. Phys.* **108**:8146–8154.

Zhou, Y., T.M. Duncan, and R.L. Cross. 1997. Subunit rotation in *E. coli* F0F1-ATP synthase during oxidative phosphorylation. *Proc. Natl. Acad. Sci. USA* **94**:10,583–10,587.

Zhou, H.X., S.T. Wlodek, and J.A. McCammon. 1998. Conformation gating as a mechanism for enzyme specificity. *Proc. Natl. Acad. Sci. USA* **95**:9280–9283.

12

What Is the Relationship between Dynamical Effects and Biological Functions?

A. WARSHEL

Dept. of Chemistry, University Park, University of Southern California,
Los Angeles, CA 90089–1062, U.S.A.

Keywords: enzyme dynamics, photosynthesis, photoisomerization

ABSTRACT

The role of dynamical effects in biological reactions is examined by computer simulation approaches. It is concluded that enzyme catalysis does not involve a major change in the transmition factor relative to the corresponding reference reaction in solution. It is also concluded that the nature of productive trajectories is similar in the enzyme and solution reaction, once the activation barriers are similar. Some differences in dynamical effects may appear as a result of changes in activation barriers, but this is the result of the change in barrier height rather than the reason for the change in the rate constant. The role of dynamical effects in light-induced photobiological reactions is also explored. Here, one finds effects that can be qualified as dynamical effects but they do involve major changes in rate constants.

INTRODUCTION

The revolution in the elucidation of the structure of proteins has paved the way for quantitative structure function correlation in biology. Such correlations should allow one, at least in principle, to explore the role of different factors. Computer simulation studies have indicated that electrostatic free energies are the primary correlation between structure and catalysis (Warshel 1981, 1991). However, it is interesting to examine whether other effects may play a significant role in determining the rate constants of biological processes. For example, it is interesting to see whether dynamical effects can contribute to the optimization of biological rate constants. In addressing this issue

Simplicity and Complexity in Proteins and Nucleic Acids
Edited by H. Frauenfelder, J. Deisenhofer, and P.G. Wolynes

it is important to be clear about the meaning of "dynamical effects," i.e., every reaction is a time-dependent process but this does not classify the corresponding rate constant as a dynamical quantity. In fact, rigorous studies (see below) divide the rate constant to a probabilistic factor that determines the chance of reaching the transition state and to a "transmition factor" that reflects the chance that the molecule will continue from the transition state to the product region. This transmition factor contains the true dynamical effects. On the other hand, the probabilistic factor depends on the activation free energy, $\Delta g^{\neq}$, and this quantity can be determined by configuration averaging without any dynamical considerations. Thus, only the deviations from the transition state theory and the possibility that the rate constant will not be proportional to $\exp\{-\Delta g^{\neq\beta}\}$ (where $\beta = 1/k_B T$) can be classified as dynamical effects. Note in this respect that all chemical reactions involve fluctuations but the chance that these fluctuations will reach the transition state, $\exp\{-\Delta g^{\neq\beta}\}$, is usually not a dynamical effect.

In this chapter I examine the role of dynamical effects in several classes of biological processes. These processes include enzymatic reactions, light-induced photobiological reactions, and diffusion processes. I argue that the contribution of dynamical effects is relatively small in all of these processes.

DYNAMICAL EFFECTS IN ENZYMATIC REACTIONS

The assumption that dynamical effects play a significant role in enzyme catalysis has been invoked repeatedly (Careri et al. 1979; Gavish and Werber 1979; McCammon et al. 1979). Before illustrating that such effects are small, it is crucial to clarify what is meant by "important role in catalysis" and what are true dynamical effects.

As has been pointed out (Warshel 1991, 1978), enzyme catalysis is determined by the difference between the activation barrier in the enzyme and the corresponding barrier in a reference solvent cage. Thus, the question of whether dynamical effects are important in enzyme catalysis boils down to the question of whether the reactive fluctuations are fundamentally different in enzymes and solution (Warshel 1984). Before addressing this question in a rigorous way, it is useful to examine how the thermal fluctuations of an enzyme lead to its chemical activity. This is done in Figure 12.1, which describes the proton transfer step from Glu35 to the sugar oxygen in the catalytic reaction of lysozyme (for further details, see Warshel 1984). This reaction is described by considering two valence bond resonance structures:

$$\begin{aligned} \Psi_1 &= \left(\mathrm{O_A} - \mathrm{H} \quad \mathrm{O_4}\!\!<\right) \\ \Psi_2 &= \left(\mathrm{O_A^-} \quad \mathrm{H} - {}^{+}\mathrm{O_4}\!\!<\right) \end{aligned} \tag{12.1}$$

where O_A and O_4 are the oxygens of Glu35 and the glycosidic bond, respectively. Figure 12.1 represents the reaction by considering the barrier for proton transfer from O_A to O_4 as a function of time. We find that the barrier oscillates with the fluctuations of

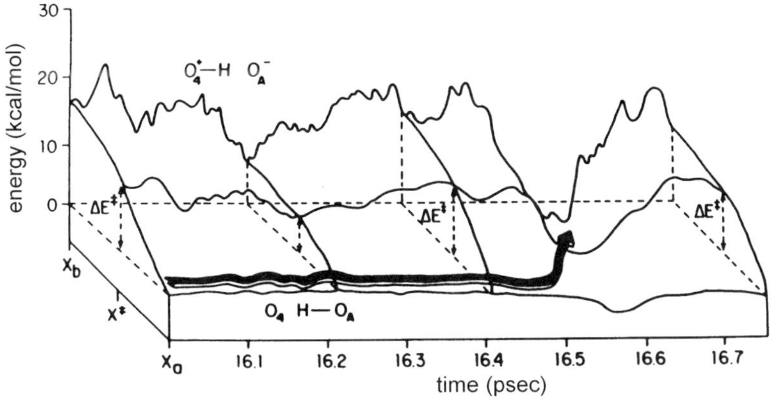

Figure 12.1 Time-dependent changes in the barrier for proton transfer, $\Delta E^{\neq}$, along a trajectory of the catalytic reaction of lysozyme. The trajectory involves a proton transfer from the oxygen of Glu 35 (O_A) to the oxygen of the glycosidic bond (O_4). This figure demonstrates that the effective barrier $\langle \Delta E^{\neq} \rangle$ is determined by the electrostatic fluctuations of the ionic state O_4^+–H O_A^- (see Warshel 1984 for further details).

the surrounding polar environment. Once in a while the environment reaches a configuration where the barrier is small and the proton can be transferred. This figure clarifies that the proton transfer probability is strongly related to the fluctuations of the surrounding active site and the resulting fluctuations of the potential energies ε_1 and ε_2, which correspond to ψ_1 and ψ_2, respectively. These environmental changes can be viewed as the fluctuations of the "reaction field" from the solvent–protein system. These fluctuations can be evaluated on a microscopic level by considering the contributions of the environment to the gap between the energies ε_1 and ε_2, respectively. The relationship between the fluctuations of the environment and the rate of proton transfer (PT) was pointed out in earlier works (Warshel 1984, 1982) and later by others (Zichi et al. 1989). However, the fact that we find an instructive relationship between the fluctuations of the reaction field and the chance of PT does not mean that we are dealing with a dynamical effect. That is, any chemical process involves thermal fluctuations that reach the transition state, but if the chance for such fluctuations can be obtained from the corresponding Boltzmann probability, then we have a statistical rather than a dynamical effect.

To examine whether or not we have significant dynamical contributions, we must consider their effect on the corresponding rate constant. Our starting point is the fact that the rate constant for chemical reactions in condensed phase can be defined in a rigorous way by (Keck 1966; Anderson 1973; McCammon and Karplus 1979; Bennett 1977; Grimmelmann et al. 1981):

$$k = \langle \dot{x} H(\dot{x}) \xi \rangle \exp\left\{-\Delta g\left(x^{\neq}\right)\beta\right\} / Z_R , \tag{12.2}$$

where $\beta = 1/k_{BT}$ (k_B is the Boltzmann constant and T the absolute temperature), x is the reaction coordinate, $x^{\neq}$ is the value of x at the transition state, $Z_R = \int_{-\infty}^{x\neq} \exp\{-g(x)\beta\}dx$, $\dot{H}(x)$ is 1 and zero for positive and negative $\dot{x}$, respectively, ξ is a factor that counts each productive trajectory only once and is taken as 1/m for a productive trajectory that crosses the transition state m times (Bennett 1977). Alternatively, one may use the reactive flux formulation (Chandler 1978).

Equation 12.2 can be written as (Hwang et al. 1988):

$$k = \left(\langle \dot{x}_+ \rangle / \Delta x^{\neq}\right) \exp\left\{-\Delta g\left(x^{\neq}\right)\beta\right\}, \tag{12.3}$$

where $\Delta x^{\neq}$ is the width of the transition state region and $\langle \dot{x}_+ \rangle$ is the average $\langle \dot{x} H(\dot{x})\xi \rangle$. The preexponential factor includes all dynamical effects while the $\Delta g(x^{\neq})$ term reflects the nondynamical probabilistic effects. Equation 12.3 can also be written as:

$$k = F\left(k_B T / h\right) \exp\left\{-\Delta g^{\neq}\beta\right\} = F k_{TST}\,. \tag{12.4}$$

where k_{TST} is the transition state theory rate constant, which does not consider dynamical effects. F is the so-called the "transmition factor" that tells us how many times a trajectory that passes the transition state from the reactant state moves back and forth before falling to the product state. This recrossing process is our dynamical effect.

As stated in the **INTRODUCTION**, all chemical reactions involve thermal fluctuations but the chance that these fluctuations will reach the transition state is usually not a dynamical effect but simply a probability factor that can be determined by nondynamical approaches (e.g., Monte Carlo simulations). In fact, we may define here "dynamical effects" as those effects that can only be determined by molecular dynamics simulations and not by Monte Carlo simulations. Such a dynamical effect could be invoked, for example, if it was true that the energy liberated in substrate binding might not be dissipated before it can be used for the enzymatic reaction (McCammon et al. 1979).

The formulation of Equation 12.2 can be related directly to the fluctuation of the above mentioned energy gap ($x(t) = \varepsilon_2(t) - \varepsilon_1(t) = \Delta\varepsilon(t)$). That is, using the linear response approximation, we can relate the effective solvent reaction coordinate to the dipole, μ, of the solute by (Hwang et al. 1988)

$$\langle x(t)_+ \rangle = \langle \Delta\varepsilon(t)_+ \rangle = \left(\frac{\Delta\varepsilon_{max}}{\mu_{max}}\right) \int_0^t \left(\frac{\langle \Delta\dot{\varepsilon}(0)\Delta\varepsilon(t-t')\rangle}{\langle \Delta\dot{\varepsilon}(0)\Delta\varepsilon(0)\rangle}\right) \langle \mu(t')_+ \rangle dt'. \tag{12.5}$$

Equation 12.5 is found to give quite a reliable estimate of the corresponding results obtained by averaging many downhill trajectories (Hwang et al. 1988). The response function $\langle \Delta\dot{\varepsilon}(0)\Delta\varepsilon(t)\rangle$ is, of course, related to the autocorrelation of $\Delta\varepsilon(t)$ by:

$$\frac{d}{dt} C(t) = \frac{d}{dt} \langle \Delta\varepsilon(0)\Delta\varepsilon(t)\rangle = -\langle \Delta\dot{\varepsilon}(0)\Delta\varepsilon(t)\rangle. \tag{12.6}$$

Thus, the dynamical effects are determined by the autocorrelation of the energy gap.

In view of the above point, one can examine the role of dynamical effects in enzymes by comparing $C(t)$ for enzymes to the corresponding $C(t)$ of the reference reaction in solution. *If these $C(t)$ are similar, then dynamical effects do not contribute to enzyme catalysis.* In Figure 12.2, we present a comparison of the $C(t)$ of the catalytic reaction of subtilisin to the corresponding $C(t)$ in solution. Both reactions were modeled using the empirical valence bond (EVB) potential surfaces (Hwang et al. 1988). As seen from the figure, the two $C(t)$ are similar and therefore the dynamical effect for the enzyme and solution reaction are similar. The same conclusions were obtained by comparing the $C(t)$ of the native subtilisin and the Asn155→Ala mutant (Warshel et al. 1988). Here again the $C(t)$ is similar in the native and mutant enzyme cases.

As is clear from Equations 12.4–12.6, $C(t)$ and F are proportional to each other and thus the transmition factors are similar in enzyme and solutions (at least for reactions with significant activation barriers) and therefore dynamical effects do not provide an important catalytic factors. Of course, instead of calculating $C(t)$ one may calculate F. (This can be done by using Equation 12.4b, running downhill trajectories [e.g., Hwang et. al 1988], or by calculating the flux autocorrelation [Chandler 1978]). However, comparing the $C(t)$s in enzymes and solutions leads to the same conclusion that would be obtained by comparing the transmition factors for the enzyme and the solution reaction.

A recent EVB study (Neria and Karplus 1997) also concluded that the transmition factor is not an important factor in the catalytic reaction of triosephosphate isomerase (TIM). However, it was implied in this work that the rate acceleration of TIM is still associated with dynamical effects. In particular, it was suggested that the nonequilibrium solvation effects are fundamentally different in the enzyme and in water, implying that in enzymes (in contrast to the situation in solution), the barrier is not modulated by the effect of the fluctuating environment. It was then suggested that in the enzyme, barrier height is determined by the intramolecular configurations of the substrate. Unfortunately, these conclusions were reached without attempting to calculate the actual activation free energy or to compare the enzyme simulations to the corresponding simulations in a solvent cage. In fact, the activation barriers in TIM and in the reference solvent reaction were calculated by Åqvist and Fothergill (1996), who determined quantitatively that the observed difference between these barriers (~13 kcal/mol) is due to the electrostatic effects of the preorganized enzyme environment. This means that other effects cannot contribute significantly. Furthermore, early simulations (Warshel 1984, 1982) have demonstrated that the barrier modulation by the fluctuation of the surrounding environment is similar in enzymes and solution as long as one considers the same reaction in both environments (in both cases we have the situation described in Figure 12.1). As to the proposed intramolecular effect of the substrate, one should note that the intramolecular potential of the substrate behaves in a very similar way in the enzyme and in solution. The main difference is in the interaction between the substrate and its surrounding. It is also useful to note that a recent simulation study (Warshel and Bentzien 1998) established that the nature of

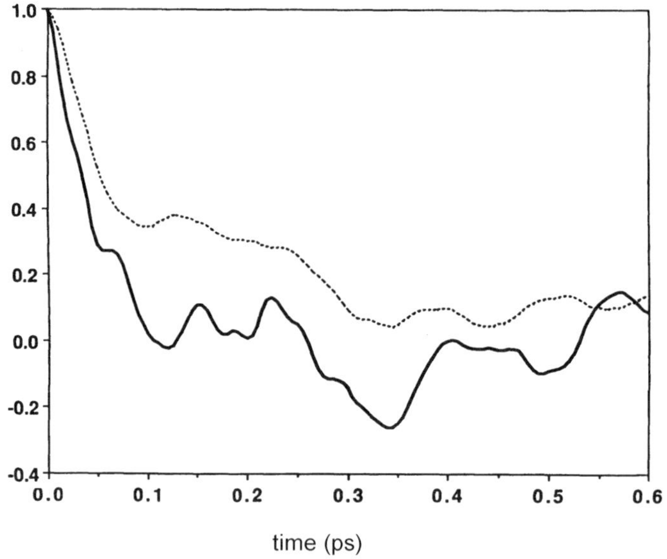

Figure 12.2 The autocorrelation, *C*(*t*), of the time-dependent energy gap for the nucleophilic attack step in the catalytic reaction of subtilisin (solid line) and for the corresponding reference reaction in water (dotted line). The similarity of these autocorrelations indicates that dynamical effects do not contribute significantly to catalysis.

nonequilibrium solvation effects is similar in enzymes and in solutions. This work pointed out that the difference in nonequilibrium effects reflects differences in reorganization free energies rather than dynamical effects. In fact, the concept of "nonequilibrium" solvation effects is an artifact of methods that do not include the solvent coordinate in the search for the transition state (Warshel and Bentzien 1998), and it is mainly a probabilistic factor.

One of the reasons why enzymes do not use dynamical effects is associated with the magnitude of the free-energy barrier in enzymatic reactions. That is, enzymes do not have any incentive to increase k_{cat} over the diffusion limit. Thus, relevant activation barriers in enzymes are usually larger than 7 kcal/mol. In general, once the barrier is high, the recrossing phenomenon is not so significant and a trajectory that reaches the transition state usually proceeds directly to the product state.

DYNAMICAL EFFECTS IN OTHER SYSTEMS

In view of the discussion in the previous section, it seems that dynamical effects do not contribute to enzyme catalysis. The question is, then, whether such effects play a substantial role in other systems. In addressing this question, we will consider light-induced reactions that occur on a very short time scale.

Photoisomerization in Rhodopsin and Bacteriorhodopsin

The photoisomerization reaction of rhodopsin and bacteriorhodopsin is a barrierless process where the system evolved to optimize the quantum yield. This is done by reaching the twisted configuration sufficiently fast so that the system does not relax back to its initial state by the fluorescent channel (see Figure 12.3). This process is likely to involve dynamical effects (in fact, the earliest simulation of the dynamics of a biological process involved a study of the excited state dynamics of a minimal model of rhodopsin [Warshel 1976]). However, the competition between the torsional motion of the chromophore and the fluorescent process is not really a dynamical effect, but an interplay between the two rate constants k_{fl} and k_i of Figure 12.3. Before examining whether k_i involves large dynamical effects, we should recognize that the main issue is again the difference between the given process in the protein and in solution. Apparently, the photoisomerization of the 13-cis protonated Schiff base of retinal (PSBR) in solution is not much slower than the corresponding process in proteins (~3 ps in solution (Logunov et al. 1996) and 0.2 ps in rhodopsin). Thus, the contributions of dynamical effects to the excited state catalysis must be smaller than the overall factor of ~20. Simulations of the photoisomerization process in bacteriorhodopsin (Warshel et al. 1991), which are summarized in Figure 12.4, indicated that the overall isomerization motion is somewhat overdamped with a significant diffusive character. Thus, trajectories that reach the 90° region lose their kinetic energy and stay at that region, rather than oscillate back and forth with a large amplitude in an inertial motion. This, however, does not tell us what is the origin of the different kinetics in the solution and protein reaction. On one hand, it is possible that the solution reaction involves a small barrier and that this leads to the slower isomerization time. Another more exotic option is provided by our recent simulation study (A. Warshel, unpublished). This study

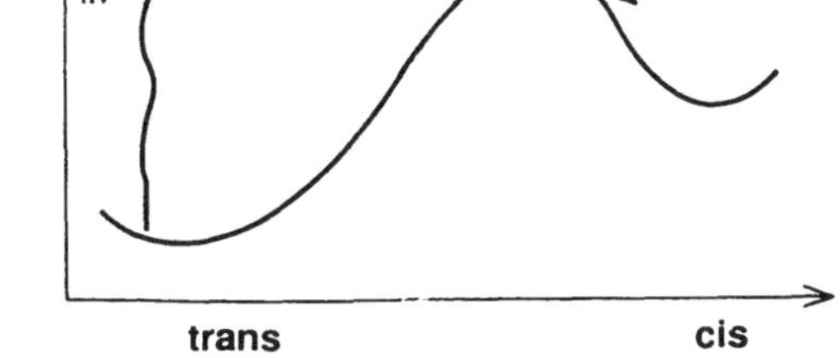

Figure 12.3 Pathways for the photoisomerization reaction in rhodopsin and bacteriorhodopsin. The competition between the forward rate, k_i, and the fluorescence-excited rate, k_{fl}, determines the number of molecules that arrive at the crossing region.

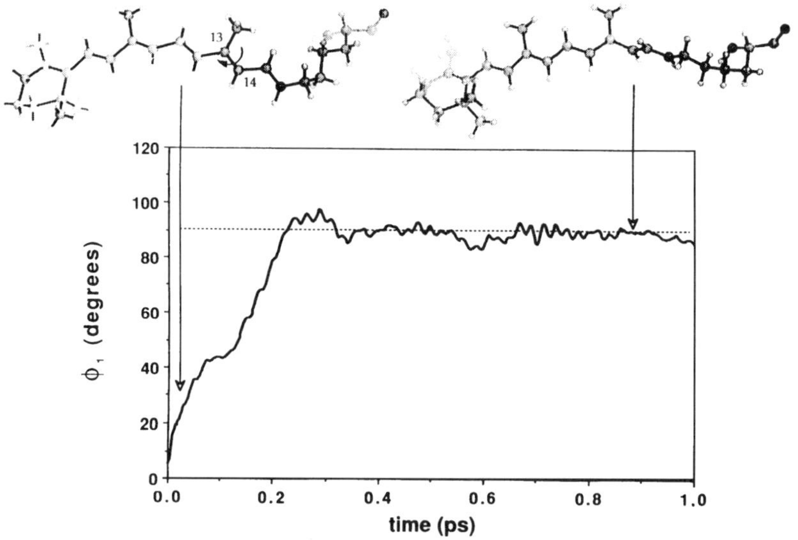

Figure 12.4 The time dependence of the $\phi_{13} - \phi_{14}$ angle (ϕ_1) for an excited state trajectory of the chromophore in the active site of bacteriorhodopsin (for further details, see Warshel et al. 1991). Two snapshots of the chromophore (at the indicated times) are depicted in the upper part of the figure. As can be seen, the overall Cartesian displacements are quite small despite the change of ϕ_1 (this reflects the fact that the isomerization around ϕ_1 involves compensating partial rotations around other bounds). As seen from the figure, the overall motion is overdamped rather than inertial, and trajectories that reach the 90° region lose their kinetic energy and stay at this region.

compared the excited state isomerization time of 11-cis PSBR in water to that in a more restricted active site (modeled by placing position constraints on the water molecules). It was found that the isomerization time is faster in the restricted environment than in the unrestricted. The results of these simulations are not yet conclusive, since it is not clear how accurate our excited state potential surface is. At any rate, the role of a more restricted environment in providing some rate acceleration is of interest and a related issue was examined recently (Leitner and Wolynes 1999).

Photosynthesis

The primary steps in bacterial photosynthesis involve light-induced electron transfer processes. Such processes might involve some dynamical effects and the question is how important these effects are. Semiclassical and quantum mechanical simulations of the dynamics of electron transfer (ET) processes in solutions and proteins were introduced by Warshel and coworkers (e.g., Warshel 1982; Creighton et al. 1988; Warshel and Parson 1991) and related studies were subsequently advanced by others (Webster et al. 1991; Treutlein et al. 1988). These studies indicated that the rate constants of regular ground state ET reactions are determined by the activation free energy and not by dynamical effects. In the case of excited state ET, it is possible that dynamical effects play some role. In exploring this question, we may examine the autocorrelation of the relevant electronic energy gap (see section on **DYNAMICAL EFFECTS IN ENZYMATIC REACTIONS**). Simulation of ET in bacterial reaction centers (RC) (Warshel and Parson 1991; Treutlein et al. 1992) indicated that initial relaxation of the autocorrelation function occurs within 0.2 ps. This fast protein relaxation is not exceptional and similar rapid relaxations have been obtained in previous simulations of charge transfer (CT) processes in solution (Warshel and Hwang 1986) and proteins (Warshel et al. 1988). It might be assumed that the femtosecond relaxation of the environment upon formation of the charge-separated state destroys the energy match needed for electron transfer reaction and thus prevents the back reaction. Such a suggestion, however, appears to overlook the fact that a similar relaxation would occur in trajectories of the reverse process, and indeed in any electron-transfer reaction, reversible or not. That is, any suggestion that a fast relaxation in the forward direction of the reaction will prevent the back reaction is inconsistent with the fact that the same fast relaxation would also occur in the backward direction. Using the same incorrect argument for the backward reaction will lead one to conclude that the forward reaction would not occur. In fact, this argument can be used to argue that all electron transfer reactions are irreversible (electron transfer in solution involves usually a very fast solvent relaxation time). Apparently, trajectories that cross in a given direction are trapped for some time, but trajectories that crossed in a previous event can use the thermal energy (according to the corresponding Boltzmann probability) and cross back in the opposite direction. Thus, the irreversibility of the initial step in the photosynthetic reaction is controlled by the proper drop in free energy (ΔG_0) and the corresponding reorganization energy, λ, and not by dynamical effects (Warshel and Schlosser 1981). It is still interesting to examine the significance of dynamical effects in photosynthetic processes. In addressing this question, one should note that the transmition factor for ET reactions (Warshel and Hwang 1986; Calef and Wolynes 1983) includes factors such as the electronic coupling matrix element, which is not dynamical in nature. The dynamical part of the preexponential factor is not expected to be much different for ET reactions in solution and protcins, as long as the reorganization energy and the reaction free energy are similar in the two environments.

In the case of ultrafast biological reactions, we might have partially frozen relaxation effects that could be qualified as time-dependent effects. That is, when the reaction is very fast, the environment might not be allowed to undergo full equilibrium relaxation (Warshel 1979). The question, however, is how important are such effects and how much can they change the rate constant obtained when considering the entire configuration space. This issue has been examined recently (Parson et al. 1998) in a simulation of the primary ET in a bacterial RC. The study was conducted by comparing the λ and ΔG_0 calculated using a full umbrella sampling/FEP approach (where the system is moved gradually from the reactant to the product state) to the corresponding calculations obtained from trajectories on the reactant and product state only. The first

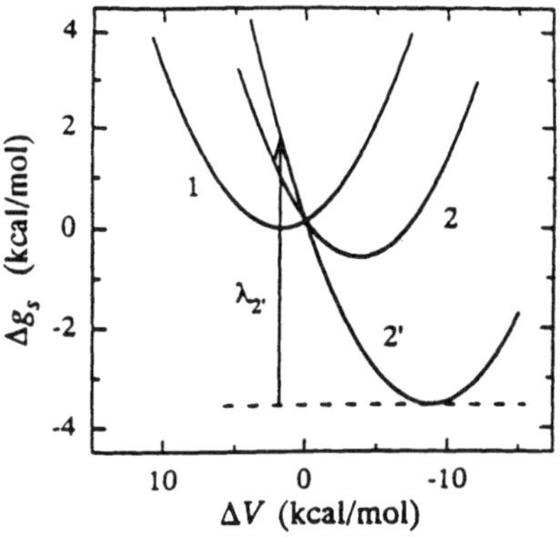

Figure 12.5 Curves 1 and 2 depict the free-energy functions for the primary charge separation process in the bacterial RC evaluated by running trajectories on the P^*B and P^+B^- states (P is the special pair and B is the neighboring bacteriochlorophyll). These curves reflect only the configuration sample during a 25 ps trajectory. Curve 2′ represents the free-energy function obtained by assuming a complete sampling, where $\Delta g_1(\Delta V) = \Delta g_2(\Delta V) + \Delta V$(for further details, see Parson et al. 1998).

method is expected to force the system to sample all the relevant configurational space while the second mainly considers configurations that are available during a relatively short time. As shown in Figure 12.5, it was found that both λ and ΔG_0 depend on the sampling approach or, in other words, on the configurations sampled. Thus, one can conclude that the effective parameters in the Marcus formula are a function of time and presumably also temperature. In the case of bacterial RC the nonequilibrium λ is probably more pertinent to the ET kinetics than the value obtained for a fully equilibrated system. Conformational relaxation that occurs on a time scale of more than 10 ps will have little bearing on the kinetics (since the reaction is completed in ~3 ps). Attempts to evaluate the rate using Marcus theory with the "rigorous" FEP calculation will therefore be inappropriate. As long as the electronic coupling is sufficiently weak, it is best to evaluate the rate constant using the semiclassical trajectory method used to evaluate the rate of surface hopping (Creighton et al. 1988) in this system. Interestingly and significantly, however, the activation barriers and rate constants obtained by using the two sets of free-energy function (and the corresponding ΔG_0 and λ), depicted in Figure 12.5 as well as those obtained by semiclassical trajectory studies are very similar.

Diffusion

Diffusion processes do involve dynamical effects since the transmition factor is reduced as a result of recrossing events. However, the effects found in biological processes are quite small. Simulations of an idealized model of an ion channel (Roux and Karplus 1993) produced a transmition factor of ~0.1. The corresponding effect on rate is quite trivial relative to the enormous reduction in activation barrier obtained by the channel's polar environment (as compared to nonpolar channels) or to the barrier of penetrating through the membrane in the absence of a channel. Furthermore, the error in estimating the activation barrier by most current models is much larger than the possible 0.6 kcal/mol correction due to dynamical effects. Diffusion processes are, of course, involved in diffusion-controlled enzymatic reactions. Thus, a significant amount of recent studies have been dedicated to this issue (Tan et al. 1993; Fisher et al. 1997). However, the catalytic effect of the diffusion processes is quite unimpressive, reaching at most one order of magnitude (larger values are not supported by experimental studies). Such effects are, of course, much smaller than the enormous effect of enzymes in reducing activation barriers. A case of point is the catalytic reaction of acetylcholinesterase. A recent study (Fuxreiter and Warshel 1998) has established that the enzyme reduces the activation barrier by ~10 kcal/mol or 7 orders of magnitudes, which, of course, is much larger than any dynamical effect.

CONCLUDING REMARKS

Although all biological processes involve atomic motions, it does not mean that these motions were optimized or used by evolution. The kinetics of thermally activated processes (which applies to all rate-determining enzymatic reactions) is largely determined by the probability of reaching the relevant transition states and this probability does not seem to reflect dynamical effects (e.g., Warshel 1991).

The fact that pure dynamical effects do not play an important role in enzyme catalysis can be easily rationalized by noting that it is much simpler to change rate constants in condensed phases by changing the activation barrier (this has an exponential effect on the rate) than by changing the preexponential transmition factor. In cases of reactions that are not controlled by diffusion, it is hard to envision a change of more than a factor of 10 in the transmition factor in the physically available range of local friction. However, it is very common to see enzymatic reactions with a rate acceleration of ten orders of magnitude, due to the reduction of the relevant activation free energies.

Light-induced photobiological reactions do occur on a short time scale where slow relaxation channels are practically frozen. This, however, might not reflect any true dynamical advantage since similar pictures occur in nonbiological noncatalyzed processes.

In summary, studies of the time dependence of biological processes provide powerful insight about such processes. However, such studies reflect to a large extent probabilistic rather than dynamical features.

REFERENCES

Anderson, J.B. 1973. Statistical theoreticals of chemical reactions. Distributions in the transition region. *J. Chem. Phys.* **58**:4684–4692.

Åqvist, J., and M. Fothergill. 1996. Computer simulation of the triosephosphate isomerase catalyzed reaction. *J. Biol. Chem.* **271**:10,010–10,016.

Bennett, C.H. 1977. Molecular dynamics and transition state theory: The simulation of infrequent events. In: Algorithms for Chemical Computations, ed. R.E. Christofferson. pp. 63–97. Washington, D.C.: ACS.

Calef, D.F., and P.G. Wolynes. 1983. Classical solvent dynamics and electron transfer. 1. Continuum Theory. *J. Phys. Chem.* **87**:3387–3400.

Careri, G., P. Fasella, and E. Gratton. 1979. Enzyme dynamics: The statistical physics approach. *Ann. Rev. Biophys. Bioeng.* **8**:69–97.

Chandler, D. 1978. Statistical mechanics of isomerization dynamics in liquids and the transition state approximation. *J. Chem. Phys.* **68**:2959.

Creighton, S., J.-K. Hwang, A. Warshel, W.W. Parson, and J. Norris. 1988. Simulating the dynamics of the primary charge separation process in bacterial photosynthesis. *Biochemistry* **27**:774–781.

Fisher, C.L., D.E. Cabelli, R.A. Hallewell, P. Beroza, E.D. Getzoff, and J.A. Tainer. 1997. Computational, pulse-radiolytic, and structural investigations of lysine-136 and its role in the electrostatic triad of human Cu, Zn superoxide dismutase. *Proteins* **29**:103–112.

Fuxreiter, M., and A. Warshel. 1998. Origin of the catalytic power of acetylcholinesterase: Computer simulation studies. *J. Am. Chem. Soc.* **120**:183–194.

Gavish, B., and M.M. Werber. 1979. Viscosity-dependent structural fluctuation in enzyme catalysis. *Biochemistry* **18**:1269–1275.

Grimmelmann, E.K., J.C. Tully, and E. Helfand. 1981. Molecular dynamics of infrequent events: Thermal desorption of xenon from a platinum surface. *J. Chem. Phys.* **74**:5300–5310.

Hwang, J.-K., G. King, S. Creighton, and A. Warshel. 1988. Simulation of free energy relationships and dynamics of S_N2 reactions in aqueous solution. *J. Am. Chem. Soc.* **110**:5297–5311.

Keck, J.C. 1966. Variational theory of reaction rates. *Adv. Chem. Phys.* **13**:85–121.

Leitner, D.M., and P.G. Wolynes. 1999. Quantum energy flow during molecular isomerization. *Chem. Phys. Lett.* **280**:411–418.

Logunov, S.L., L. Song, and M.A. El-Sayed. 1996. Excited-state dynamics of a protonated retinal Schiff base in solution. *J. Phys. Chem.* **100**:18,586–18,591.

McCammon, J.A., and M. Karplus. 1979. Dynamics of activated processes in globular proteins. *Proc. Natl. Acad. Sci. USA* **76**:3585–3589.

McCammon, J.A., P.G. Wolynes, and M. Karplus. 1979. Picosecond dynamics of tyrosine side chains in proteins. *Biochemistry* **18**:927–942.

Neria, E., and M. Karplus. 1997. Molecular dynamics of an enzyme reaction: Proton transfer in TIM. *Chem. Phys. Lett.* **267**:23–30.

Parson, W.W., Z.T. Chu, and A. Warshel. 1998. Reorganization energy of the initial electron-transfer step in photosynthetic bacterial reaction centers. *Biophys. J.* **74**:182–191.

Roux, B., and M. Karplus. 1993. Ion transport in the gramicidin channel: Free energy of the solvated right-handed dimer in a model membrane. *J. Am. Chem. Soc.* **115**:3250–3260.

Tan, R.C., T.N. Truong, J.A. McCammon and J.L. Sussman. 1993. Acetylcholinesterase: Electrostatic steering increases the rate of ligand binding. *Biochemistry* **32**:401–403.

Treutlein, H., K. Schulten, A.T. Brunger, M. Karplus, J. Deisenhofer, and H. Michel. 1992. Chromophore-protein interactions and the function of the photosynthetic reaction center: A molecular dynamics study. *Proc. Natl. Acad. Sci. USA* **89**:75–79.

Treutlein, H., K. Schulten, C. Niedermeier, J. Deisenhofer, H. Michel, and D. DeVault. 1988. Electrostatic control of electron transfer in the photosynthetic reaction center of *Rhodopseudomonas viridis*. In: The Photosynthetic Bacterial Reaction Center, ed. J. Breton and A. Verméglio. pp. 369–377. New York: Plenum.

Warshel, A. 1976. Bicycle-pedal model for the first step in the vision process. *Nature* **260**:679–683.

Warshel, A. 1978. Energetics of enzyme catalysis. *Proc. Natl. Acad. Sci. USA* **75**:5250–5254.

Warshel, A. 1979. Conversion of light energy to electrostatic energy in the proton pump of halobacterium halobium. *Photochem. Photobiol.* **30**:285–290.

Warshel, A. 1981. Electrostatic basis of structure–function correlation in proteins. *Accts. Chem. Res.* **14**:284–290.

Warshel, A. 1982. Dynamics of reactions in polar solvents. Semiclassical trajectory studies of electron-transfer and proton-transfer reactions. *J. Phys. Chem.* **86**:2218–2224.

Warshel, A. 1984. Dynamics of enzymatic reactions. *Proc. Natl. Acad. Sci. USA* **81**:444–448.

Warshel, A. 1991. Computer modeling of chemical reactions in enzymes and solutions. New York: John Wiley.

Warshel, A., and J. Bentzien. 1999. Energetics and dynamics of transition states of reactions in enzymes and solutions. In: Transition State Modeling for Catalysis, ed. D.G. Truhlar and K. Morokuma, pp. 489–498. ACS Symp. Series 721.

Warshel, A., Z.T. Chu, and J.-K. Hwang. 1991. The dynamics of the primary event in rhodopsins revisited. *Chem. Phys.* **158**:303–314.

Warshel, A., and J.-K. Hwang. 1986. Simulation of the dynamics of electron transfer reactions in polar solvents: Semiclassical trajectories and dispersed polaron approaches. *J. Chem. Phys.* **84**:4938–4957.

Warshel, A., and W.W. Parson. 1991. Computer simulations of electron transfer reactions in solution and photosynthetic reaction centers. *Ann. Rev. Phys. Chem.* **42**:279–309.

Warshel, A., and D.W. Schlosser. 1981. Electrostatic control of the efficiency of light-induced electron transfer across membranes. *Proc. Natl. Acad. Sci. USA* **78**:5564–5568.

Warshel, A., F. Sussman, and J.-K. Hwang. 1988. Evaluation of catalytic free energies in genetically modified proteins. *J. Mol. Biol.* **201**:139–159.

Webster, F., P.J. Rossky, and R.A. Friesner. 1991. Nonadiabatic processes in condensed matter: semi-classical theory and implementation. *Comp. Phys. Comm.* **63**:494–522.

Zichi, D.A., G. Ciccotti, J.T. Hynes, and M. Ferrario. 1989. Molecular dynamics simulation of electron-transfer reactions in solution. *J. Phys. Chem.* **93**:6261–6265.

13

Averaging versus Zooming-in Complexity

Contraria sunt Complementa?

F.G. PARAK[1] and M.U. PALMA[2]

[1] Physik-Department E17, Technische Universität München, James-Franck-Straße 85748 Garching, Germany
[2] Dipartimento di Scienze Fisiche ed Astronomiche, Università di Palermo, via Archirafi 36, 90123 Palermo, Italy

Keywords: hemoglobin, protein dynamics, protein–solvent interaction, coagulation, energy landscape

> Russell's "complexes" were to have the useful property of being compounded, and were to combine with this the agreeable property that they could be treated like "simples." But this alone makes them unserviceable as logical types [....], since there would then have been significance in asserting of a simple that it was complex. But a property cannot be a logical type.
>
> —Ludwig Wittgenstein (Notes on Logic, In: Notebooks, 1914–16).

INTRODUCTION

In this chapter, we consider complexity and simplicity as properties to avoid logical inconsistency in discussing how can they possibly coexist and be complementary. Phenomena in our physical world cover a remarkable span of scale lengths. Explaining and predicting them in terms of a small number of universal physical laws is a longstanding, although philosophically unwarranted quest. The variety of goals and scale lengths characterizing different disciplines may imply averaging-out and zooming-in down to the shortest relevant scale. A well-known example of the productivity of this approach is historically provided by the successful attempt of using the laws of mechanics to view notions such as heat and temperature at the atomic scale.

Simplicity and Complexity in Proteins and Nucleic Acids
Edited by H. Frauenfelder, J. Deisenhofer, and P.G. Wolynes

Macroscopically, these are simple concepts. At the atomic scale, however, they appear related to bewildering varieties of configurational and dynamical states which hardly convey the idea of steady state associated with the notion of temperature. Macroscopic simplicity and the idea of steady state are recovered by appropriate averaging of microscopic scale complexity. Historically, this process resulted in the birth of new branches of physics. This example illustrates how the simplicity of apparently steady conditions and the complexity of apparently random unsteadiness can coexist and complement each other.

Here we deal with the fascinating world of biomolecules. In many cases they show functional simplicity if taken as elemental units, and complexity when viewed at the microscopic scale. To gain a better understanding, we shall discuss some results concerning model systems as well as complex ones, such as myoglobin and hemoglobin, and interactions with water. We shall see how macroscopic functional simplicity results from complex interactions on the atomic scale and how extending our view so as to include the role of the solvent can reveal, even in the case of simple model solutes, a subtle interplay of simplicity and complexity. Sometimes, however, the apparently simple behavior on the macroscopic scale proves to be an oversimplification if one investigates the complexity on the microscopic scale.

PROTEIN DYNAMICS AND COMPLEXITY

Human beings depend on oxygen. If the brain lacks oxygen for more than five minutes life is finished. Oxygen binds to hemoglobin and is transported by the blood circulation to all places in the body where it is needed. In the muscles another protein, myoglobin, takes the oxygen and stores it. The storing and releasing of oxygen is the simple function of myoglobin (Mb) and hemoglobin (Hb). This function has been studied in great detail. In many studies, oxygen has been replaced by carbon monoxide, CO, to simplify experiments. A typical experiment for characterizing the function is the measurement of the number of CO-ligated Mb molecules, MbCO, as a function of the partial pressure of CO. CO-binding kinetics has been studied by flashing off the ligand and watching it rebind to the iron center within the Mb as a function of time. Until 1972, the state of the art was well described by: "The results on myoglobin present no problem; indeed, they are just what would be expected for simple one site molecule that undergoes no conformational change on ligand binding, has no Bohr effect, and, of course, no cooperativity, and whose reaction with CO under all conditions can be described as a simple one step process" (Brunori et al. 1972).

It soon became obvious that this evident simplicity is the result of an averaging of many processes, which even at present are not fully understood. Frauenfelder and coworkers investigated the rebinding of CO after flash photolysis at temperatures down to 40 K (first basic paper: Austin et al. 1975). At low temperatures there is no indication for the exponential rebinding occurring at room temperature. On the molecular scale this can be viewed as follows: at low temperatures, the environment of the Mb

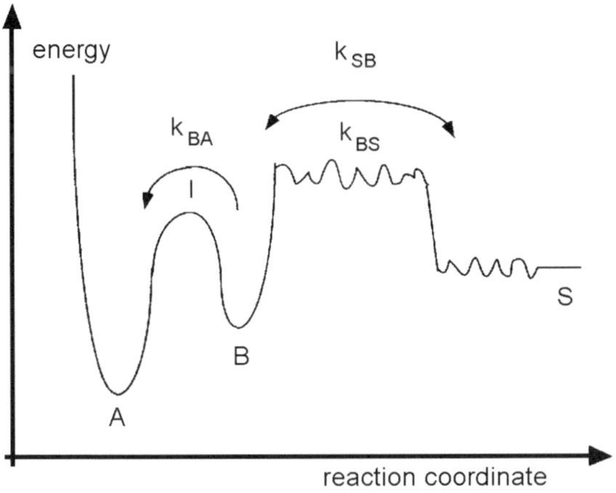

Figure 13.1 Potential energy of a single MbCO molecule as a function of the reaction coordinate. State A: CO ligated to the heme iron; state B: FeCO bond broken, CO within the heme pocket; state S: Mb deoxy where the CO is in the solution. The barrier height *I* differs in different Mb molecules; *k*: transition rates between the states given as lower index.

and, maybe the molecules themselves, behave like a glass. If a light flash breaks the bond between the heme iron and the CO, the latter moves away about 3.7 Å from the iron, still remaining within Mb in a hole called the heme pocket. The rebinding should be an exponential process with a temperature-dependent rate, k_{BA}, determined by the barrier height *I*, which the CO molecule has to overcome along the reaction coordinate (cf. Figure 13.1).

However, a power law has been experimentally observed. A simple explanation is that individual Mb molecules have a different barrier height *I*. Thus, the rebinding kinetics can be obtained by a superposition of many exponentials with different rates. Individual molecules are no longer identical; they are frozen in different "conformational substates." The picture of rebinding has become more complicated, but is it already complex?

Exponential rebinding at room temperature requires one common barrier height for all molecules. This implies thermodynamic averaging over all accessible subconformations of the Mb. This averaging starts to be effective only above 200 K, and it becomes possible because the molecules are no longer frozen in individual substates. There is a rapid conversion among substates. The barrier along the reaction coordinate is now an enthalpy barrier. While it was believed for a long time that the barrier height distribution is temperature independent up to 200 K, it is now assumed that it is shifted to higher energies between 160 K and 180 K (Nienhaus et al. 1992). The shifting of the barrier height distribution suggests that the Mb molecules relax into another set of conformational substates above 160 K, where the rebinding to the iron becomes less favorable (Mb*CO(B_1) → Mb*CO(B_2) in Figure 13.2). Does this come

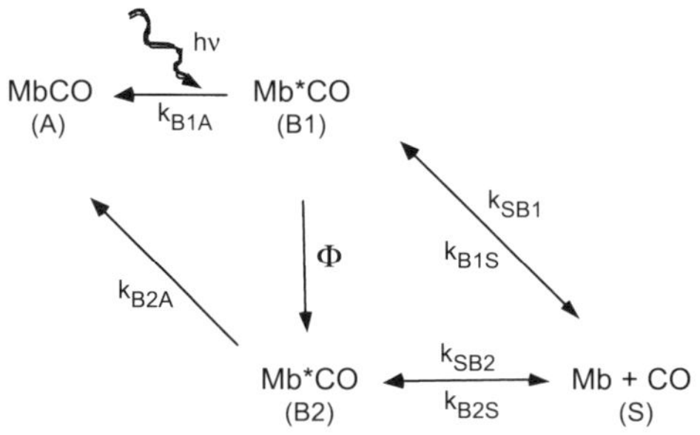

Figure 13.2 Reaction scheme of myoglobin after flash photolysis. The states correspond to Figure 13.1; however, state B is split into two states, B1 and B2. After photolysis the myoglobin is in state B1. Above 160 K the molecule may structurally relax into state B2 before the CO rebinds. The rebinding rate k_{B2A} is smaller than k_{B1A}.

from the fact that the energy of the molecule with an unbound CO in the heme pocket decreases with respect to MbCO (state A in Figure 13.2)? Can one really exclude that the relative energies of MbCO and Mb*CO remain constant, while the barrier height I between these two states increases (Goldanskii and Parak 1994)? Even below 160 K the situation may be more complex. It was shown that the rebinding kinetics can be explained by assuming in part a temperature dependence of the barrier height distribution even at low temperatures (Ober et al. 1997). This type of data analysis allowed a scaling of the kinetic data between 40 K and 180 K.

Averaging discussed above requires intramolecular motions, which have been studied in many experiments. Mean-square displacements of atoms in the protein have been obtained from X-ray structure analysis, incoherent neutron scattering, and Mössbauer spectroscopy (cf. Parak and Frauenfelder 1993 and Figure 13.3). Below about 200 K most intramolecular motions coupled to the iron can be described by a normal mode analysis (Melchers et al. 1996). Above about 200 K a new type of motion reveals itself in Mössbauer spectroscopy and neutron scattering, although these methods explore very different time windows. This motion can be described as a diffusion between different conformational substates. Just when these protein-specific motions are present, the CO molecule can leave the heme pocket. Here again a very different time scale is involved.

Protein complexity and the hierarchical organization also reveals itself at very low temperatures. Zooming into the energy valley, for instance, of state A in Figure 13.1 shows a rough energy landscape, which has mainly been explored by hole-burning techniques (cf. Fritsch et al. 1996), specific heat measurements, and microwave absorption techniques. The similarity with glass is striking. Nevertheless, it has been shown that the "two-level tunneling system" (TLS) is not good enough to explain the data.

All in all, the simple functional behavior described by Brunori et al. (1972) comes very close to the unstructured effective barrier height that results at physiological

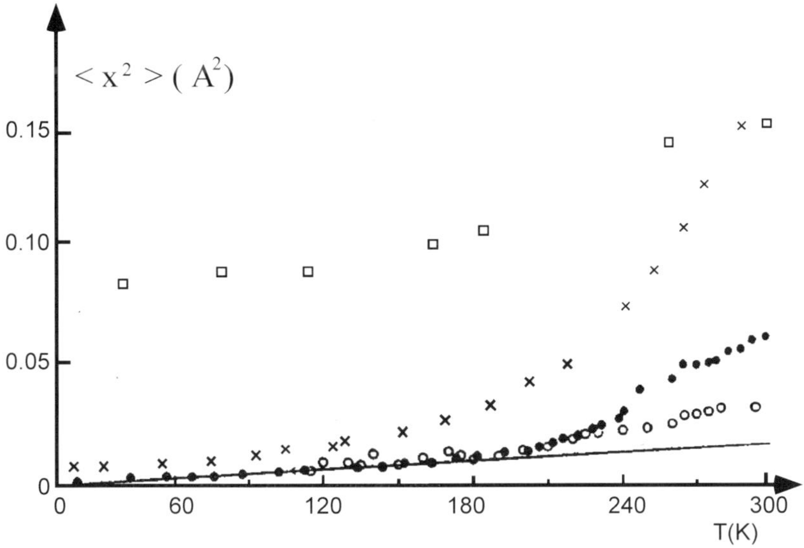

Figure 13.3 Mean-square displacements, $<x^2>$, in myoglobin as function of temperature. Circles: $<x^2>$ of the heme iron as determined by Mössbauer spectroscopy in deoxymyoglobin crystals (full circles) and freeze-dried Mb (open circles); the solid line gives the iron displacements calculated by normal mode analysis; crosses: displacements obtained by incoherent neutron scattering; squares: $<x^2>$ values averaged over all nonhydrogen atoms of myoglobin obtained from X-ray structure analysis.

temperatures from thermodynamical averaging over fast motions among multiplicities of substates. Simplicity is thus recovered by "zooming out," yet we now have a better understanding of the microscopic picture that underlies thermodynamics. Moreover, thermodynamical averaging is often fulfilled only in part, which makes a detailed understanding of the energy landscape inevitably for an understanding of functional properties.

A large number of open questions remains: Are the harmonic motions and the quasi-diffusive motions independent, or do they couple to each other? Do very low-frequency modes exist that have not yet been characterized? How good is the concept of temperature-independent and hierarchically arranged barrier heights, and on what time scale is this concept no longer able to explain functions such as ligand rebinding?

The complexity of protein molecules often appears under unphysiological conditions, perhaps because specific experimental conditions have no functional significance. This is, however, very unlikely. There seems to be a border temperature around 200 K that separates more or less functionally dead molecules from really working molecules. Nevertheless, there is no sharp transition, only a smooth connection between different hierarchies of motions. Just as anatomists used cadavers to gain knowledge in medicine, so does it seem appropriate to analyze Mb below 200 K.

A more complicated case is that of Hb, as discussed in depth by Eaton (this volume). Here it is enough to recall that Hb contains four subunits, each of which is very similar to Mb and can bind an oxygen molecule per subunit. The uptake and release of oxygen is optimized by interactions among the subunits. This is related to a conformational transition between the so-called T- and R-states, where up to half or more than half of the subunits are occupied by oxygen, respectively. As determined experimentally, the free energy of the R-state exceeds that of the T-state by about 35.5kJ/mol and it is balanced by the oxygen binding. This is sufficient to describe the function of Hb. Complexity arises when one tries to understand the physical reason for the T–R free-energy difference. Chothia et al. (1976) have shown that in the R-form the hydrophobic surface exposed to the solvent increases by about 1000 A^2. In the usual understanding of hydrophobic interactions, this implies a free-energy increase. Detailed experiments have shown that the increase is due to about 75 water molecules statistically transferred from bulk water to hydrophobic hydration. The related free-energy contribution was determined to be about 60kJ mol^{-1} (Bulone et al. 1993; Palma et al. 1994). Clearly, this contribution absolutely dominates the function due to interaction with the solvent. The intramolecular contribution is, by contrast, –24.5kJ mol^{-1}, so that it even has the wrong sign. This suggests zooming-in the complexity of water for a deeper scrutiny of biomolecule–solvent interactions.

COMPLEXITY OF PROTEIN AND SOLVENT INTERACTION

So far, we have only considered the energy landscape relative to the internal degrees of freedom of a protein. Now we include the solvent in our view. Consider the very complex potential energy landscape of a system of N water molecules in the state of bulk liquid. With reference to this landscape, the total free energy of the liquid, G_w, can be given in terms of depth, Φ_m, and logarithmic multiplicity, σ (Φ_m), of the subset of basins occupied with overwhelming probability, and of their related vibrational contributions f (T, Φ_m), that is: $G_w = N[\Phi_m - k_B T \, \sigma(\Phi_m) + f(T, (\Phi_m)]$. This expression (Stillinger 1988) is the result of thermodynamic averaging over the water energy landscape. Interaction with solutes alters the originally unperturbed landscape and its populated basins, causing a free-energy change ΔG_{SW}, which is the hydration free energy. The thermodynamic drive towards more favorable ΔG_{SW} is expressed by solvent-induced forces (SIFs) defined as SIF = $-\text{grad}\, \Delta G_{SW}$. The differential nature of SIFs makes them particularly suited (at variance with free energies) to studying important space-resolved, i.e., local details of solute–water interactions. The relevance of SIFs to biomolecular conformation and interactions is appreciated by noting that the work of a 70 pN force across 1 Å is equivalent to about 4 kJ mol^{-1}, and that typical SIF values are in the 20–500 pN range. Consequently, the effects due to perturbations caused by solutes to the energy landscape of the solvent cannot be ignored.

A simple and significant example is shown in Figure 13.4. The solute is represented by six fixed Lennard–Jones (LJ) spheres of the same size as the water oxygen. Sphere 4

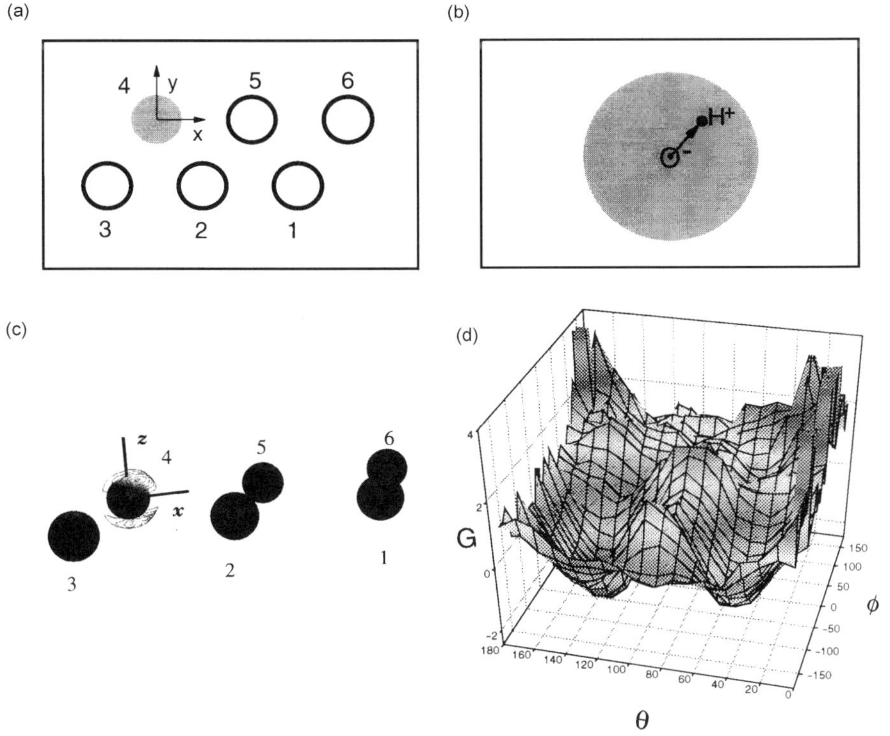

Figure 13.4 Rugged free-energy landscape in the two-dimensional configurational space of a single-model solute resulting from interaction with the solvent. The planar model solute shown in (a) is made of six Lennard–Jones spheres, with the same parameters of TIP4P water. Element 4 bears a freely orientating electric dipole with the negative charge in the center as shown in (b), so as to model an OH group, as in the TIP4P model. *In vacuo* the energy landscape seen by the OH group is obviously flat. This is no longer the case in a bath of explicit water molecules. Contour maps of the probability distribution of dipole orientation are shown in (c). The corresponding free-energy landscape is shown in (d). This is generated by interaction of the OH group with water. The *x, y* plane is that of the planar composite solute. The related angular coordinates of the dipole orientation are Θ and Φ. (Molecular dynamic simulations with explicit water by San Biagio et al. [1999a,b].)

bears in its center an orientationally free dipole moment. Its orientation is defined by the Θ and Φ polar angles. These are the only two degrees of freedom that the composite solute possesses. *In vacuo*, the energy is obviously independent of Φ and Θ. Consequently, the energy landscape of our model system is just flat. Now we add the bath of water molecules (explicit water). As discussed above, the interaction of our composite solute with water, averaged over water configurations, results in a free energy that is now strongly dependent upon dipole orientation. Thus, our composite solute will now experience, with respect to its two degrees of freedom Φ and Θ, a landscape that is no longer flat and represents free energy. Dipole orientations are no longer equally probable and show two well-defined probability maxima corresponding to two free-energy minima in Figure 13.4. Away from one of these minima, the orientation-restoring SIF on H^+ reaches a peak value of about 250 pN (San Biagio et al. 1999a,b). This is a remarkable value: the work of this force across 1 Å would amount to 15 kJ mol^{-1}. We see that, even in such a simple case, interaction with solvent water can turn a flat intrasolute potential energy landscape into a complex free-energy landscape, containing features of high potential relevance to protein folding and function. It should be remembered that the rugged surface shown in the Figure 13.4 is dictated by the totality of (local and nonlocal) constraints imposed to the solvent by the entire composite solute in the given specific configuration, i.e., by the conformational context. These specific constraints dictate which solvent configurations are energetically favored and therefore statistically relevant and responsible for ΔG_{SW} and SIFs. In turn, these two quantities possess a marked complexity of their own, due to their inherently strong nonadditivity and related strong many-body character of ΔG_{SW} (Brugé et al. 1996; Martorana et al. 1997; Bulone et al. 1997; San Biagio et al. 1998). A subtle interplay of complexity–simplicity emerges, therefore, even in the case of a very simple model solute in water.

Another aspect of the strong nonadditivity and context dependence of SIFs is shown in Figure 13.5. In (a) we have a model system similar to that of Figure 13.4. However, we now only have apolar LJ spheres, and element 5 is missing. Averaging over all of the configurations imposed on water by this new solute context causes a SIF that drives element 6 towards the solvent, despite its hydrophobic character. Note the absence of simple symmetry relations between the direction of the SIF and the solute configuration. A similar effect (shown in Figure 13.5b) occurs on the hydrophobic Pro2 residue of bovine pancreatic trypsin inhibitor (Martorana et al. 1998). This very sensitive dependence upon the conformational context is not accidental, and it endows SIFs with an unsuspected character of exquisite specificity, obviously relevant to biomolecular folding, recognition, and function.

Further averaging on coarser scales can also be useful, even if the predictive power cannot be expected to cover exquisitely specific and possibly very important details. It should be noted that even minute inaccuracies in predicting protein folding or recognition can miss all-important functional details. Averaging can be done in terms of a single solute–water interaction parameter X (Flory 1953), which describes in a mean field approximation microscopic collapse/unfolding and mesoscopic demixing/remixing. The two classes of phenomena are described in terms of the same interaction parameter and have in common a thermodynamic drive towards more favorable interactions within the (solute + solvent) system. Therefore, a link between them is expected. Accordingly, experiments on proteins, polypeptides, and a polysaccharide show that mesoscopic demixing promotes, as a consequence of local concentration enhancements, important conformational changes. Reciprocally, even slight conformational

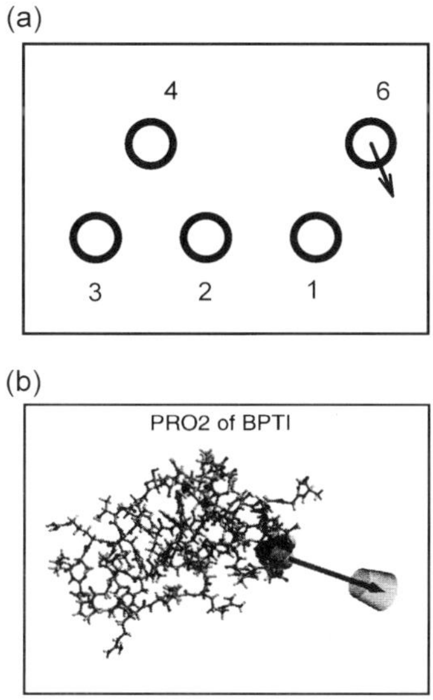

Figure 13.5 Examples of the many-body character of ΔG_{SW} and SIFs, from molecular dynamic simulations in explicit water. (a) Here the model is similar to that of Figure 13.4; however, element 5 is now missing and element 4 does not bear a dipole. This new context of solute elements interacts with the solvent and causes a solvent-induced force to drive element 6 towards the solvent, despite its hydrophobic character. Solvent-induced forces on all other solute elements (not shown) concur instead towards their hydrophobic collapse (Martorana et al. 1997; San Biagio et al. 1998). (b) SIFs acting on the hydrophobic Pro2 residue of bovine pancreatic trypsin inhibitor (BPTI). As in (a), spatial context and nonadditivity are capable of reversing the classical hydrophobic force. Data from Martorana et al. (1998).

changes cause remarkable shifts in the solution instability region as a result of related changes of protein–solvent interactions. These experiments also show that demixing promotes coagulation/gelation in high concentration regions. A general feature emerges, whereby demixing mediates between processes of conformational change and coagulation. The picture can be made more complex by kinetic competitions (Manno and Palma 1997; Manno et al. 1999; San Biagio et al. 1999a,b). Examples are given in Figure 13.6. Two of them are related to pathological coagulation of proteins, the other to chemio- and thermodynamic energy transduction.

CONCLUSIONS

Sunt contraria complementa? Thermodynamic averaging helps understanding of phenomenologically simple behaviors. Averaging, however, requires a clear picture of the complex microscopic states of protein and water, their mutual perturbations, and relevant time scales. It corresponds to a marked expansion of the dimensionality of the functional region of the conformational space and related entropic stabilization of

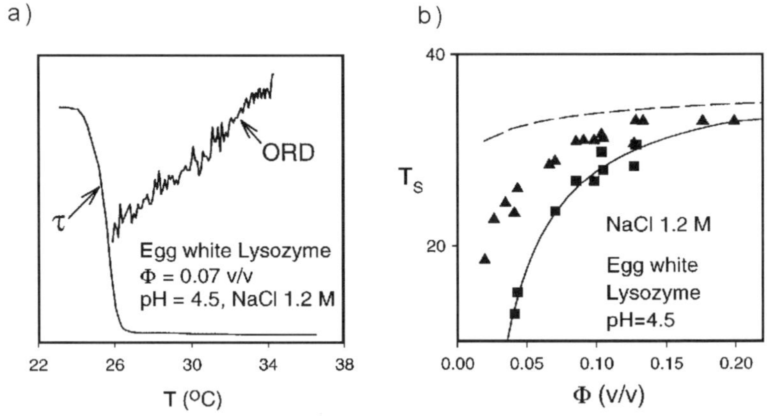

Figure 13.6 Interdependent (microscopic) conformational changes and (mesoscopic) coagulation of proteins. First Part: hen egg lysozyme. This small protein shows a complex behavior. Two mutants of its human variant cause systemic amyloidosis (Booth et al. 1997). Lysozyme is known to coagulate on heating. However, its coagulation from the native fold is also observed upon cooling, concurrently with some amount of conformational change. (a) Coagulation is monitored by the rapid rise of turbidity, τ. Concurrently, some conformational change is detected by the optical rotation signal (ORD) and by a change of the hydrodynamic radius, revealed by dynamic light scattering (not shown). (b) Spinodal and binodal lines of the same system. Experimental spinodal points (squares) are determined from mean-field extrapolation of critically divergent fluctuations. The continuous line represents the corresponding Flory–Huggins best fit. The binodal line obtained by using the same best-fitting parameters is also shown as a broken line. Discrepancy with glow-point data (triangles) is evident. The theoretical binodal can be reconciled with experimental glow-point data only by changing appreciably the P-value measuring the protein flexibility, in agreement with the detected conformational changes. This agrees with the expected interplay of conformational changes and demixing (San Biagio et al. 1999a,b; Manno et al. 1999). *(Continued on next page)*

functional conformations. Interaction of a relatively simple system, such as in Figure 13.4, with a complex one, such as water, can render the simple system unexpectedly complex, despite averaging simplifications. Averaging over increasingly longer time and length scales provides increasingly simple, but incomplete views. The two complementary views converge, therefore, in providing deeper understanding and additional predictive power, as shown in the examples we have discussed, concerning taxonomic states of MbCO, the role of water in the function of Hb, and the interweaving of conformational changes, demixing, and aggregation.

Viewing the complexity of protein solvent and protein–protein interaction on different scales allows the role of SIFs and ΔG_{SW} to be revealed in specificity and in the configurational landscape of the appropriate energy. The picture in terms of hydrophobic and hydrophilic interactions is oversimplified and can lead to wrong conclusions.

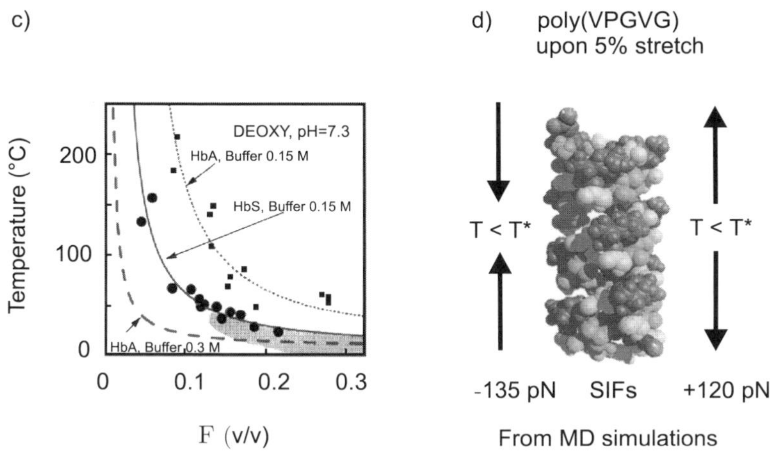

Figure 13.6 *(continued)* (c) Two mutated residues out of 574 cause the gelation of deoxy human Hb at physiological temperatures (Eaton and Hofrichter 1990). This mesoscopic process is triggered (San Biagio and Palma 1991) by the shift of the instability (spinodal) region of Hb in its T conformation caused by the mutation. (d) Poly(VPGVG). Its solutions in water undergo simultaneous microscopic and mesoscopic changes across the spinodal line (Sciortino et al. 1993; Urry 1993). The coagulated material (crosslinked by gamma irradiation) can be cycled repeatedly across the transition, by either temperature or chemical changes. This causes shrinkage/relaxation cycles. Related ponderomotive effects perform thermal and chemiomechanical energy conversions (Urry 1993). Molecular dynamic simulations show that shrinking/relaxation are caused by a change of sign of SIFs across the transition, as indicated by arrows in the Figure. Accordingly, the energy conversion is accomplished by the work performed by SIFs on the microscopic scale of single residues (ongoing work by Martorana et al.).

Given the size of SIFs and ΔG_{SW}, nonaddivity and multiple path feedback between micro and macroscopic processes can have a crucial role in pathways of funneling, folding, recognition, and coagulation.

Can zooming be continued forever? Zero point vibrations and, probably, zero point entropy set a limit. Beyond this border, the picture of a protein molecule disappears. Why did nature develop such complex systems like proteins? Could smaller molecules also do the job? Structural fluctuations between conformational substates may be the lubricant allowing conformational relaxation. Complex systems offer "possibilities" crucial to evolution. Borstnik and Hofacker (1985) have shown that the protein design covers function as well as evolutionary flexibility. Molecules in conformational substates at the wing of a Gaussian distribution may better fulfill their function after a change of the environment. A point mutation may easily make these substates the center of a new Gaussian distribution. However, even without this mutation, the molecule is still active.

REFERENCES

Austin, R.H., K.W. Beeson, L. Eisenstein, H. Frauenfelder, and I.C. Gunsalus. 1975. Dynamics of ligand binding to myoglobin. *Biochemistry* **4**:5355–5372.

Booth, D.R., M. Sunde, V. Bellotti, C.V. Robinson, W.L. Hutchinson, P.E. Fraser, C.M. Dobson, S.E. Radford, C.C.F. Blake, and M.B. Pepys. 1997. Instability, unfolding, and aggregation of human lysozyme variants underlying amyloid fibrillogenesis. *Nature* **385**:787–793.

Borstnik, B., and G.L. Hofacker. 1985. Functional aspects of the neutral patterns in protein evolution. In: Structure and Motion: Membranes, Nucleic Acids, and Proteins, ed. E. Clementi, M. Corongiu, M.H. Sarma, and R.S. Sarma, pp. 277–92. Guilderland, NY: Adenine Press.

Brugé, F., S.L. Fornili, G.G. Malenkov, M.B. Palma-Vittorelli, and M.U. Palma. 1996. Solvent-induced forces on a molecular scale: Non-additivity, modulation, and causal relation to hydration. *Chem. Phys. Lett.* **254**:292–301.

Brunori, M., J. Bonaventura, C. Bonaventura, E. Antonini, and J. Wyman. 1972. Carbon monoxide binding by hemoglobin and myoglobin under photodissociating conditions. *Proc. Natl. Acad. Sci. USA* **69**:868–871.

Bulone, D., V. Martorana, P.L. San Biagio, and M.B. Palma-Vittorelli. 1997. Effects of the electric charges on hydrophobic forces. *Phys. Rev. E-Rapid Comm.* **56**:4939–4942.

Bulone, D., P.L. San Biagio, M.B. Palma-Vittorelli, and M.U. Palma. 1993. On the role of water in hemoglobin function and stability. *Science* **259**:1335–1336.

Chothia, C., S. Wodak, and J. Janin. 1976. Role of subunit interfaces in the allosteric mechanism of hemoglobin. *Proc. Natl. Acad. Sci. USA* **73**:3793–3797.

Eaton, W.A., and J. Hofrichter. 1990. Sickle cell hemoglobin polymerization. *Adv. Prot. Chem.* **40**:63–279.

Flory, P.J. 1953. Principles in Polymer Chemistry. Ithaca, NY: Cornell Univ. Press.

Fritsch, K., J. Friedrich, F. Parak, and J.L. Skinner. 1996. Spectral diffusion and energy landscape of a protein. *Proc. Natl. Acad. Sci. USA* **93**:15141–15145.

Goldanskii, V.I., and F. Parak. 1994. Some still open questions of relaxational CO-heme rebinding. *Chem. Phys. Lett.* **219**:379–383.

Manno, M., A. Emanuele, V. Martorana, D. Bulone, P.L. San Biagio, M.B. Palma-Vittorelli, and M.U. Palma. 1999. Multiple interactions between molecular and supramolecular ordering. *Phys. Rev. E.* **59**:2222–2230.

Manno, M., and M.U. Palma. 1997. Fractal morphogenesis and interacting processes in gelation. *Phys. Rev. Lett.* **79**:4286–4289.

Martorana, V., D. Bulone, P.L. San Biagio, M.B. Palma-Vittorelli, and M.U. Palma. 1997. Collective properties of hydration: Long-range and specifity of hydrophobic interactions. *Biophys. J.* **73**:31–37.

Martorana, V., G. Corongiu, and M.U. Palma. 1998. Interaction of explicit solvent with hydrophobic/philic-charged residues of a protein: Residue character vs. conformation context proteins. *Proteins* **32**:129–135.

Melchers, B., E.W. Knapp, F. Parak, L. Cordone, A. Cupane, and M. Leone. 1996. Structural fluctuations of myoglobin from normal-modes, Mössbauer, Raman, and absorption spectroscopy. *Biophys. J.* **70**:2092–2099.

Nienhaus, G.U., J.R. Mourant, and H. Frauenfelder. 1992. Spectroscopic evidence for conformational relaxation in myoglobin. *Proc. Natl. Acad. Sci. USA,* **89**:2902–2906.

Ober, C., M. Burkhardt, H. Winkler, A.X. Trautwein, A.A. Zharikov, S.F. Fischer, and F.G. Parak. 1997. Low temperature study of myoglobin-ligand rebinding kinetics with Mössbauer spectroscopy. *Eur. Biophys. J.* **26**:227–237.

Palma, M.U., D. Bulone, P.L. San Biagio, and M.B. Palma-Vittorelli. 1994. Biological relevance of solvent-induced forces. In: Hydrogen Bond Networks, ed. M.C. Bellissent-Funel and J.C. Dore. Dordrecht: Kluwer Academic.

Parak, F., and H. Frauenfelder. 1993. Protein dynamics. *Physica A* **201**:332–345.

San Biagio, P.L., D. Bulone, V. Martorana, M.B. Palma-Vittorelli, and M.U. Palma. 1998. Physics and biophysics of solvent-induced forces: Hydrophobic interactions and context-dependent hydration. *Eur. Biophys. J.* **27**:183-196.

San Biagio, P.L., V. Martorana, D. Bulone, M.B. Palma-Vittorelli, and M.U. Palma. 1999a. Solvent-induced free energy landscape and solute–solvent dynamic coupling in a multielement solute. *Biophys. J.*, in press.

San Biagio, P.L., V. Martorana, A. Emanuele, S.M. Vaiana, M. Manno, D. Bulone, M.B. Palma-Vittorelli, and M.U. Palma. 1999b. Interacting processes in protein coagulation. *Proteins* **37**:116–120

San Biagio, P.L., and M.U. Palma. 1991. Spinoidal lines and Flory-Huggins free-energies for solutions of human hemoglobins HbS and HbA. *Biophys. J.* **60**:508–512.

Sciortino, F., K.U. Prasad, D.W. Urry, and M.U. Palma. 1993. Self-assembly of bioelastomeric structures from solutions: Mean-field critical behavior and Flory-Huggins free-energy of interaction. *Biopolymers* **33**:743–752.

Stillinger, F.H. 1988. Supercooled liquids, glass transition, and the Kauzman paradox. *J. Chem. Phys.* **88**:7818–7825.

Urry, D.W. 1993. Molecular machines: How motion and other functions of living organisms can result from reversible chemical changes. *Angew. Chem. Intl. Ed.* **32**:819–841.

Wittgenstein, L., Notes on Logic (Notebooks 1914–16). English edition by Blackwell, Oxford 1979. For a thorough discussion of Russell's and Wittgenstein's views on this important subject, see: D. Pears. 1977. The relation between Wittgenstein's Picture Theory of Propositions and Russell's Theory of Judgement. *Philos. Rev.* **86**:177–196.

14

From Simplicity to Complexity and Back*

Function, Architecture, and Mechanism of Light-harvesting Systems in Photosynthetic Bacteria

K. SCHULTEN

Beckman Institute and Department of Physics, University of Illinois at Urbana-Champaign, Urbana, IL 61801, U.S.A.

Keywords: pigment organization, molecular aggregates, excitons, energy transfer

*"... quod verum simplex sincerumque sit, id esse naturae hominis aptissimum."
—Cicero (De Officiis, 1,13)
[... what is true, simple, and pure is most fitted to the nature of man.]

ABSTRACT

Photosynthetic organisms fuel their metabolism with light energy and have developed for this purpose an efficient apparatus for harvesting sunlight, key features of which were conceptually established long ago. Recently, the atomic structure of a main protein constituent of the apparatus, as it evolved in purple bacteria, has been solved through a combination of modeling, X-ray crystallography, and electron microscopy. This permitted the modeling of the entire light-harvesting system, a complex nanometric aggregate of transmembrane proteins. This chapter discusses how an ongoing analysis has wrested an explanation of the light-harvesting function from an atomic level structure, based on quantum physics. The investigations of the light-harvesting system of purple bacteria demonstrate very clearly the course typical for research in biological physics, starting from a simply stated, known function and proceeding through experimental and theoretical investigations carried out at more and more refined levels of molecular reality: First, the macromolecular components of the system underlying a function are identified and their role characterized, e.g., through spectroscopy. Second, the complex

Simplicity and Complexity in Proteins and Nucleic Acids
Edited by H. Frauenfelder, J. Deisenhofer, and P.G. Wolynes

structures of these components are established at atomic resolution and functionally relevant architectural elements are recognized. Finally, through refined observation and theoretical analyses of these elements, the physical mechanisms exploited by the organism to achieve the function are determined.

INTRODUCTION

Living organisms are characterized through a remarkable set of facilities, for example, heredity and evolution, maintenance of a state of order through metabolic energy consumption, or information processing through chemical and electric signaling. These and other facilities were discovered initially on the basis of macroscopic observation or physiological experiments, and often an impressive degree of understanding was achieved at this level. Advances in crystallography, X-ray scattering, NMR, and electron microscopy in combination with arduous biochemical and physicochemical investigations have pushed the boundary of knowledge on the cell's machinery involved in the above-mentioned facilities from the level of organisms and cells to molecules, and ultimately to atomic level structures. However, such structures, welcomed and often heralded as great breakthroughs, are in themselves extremely complex and do not result promptly in deeper understanding. The step from often clearly and succinctly characterized, i.e., simple, cell functions to intrinsically complex protein structures needs to be succeeded by an attempt to regain clear and succinct descriptions of the mechanisms underlying the particular functions, which has often proven to be a long and difficult process.

Other chapters in this volume describe this process for the protein myoglobin, the structure of which was solved long ago. In this chapter, I investigate the journey from simplicity to complexity and back for a cellular function only recently characterized through atomic level structures: light harvesting in photosynthesis.

It is important to realize that this journey might pass across several levels of resolution of a biological organism. The least resolved level might be concerned with the establishment of cellular function. On the next level, the macromolecular components of the underlying system might be identified and their role characterized, e.g., through spectroscopy; still on a further level, the complex structures of these components might be established at atomic resolution and functionally relevant architectural elements might be recognized. On the final level, refined observation and theoretical analyses of these elements might determine the physical mechanisms exploited by an organism to achieve the cellular function.

The cellular apparatus involved in light harvesting will allow us to discuss how explanations of the mechanisms underlying the well-established function are wrested from complex protein structures and a conceptual foundation of cell biology grounded on the laws of physics is achieved. Since the proteins of the light-harvesting system furnish, in the first place, simply a scaffold for chromophores, the structure–function relationship is relatively more easy to state than in proteins undergoing dynamic processes and conformational transitions. However, the light-harvesting system studied involves a nanometric aggregate of proteins, the size of which establishes a new level of complexity that is characteristic of many recently discovered biomolecular structures. In this regard, this chapter complements the focus on smaller proteins presented elsewhere in this volume. The light-harvesting system is also a harbinger of progress in research on bioelectronic systems, founded necessarily on quantum mechanical concepts.

In this chapter I initially outline the light-harvesting function as it was characterized many decades ago. I then describe the characterization of the components of the light-harvesting system, the establishment of the structures of the components as well as the whole photosynthetic unit (PSU), and recognition of the functionally relevant architectural elements. Finally, I outline the quantum mechanical description of these elements and the present perception of the physical mechanisms exploited by purple bacteria. This chapter follows closely the presentation in Hu et al. (1998).

FUNCTION AND PHYSICAL PRINCIPLES OF LIGHT HARVESTING

Photosynthetic life forms reduce CO_2 to carbohydrates by utilizing the energy of sunlight. The prevalent color of green in the Earth's biosphere is testimony to the important role that chlorophylls play in this regard. Chlorophylls are assisted in their role by carotenoids, also widely known through their coloration of petals and fruits in plants. Emerson and Arnold (1932) demonstrated that it requires hundreds of chlorophylls to reduce one molecule of CO_2 under saturating flashlight intensity. To explain the cooperative action of these chlorophylls, these authors postulated that only very few chlorophylls in the primary reaction site, termed the photosynthetic reaction center (RC), take part directly in photochemical reactions; most chlorophylls serve as light-harvesting antennae, capturing the sunlight and funneling electronic excitation towards the RC. This notion gave rise to the definition of the PSU as an ensemble of an RC with associated light-harvesting complexes containing up to 250 chlorophylls. The PSU absorbs sunlight, fuels its energy in the form of electronic excitation into a system of chlorophylls, which acts as a funnel that conducts the excitation to the RC. The suggestion became widely accepted when Duysens carried out a critical experiment in which excitation transfer between different chlorophylls was observed (Duysens 1952).

The notion of 250 chlorophylls absorbing sunlight and funneling its energy in the form of electronic excitation through the chlorophyll ensemble to the RC caught the attention of the physicists R. Oppenheimer and T. Förster. Half a century ago they demonstrated independently, through application of quantum mechanical perturbation theory, that molecules like chlorophylls can transfer among each other electronic excitation by means of induced dipole–induced dipole interaction. Förster's respective formula for the rate of transfer from donor D to acceptor A is widely used in spectroscopy today in the form:

$$k_{D\to A} = \frac{1}{\tau_D}\left(\frac{R_0}{R}\right)^6 , \qquad (14.1)$$

where τ_D is the radiative lifetime of D^* and R_0 is the so-called Förster radius, the length scale over which excitation transfer is efficient. R_0 is defined through

$$R_0^6 = \frac{c\kappa^2\phi_D}{n^4}\int d\nu\, \nu^{-4} S_{D^*}(\nu) S_A(\nu) . \qquad (14.2)$$

Here, c is a system-independent constant, κ is a parameter dependent on the relative orientation of A and D, ϕ_D is the fluorescence quantum yield of D^*, n is the environmental optical density, and the spectral line shape functions $\widetilde{S}_{D^*}(\nu), S_A(\nu)$ account for the transitions $D^*\to D$, $A\to A^*$; $\widetilde{S}_{D^*}(\nu)$ is normalized as $\int \widetilde{S}_{D^*}(\nu)d\nu = 1$, whereas $S_A(\nu)$ represents an extinction coefficient.

Equation 14.1 is based on a multipole expansion, accounting for dipolar contributions, and actually applies only when the distance between the edges of A and D is much larger than the overall size of the molecules themselves. For such geometry, the Förster formula predicts that excitation transfer arises only when the transitions $D^* \to D, A \to A^*$ are both optically allowed. When the stated geometrical criteria are not met, the latter is not a precondition for excitation transfer and even optically forbidden transitions can be participants in the process. For a general description one cannot, therefore, rely on a multipole expansion, but rather one needs to account in a numerical calculation for the complete Coulomb coupling between donor and acceptor electronic states. Below I refer to the respective situation as being governed by the Coulomb mechanism.

In 1953, Dexter pointed out that a second mechanism exists for excitation transfer, which is based on electron exchange between donor D and acceptor A (Dexter 1953) and which has been described in detail for light harvesting in purple bacteria by Damjanovic et al. (1999). The Coulomb and exchange mechanisms for excitation transfer differ significantly in their operative range. While the Coulomb mechanism is effective over distances of typically 20–50 Å, the exchange mechanism is effective only when there is sufficient overlap of the wave functions of D^* and A, i.e., for distances of a few Å. The Coulomb mechanism, for optically allowed transitions, has an R^{-6} distance dependence as described by Equation 14.1, with a typical operational range R_0 of 30 Å. The exchange mechanism decays exponentially with distance; for it to be effective, donor and acceptor must be in direct contact.

As a consequence of selection rules, the Coulomb mechanism applies only to the transfer of singlet excitations (assuming singlet ground states of A and D), whereas the exchange mechanism is applicable also to the transfer of triplet excitations (Damjanovic et al. 1999). Using the Dexter mechanism, the carotenoids can protect the bacteriochlorophylls (BChls) and the entire bacterium against the detrimental effects of BChl triplet states which arise with a small but finite probability and can generate excited oxygen according to the reaction $^3O_2 + {}^3\mathrm{BChl}^* \to {}^1O_2^* + {}^1\mathrm{BChl}$. The carotenoids quench the BChl triplet states and, as a result, electronic excitation of O_2 is prevented and with it the damaging properties of this compound in its reactive singlet form.

The successful implication of the basic laws of physics in the suggested light-harvesting function by the mentioned researchers and others (e.g., Knox 1963) posed a clear challenge to another generation of scientists to discover the architecture of the PSU at a level of detail that would permit the application of quantum physics to deduce observed properties and understand its mechanism.

LIGHT HARVESTING IN PHOTOSYNTHESIS IN PURPLE BACTERIA

In the following, we jump to the discovery of the organization of the photosynthetic unit of one particular photosynthetic species: purple bacteria. Not mentioned are decades of research leading to the preparation of the system, to spectroscopic observations of excitation transfer down to the femtosecond time scale, and to a more complete theoretical understanding of molecular energy transfer. Figure 14.1 depicts schematically the intracytoplasmic membrane of purple bacteria with its primary photosynthetic apparatus. In the PSU, an array of light-harvesting complexes captures light and transfers the excitation energy to the photosynthetic RC. The latter is used to separate electron/hole pairs and to attract protons, reducing quinone, Q, to hydroquinone, QH_2. For this purpose, Q needs to shuttle into the RC and QH_2 out of the RC. The holes are replaced by electrons furnished by cytochrome c_2. The remaining part of the primary photosynthetic apparatus utilizes QH_2 to generate, by means of the bc_1 complex, a proton gradient which drives the synthesis of ATP.

The PSU of most purple bacteria contains two types of light-harvesting complexes: light-harvesting complex I (LH-I) and light-harvesting complex II (LH-II) (Zuber and Brunisholz 1991). LH-I is found directly surrounding the RCs (Miller 1982; Walz and Ghosh 1997), while LH-II is not in direct contact with the RC but transfers energy to the RC via LH-I (Monger and Parson 1977; van Grondelle et al. 1994). For some bacteria, such as *Rhodopseudomonas* (*Rps.*) *acidophila* and *Rhodospirillum* (*Rs.*) *molischianum* strain DSM 120 (Germeroth et al. 1993), there exists a third type of light-harvesting complex, LH-III. A 1:1 stoichiometry is found between RC and LH-I (Walz and Ghosh 1997); the number of LH-IIs and LH-IIIs varies according to growth conditions, such as light intensity and temperature (Aagaard and Sistrom 1972).

Figure 14.2 shows the energy levels for the key electronic excitations in the PSU of purple bacteria. The energy levels reveal two types of components of light harvesting: components involving carotenoids and BChls, which fuel photons into the PSU as electronic excitations of BChl, and components involving solely the BChls that funnel the latter excitations to the RC. The photon-fueling components absorb light mainly at wavelengths of about 500 nm through carotenoids and above 800 nm through BChls, transferring the resulting excitations to the BChls forming the excitation funnel. LH-I, LH-II, and LH-III contain carotenoids, but only LH-II and LH-III contain a set of

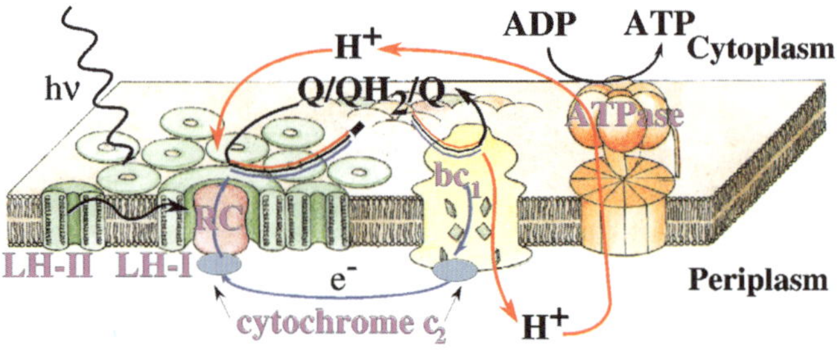

Figure 14.1 Schematic representation of the photosynthetic apparatus in the intracytoplasmic membrane of purple bacteria. The reaction center (RC, red) is surrounded by the light-harvesting complex I (LH-I, green) to form the LH-I–RC complex, which is surrounded by multiple light-harvesting complexes LH-II (green), forming altogether the photosynthetic unit (PSU). Photons are absorbed by the light-harvesting complexes and excitation is transferred to the RC initiating a charge (electron-hole) separation. The RC binds quinone Q_B, reduces it to hydroquinone Q_BH_2, and releases the latter. Q_BH_2 is oxidized by the bc_1 complex, which uses the exothermic reaction to pump protons across the membrane; electrons are shuttled back to the RC by the cytochrome c_2 complex (blue) from the ubiquinone-cytochrome bc_1 complex (yellow). The electron transfer across the membrane produces a large proton gradient that drives the synthesis of ATP from ADP by the ATPase (orange). Electron flow is represented in blue, proton flow in red, and quinone flow — likely confined to the intramembrane space — in black.

BChls absorbing at B800. The excitation funnel BChls are characterized through a pronounced energetic hierarchy: the respective BChls in LH-III at the periphery of the PSU absorb at 820 nm, the ones in the more central LH-II absorb at 850 nm, and the ones in LH-I in contact with the RC absorb at 875 nm (van Grondelle et al. 1994). This energy cascade, according to the principles of thermodynamcis, serves to funnel electronic excitations from the LH-IIIs and LH-IIs through LH-I to the RC. Time-resolved picosecond and femtosecond spectroscopy revealed that excitation transfer within the PSU occurs on a subpicosecond (in the photon-fueling components) to picosecond (in the excitation funnel components) time scale and altogether at near unit (95%) efficiency (Pullerits and Sundstrom 1996; Fleming and van Grondelle 1997).

The photon fueling/excitation funnel system in Figure 14.2 constitutes a detailed, yet conceptually simple, picture of light harvesting in purple bacteria. The picture combines knowledge of the constituents of the system with measured transfer pathways and rates (not shown) in a physically sensible scheme. One could consider this picture a satisfactory explanation for light harvesting, but it still leaves unanswered the questions: How exactly are the chlorophylls and carotenoids organized in the PSU, and how does the system achieve its high efficiency? These questions become more serious if one compares light-harvesting systems of different photosynthetic life forms which show little resemblance. Is there a common ancestor to the PSUs of all photosynthetic life, and what factors might have caused evolutionary divergence? Such answers require a deeper knowledge of the molecular machinery behind Figure 14.1.

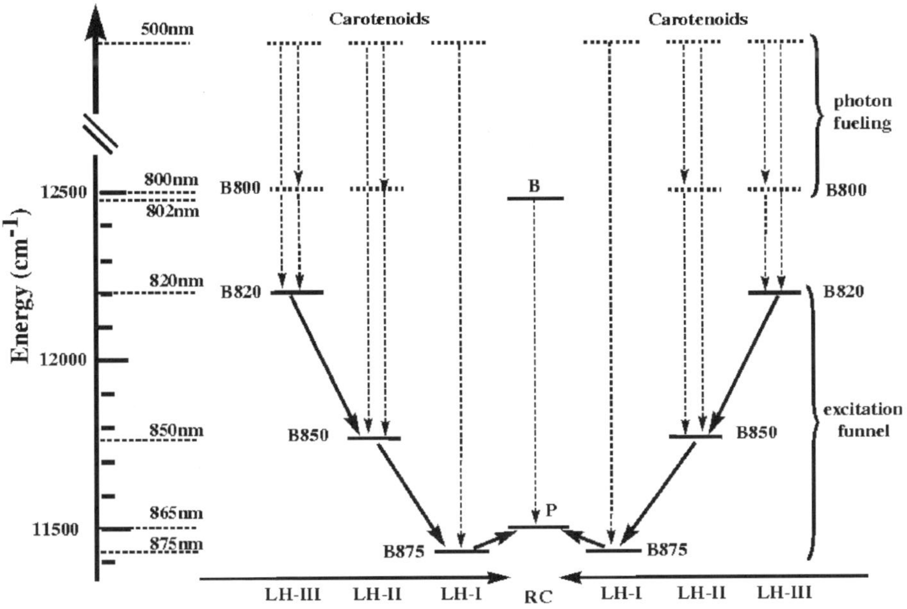

Figure 14.2 Energy levels of the electronic excitations in the PSU of BChl *a* containing purple bacteria. The diagram illustrates a funneling of excitation energy towards the photosynthetic reaction center (RC). The dashed lines indicate (vertical) intra-complex excitation transfer; the solid lines (diagonal) inter-complex excitation transfer. LH-I exists in all purple bacteria; LH-II exists in most species; LH-III arises in certain species only.

STRUCTURAL MODEL OF THE PSU OF *RHODOBACTER SPHAEROIDES*

The PSU is a nanometric aggregate of many components and as such it poses an extreme challenge for structure analysis methods available today. Figure 14.3 presents a structural model of the PSU for the purple bacterium *Rhodobacter* (*Rb.*) *sphaeroides*. Shown is a top view of an RC, surrounded by a (large) ring-like protein, the light-harvesting complex I (LH-I) with a 16-fold symmetry axis, which in turn is surrounded by (small) ring-like proteins, the light-harvesting complex II (LH-II) with a ninefold symmetry axis. Only three LH-IIs are shown. The actual photosynthetic apparatus can contain up to about nine LH-IIs around each LH-I, a number derived from observed stoichiometries as well as concluded from the assumption that nine LH-IIs fit around LH-I, leaving a sufficient opening for Q/QH_2 shuttling.

All components in Figure 14.3, except the RC, have been modeled from homologous proteins as explained below. Even though only a model of the PSU is presented here, I wish to stress that for the purposes of the central mechanism of light harvesting, the model is likely satisfactory and as good as a real structure.

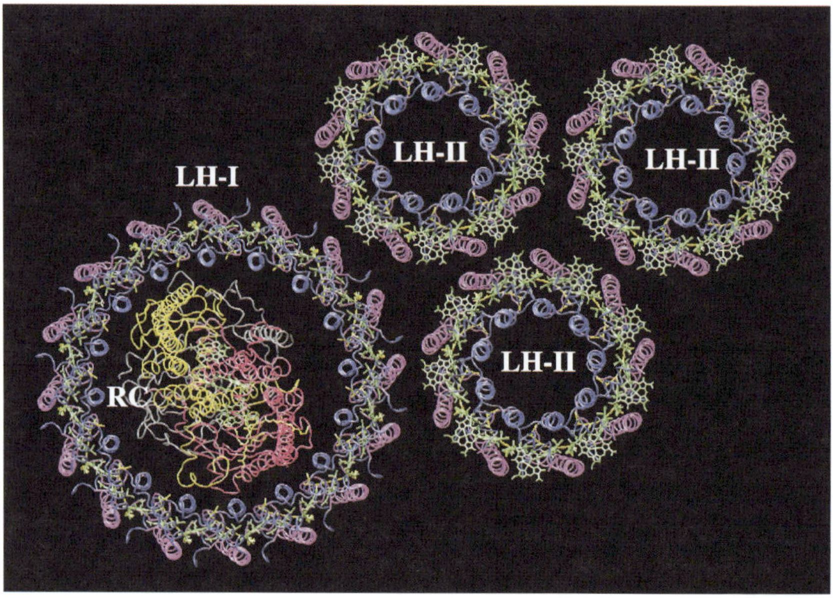

Figure 14.3 Arrangement of pigment protein complexes in the modeled bacterial photosynthetic unit (PSU) of *Rb. sphaeroides.* The α-helices are represented as C_α-tracing tubes with α-apoproteins of both LH-I and LH-II in blue and β-apoproteins in magenta, and the L, M, and H subunits of RC in yellow, red, and gray, respectively. All the BChls are in green, and carotenoids in yellow. (Produced with the program VMD [Humphrey et al. 1996]).

Molecular biologists critical of modeling are reminded that other disciplines employ models routinely and successfully (e.g., condensed matter physics) and have learned to judge the benefits as well as the shortcomings of models. Certainly, it is more desirable to base conclusions on a highly resolved structure of the PSU at below 1 Å resolution (the relevant length scale relevant for a faithful quantum mechanical description of light harvesting) and in the form of a movie with 100 fs time resolution (the physically relevant time scale); however, since this information is not available, and may never be due to the naturally disordered character of a cell membrane or a thermal protein, a computational model must suffice. Theoretical biology has invested much effort in extrapolating structure and dynamics from homologous and static structures by means of many types of model building algorithms as well as molecular dynamics; such tools, as well as bioinformatics methodologies, have been employed to derive the structure shown in Figure 14.3 (Hu and Schulten 1998; Hu et al. 1998).

In the PSU of *Rb. sphaeroides*, the structure of the photosynthetic RC is known (Ermler et al. 1994). In fact, the availability of this structure prompts us here to model the PSU of this species, rather than that of another species. Since we are presently concerned with the light-harvesting function, we will not discuss the structure of the RC further, except later in regard to its function as an acceptor of electronic excitation.

The structure of LH-II for *Rb. sphaeroides* is not known, but it has been determined for *Rps. acidophila* (McDermott et al. 1995) and for *Rs. molischianum* (Koepke et al. 1996). For the latter species the structure, determined to 2.4 Å resolution (Koepke et al. 1996), is shown in Figure 14.4a. The protein is an octameric aggregate of αβ-heterodimers; the latter contain a pair of short peptides (α- and β-apoproteins) noncovalently binding three BChl *a* molecules and one lycopene (a specific type of carotenoid). Presumably, a second lycopene exists for each αβ-heterodimer. The electron density map exhibits indeed a stretch of assignable density, but not long enough to resolve the entire lycopene (Koepke et al. 1996). Two concentric cylinders of α-helices, with the α-apoproteins inside and the β-apoproteins outside, form a scaffold for BChls and lycopenes. Figure 14.4b depicts the 24 BChl molecules and eight lycopene molecules in LH-II with all other components stripped away. Sixteen B850 BChl molecules form a continuous overlapping ring of 23 Å radius (based on central Mg atoms of BChls) with each BChl oriented perpendicular to the membrane plane. The Mg–Mg distance between neighboring B850a and B850b BChls is 9.2 Å (within an αβ-heterodimer, cf. Figure 14.4) and between B85Oa′ and B850b is 8.9 Å (between heterodimers, cf. Figure 14.4). Eight B800 BChls, forming another ring of 28 Å radius, are arranged with their tetrapyrrol rings nearly parallel to the membrane plane and exhibit a Mg–Mg distance of 22 Å between neighboring BChls, i.e., the BChls are coupled only weakly. The ligation sites for the B850 BChl are α-His-34 and β-His-35, while the B800 BChls ligate to α-Asp-6. Eight lycopene molecules are found to span the transmembrane region, each making close contact with a B800 BChl and a B850a BChl.

The structure of LH-II of *Rb. sphaeroides* has not yet been determined through crystallography. The α- and β-apoproteins of this protein exhibit a close homology to the subunits of LH-II from *Rps. acidophila*: α-apoprotein has 45% sequence identity with its *Rb. sphaeroides* counterpart, the transmembrane segment of the β-apoprotein has 65% sequence identity. Electron microscopy observations suggest that LH-II of *Rb. sphaeroides* contains nine αβ-heterodimers (Olsen, pers. comm.), instead of eight as in LH-II of *Rs. molischianum*. LH-II of *Rb. sphaeroides*, as shown in Figure 14.3, has been constructed, therefore, as a nanomer of αβ-heterodimers by means of homology modeling using the αβ-heterodimer of LH-II from *Rps. acidophila* as a template. For this purpose, the modeling protocol developed and applied successfully to LH-II of *Rs. molischianum* in Koepke et al. (1996), Hu et al. (1995a), and Hu and Schulten (1998) was utilized.

The structure of LH-I of *Rb. sphaeroides* and of its complex with RC are also not yet available. However, electron micrographs of the protein alone as well as in complex with RC (R. Ghosh, pers. comm.) exist. The protein exhibits a ring-like structure with a 16-fold symmetry axis, the degree of symmetry being apparently subject to sample preparation and possibly to data-averaging procedures. In particular, the complex with RC can show a deformation from a 16-fold symmetry, e.g., a 4-fold symmetry (R. Ghosh, pers. comm.). Actually, Jungas et al. (1999) provide electron micrographs for *Rb. sphaeroides* that reveal an aggregate of two half-circular LH-Is with two RCs. For

(a)

(b)

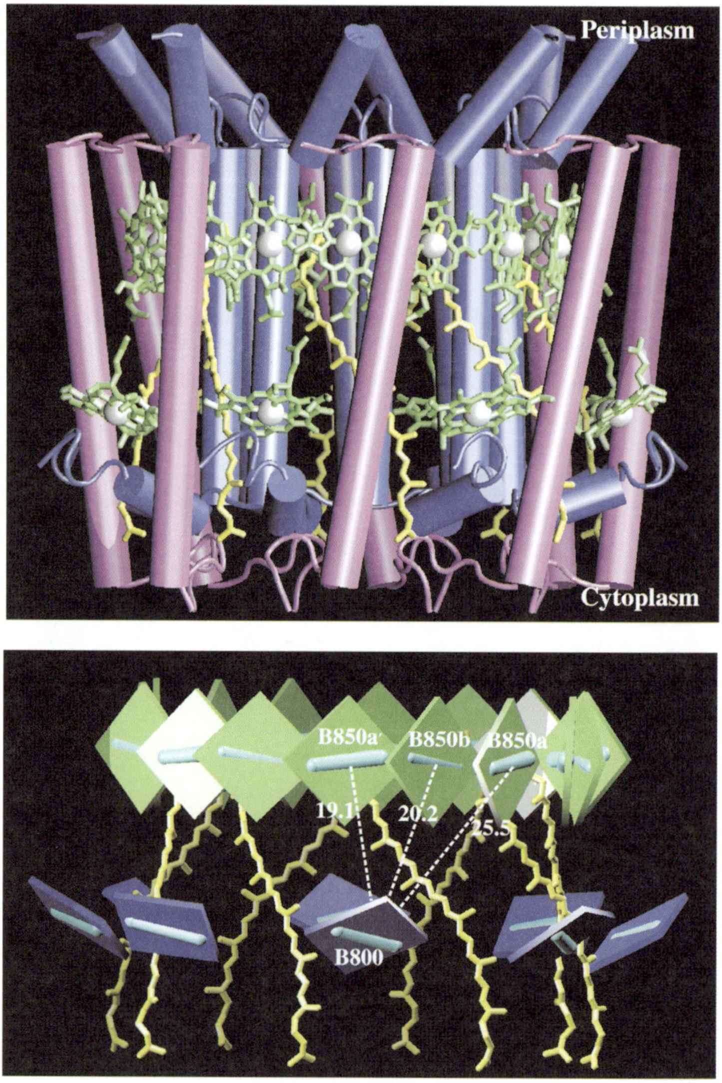

Figure 14.4 The octameric LH-II complex from *Rs. molischianum* (from Koepke et al. 1996). (a) The α-helical segments are represented as cylinders with the α-apoproteins (inside) in blue and the β-apoprotein (outside) in magenta. The BChl molecules are in green with phytyl tails truncated for clarity. The lycopenes are in yellow. (b) Arrangement of chromophores with BChls represented as squares, and with carotenoids (lycopenes) in a licorice representation. Bars connected with the BChls represent the Q_y transition dipole moments as defined by the vector connecting the N atom of pyrrol I and the N atom of pyrrol III (Gouterman 1961). Representative distances between central Mg atoms of B800 BChl and B850 BChl are given in Å. The B850 BChls bound to the α-apoprotein and the β-apoprotein are denoted as B850a and B850b, respectively; BChl B850a′ is bound to the (left) neighboring heterodimer.

the purpose of developing a description of the PSU of *Rb. sphaeroides*, we adopt the model of LH-I as derived in Hu and Schulten (1998), which involves a hexadecamer of αβ-heterodimers; the modeling exploited a close homology of these heterodimers to those of LH-II of *Rs. molischianum*. The resulting LH-I structure yields an electron density projection map that is in agreement with an 8.5 Å resolution electron microscopy projection map for the highly homologous LH-I of *Rhodospirillum rubrum* (Karrasch et al. 1995). The LH-I complex contains a ring of 32 BChls referred to as B875 BChls according to their main absorption band. The Mg–Mg distance between neighboring B875 BChls is 9.2 Å within the αβ-heterodimer and 9.3 Å between neighboring heterodimers.

The modeled LH-I has been docked to the photosynthetic RC of *Rb. sphaeroides* by means of a constrained conformational search (Hu and Schulten 1998), employing for the RC the structure reported in Ermler et al. (1994). Figure 14.5a, b presents the LH-I–RC complex. The arrangement of the BChls in the LH-I–RC complex is depicted in Figure 14.4b. One can discern the ring of B875 BChls of LH-I that surrounds the reaction center special pair (P_A and P_B) and the so-called accessory BChls (B_A, B_B). The closest distance between the central Mg atom of the RC's special pair (bacteriochlorophylls P_A and P_B) and the Mg atom of the BChls in LH-I is 42 Å. The distance between the Mg atom of the accessory BChl (bacteriochlorophylls B_A, B_B) and the LH-I BChls is shorter, the nearest distance measuring 35 Å. *Rb. sphaeroides* contains an additional *Puf*X gene. It has been suggested that the *Puf*X protein substitutes one or more αβ-heterodimers of LH-I to open up the circular ring shown in Figure 14.4a and to facilitate, thereby, the flow of quinones (Q_B/Q_BH_2) between the RC and the cytochrome bc_1 complex (see Figure 14.1) (Farchaus and Oesterhelt 1989; Cogdell et al. 1996; Walz and Ghosh 1997). A preliminary model of a *Puf*X dimer, replacing a single LH-I heterodimer, exhibits an opening suitable for the RC-bound quinone (cf. Figure 14.5a) to shuttle out of the LH-I–RC complex.

ARRANGEMENT OF CHROMOPHORES IN THE PSU

The PSU is a protein aggregate of overwhelming complexity. The task of connecting its structure to the photon fueling/excitation funnel function, as depicted in Figure 14.2, appears formidable. However, the light-harvesting proteins in the PSU serve mainly a structural role, as a scaffold for the chlorophyll and carotenoid chromophores. Discarding the peptide components of the PSU results in the arrangement of chromophores as shown in Figure 14.6. One can discern a hierarchical aggregate of the BChls, organized into rings of 18 closely coupled (B850) and nine loosely coupled (B800) BChls in the peripheral LH-IIs surrounding a large ring of 32 closely coupled (B875) BChls of LH-I which in turn surrounds four chlorophylls of RC. Close groupings, shown separately in Figure 14.7, of carotenoids with three LH-II BChls can also be recognized.

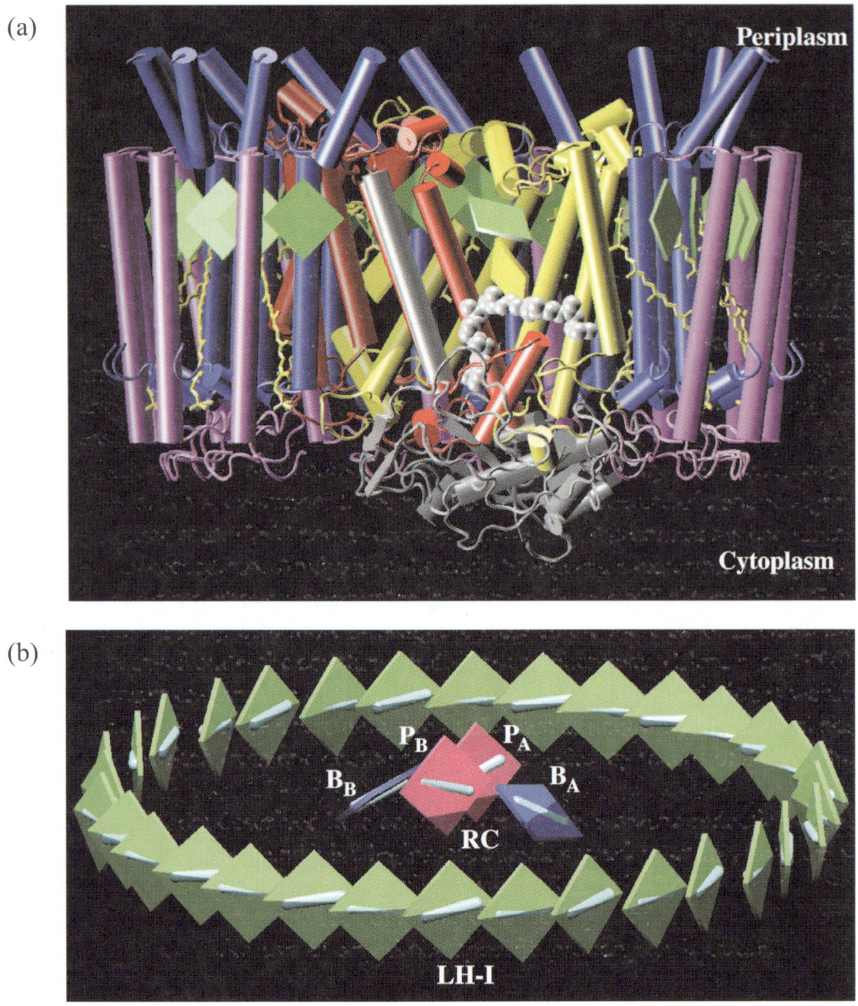

Figure 14.5 Structure of the LH-I–RC complex. (a) Side view of the LH-I–RC complex with three LH-I αβ-heterodimers on the front side removed to expose the RC in the interior. The α-helices are represented as cylinders with the L, M, and H subunits of the RC in yellow, red, and gray, and the α-apoprotein and the β-apoprotein of the LH-I in blue and magenta. BChls and bacteriopheophytins are represented as green and yellow squares, respectively. Carotenoids (spheroidenes) are in a yellow licorice representation, and quinone, Q_B, is rendered by gray van der Waals spheres. Q_B shuttles in and out (as Q_BH_2) of the LH-I–RC complex, as indicated in Figure 14.1. (b) Arrangement of BChls in the LH-I–RC complex. The BChls are represented as squares with B875 BChls of LH-I in green, and the special pair (P_A and P_B) and the accessory BChls (B_A and B_B) of the RC in red and blue, respectively; cyan bars represent Q_y transition moments of BChls. (Produced with the program VMD [Humphrey et al. 1996]).

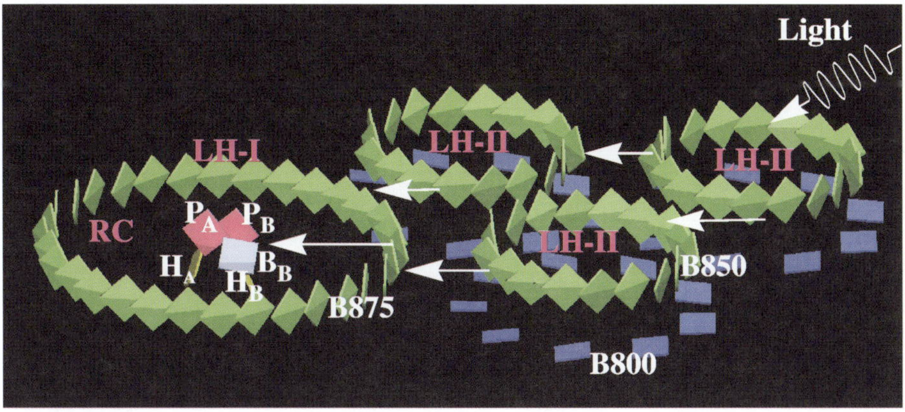

Figure 14.6 Excitation transfer in the bacterial photosynthetic unit. LH-II contains two types of BChls, commonly referred to as B800 (dark blue) and B850 (green), which absorb at 800 nm and 850 nm, respectively. BChls in LH-I absorb at 875 nm and are labeled B875 (green). P_A and P_B refer to the reaction center special pair, and B_A, B_B to the accessory BChls in the RC. The figure demonstrates the coplanar arrangement of the B850 BChl ring in LH-II, the B875 BChl ring of LH-I, and the reaction center BChls P_A, P_B, B_A, B_B. (Produced with the program VMD [Humphrey et al. 1996]).

An essential feature of the chromophore organization of the PSU is that the ring-like aggregates of BChls within LH-I and LH-II exhibit a coplanar arrangement together with the BChls in the RC. An analysis of the LH-I and LH-II structures (Hu and Schulten 1998; Hu et al. 1997) indicates that each BChl of the B850 ring of LH-II and of the B875 ring of LH-I is noncovalently bound to three side-chain atoms of the α- or β-apoprotein such that the BChls are held in a rigid orientation, underscoring the relevance of the arrangement shown in Figure 14.6.

The chromophore aggregate shown in Figure 14.6 is a structure of simplicity and beauty. One may want to conclude at this point that the structural model implies an obvious answer to the question posed by Figure 14.2, i.e., how photons are fueled into the BChl system and electronic excitation is funneled to the RC. It has long been observed that excitation transfer LH-II → LH-I → RC occurs in the PSU in less than 100 picoseconds and with about 95% efficiency (Pullerits and Sundstrom 1996). In this respect, it is interesting to note that the transition dipole moments of the Q_y excitations of the B850 and B875 BChls are all oriented in the two-dimensional plane that encompasses the ring-like BChl aggregates of LH-II, LH-I, and the RC special pair and is optimally attuned to the desired flow of electronic excitation LH-II → LH-I →RC. However, the intuitive appeal of the aggregate architecture in Figure 14.6 cannot be confused with a sound understanding, which requires that one applies the laws of physics to the chromophore aggregate shown and investigates the emergent properties as well as their relationship to the function of the PSU. The efficient transfer of electronic excitation from the periphery to the center is hardly a foregone conclusion, prompting the consideration of several issues.

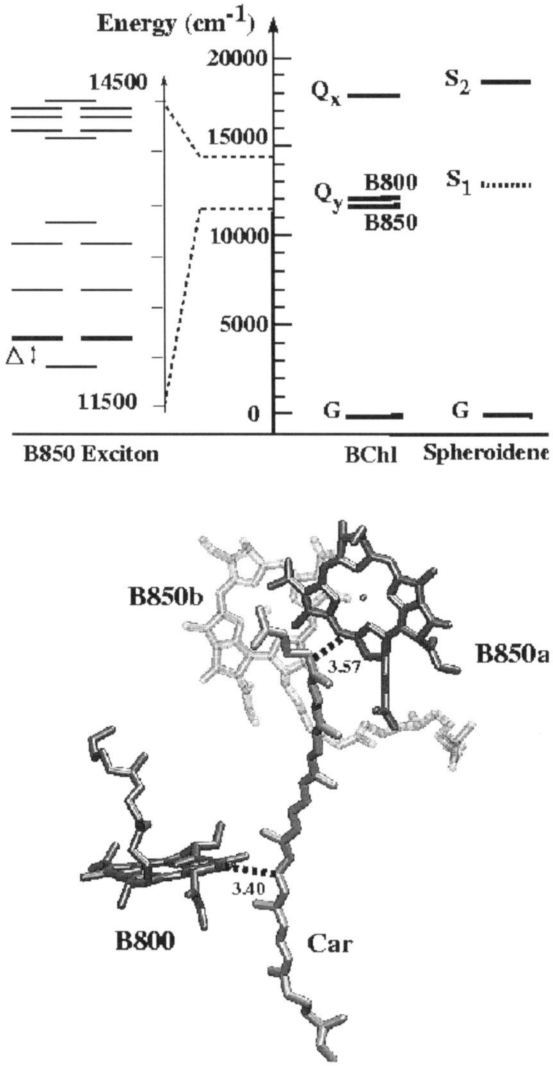

Figure 14.7 Bacteriochlorophyll–carotenoid interactions. Top left: Exciton bands of the circular B850 BChl aggregate as determined by quantum chemical (INDO/S) calculations (Zerner et al. 1998) based on coordinates of the crystal structure of LH-II from *Rs. molischianum* (Koepke et al. 1996). The degenerate states that carry all the oscillator strength are highlighted by thickened lines. Top right: excitation energies of BChl and carotenoid states in LH-II of *Rb. sphaeroides*. Solid lines represent spectroscopically measured energy levels. Dashed lines indicate the estimated (see Damjanovic et al. 1998; Nagae et al. 1993) energy for the optically forbidden S_1 state of the carotenoid spheroidene. Bottom: arrangement of spheroidene and the most proximate BChls based on the modeled structure of LH-II from *Rb. sphaeroides*. Close contacts between BChl and carotenoid spheroidene are indicated by representative distances (in Å).

First, one must concede that the system of 200–300 BChls and about 100 carotenoids is too large to be described *ab initio* on the basis of quantum physics. The system needs to be broken up into constituents, with proper justification that the necessary dissection does not essentially alter relevant properties. Fortunately, the aggregate shown exhibits a distinct hierarchy of spatial scales. BChls in the individual rings of LH-IIs, LH-I, and the RC are spatially closer and likely more tightly coupled than are BChls belonging to different units (rings, RC). One also notices that the so-called B800 BChls are themselves all spatially separated (center-to-center) by 20 Å and are likely well described as individual BChls. The carotenoids experience close contacts with certain BChls (see Figure 14.7) and the effect of the resulting interactions requires a detailed description. The dissection of the chromophore aggregate in the PSU conforms actually with quantum mechanical perturbation theory, which suggests how a hierarchy of interacting units can be composed from descriptions of its components. Perturbation theory is well tested and provides, in particular, criteria that determine its applicability.

Second, even the mentioned subcomponents are too large for a truly fundamental description. Rather, so-called semiempirical descriptions need to be invoked or so-called *ab initio* methods with a high degree of simplification, the choice of which implies also a strong empirical component. However, a satisfactory description should be one that has been previously and independently, i.e., in a separate context, applied successfully to homologous molecular systems. Since chlorophylls and carotenoids have been investigated intensely for decades by means of semiempirical methods, one treads here on fairly safe ground.

Third, one must note that the structure shown in Figure 14.6 is a static, stroboscopic view, and a highly idealized one in regard to the symmetry of the individual units. The actual system is subject to thermal motion which distorts the imposed symmetries. The LH-I ring surrounding the RC may actually be severely distorted in its shape by the RC or may be only a half-circle, as observed by Jungas et al. (1999). Furthermore, in some purple bacteria, one (or possibly more) of the sixteen units of LH-I is replaced by *Puf*X, which interrupts the ring of BChls to provide a gate through which Q_B/Q_BH_2 can be shuttled. The physical description achieved must be subjected to the mentioned distortions in order to discern which features of the description are robust against thermal motion and shape changes. The computational task involved in the suggested theoretical investigations is formidable, and the results necessarily complex. A simple finding is by no means certain.

QUANTUM MECHANICAL DESCRIPTION OF EXCITATION FUNNEL

One of the most intriguing structural features of the bacterial light-harvesting complexes is the circular organization of BChl aggregates (Hu and Schulten 1997). To understand the primary processes of light absorption and the subsequent excitation transfer from LH-IIs, through LH-I, to the RC, it is essential to characterize the electronic properties of the excited states of a circular BChl aggregate. For this purpose we have carried out INDO-CIS (Zerner et al. 1998) calculations of individual BChl, small groupings of BChls as they occur in LH-II of *Rs. molischianum*, and of an entire circular aggregate of sixteen B850 BChls. The INDO-CIS method proved earlier to be satisfactory for the description of electronic excitations of organic chromophores, in particular, chlorophylls (Thompson and Zerner 1991). The low energy part of the spectrum resulting for the ring of sixteen B850 BChls is presented in Figure 14.7. The energy levels shown are mainly due to linear combinations of BChl Q_y excitations, so-called excitons. The two bands of the spectrum arise from a dimerization of the BChls in LH-II, i.e., B850a′ and B850b BChls are closer than B850a and B850b BChls as mentioned above (cf. Figure 14.4b).

A complete analysis of the INDO-CIS results also reveals higher energy excitations corresponding to Q_x states of individual BChls as well as to so-called charge resonance states. However, these states should not play a role for the excitation funnel since they are out of resonance with other relevant electronic excitations. Nonetheless, these states may absorb light and fuel its energy into the funnel system.

The B850 BChl spectrum in Figure 14.7 arises mainly from interactions between the BChl Q_y excitations of individual BChls. One may ask how far a reduced description (in terms of just these excitations) would reproduce the INDO-CIS results in which thousands of other excitations have been identified. Such a reduced description would be formulated in terms of a 16×16 Hamiltonian matrix $\langle j|\hat{H}|k\rangle$, where $|j\rangle, j = 1,...16$ represents the Q_y excitations of individual BChls. The diagonal elements $\langle j|\hat{H}|j\rangle = \varepsilon$ account for the excitation energy of the Q_y state. For non-nearest neighbor BChls j and k, associated interactions $\langle j|\hat{H}|k\rangle$ should be well approximated by dipole–dipole coupling terms:

$$\langle j|\hat{H}|k\rangle = C\left(\frac{\vec{d}_j \cdot \vec{d}_k}{r_{jk}^3} - \frac{3(\vec{r}_{jk} \cdot \vec{d}_j)(\vec{r}_{jk} \cdot \vec{d}_k)}{r_{jk}^5}\right), \; k \neq j, \; j \pm 1, \tag{14.3}$$

where $\vec{d}_j$ are unit vectors describing the direction of the transition dipole moments of the ground state $\rightarrow Q_y$ state transition of the jth BChl and $\vec{r}_{jk}$ is the vector connecting the centers of BChl j and BChl k. C is a parameter, yet unspecified, related to the oscillator strength of the *ground state* $\rightarrow Q_y$ transition. The induced dipole–induced dipole coupling does not account well for interactions between neighboring BChls since the geometry criterion for the multipolar expansion is not met. Due to the eightfold symmetry of LH-II, the respective matrix elements $\langle j|\hat{H}|j+1\rangle$ assume only two different values, namely, v_1 (v_2) for odd (even) j. The Hamiltonian is then specified through four parameters: ε, v_1, v_2, and C. These parameters can be chosen such that the spectrum resulting for the matrix $\langle j|\hat{H}|k\rangle$ reproduces closely the spectrum of a full INDO-CIS calculation. The corresponding parameters are $\varepsilon = 13{,}242\ \text{cm}^{-1}$, $v_1 = 790\ \text{cm}^{-1}$, $v_2 = 369\ \text{cm}^{-1}$, and $C = 505{,}644\ \text{Å}^3\ \text{cm}^{-1}$.

The effective Hamiltonian reproduces the INDO-CI spectrum of the B850 BChls in LH-II well, except for the energy gap Δ (see Figure 14.6) between the lowest and second lowest excited state, which measures 422 cm^{-1}, i.e., twice the value predicted in a pure effective Hamiltonian picture. The reason for the difference is that the lowest electronic excitation experiences level repulsion with higher charge resonance states not accounted for by the effective Hamiltonian. Other calculations, less extensive than the INDO-CIS procedure in Zerner et al. (1998), ranging from alternative effective Hamiltonian representations (Dracheva et al. 1996; Hu et al. 1995b; Sauer et al. 1996) to the quantum mechanical consistent-force-field/π-electron (QCFF/PI) approach (Alden et al. 1997), yield a similar exciton band structure as INDO-CIS but differ in detailed exciton levels and band gaps.

A key feature of the electronic excitations connected with the B850 BChl spectrum in Figure 14.7 is the distribution of oscillator strength of the exciton states, i.e., the ability of the states to absorb or emit light directly. Only the energetically degenerate pair of states $|\tilde{2}\rangle, |\tilde{3}\rangle$ above the lowest energy state $|\tilde{1}\rangle$ carry nonzero oscillator strength. Due to a well-known sum rule, the addition of all oscillator strengths of the exciton band system, in the effective Hamiltonian description, must be equal to the sum of Q_y state oscillator strength D_y of the individual BChls (equal $16D_y$), i.e., $|\tilde{2}\rangle$ and $|\tilde{3}\rangle$ must each carry oscillator strength $8D_y$. This value reflects the property of the exciton states $|\tilde{2}\rangle, |\tilde{3}\rangle$ to be coherent superpositions of Q_y excitations $|j\rangle, j=1,\dots 16$ of individual BChls, just as the vanishing of oscillator strength of state $|\tilde{1}\rangle$ reflects a coherent spread of excitations around the entire BChl ring.

The distribution of oscillator strength can have important functional implications: excitation of the B850 BChl system would result, after thermal relaxation, in the preferential population of the energetically lowest exciton state $|\tilde{1}\rangle$, which is optically forbidden due to its vanishing oscillator strength and, hence, is prevented from wasteful fluorescence. The population of state $|\tilde{1}\rangle$ would depend sensitively on the actual energy difference Δ between $|\tilde{1}\rangle$ and the energetically degenerate states $|\tilde{2}\rangle, |\tilde{3}\rangle$.

One must note that the properties of the B850 BChl system outlined depend on the ideal eightfold symmetry axis of LH-II of *Rs. molischianum*. Distortions due to thermal motion or interruptions of the complete circle would alter the oscillator strength distribution. The characteristics of the exciton states due to a complete coherent spread of the excitations over the LH-II ring need to be studied in the presence of distortions. It is widely believed that the B850 BChl excited states, despite natural disorder, are delocalized; however, the extent of delocalization is debated (Fleming and van Grondelle 1997). The estimate for the number of coherently coupled BChls ranges from two BChl molecules (Jimenez et al. 1996) to the entire length of the B850 BChl aggregate (Wu et al. 1997; van Oijen et al. 1999). In principle, the relative strengths of the disorder and of the coupling between BChls determine the delocalization length. According to the effective Hamiltonian picture, the coupling between nearest neighbor BChls is 790 cm^{-1} (v_1) within the αβ-heterodimer, and 369 cm^{-1} (v_2) between the

$\alpha\beta$-heterodimers (see above). The effect of static disorder has been modeled in Hu et al. (1997) by randomizing the diagonal elements of an effective Hamiltonian and carrying out an effective Hamiltonian description numerically for a large ensemble of randomized Hamiltonians. Using a distribution consistent with the inhomogeneous broadening measured by hole-burning spectroscopy, the effect of diagonal disorder on exciton delocalization and oscillator strength distribution was found to be noticeable, but small. Static off-diagonal disorder shows similar effects (Schulten et al., unpublished results).

The treatment of dynamic disorder for an exciton system, like that of the Q_y excitons in LH-II, is technically extremely difficult and essentially impossible. One needs to describe respective systems mathematically in the product space of quantum states, Ω, and of stochastic states, Σ, i.e., in the very high-dimensional space $\Omega\otimes\Sigma$. Descriptions achieved so far restrict Σ to two states (Barvík et al. 1998). Desirable would be a sampling of an effective state space of the B850 BChl ring in LH-II through a molecular dynamics simulation, abstraction of a stochastic operator which would account for the monitored transitions between stochastic states, and a subsequent stochastic quantum mechanical description. For two-state quantum systems, such treatment has been achieved in Bittl and Schulten (1988, 1989), Schulten and Tesch (1991), and Xu and Schulten (1994).

The state of the quantum mechanical description of the B850 BChl ring in LH-II is still marred by the type of complexity that often arises in a still incomplete investigation. More refined *ab initio* quantum chemical calculations — including the effect of electron exchange between BChls, which is not accounted for in an INDO-CIS treatment nor in studies of static and dynamic randomization, all in close comparison with optical spectroscopy data — need to shed light on the issue of how localized or delocalized excitons in the B850 BChl ring actually are. Below, for the sake of argument, we assume that the description achieved in Zerner et al. (1998) and Hu et al. (1997) and the associated effective Hamiltonian description outlined are satisfactory.

EFFECTIVE HAMILTONIAN FOR EXCITATION FUNNEL

The effective Hamiltonian introduced above for LH-II of *Rs. molischianum* can be readily extended to describe the ring of 18 BChls in LH-II of *Rb. sphaeroides* as well as the ring of 32 BChls in LH-I of this species (Hu et al. 1997; Ritz et al. 1998). If one assumes that the nearest neighbor coupling in these light-harvesting complexes is the same as in LH-II of *Rs. molischianum*, i.e., v_1 and v_2, one can generalize the effective Hamiltonian without any further parameters, only the dimension of the matrix and the coordinates of the respective ring geometries, which enter Equation 14.3, need to be adjusted. The general form of the effective Hamiltonian is:

$$\hat{H} = \begin{pmatrix} \varepsilon & \nu_1 & W_{1,3} & W_{1,4} & \cdot & \cdot & \cdot & \cdot & W_{1,2N-1} & \nu_2 \\ \nu_1 & \varepsilon & \nu_2 & W_{2,4} & \cdot & \cdot & \cdot & \cdot & W_{2,2N-1} & W_{2,2N} \\ W_{3,1} & \nu_2 & \varepsilon & \nu_1 & \cdot & \cdot & \cdot & \cdot & W_{3,2N-1} & W_{3,2N} \\ W_{4,1} & W_{4,2} & \nu_1 & \varepsilon & \cdot & \cdot & \cdot & \cdot & \cdot & \cdot \\ \cdot & \cdot & \cdot & \cdot & \cdot & \cdot & \cdot & \cdot & \cdot & \cdot \\ \cdot & \cdot & \cdot & \cdot & \cdot & \cdot & \cdot & \cdot & \cdot & \cdot \\ \cdot & \cdot & \cdot & \cdot & \cdot & \cdot & \cdot & \cdot & \cdot & \cdot \\ \cdot & \cdot & \cdot & \cdot & \cdot & \cdot & \cdot & \varepsilon & \nu_2 & W_{2,2N-2,2N} \\ \cdot & \cdot & \cdot & \cdot & \cdot & \cdot & \cdot & \nu_2 & \varepsilon & \nu_1 \\ \nu_2 & \cdot & \cdot & \cdot & \cdot & \cdot & \cdot & W_{2N,2N-2} & \nu_1 & \varepsilon \end{pmatrix} \quad (14.4)$$

Here, ε represents the excitation energy of the Q_y state. The quantities $W_{j,k}$ correspond to the elements $\langle j|H|k\rangle$ and are given by Equation 14.3. In case of LH-II of *Rb. sphaeroides*, one chooses $N = 9$ and in case of LH-I, $N = 16$. In both cases this description yields the same characteristics of the exciton bands as for the B850 BChl aggregate of LH-II of *Rs. molischianum*: the second and the third exciton states carry all the oscillator strength, the lowest energy exciton state being optically forbidden (Hu et al. 1997).

One can even extend the effective Hamiltonian to describe the BChls of the entire PSU, including the BChls in RC. The form of the Hamiltonian is:

$$\hat{H} = \begin{pmatrix} \hat{R}_{11} & \hat{R}_{12} & \cdots & \hat{R}_{1M} \\ \hat{R}_{21} & \hat{R}_{22} & \cdots & \hat{R}_{2M} \\ \vdots & \vdots & \ddots & \vdots \\ \hat{R}_{M1} & \hat{R}_{M2} & \cdots & \hat{R}_{MM} \end{pmatrix} \quad (14.5)$$

Here, the elements $\hat{R}_{jk}$ are themselves matrices. $\hat{R}_{jj}$ describes the subcomponents of the excitation funnel, i.e., several 18×18 matrices accounting for the B850 BChl rings, a 32×32 matrix describing the B875 BChl ring, and a 4×4 matrix describing the BChls in RC. The elements of the latter matrix would need to be properly adjusted, since the nearest neighbor couplings in RC differ from those in LH-I and LH-II. All non-nearest neighbor couplings as well as the elements of the matrices $\hat{R}_{jk}, j \neq k$, describing couplings between the BChl rings or between the B875 BChl ring and the RC BChls, are furnished through Equation 14.3.

The effective Hamiltonian (Equation 14.5) furnishes a simple description of the relevant electronic properties of the light-harvesting system that hides the complexity of the underlying system of hundreds of partaking electrons. The Hamiltonian has been employed in Hu et al. (1997) and Ritz et al. (1998) to determine the rates of transfer of electronic excitation between LH-IIs, between LH-II and LH-I, and between LH-I and RC, i.e., to describe the excitation funnel component of the PSU shown in Figure 14.7. The calculated time constants of 3.3 and 65 ps for the excitation transfer processes

LH-II → LH-I and LH-I → RC in *Rb. sphaeroides*, respectively, are in agreement with experimental values of 3 ~ 5 ps and 35 ps (Hess et al. 1995; Visscher et al. 1989). A startling result from these calculations has been a suggested role of the accessory BChls as mediators of the excitation transfer from LH-I to the RC special pair: the calculated time for LH-I → RC transfer, in the absence of accessory BChls, is about 600 ps, which is an order of magnitude too long compared with observations; the accessory BChls in RC provide a tunneling path for the excitation transfer which bridges the large distance of 42 Å or longer between LH-I BChls and the RC special pair.

ROLE OF B800 BCHLS AND CAROTENOIDS IN PHOTON FUELING

The energy of photons is fueled into the light-harvesting system through primary absorption by the B800 BChls and by the carotenoids. First, I will discuss the role of the B800 BChls and then that of the carotenoids.

The B800 BChls are oriented such that they absorb light in a direction perpendicular to that of the B850 BChls. The individual B800 BChls can transfer the resulting excitation energy to the B850 ring through the Förster mechanism (Oppenheimer 1941; Förster 1948). The transfer proceeds within 700 fs (Shreve et al. 1991). Quantum chemical calculations in Zerner et al. (1998) have demonstrated that the B800 BChls are only weakly coupled with each other and with the B850 BChls, a property also revealed by single-molecule spectroscopy (van Oijen et al. 1999). Calculations of excitation transfer in Damjanovic et al. (1999) reveal that the short transfer time, to a large degree, results from the exciton splitting of the accepting B850 exciton levels shown in Figure 14.6; the exciton splitting greatly improves the resonance of the excitations of B800 and B850 BChls.

Carotenoids absorb light at 500 nm into a strongly allowed state and transfer the excitation energy within 200 fs and with nearly 100% efficiency to the Q_y exciton states of the B850 ring (Chadwick et al. 1987). The question arises by which pathways and by which mechanism such an efficient excitation transfer is achieved. Figure 14.7 presents the energies of the excited states of spheroidene, the carotenoid in LH-II of *Rb. sphaeroides*, and of the BChls in LH-II of *Rb. sphaeroides*. Spheroidene features a strongly allowed singlet-excited state absorbing at 500 nm, labeled S_2, and a lower lying optically forbidden state labeled S_1. The latter state plays a prominent role and has been characterized by Hudson et al. (1982) and Tavan and Schulten (1987). In fact, the state S_2, which initially absorbs the 500 nm photons, decays within less than 200 fs into S_1. The S_1 state is in resonance with the accepting Q_y exciton states and, thus, provides a gateway for transfer to the B850 ring.

The optically forbidden character of the S_1 state of spheroidene precludes its coupling to the B850 ring through the Förster mechanism, thus limiting potential mechanisms to coupling through Coulomb interaction including higher-order multipoles or coupling through electron exchange (Dexter mechanism; Dexter 1953). Both transfer routes require a close proximity between spheroidenes and BChls. Spheroidene and BChls are indeed found in close contact, as shown in Figure 14.7. Calculations based on the geometric arrangement of carotenoids and BChls in LH-II (Figure 14.7) and on CI expansions of the electronic states of carotenoids and chlorophylls, reported in Nagae et al. (1993) and Damjanovic et al. (1998), however, suggest that the exchange mechanism is largely ineffective since the respective coupling strength is very small. These calculations showed that the Coulomb mechanism results in a transfer of singlet excitations through the S_1 (carotenoid) $\rightarrow$ B850 (exciton states) pathway. This transfer is also strongly accelerated by the splitting of the B850 exciton levels (Damjanovic et al. 1999).

In addition to transfer through the forbidden S_1 state of spheroidene, the absorbing S_2 state of carotenoids transfers excitation directly to the Q_x state of BChl as suggested by the calculations in Damjanovic et al. (1999) and the shortened (60 fs) *in vivo* lifetime of the S_2 state in LH-II (Ricci et al. 1996).

Besides the light-harvesting function, carotenoids protect the light-harvesting system from the damaging effect of BChl triplet states as pointed out above. Carotenoids quench the BChl triplet states through triplet excitation transfer from the BChls. This transfer involves a spin change and can only proceed through the electron exchange mechanism (Dexter 1953). The triplet excitation transfer in LH-II of *Rs. molischianum* has been described in detail (Damjanovic et al. 1999). The calculations showed that B850a is well protected by one of the eight lycopenes seen in the crystal structure of LH-II of *Rs. molischianum* (see Figures 14.4 and 14.7), while B850b is not directly protected, but can transfer triplet excitation within a few picoseconds to the well-protected B850a BChl. The motif that BChls are protected via a bridge involving a further BChl has been found also in the RC (Laible et al. 1998) and in the light-harvesting complex of plants LHC-II (Kuhlbrandt 1994). B800 may need not be protected since it transfers its singlet excitation within ~1 ps to the B850 BChls.

HOW IS THE PSU ASSEMBLED *IN SITU*?

The remarkable overall architecture of the PSU is matched by the equally remarkable architecture of the individual LH-I and LH-II complexes, which in the case of LH-II of *Rs. molischianum* is composed of sixteen independent peptides (eight identical α-apoproteins and eight identical β-apoproteins), 24 chlorophylls, and eight carotenoids. These components self-aggregate in the cytoplasmic membrane, most likely first into heterodimers as defined above, and then, using the heterodimers as building blocks, into a ring of eight heterodimers. In the case of LH-II of *Rps. acidophila* (Duysens 1952) and *Rb. sphaeroides*, nine heterodimers aggregate into LH-II. The LH-I complex consists of sixteen heterodimers, each of which contain two chlorophylls and one carotenoid. The self-aggregation of the individual light-harvesting complexes LH-I and LH-II and of the whole PSU from its main building blocks, the heterodimers, poses a great challenge. Several questions regarding the self-aggregation process arise: How do geometry and energetics contribute to the aggregation of two heterodimers? How is this aggregation guided in the cell membrane? How do the single heterodimer building blocks control the overall ring size? How do changes of ring size affect the stability of an overall complex? These questions might be pursued through modeling calculations; however, due to the large size of the overall ring systems and the qualitative nature of the phenomenon of self-aggregation, i.e., one does not know *a priori* what observables to monitor, I chose an approach through mechanical models of the RC, LH-I composed of sixteen heterodimers, and LH-II composed of 7–9 heterodimers. These models have been described in Bailey et al. (1998) and were used to assemble LH-II from eight heterodimer units, a most revealing experience: it turned out that the aggregation of the building blocks did not require any flexibility of the dimer surfaces; the individual components naturally fit together tightly without the need of a flexible adjustment of their surfaces. This feature implies a favorable entropy of aggregation, i.e., the heterodimers do not need to explore much conformation space to interlock. Viewing the resulting complex strikes the observer with the realization that the heterodimers have evolved purposefully corrugated self-complementary surfaces to steer the aggregation process and to stabilize the eventual aggregates. One notices that any coarse-graining of atomic level surface features of the heterodimers would deteriorate adhesion properties. In fact, the manufacturing of the models required about 30,000 triangles for the proper representation of surface features of each heterodimer; this corresponds to 1 Å resolution and proved necessary to represent properly the relevant complementarity of surfaces. The heterodimers are composed of two, rather short, transmembrane helices, a feature which requires the heterodimers to be oriented strictly normal to the membrane plane; a tilt would immerse polar side groups into the hydrophobic phase of the lipid bilayer. A pronounced top-bottom asymmetry of the heterodimer favors a vectorial orientation in the membrane. Manipulation of two heterodimer models properly juxtaposed and rotated relative to each other reveals readily the proper interlocking geometry of a pair. A similar one-dimensional (rotation around the membrane normal) search process may also govern the rapid self-assembly of LH-II in the cytoplasmic membrane. One can readily build rings of slightly "wrong" sizes, e.g., an LH-II ring of only seven or nine units, i.e., one too few or one too many. In both cases the mechanical models reveal a loose packing with wobbling heterodimer units. This is certainly a feature which results from the fact that the LH-II heterodimer models stem from the eight-ring structure of *Rs. molischianum*, i.e., a structure optimized for that particular ring size. However, the model demonstrates clearly that complementarity and orientation of the corrugated surfaces of the individual heterodimers are precise enough to control the overall ring size of eight units. I have built also the LH-I complex from sixteen models of the respective heterodimers as well as a model of the RC. This allowed me to build a key section of the PSU. The models revealed that one can readily clap the LH-I ring over the RC leading to a unique geometry of the complex. In this complex, LH-I forms an atrium-type space adjacent to the reaction center surface with the entrance of the functionally important quinone binding site. One also notices that the sixteen-unit

LH-I ring is considerably more flexible in its shape than is the eight-unit LH-II ring; it does not appear likely that the LH-I ring would assume its proper size and shape without the RC template.

SUMMARY

From a few common components, photosynthetic organisms have developed a rather divergent set of antenna systems. The evolutionary divergence has been discussed comparing the antenna systems of purple bacteria, green bacteria, cyanobacteria, dinoflagellates, and green plants in Hu et al. (1998), which contains other references on the subject. The properties of the antenna systems of purple bacteria are likely relevant for an understanding of the mechanisms common to all the mentioned light-harvesting systems. These properties are summarized here in general terms to serve potentially an understanding of a broader class of photosynthetic life forms.

The chromophores of purple bacteria, i.e., BChls and carotenoids, are attuned to their ambient light. In the case of lycopene/spheroidene and B800/B850 BChls, the combined absorption spectrum is complementary to that of chlorophyll a or b in green plants, i.e., adjusted to a habitat below plants. The purple bacteria exploit the optically allowed as well as the optically forbidden low-lying excited states of polyenes (Hudson et al. 1982; Tavan and Schulten 1987) to couple the carotenoid excitations to BChls. The carotenoids are entrusted with the excitation energy for only a few hundred femtoseconds, after which time BChls are the wardens of the energy.

The spectra of BChls are tuned only to a limited degree though interaction with the protein environment, e.g., through formylmethionine–Mg ligation in case of B800 of LH-II from *Rps. acidophila* (McDermott et al. 1995) or through an Asp–Mg ligation in case of B800 of LH-II from *Rs. molischianum* (Koepke et al. 1996); the observed spectra result mainly from intrinsic properties of BChls and excitonic interactions (Hu et al. 1997; Zerner et al. 1998). Excitonic coupling splits the excited state energies, thus improving the overlap between donor and acceptor spectra in the excitation cascade (Hu et al. 1997; Zerner et al. 1998; Damjanovic et al. 1998).

BChls have the disadvantage that their lowest energy triplet state lies high enough to excite molecular oxygen. Their companion carotenoids quench the triplet excitations of BChls.

The efficient flow of excitation through the chromophore system requires highly ordered aggregates, the geometry of which is adapted to the needed interactions; carotenoids must be in close (van der Waals) contact with BChls for triplet quenching and must be proximate within a few angstroms for transfer of optically forbidden excitations. In order to achieve significant exciton splitting, chlorophylls must have Mg–Mg distances of about 10 Å; for energy transfer on a picosecond time scale, Mg–Mg distances must be of the order of 20 Å. It is possible that BChls form aggregates to achieve coherence over many chromophores, such that the lowest energy state becomes optically forbidden, increasing its lifetime.

A multi-protein architecture is necessary to provide a large enough scaffold for the number of chromophores employed in light harvesting. Due to this architecture, antenna systems employ a hierarchy of chromophore aggregates; the chromophores are closer and more tightly coupled in the individual pigment–protein complex, e.g., in LH-II, and more loosely coupled between different pigment–protein complexes. The control of the overall aggregation of the multi-protein system is in itself an impressive achievement worthy of study.

To direct flow of excitation to the RC, the antenna system of purple bacteria assumes a spatial organization in which the BChls with lower-energy excitations are closer to the RC. Such arrangement, as shown in Figure 14.2, yields an energy funnel that prevents detours in the excitation flow, enhancing the overall efficiency of light harvesting as measured by the quantum yield for a photon absorbed to reach an RC.

The multi-protein photosynthetic apparatus as shown in Figure 14.1 poses the challenge for eventually modeling the conversion of light into ATP in its entirety. Few would have predicted that the protein constituents of the photosynthetic apparatus would be structurally known in principle already today; however, many expect that biologists will more frequently see entire protein systems engaged in complex overall functions resolved at atomic resolution. The questions posed by the photosynthetic apparatus will then be typical for biology in the 21st century: How are multiprotein systems genetically controlled? How do they physically aggregate? How did they evolve and how do they compare among species? The PSU constitutes an ideal subsystem of the photosynthetic apparatus which, due to its smaller size, is more amenable to study while posing the same principal challenges: How do LH-I and LH-II form from their many independent components? What determines the ring size and stability? How do the completed LH-IIs aggregate around the LH-I–RC complex? The function of the PSU emerges as a true system property, all components being designed to cooperate in absorbing light effectively and channel its energy to the RC. The common origin of photosynthetic, respiratory, and other organisms makes the PSU and the photosynthetic apparatus a valuable model for understanding, at the level of multiprotein systems, not only photosynthesis, but life in general.

ACKNOWLEDGMENTS

I would like to thank H. Michel, Xiche Hu, A. Damjanovic, and T. Ritz for their collaboration on light-harvesting protein function and mechanism, and acknowledge financial support from the National Institutes of Health [P4IRR059691, the National Science Foundation [NSF BIR 9318159 and NSF BIR-94-23827(EQ)], as well as from the Carver Charitable Trust.

REFERENCES

Aagaard, J., and W.R. Sistrom. 1972. Control of synthesis of reaction center bacteriochlorophyll in photosynthetic bacteria. *Photochem. Photobiol.* **15**:209–225.

Alden, R.G., E. Johnson, V. Nagarajan, W.W. Parson, C.J. Law, and R.J.Cogdell. 1997. Calculations of spectroscopic properties of the LH2 bacteriochlorophyll-protein antenna complex from *Rhodopseudomonas acidophila. J. Phys. Chem. B* **101**:4667–4680.

Bailey, M., K. Schulten, and J.E. Johnson. 1998. The use of solid physical models for the study of macromolecular assembly. *Curr. Opin. Struct. Biol.* **8**:202–208.

Barvík, I., C. Warns, and P. Reineker. 1998. Exciton transfer in disordered molecular aggregates: Computer modeling of optical line shapes in ring bacteria antenna systems. *J. Luminesc.* **76/77**:331–334.

Bittl, R., and K. Schulten. 1988. Length dependence of the magnetic field modulated triplet yield of photogenerated biradicals. *Chem. Phys. Lett.* **146**:58–62.

Bittl, R., and K. Schulten. 1989. A static ensemble approximation for stochastically modulated quantum systems. *J. Chem. Phys.* **90**:1794–1803.

Chadwick, B.W., C. Zhang, R.J. Cogdell, and H.A. Frank. 1987. The effect of lithium dodecyl sulfate and sodium borohydride on the absorption spectrum of the B800–850 light-harvesting complex from *Rhodopseudomonas acidophila 7750. Biochim. Biophys. Acta* **893**:444.

Cogdell, R.J., P.K. Fyfe, S.J. Barrett, S.M. Prince, A.A. Freer, N.W. Isaacs, P. McGlynn, and C.N. Hunter. 1996. The purple bacterial photosynthetic unit. *Photosynth. Res.* **48**:55–63.

Damjanovic, A., T. Ritz, and K. Schulten. 1999. Energy transfer between carotenoids and bacteriochlorophylls in a light-harvesting protein. *Phys. Rev. E* **59**:3293–3311.

Dexter, D.L. 1953. A theory of sensitized luminescence in solids. *J. Chem. Phys.* **21**:836–850.

Dracheva, T.V., V.I. Novoderezhkin, and A. Razjivin. 1996. Exciton delocalization in the antenna of purple bacteria — Exciton spectrum calculations using x-ray data and experimental site inhomogeneity. *FEBS Lett.* **387**:81–84.

Duysens, L.N.M. 1952. Transfer of Excitation Energy in Photosynthesis. Ph.D. thesis, Utrecht.

Emerson, R., and W. Arnold. 1932. The photochemical reaction in photosynthesis. *Gen. Physiol.* **16**:191–205.

Ermler, U., G. Fritzsch, S.K. Buchanan, and U. Michel. 1994. Structure of the photosynthetic reaction center from *Rhodobacter sphaeroides* at 2.65 Å resolution: Cofactors and protein–cofactor interactions. *Structure* **2**:925–936.

Farchaus, J.W., and D. Oesterhelt. 1989. A Rhodobacter sphaeroides *puf*L, M, and X deletion mutant and its complementation in trans with a 5.3 kb *puf* operon shuttle fragment. *EMBO J.* **8**:47–54.

Fleming, G.R., and R. van Grondelle. 1997. Femtosecond spectroscopy of photosynthetic light-harvesting systems. *Curr. Opin. Struct. Biol.* **7**:738–748.

Förster, T. 1948. Zwischenmolekulare Energiewanderung und Fluoreszenz. *Ann. Phys. (Leipzig)* **2**:55–75.

Germeroth, L., F. Lottspeich, B. Robert, and H. Michel. 1993. Unexpected similarities of the B800–850 light-harvesting complex from Rhodospirillum molischianum to the B870 light-harvesting complexes from other purple photosynthetic bacteria. *Biochemistry* **32**:5615–5621.

Gouterman, M. 1961. Spectra of porphyrins. *J. Mol. Spectrosc.* **6**:138–163.

Hess, S., M. Chachisvilis, K. Timpmann, M.R. Jones, G.J.S. Fowler, C.N. Hunter, and V. Sundstrom. 1995. Temporally and spectrally resolved subpicosecond energy transfer within the peripheral antenna complex (LH2) and from LH2 to the core antenna complex in photosynthetic purple bacteria. *Proc. Natl. Acad. Sci. USA* **92**:12,333–12,337.

Hu, X., A. Damjanovic, T. Ritz, and K. Schulten. 1998. Architecture and function of the light-harvesting apparatus of purple bacteria. *Proc. Natl. Acad. Sci. USA* **95**:5935–5941.

Hu, X., T. Ritz, A. Damjanovic, and K. Schulten. 1997. Pigment organization and transfer of electronic excitation in the purple bacteria. *J. Phys.Chem. B* **101**:3854–3871.

Hu, X., and K. Schulten. 1997. How nature harvests sunlight. *Phys. Today* **50**:28–34.

Hu, X., and K. Schulten. 1998. A model for the light-harvesting complex I (B875) of *Rhodobacter sphaeroides*. *Biophys. J.* **75**:683–694.

Hu, X., D. Xu, K. Hamer, K. Schulten, J. Koepke, and H. Michel. 1995a. Predicting the structure of the light-harvesting complex II of *Rhodospirillum molischianum*. *Protein Sci.* **4**:1670–1682.

Hu, X., D. Xu, K. Hamer, K. Schulten, J. Koepke, and H. Michel. 1995b. Prediction of the structure of an integral membrane protein — The light-harvesting complex II of *Rhodospirillum molischianum*. In: Biological Membranes: A Molecular Perspective from Computation and Experiment, ed. K.M. Merz and B. Roux, pp. 503–533. Cambridge, MA: Birkhäuser.

Hudson, B.S., B.E. Kohler, and K. Schulten. 1982. Linear polyene electronic structure and potential surfaces. In: Excited States, ed. E.C. Lim, Vol. 6, pp. 1–95. New York: Academic Press.

Humphrey, W.F., A. Dalke, and K. Schulten. 1996. VMD — Visual Molecular Dynamics. J. Mol. Graphics **14**:33–38.

Jimenez, R., S.N. Dikshit, S.E. Bradforth, and G.R. Fleming. 1996. Electronic excitation transfer in the LH2 complex of *Rhodobacter sphaeroides*. *J. Phys. Chem.* **100**:6825–6834.

Jungas, C., J.L. Ranck, J.L. Rigaud, and A. Vermeglio. 1999. Supramolecular organization of the photosynthetic apparatus of *Rhodobacter sphaeroides*. *EMBO J.* **18**:534–542.

Karrasch, S., P.A. Bullough, and R. Ghosh. 1995. 8.5 Å projection map of the light-harvesting complex I from *Rhodospirillum rubrum* reveals a ring composed of 16 subunits. *EMBO J.* **14**:631.

Knox, K. 1963. Theory of Excitons. New York: Academic Press.

Koepke, J., X. Hu, C. Münke, K. Schulten, and H. Michel. 1996. The crystal structure of the light-harvesting complex II (B800–850) from *Rhodospirillum molischianum*. *Structure* **4**:581–597.

Kuhlbrandt, W. 1994. High-resolution electron crystallography of membrane proteins. In: Membrane Protein Structure: Experimental Approaches, ed. S.H. White, pp. 206–223. New York: Oxford Univ. Press.

Laible, P.D., V.C. Chynwat, M.C. Thurnauer, M. Schiffer, D.K. Hanson, and H.A. Frank. 1998. Protein modifications affecting triplet energy transfer in bacterial photosynthetic reaction centers. *Biophys. J.* **74**.2623–2637.

McDermott, G., S.M. Prince, A.A. Freer, A.M. Hawthornthwaite-Lawless, M.Z. Papiz, R.J. Cogdell, and N.W. Isaacs. 1995. Crystal structure of an integral membrane light-harvesting complex from photosynthetic bacteria. *Nature* **374**:517–521.

Miller, K.R. 1982. Three-dimensional structure of a photosynthetic membrane. *Nature* **300**:53–55.

Monger, T.G., and W.W. Parson. 1977. Singlet-triplet fusion in *Rhodopseudomonas sphaeroides* chromatophores. A probe of the organization of the photosynthetic apparatus. *Biochim. Biophys. Acta* **460**:393–407.

Nagae, H., T. Kakitani, T. Katoh, and M. Mimuro. 1993. Calculation of the excitation transfer matrix elements between the S_2 or S_1 state of carotenoid and the S_2 or S_1 state of bacteriochlorophyll. *J. Chem. Phys.* **98**:8012–8023.

Oppenheimer, J.R. 1941. Internal conversion in photosynthesis. *Phys. Rev.* **60**:158.

Pullerits, T., and V. Sundstrom. 1996. Photosynthetic light-harvesting pigment–protein complexes: Toward understanding how and why. *Acc. Chem. Res.* **29**:381–389.

Ricci, M., S.E. Bradforth, R. Jimenez, and G.R. Fleming. 1996. Internal conversion and energy transfer dynamics of spheroidene in solution and in the LH-1 and LH-2 light-harvesting complexes. *Chem. Phys. Lett.* **259**:381–390.

Ritz, T., X. Hu, A. Damjanovic, and K. Schulten. 1998. Excitons and excitation transfer in the photosynthetic unit of purple bacteria. *J. Luminesc.* **76/77**:310–321.

Sauer, K., R.J. Cogdell, S.M. Prince, A. Freer, N.W. Isaacs, and H. Scheer. 1996. Structure-based calculations of the optical spectra of the LH2 bacteriochlorophyll-protein complex from *Rhodopseudomonas acidophila. Photochem. Photobiol.* **64**:564–576.

Schulten, K., and M. Tesch. 1991. Coupling of protein motion to electron transfer: Molecular dynamics and stochastic quantum mechanics study of photosynthetic reaction centers. *Chem. Phys.* **158**.421–446.

Shreve, A.P., J.K. Trautman, H.A. Frank, T.G. Owens, and A.C. Albrecht. 1991. Femtosecond energy-transfer processes in the B800–850 light-harvesting complex of *Rhodobacter sphaeroides* 2.4.1. *Biochim. Biophys. Acta* **1058**:280–288.

Tavan, P., and K. Schulten. 1987. Electronic excitations in finite and infinite polyenes. *Phys. Rev. B* **36(8)**:4337–4358.

Thompson, M., and M. Zerner. 1991. A theoretical examination of the electronic structure and spectroscopy of the photosynthetic reaction center from *Rhodopseudomonas viridus. J. Am. Chem. Soc.* **113**:8210–8215.

van Grondelle, R., J. Dekker, T. Gillbro, and V. Sundstrom. 1994. Energy transfer and trapping in photosynthesis. *Biochim. Biophys. Acta* **1187**:1–65.

van Oijen, A.M., M. Ketelaars, J. Kohler, T.J. Aartsma, and J. Schmidt. 1999. Unraveling the electronic structure of individual photosynthetic pigment-protein complexes. *Science* **285**:400–402.

Visscher, K.J., H. Bergstrom, V. Sundstrom, C.N. Hunter, and R. van Grondelle. 1989. Temperature dependence of energy transfer from the long wavelength antenna BChl-896 to the reaction center in *Rhodospirillum rubrum*, *Rhodobacter sphaeroides* (w.t and m21 mutant) from 77 to 177 k, studied by picosecond absorption spectroscopy. *Photosynth. Res.* **22**:211–217.

Walz, T., and R. Ghosh. 1997. Two-dimensional crystallization of the light-harvesting I-reaction centre photounit from *Rhodospirillum rubrum. J. Mol. Biol.* **265**:107–111.

Wu, H.-M., N.R.S. Reddy, and G.J. Small. 1997. Direct observation and hole burning of the lowest exciton level (B870) of the LH2 antenna complex of *Rhodopseudomonas acidophila* (strain 10050). *J. Phys. Chem. B* **101**:651–656.

Xu, D., and K. Schulten. 1994. Coupling of protein motion to electron transfer in a photosynthetic reaction center: Investigating the low temperature behaviour in the framework of the spin-boson model. *Chem. Phys.* **182**:91–117.

Zerner, M.C., M.G. Cory, X. Hu, and K. Schulten. 1998. Electronic excitations in aggregates of bacteriochlorophylls. *J. Phys. Chem.* B **102**:7640–7650.

Zuber, H., and R.A. Brunisholz. 1991. Structure and function of antenna polypeptides and chlorophyll-protein complexes: Principles and variability. In: Chlorophylls, ed. H. Scheer, pp. 627–692. Boca Raton: CRC Press.

Standing, left to right: Helmut Grubmüller, Ugo Palma, Arieh Warshel, Uli Nienhaus
Seated, left to right: Don Crothers, Keith Moffat, Klaus Schulten, Fritz Parak
Not pictured: Jean-Pierre Changeux

15

Group Report: How Does Complexity Lead to an Apparently Simple Function?

K. MOFFAT, Rapporteur

J.-P. CHANGEUX, D.M. CROTHERS, H. GRUBMÜLLER, G.U. NIENHAUS, M.U. PALMA, F.G. PARAK, K. SCHULTEN, and A. WARSHEL

Keywords: macromolecular dynamics, thermodynamic averaging, energy landscapes, biological selection, quantum treatments, simplification, molecular environments, macromolecular assemblies, common rules

INTRODUCTION: FROM COMPLEXITY TO SIMPLICITY

A bewildering complexity is revealed by even a first, casual glance at the information available on macromolecular structure and function, on integration of these structures into higher-order molecular and cellular assemblies, and on the more elaborate functions and modes of regulation displayed by such assemblies. It appears as though each entity has to be considered alone, its uniqueness more important than its similarities to its neighbors. If this were true, then the study of biology at the molecular and cellular level by the techniques of physics and chemistry would be daunting indeed. In this chapter we consider to what extent, and how, this complexity may be reduced to a simpler view of both structure and function. Can "conceptual averaging" capture the essential properties? How may apparently simpler functions, all of which are associated at some level with flexibility and the motion of atoms, arise from appropriate statistical averaging over vastly more complicated, underlying motions? With such a reductionist viewpoint, there is always the danger that injudicious simplification may lose the key elements of the problem; one ends up with naïveté rather than simplicity.

We choose not to venture a formal definition of the word "complexity." There are too many pitfalls in potential conflict with existing definitions in physics and

Simplicity and Complexity in Proteins and Nucleic Acids
Edited by H. Frauenfelder, J. Deisenhofer, and P.G. Wolynes

mathematics. In this chapter, we use the word primarily in the sense of our inability to understand intuitively those systems that are properly described by large numbers of variables, for example in coordinate and momentum space. Such highly multidimensional systems are likely to be complex and yet, reduction of dimensionality by averaging may lead to a description of a macromolecule and its functions that exhibits simplicity. Although our approach is largely that of the physical chemist, we illustrate our points with both simpler and more complex biological examples.

CLASSICAL AND QUANTUM DYNAMICS, THERMODYNAMICS, LANDSCAPES, AND KINETICS

Simplification through Thermodynamic Averaging

The motion of biopolymers is governed by potential surfaces with many thermally accessible minima. Averaging over the resulting heterogeneities may produce a simpler effective behavior. Computational execution of such averaging includes integration over phase space to yield the partition function from which the thermodynamic parameters are obtained, or reduction of a complex reaction profile to a low-dimensional reaction coordinate. A protein's structural change may be captured by a two-state model with a single rate constant connecting the states, even though there may be billions of substates in each "state" and a vast number of trajectories that connect the two states.

An example is the R to T transition in hemoglobin. In a simple model, one may consider two reaction coordinates: a metal out-of-plane coordinate and the distance between a ligand and the heme iron. One can then calculate the corresponding free-energy barrier for moving from the R to T minima in this two-dimensional diagram (see, e.g., Warshel and Weiss 1981). Simplification, however, inevitably leads to a loss of information, and it is possible to carry the process too far and lose the key features. Indeed, some systems may be irreducible, in that any averaging comes at the price of loss of fundamental features.

Many processes in proteins acquire simplicity since they can be dealt with in terms of first-order kinetics, in contrast to cases in which the detailed dynamics of the system must be considered in order to explain the observed behavior. This question can be reformulated in terms of the permitted order of averaging over the substates of a system. If, for example, the substates of states A and B equilibrate rapidly compared to the rate of the A to B transition, then the system should be adequately described by a single rate constant, and a single decay time in the relaxation spectrum. If, however, equilibration over substates is only as fast as the A to B transition or is slower, the detailed dynamics is important and will be reflected in a multiexponential or even continuous relaxation spectrum. In the first of these cases one can, in principle, get a reasonable estimate of the rate constant by calculating the probability of the transition state, whereas the second case requires an explicitly time-dependent treatment. However, at the limit of slow equilibration we may average over rate constants of reactions that start from different substates.

Biological Selection Acting on Simple Function

The "apparently simple function" that characterizes proteins and nucleic acids is a consequence of natural selection of polymers from a highly multidimensional sequence space in a self-replicating system. Since each polymer sequence is also highly multidimensional in coordinate and momentum space, the selection of a function that confers reproductive advantage reflects an extraordinary degree of simplification. The basis for selection of simple function can be viewed in simple terms. Two obvious properties are the stability of a protein or nucleic acid, and the rate at which it can carry out a process. It is obvious that a protein must be selected for stability of the native structure in the environment where the organism lives. On the other hand, maximizing stability is clearly not an objective, since the unusual stability of proteins from extremophiles demonstrates that much higher stabilities are possible than seen in mesophiles. Furthermore, more stable mutants frequently lead to less efficient enzymes (Shoichet et al. 1995).

Selection for kinetic properties is obvious in the case of enzymes, since organisms may benefit from a substantial acceleration of a chemical reaction rate. On the other hand, maximizing all reaction rates is not a viable general objective, since the metabolism of an organism must retain a steady state balance and evolution actually operates at the level of metabolic pathways. Selection of a specific relaxation time for successful function is illustrated by neurotransmitter proteins, which display fast and slow conformational changes as a basis for developing a "memory effect" (see below).

An illustration of the balance between kinetic and stability selection can be found in the interaction of two RNA loops which is the first step in a sense–antisense interaction that leads to control of copy number of the *ColE1* plasmid in *Escherichia coli* (Polisky 1998). The sequences of the complementary 7-nucleotide loops are not optimized for maximum stability, but rather for a high reaction rate constant to form the loop–loop interaction (Eguchi and Tomizawa 1991).

Selection for explicitly dynamic properties can also be identified. It is obvious that the polymers on which life is based must be able to search configuration space rapidly enough to solve the folding problem. As a more specific example, protein dynamics must be sufficiently rapid to open a channel to the active site on a time scale appropriate to the needed turnover rate of the enzyme.

Energy Landscapes

In the case of a protein, a description in terms of energy landscapes has proved useful. Contact between a microscopic model and measurable macroscopic quantities is provided by weighted thermodynamic averaging among a huge variety of configurations. This concept greatly helps our understanding of how a protein works at physiological temperatures, if we average thermodynamically accessible configurations on a time scale short compared to the time scale of protein function. This averaging over "fast" variables is expected to smooth out the energy landscape as described through the potential energy function. The averaging gradually transforms the energy landscape into an effective free energy.

The "energy landscape" picture of proteins has been introduced to describe a potential energy surface in an essential subset of degrees of freedom that is structured over wide ranges of energy (barriers) and conformational length scales (see Eaton et al., this volume). This potential energy landscape turns into a free-energy surface for temperatures high enough or times of conformational interconversions long enough to ensure thermodynamic averaging within quasi-equilibrium intermediate states. In this limit, the landscape picture is equivalent to a kinetic scheme with multiple states; an example is provided by the bacteriorhodopsin photocycle (Oesterhelt et al. 1992; Schulten et al. 1995).

However, experiments have shown that complete thermodynamic averaging is more the exception than the rule for actual protein processes so that simple schemes give only a zero-order description. For example, in reaction centers for photosynthetic bacteria, sequential electron transfer over several steps occurs with transfer rate coefficients spanning the time range between picoseconds and seconds. Even at room temperature, only a small fraction of averaging over protein configurations occurs on the picosecond time scale so that the fast events occur in a (largely) heterogeneous ensemble of proteins (Parson et al. 1998; McMahon et al. 1998).

A second example of nontrivial kinetic behavior is ligand binding to heme proteins (see chapters by Nienhaus and Eaton, both this volume). After breaking of the covalent bond between the heme iron and the oxygen, the ligand either rebinds in a geminate process or exits. The nonexponentiality of the geminate process, even at physiological temperature, shows that fluctuational averaging is not complete at the time the ligand decides to either rebind or exit. Consequently, a simple three-well description of the process of ligand dissociation (and association) cannot adequately describe this process either.

These are just two illustrations of a problem that is general for biomolecules and arises from the fact that proteins are complex structures with relaxations stretched out over the entire time range, starting with picoseconds and extending beyond kiloseconds.

The energy landscape picture is clearly necessary to adequately describe experimental results. A quantitative and predictive use of this concept in theory seems to be very difficult, but the question of how to reduce the complexity of an energy landscape to a few relevant coordinates able to describe protein function is posed as a challenge to the theorists among us. Following pioneering work by Go and Scheraga (1969), initial steps have been taken (see, e.g., Karplus and Kushick 1981; Garcia 1992; Amadei et al. 1993; Orešic and Shalloway 1994; Grubmüller 1995; Balsera et al. 1996).

In modeling chemical processes in enzymes, one usually starts the evaluation of the activation free energy by performing a free-energy perturbation/umbrella sampling calculation and forcing the system to move from the reactant to the intermediate (or product) state. Such calculations are done starting from one protein structure and sometimes averaging over several initial coordinates and utilizing drastically simplified solvent models, by simulated annealing. The brute force approach will involve generating more and more coordinates. In principle, one may look for landscape concepts to determine how much averaging should be done in the direction orthogonal to the reaction coordinate, if suitable algorithms can be provided (Eaton et al., this volume). By sampling a (possibly small) fraction of (relative) energy minima and properly employing energy landscape concepts, one may obtain an estimate of the global energy minimum without actually having to visit all minima.

Since the pioneering work of Frauenfelder and coworkers, it is known that substates can play an important role in protein dynamics as a function of temperature (Austin et al. 1975). It is often stated that the substates occur in a hierarchical fashion with a series of separate tiers characterized by markedly different heights of barriers that separate the substates within each tier (Ansari et al. 1985; Frauenfelder et al. 1991). This view seems to arise at least in part from the limited measurements that can be made. In myoglobin, for example, each tier is characterized by a single experimental marker. Simulations suggest, however, that there is a continuum of substates, at least in the range of fluctuations that can be studied. The concept of substates is clearly important and one in which additional research is needed.

One may question, however, the extent to which the landscape view offers more than the potential energies employed in simulations, when the latter are combined with straightforward thermodynamic analyses. A method for selection of essential coordinates has not been provided by the advocates of the landscape picture and there appear to be few, if any, examples where the landscape picture has yielded insight into protein mechanisms that could not have been gained otherwise.

Determining Essential Degrees of Freedom

One way to depart from the overwhelming complexity of 3N-6 degrees of freedom of a protein (N being the number of atoms of the protein molecule) to simpler — yet sufficiently accurate — descriptions is dimension reduction, which is considered a particularly promising route by which to base the landscape concept on microscopic models. Success here rests on the ability to identify the essential degrees of freedom and to eliminate the irrelevant ones. Normal mode analysis or the more general tool of principal component analysis (applied to Boltzmann structure ensembles) can provide such essential degrees of freedom (Amadei et al. 1993; Hayward et al. 1994), but only after sufficient sampling and not in a predictive manner (Balsera et al. 1996). In particular, principal component analysis has been used to predict slow collective conformational motions (Grubmüller 1995) that cannot otherwise be predicted by, for example, conventional brute force, nanosecond molecular dynamics simulations.

Once essential degrees of freedom are defined, by integration over the orthogonal subspaces, a free-energy landscape for the essential degrees can be derived from the potential (enthalpic) energy landscape for all atomic coordinates (for a good review, see Elber 1996). It is assumed that such a dimension-reduced description of the dynamics of the essential degrees of freedom generally will have to be corrected for memory effects arising from the neglect of degrees of freedom considered irrelevant. Here the crucial question is how many "essential" coordinates suffice for a proper description of protein dynamics.

Nucleic Acids

Nucleic acids in their unfolded or single-stranded state are just as complex in coordinate and momentum space as are proteins. However, they differ from proteins in that simple base-pairing and base-stacking interactions dominate the structural organization. The "folding problem" for double helical DNA, whose logic was solved by Watson and Crick, proceeds by linear helix growth from a (randomly chosen) double helical nucleus. Folding of RNA molecules generally nucleates from hairpin helices, followed by double helix formation between more distant sequences, and formation of tertiary structural interactions. The secondary structure of complex RNAs, such as those in the ribosome, has been established by phylogenetic comparisons, in which it is observed that a sequence change at one site is accompanied by a complementary change at the site to which it is base paired. Thus, because of the dominance of strongly polar interactions in nucleic acids, the logic and pathway of their folding is much simpler than for proteins for which nonpolar interactions with their short range and nondirectional character are of great importance (Zarrinkar and Williamson 1994; Pyle and Green 1995; Doudna and Doherty 1997; Batey and Doudna 1998).

Is Dynamics Essential?

In many cases, and especially when the barrier for a reaction is sufficiently high, one finds that the reaction rate is determined by the probability of reaching the transition state (the activation free energy, $\Delta G^{\neq}$) rather than by pure dynamical effects. To address this point, we first need to define "dynamical effects." Several definitions are acceptable and we can, for example, state that "if we can determine a rate constant by using configurational sampling approaches (e.g., Monte Carlo simulation) then dynamical effects do not play a major role." This will be the case if the transmission factor in transition state theory is near unity and the prefactor does not vary too much in protein-solvent systems (Warshel 1991). The latter assumption is, however, the subject of controversial discussions. In other words, all processes involve motion but in many cases we can determine the probability of a process from the relevant activation energy.

In describing chemical reactions one has to account for the fact that a molecule fluctuates for a long time in the reactant state until the rare event occurs in which a fluctuation carries the molecule to the transition state and to the product state. The chance for such fluctuation is proportional to $\exp\{-\Delta G^{\neq}/RT\}$. The rate constant is then given by this expression multiplied by a preexponential factor that reflects the time that a trajectory needs to pass the transition state region. This time can be determined by running downhill trajectories and monitoring how long it takes to fall from the transition state to the reactant or product state minima. One can discern from such simulations how many times a system goes back and forth at the transition state, thereby providing the transmission factor. Once the barrier is sufficiently high, the time of the downhill trajectories is similar for barriers of different heights and, thus, we have a similar prefactor (usually not much different than unity) for many processes with moderate or large activation barriers (Warshel 1991).

A space and time-average structure in the native state, presumably at the global minimum of free energy, is revealed by such techniques as X-ray crystallography and magnetic resonance. The structure itself does not reveal directly the energy or the free energy; these must be derived from the structure by the application of suitable potential functions and estimates of the entropic contributions. If this can be accurately performed, then a comparison of two structures will yield a measure of the free-energy difference between them and hence (if appropriate) the equilibrium constant that governs their interconversion. This procedure is the more accurate, the more closely related the structures are; the free-energy perturbation method has proved effective in such cases (see, e.g., Radmer and Kollman 1997).

In relating structure to kinetics, the additional difficulty arises that the structural comparison is between structures at a free-energy minimum (a valley in the free energy vs. reaction coordinate cartoon) and a maximum (a peak or transition state). Although the minima may be directly accessible to experiment via either a conventional, static structure determination in the case of reactants or products, or via a time-resolved structure determination in the case of transient intermediates, the maxima will always be experimentally inaccessible (see Dill et al., this volume). Transition state geometries can only be inferred by suitable extrapolation from accessible structures, based on further chemical and enzymological knowledge. Such inferences, however, may be hard to draw. The prediction of rate constants from structure, or the understanding of observed rate constants in terms of structure, is therefore a hazardous process.

A first example may be provided by the reactivity of the α-and β-chains of deoxyhemoglobin in the T quaternary structure towards oxygen. The complete structure of deoxyhemoglobin reveals that the prospective binding pocket is open in the α-chain, but obstructed by the side chain of a valine residue in the β-chain. One might, therefore, predict that the α-chain would react with ligands more rapidly than the β-chain (Perutz 1970). However, kinetic experiments yielded the opposite result (Olson and Gibson 1972). This discrepancy may be accounted for by noting that the free energies of activation associated with transfer of ligand from bulk solvent to the ligand pocket of the α- and β-chains, adjacent to the iron, are required. Thus, the free energies of activation involved in the percolation of ligand through the α- and β-chains towards the ligand pocket, and the free energy required to displaced the valine are needed — and these are by no means obvious from a simple inspection of the deoxyhemoglobin structure.

Superoxide dismutase (SOD) provides a second example for a clear, yet surprising, structure–kinetics relationship. The very high binding rate relies on the electrostatic field surrounding the protein dimer. This field is shaped through the dielectric inhomogeneity of the protein–solvent interface as well as through salt effects; its theoretical description requires solution of the Poisson–Boltzmann equation. Brownian dynamics of the ligand–enzyme system in the force field along both translational and rotational degrees of freedom can explain the observations of Getzoff et al. (1992) on the action of human SOD and its mutants, binding rates of which differ by factors of 2–4. However, in the case of a mutant SOD from *Propionibacterium shermanii*, a reduction of the activity by 90% was observed, but only 17% could be accounted for by electrostatic effects (Gabbianelli et al. 1997). Nevertheless, no structural differences between the wild type and the mutant could be seen (M. Schmidt, F. Parak, and A. Desideri, unpublished results).

More challenging is an attempt to determine the rate constants for an enzyme mechanism by knowing the structures of the free enzyme and an enzyme-substrate complex. Here, one can take the initial structure, define potential surfaces for reactants and intermediate states in an assumed chemical mechanism, and perform free-energy perturbation studies of the relevant activation barriers. Such studies may be quite reliable when implemented with proper theoretical methods (Warshel 1991; Åqvist and Warshel 1993). Such calculations would be even more effective if the structures of the enzyme-product complex and of short-lived intermediates were also available (Dill et al., this volume).

In cytochrome c, the structure of the reduced and the oxidized form, cyt_{red} and cyt_{ox} respectively, has been determined by X-ray structure analysis and by NMR. Within the experimental error it is difficult to see any structural difference. Is it nevertheless possible to calculate from these structures the barrier height between the oxidized and the reduced form? This issue has been addressed by taking the two relevant structures and performing molecular dynamic simulations with the reduced system around the reduced structure and the oxidized system around the oxidized structure (Muegge et al. 1997). One can determine from these calculations the Marcus theory parabolas and, hence, the reaction barrier at the crossing point of the two parabolas.

Need for Quantum Treatment

Simulation tools currently employed for the dynamics of biomolecular systems are based mainly on classical mechanics. The use of those tools has fueled a remarkable progress in computational biology and led to many successes in the interpretation of experimental data. However, for systems having three to four atoms, say, quantum mechanical descriptions are required in order to make quantitative comparisons with experiments that yield detailed trajectories. Relatively accurate quantum mechanical calculations may also be required for many-atom biopolymer systems. We mention a few quantum effects which underscore this requirement.

So-called zero point quantum effects arise from the difference in classical and quantum mechanics in the low energy behavior of vibrational modes. In quantum mechanics there exists a finite, nonvanishing minimal energy content in such modes which leads to the specific heat approaching zero at temperatures less than $h\omega/k_B$, i.e., for many modes at physiological temperatures. This implies that classical mechanics overestimates energy transfer rates.

Tunneling is a second characteristic quantum process without a classical equivalent. Tunneling reactions have been widely reported, often reflected in strong isotope effects. The geminate rebinding of CO to myoglobin at helium temperature is a classical example for the tunneling of entire molecules (Alben et al. 1983). Recent studies by Klinman and coworkers on tunneling of hydrogen in the glucose oxidase reaction have attracted much attention (Bahnson et al. 1997). Strong isotope effects have also been observed for the kinetics of the proton pump cycle of bacteriorhodopsin.

Interpretation of spectroscopic observations, in particular in regard to force fields, needs to be based on a quantum mechanical description of the underlying vibrational motion, even in cases where the dynamical motions of the atoms involved are not subject to large quantum effects. Intensities of infrared and, in particular, resonance Raman spectra require accurate quantum mechanical descriptions. Vast amounts of vibrational spectra are available for certain protein systems, such as bacteriorhodopsin.

Nonradiative transitions between different electronic states occur in the course of many protein reactions, e.g., in electron transfer or in primary photobiological processes. The transition is an inherently nonclassical effect which typically takes place at intersections or near intersections of potential energy surfaces corresponding to different electronic states. Important photobiological processes where nonadiabatic transitions are known or believed to play a major role are the photoisomerization of retinal in rhodopsin and bacteriorhodopsin (Ebrey 1993; Ben-Nun et al. 1998), of the 4-hydroxycinnamic acid chromophore in photoactive yellow protein, PYP (Perman et al. 1998) and CO binding to myoglobin and hemoglobin (Buhks and Jortner 1985; Frauenfelder and Wolynes 1985).

Following the Born–Oppenheimer approximation, nuclear and electronic degrees of freedom are separated in most present-day simulations of biopolymers: establishment of a force field based on electron properties is succeeded by a description of nuclear motion. However, the increased power of computers already permits electronic and nuclear degrees of freedom to be treated simultaneously for systems of less than about hundred atoms, e.g., using semi-empirical electron Hamiltonians (Warshel and Levitt 1976; Field et al. 1990; Warshel 1991), density functional methods (Stanton et al. 1993; Wesolowski et al. 1996), or the Car–Parrinello method (Molteni and Parrinello 1998).

The described shortcomings will be overcome through a new generation of quantum simulation methods. The feasibility of such simulations has already been documented in a few cases, e.g., for a full quantum mechanical treatment of protein vibrations (Roitberg et al. 1996), for a quantum simulation of all nuclear motions in bacteriorhodopsin during the primary photoprocess (Ben-Nun et al. 1998), for a full description of electronic-nuclear degrees of freedom for an oligonucleotide crystal (Parrinello 1997), and for quantum mechanical tunneling in enzymes by path integral approaches (Hwang and Warshel 1996).

Quantum Mechanics as an Example of Simplification

The quintessential reaction process *reactant* → *product* in chemistry involves electrons switching between two states. Examples are electron transfer reactions in proteins, when an electron moves from an orbital on the donor moiety D to an orbital on the acceptor moiety A, and bond formation or bond breaking in an enzyme, when electrons shift from a nonbonding state to a bonding state or vice versa. When the reaction proceeds in a protein environment, systematic interactions with the protein environment influence the reaction. For example, electrostatic forces stabilize the transition state, but there arise also unsystematic contributions due to the thermal motion of the protein. The latter contribution is not necessarily unwanted. Rather like a diffusion controlled reaction driven by Brownian motion due to collisions with a condensed environment, a two-state quantum system in a protein can be driven by thermal vibrations, as demonstrated in Schulten (1995) and earlier work cited therein.

In fact, monitoring the time-dependent energy gap between two quantum states, $\Delta(t) = E_1(t) - E_2(t)$, provides a useful way of modeling chemical reactions in solution and in proteins (Warshel 1984; Treutlein et al. 1988). Fluctuations in this energy gap are usually determined by the fluctuation of the field from the protein and solvent, arising from nearby dipoles as well as charged residues. These fluctuations occur along structural coordinates $q_j, j = 1, 2, \dots 3N$ (N = number of protein atoms). Fluctuations across essentially the entire protein are coupled. In the case in which only $\Delta(t)$ is affected and the coupling to the protein is linear, i.e., $\Delta(t) = \Sigma_j c_j q_j$, the influence of the protein on the process can be captured succinctly through stochastic quantum mechanics as outlined in Warshel and Hwang (1986) and Leggett et al. (1987), and applied to proteins by Xu and Schulten (1994); see also Warshel et al. (1989). The theory requires knowledge of the weighted spectral density $J(\omega) \sim \Sigma j\ (c_j^2/\omega_j)\ \delta(\omega - \omega_j)$, where ω_j are the vibrational frequencies of the protein. However, one can express $J(\omega)$ through the correlation functions $<\Delta(t)\,\Delta(0)>$ and RMSD values of $\Delta(t)$ of the protein such that the theory does not rest on the assumption of ad hoc parameters (Xu and Schulten 1994). This electron transfer theory has been successfully applied to experimental data by McMahon et al. (1998). The stated description reduces to the Marcus theory in the limit of high temperature.

The dynamical effects of the protein fluctuations may be converted to activation barriers at the semiclassical high temperature limit. That is, integration of $J(\omega)$ gives the so-called Marcus reorganization energy (Warshel and Hwang 1986) and this can be used in the evaluation of the Marcus activation barrier. If the integrated $J(\omega)$ gives the same rate constant as the more rigorous treatment with the detailed $J(\omega)$, we may conclude that dynamical effects are not critical. This implies that quantum effects do not play a role for all vibrations involved in electron transfer ($h\omega << kT$).

The development of a potential stochastic quantum mechanics description for more general quantum systems in proteins is an active field of research and molecular biophysics could benefit greatly from advances (Schulten 1995). An example for a quantum system that requires a more general description is electronic excitations in chlorophyll aggregates in the photosynthesis apparatus (as discussed in Hu et al. 1997), where thermal fluctuations introduce incoherencies and localization of quantum states.

WATER MOLECULES TO ORGANELLES: FROM RELATIVE SIMPLICITY BACK TO COMPLEXITY

Molecules and Their Environment

The structure, dynamics, and function of biological macromolecules depend crucially on the properties of their environment. Macromolecules therefore cannot be treated separately from their environment; they form an inseparable system. The environment may differ for different regions of a single protein. For example, an integral membrane protein such as the acetylcholine receptor has three principal domains: a transmembrane domain in which the environment is that of the hydrophobic interior and the more hydrophilic head groups of the lipid bilayer, an extracellular domain which binds the neurotransmitter, and a cytoplasmic domain. The environments of the latter two domains are hydrophilic, aqueous solution. The environment may vary in time; for example, a protein may transiently associate with a membrane surface, or with other macromolecules as part of a multimeric complex, such as a transcription initiation complex or a spliceosome. The environment may be chosen as part of the experimental setup. For example, rarely does one investigate structure–function relations in natural environments. Normally, the protein is isolated from its native environment, purified, and (for example) caused to aggregate in the form of a crystal for structure determination. The crystal structure then provides the basis for a mechanistic explanation of function, although the function itself may have been examined in quite different solvents and environments. For a few systems like hemoglobin and myoglobin that have been studied in more detail, reaction rates in the crystal are similar to those in solution (Rivetti et al. 1993; Nienhaus et al. 1998) which is encouraging, but this fact should not encourage generalization of this observation. Indeed, ligand binding to hemoglobin crystals does not display the key functional feature of cooperativity (Rivetti et al. 1993).

Proteins are often embedded in viscous solvents like glycerol in experiments that study such protein functions as ligand binding, electron transfer, and proton transfer over a wide range of temperature. In metmyoglobin, even the electronic state of the heme iron is modified by surroundings (Bizzari et al. 1995). In many experiments, it has become obvious that protein motions are strongly affected by the dynamics of the surrounding glassy matrix, and the term "slaved glass transition" has been coined to described this situation (Ansari et al. 1987). However, the details are less clear. Which motions are arrested, and which are not affected? In small globular proteins like bacteriorhodopsin, myoglobin and blue copper proteins, most motions at the active site in the interior are frozen in upon vitrification of the solvent. However, for larger proteins, lesser constraints of the dynamics in the interior by the arrested surface atoms are expected.

One may distinguish two temperature regimes separated by the glass transition temperature, T_g, of the solvent. This separation is by no means sharp. Below T_g, the dynamics can be described essentially by a normal mode analysis. Jumps between conformational substates are rare; a great part of the functional properties is absent; and the molecule is nearly dead. Above T_g, all functions are present although the rates depend on temperature and slow down as T approaches T_g of the surrounding medium. In this physiological temperature region, practically all modes of motion present below T_g are retained but an additional channel of motions is activated. These motions may be described by a diffusion of parts of the molecule with respect to each other, limited in space by the forces which try to preserve the average protein structure. For example, the efficiency of the electron transfer Q_A to Q_B in the reaction center of *Rhodospirillum rubrum* (Parak et al. 1980; Novella and Schulten 1991) or photosystem II of spinach (Garbers et al. 1998) correlates with these motions. In this picture, the glass transition of the environment suppresses functionally important motions. However, it is not necessarily enforcing a glass transition in the protein. A "slaving" of protein dynamics by the environment of the surrounding medium occurs only below T_g (see Parak and Palma, this volume).

Proteins can be embedded in rigid polymer or sugar matrices (see, e.g., Cordone et al. 1998) that restrict or suppress fluctuations between conformational substates even at room temperature, or incorporated into other non-native environments like surfaces or interfaces. Altered, controllable environments are of great interest not only for certain types of experiments, but also for technological applications to the preservation of biomolecules in the food and pharmaceutical industries, and to the development of biocatalysts and biosensors. A better understanding of the interactions between proteins and their environments could greatly assist in their technological application.

Solvent-induced Forces

The known flexibility of water molecules in making hydrogen-bonded structures is the reason for the particular complexity and ruggedness of the energy landscape of liquid water. Solutes such as macromolecules alter this energy landscape and related free energy of water. Consequently, a thermodynamic force arises that attempts to bring solutes toward the nearest spatial arrangement which minimizes the solute-solvent contributions to free energy. The gradient of the latter expresses the pertinent solvent-induced force (SIF).

An example of the interaction between the solvent and a protein is provided by hemoglobin. In its transition from the R (oxy) state to the T (deoxy) state, hemoglobin exposes some 1,000 Å^2 of additional hydrophobic surface to the solvent (Chothia et al. 1976). Experiments estimate the free-energy cost of the additional hydration of this surface at about 60 kJ mol^{-1} (Parak and Palma, this volume). It involves the statistical transfer of about 75 water molecules from the bulk to the hydrophobic surface (Bulone et al. 1993). This solvent-based contribution is essential to the structural transition and to the associated functional properties of hemoglobin. The simplifying view is reached by averaging over the relatively large hydrophobic surface. At single residue resolution, however, local effects of the solvent are best elicited by SIFs due to their intrinsically differential nature. This allows the notable simplification of expressing the local effects of interactions between the two highly complex systems of protein and water with a limited set of force vectors. Complexity reappears, nevertheless, under the aspect of a strong non-pair additivity of SIFs and related strong many-body character of hydration free energy. Strong evidence for such complexity has been obtained recently (Brugé et al. 1997; Martorana et al. 1996, 1997, 1998; Bulone et al. 1997; San Biagio et al. 1998). As a consequence, SIFs are characterized by a marked dependence upon the specific spatial arrangements (or "context") of solutes. This causes highly selective interactions and unexpected modifications of the free-energy landscape of solutes (Parak and Palma, this volume).

This view can be integrated over a scale even larger than the 1000 Å^2 mentioned above, e.g., over entire proteins. In this case, rewarding simplification is obtained in terms of a mean field approximation (Bulone et al. 1993). Simultaneous consideration over two scales allows eliciting interwined simplicity/complexity effects (Parak and Palma, this volume).

Macromolecular Assemblies

Many cellular functions are achieved by assemblies of biopolymers, for example, multi-enzyme complexes, ribonucleoproteins, individual proteins or groups of proteins embedded in membranes, viruses, and complexes of proteins with DNA. Through advances in crystallography and electron microscopy, structures of these large systems are being solved today with increasing frequency. Due to their important functions, the extension of structural biology towards cell biology through study of such assemblies will assume center stage during the next decade.

This will pose new conceptual and technical challenges to researchers. For example, computational methodologies need to be developed further, based on the general increase of processor power and the advent of massively parallel computing (Heller et al. 1990) as well as highly efficient molecular dynamics algorithms (Grubmüller et al. 1991; Eichinger et al. 1995, 1997; Luty and Van Gunsteren 1996; Schlick et al. 1997; Grubmüller and Tavan 1998; Schlick et al. 1999; Kale et al. 1999). Molecular graphics must be adapted to the study of very large assemblies, with the main goal being support in navigating through large structures, up and down the spatial resolution scale. New conceptual (cartoon) and detailed (physical) representations may combine geometrical and physical data. For example, mechanical models of protein domains as the building blocks of aggregates are valuable for examining virus capsids and light-harvesting systems (Bailey et al. 1998).

Simulation tools must be capable of dealing with the numerical task posed by accurate many-atom simulations (10^6 atoms), and of manipulating individual components, e.g., pulling components apart while adhesion forces are monitored. For the latter purpose new simulation tools, so-called steered (targeted) molecular dynamics methods, are presently being developed and have been successfully applied (Grubmüller et al. 1996; Izrailev et al. 1997; Lu et al. 1998; Marrink et al. 1998; Marszalek et al. 1999; see also Dill et al., this volume). The main challenge at present is the short time scale of computational manipulations which implies a large dissipative component to all forces measured (Grubmüller et al. 1996; Izrailev et al. 1997). This component must be discounted in order to furnish the effective force fields relevant for cellular processes (Gullingsrud et al. 1999; Heymann and Grubmüller 1999).

An example that demonstrates the necessity for modeling and simulation on a larger scale is provided by the problem of domain mobility in the cytochrome bc_1 complex. The cytochrome bc_1 complex from mitochondria consists of 11 protein subunits and four cofactors. As part of the respiratory chain, it oxidizes ubiquinol (QH_2) to quinone (Q) and pumps four protons from the matrix to the intermembrane space. It has two major quinone binding sites, Q and Q_0. At the quinone oxidation site Q_0, one electron of QH_2 is transferred to the iron sulfur protein (ISP) in a strongly exothermal reaction ($\Delta E \sim 300$ mV); the second electron of QH_2 is transferred via the two hemes of cytochrome b to the quinone reduction site Q in a much less exothermal reaction ($\Delta E \leq$ 100 mV) (Brandt and Trumpower 1994). X-ray structure analyses (Xia et al. 1997; Kim et al. 1998; Zhang et al. 1998; Iwata et al. 1998) of bovine bc_1 in the presence and absence of different Q_0 inhibitors showed that the extramembrane domain of the ISP can switch between a fixed state, where it is bound to the surface of cytochrome b, and a loose state, where it does not populate a structurally well-defined state. In the loose state, it appears to be free to undergo a hinged motion that could achieve two goals. It could assist electron transfer from the ISP to cytochrome c_1 by allowing transient contacts between these proteins; or, transition from the fixed to the loose state could prevent the transfer of both electrons to the iron-sulfur protein which would halve the enzymatic efficiency. Different conformations of the ISP were also reported in crystal structures of chicken bc_1 complex in the presence and absence of a Q_0 inhibitor and of the complex from rabbit (Zhang et al. 1998)

The nature and kinetics of the ISP motion can be simulated by large-scale computational modeling (Izrailev et al. 1999) and may eventually explain the electron transfer kinetics between QH_2 and cyt c_1.

Levels of Organization

In the biological sciences, the simplifying concept of level of organization has been found rather useful. In the brain, for instance, one may distinguish the molecular from the cellular level, the "small networks" level from the level of neuronal assemblies, and the level of "assemblies of assemblies." Theoretical and experimental work will attempt to relate a particular level of organization to a relevant function, knowing *a priori* that these hierarchical levels are "nested" and interdependent. The Hodgkin and Huxley model for transmission of the nerve impulse was proposed at the cellular level without knowing the actual structure of ion channels, and the Hebb rule for learning was proposed at the neuron assembly level without knowing what neurotransmitter receptors are as molecular entities.

One may try to apply such views to protein molecules themselves, where hierarchical levels of organization may also be recognized. For example, individual subunits of the acetylcholine receptor are split into spatially distinct, functional domains; the extracellular N-terminal domain contributes to ligand binding, the transmembrane domain contributes to the ion channel, and the cytoplasmic domain to regulation. These domains display both folding and functional autonomy. Indeed, one may build a functional receptor by linking together the binding domain of the $\alpha 7$ neural nicotinic receptor and the transmembrane or channel domain of the $5HT_3$ (serotonin) receptor, despite considerable sequence differences (Eiselé et al. 1993). In the case of the glutamate receptor, the evolutionary origin of these two distinct domains can be traced back to bacteria (Paas 1998). The glutamate-binding domain strikingly resembles the bacterial periplasmic lysine/arginine/ornithine-binding protein or the leucine/isoleucine /valine-binding protein. On the other hand, the channel domain shares such structural features with the bacterial potassium channel as its tetrameric organization and presence of a P loop. One may say that "tinkering" with two bacterial proteins suffices to create a brain receptor!

At the next highest structural level, that of quaternary structure, "emergent" properties may be recognized which are directly dependent upon specific features of that structure. To build a binding site, the single subunit level suffices. On the other hand, to achieve cooperativity between binding sites seems to require the assembly of subunits (protomers) into limited, closed, symmetrical assemblies referred to as oligomers. This is the case for hemoglobin, for many regulatory enzymes such as aspartate transcarbamoylase, phosphorylase b and L-lactate dehydrogenase, for receptors of neurotransmitters, and so on. Without a doubt, the cooperative binding of specific ligands to these flexible oligomers is dependent upon their quaternary organization. As proposed in the Monod–Wyman–Changeux (MWC) model, their equilibrium binding and kinetic data can be accounted for by a "concerted" or cooperative conformational transition of the quaternary assembly (Eaton, this volume). In other words, the "simple" functional feature of responding to the threshold concentration of a regulatory ligand is "determined" by the more complex assembly of subunits into an oligomeric structure – a common feature of many regulatory, allosteric proteins. From an evolutionary point of view, only a few surface mutations may suffice to confer on an independent subunit the ability to associate with itself, to acquire quaternary structure, and thus to form a cooperative oligomer.

In the case of the acetylcholine receptor, the situation is even more complex. The receptor prepared from fish electric organ (Changeux 1990) is a 300 kDa heteropentamer made up of 4 homologous subunits in the clockwise order $\alpha\gamma\alpha\delta\beta$ (Hucho et al. 1996). It carries two acetylcholine binding sites at the boundaries between subunits and possesses an ion channel located along the fivefold rotation axis of pseudosymmetry of the molecule. The assembly of a fivefold pseudosymmetrical oligomer from four chemically different subunits remains a challenge for the physical chemists as well as for the molecular biologists. Early work suggests that: (1) the assembly of the subunits takes place through initial formation of an $\alpha\gamma$ and an $\alpha\delta$ dimer, their association together, and inclusion of the β subunit to form the functional pentameric oligomer; (2) folding and assembly of the subunits occurs in the endoplasmic reticulum and is a very slow process, requiring hours; and (3) many molecules do not reach the correct conformation and are destroyed (only about 20% of adequately assembled molecules "survive" in the cytoplasmic membrane). The folding and assembly processes of such complex molecules include a long series of molecular and cellular processes which result in drastic final selection of the functional molecule. It would be a mistake to view the assembly process as a strictly computational problem.

The distance between acetylcholine binding sites and the ion channel is in the range of 20–30 Å (Herz et al. 1989). Interaction typically takes place between topographically distinct sites; it is, by definition, allosteric (Monod et al. 1963). The signal transduction mechanism thus emerges at the level of the quaternary organization of the protein.

From a kinetic point of view, the receptor oligomer undergoes a cascade of conformational transitions upon binding the neurotransmitter: a fast opening of the ion channel in the millisecond time scale (activation transition), and a slower "desensitization" transition in the 100 ms to second (or even minute) time scale which leads to a high affinity "refractory state." A simple representation of the transition includes four states denoted R, A, I, and D (Figure 15.1) which, in agreement with the MWC scheme, may preexist in the absence of ligand binding (Changeux 1990; Edelstein et al. 1996).

Many, if not all, ligand-gated ion channels exhibit slow desensitization transitions. Their functional significance is evident. They regulate the "efficiency" of the receptor response to the neurotransmitter. They may serve as elementary "memory" devices for the storage of short-term memories in neural networks (see Heidmann and Changeux 1982; Dehaene and Changeux 1997). Moreover, the transmembrane organization of the receptor is such that they may "integrate" multiple extracellular as well as intracellular signals through such transitions, thus serving as time coincidence detectors (see Edelstein and Changeux 1998). A more complicated structure thus allows for a more diverse range of function and much more sophisticated regulation.

Selection may not exert its action solely on fast reactions. Slow transitions may have important physiological functions in the brain, in photoreceptors and in other systems, and thus be directly selectable by evolution. Indeed, as a rule of thumb the larger a molecular entity or complex, the slower are certain of its structural transitions. Quaternary structural transitions are usually slower than (and preceded by) tertiary structural transitions.

Further structural and functional complexities arise at the level of the whole cell. Proteins are not necessarily dispersed within the cytoplasm in a random, nonorganized manner. Membrane proteins may, for instance, cluster in small-scale assemblies or

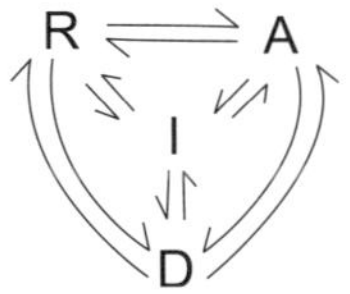

Figure 15.1 Conformational transitions in the acetylcholine receptor (Changeux 1990; Edelstein et al. 1996). For acetylcholine, the R to A activation equilibrium constant has a value of around 10^{-5} M; the I to D desensitization equilibrium constant has a value around 10^{-8} M.

even in well-defined edifices limited, on the outside, to the basal lamina or to the connective tissue environment; or, on the inside, to the cytoskeleton. A particularly remarkable example is the very restricted topological distribution of neurotransmitter receptors at the synapse (Duclert and Changeux 1995). At the neuromuscular junction, the surface density of acetylcholine receptor molecules under the nerve ending where neurotransmitter release takes place is near to close packing (several thousand receptors per μm^2); the lateral diffusion of the receptors is negligible; and the receptor protein is metabolically stable with a half-life of about 11 days. The stability of the postsynaptic aggregate of receptor molecules relies upon multiple interactions with other extracellular structures (agrin, collagen, "cleft substances"...) and with intracellular, cytoplasmic, structures (43K-rapsyn, cytoskeletal proteins...). The receptor molecules traverse the lipid bilayer and in the aqueous environment on both sides of the membrane, they may interact with other protein or polysaccharide species through specialized binding sites.

We emphasize that this rather static picture of the higher-order assemblies of macromolecules does not fully suffice to understand their properties and stability. In particular, we must take into account the developmental aspects of synapse formation, in particular the contribution of local regulation of gene transcription in this process. Indeed, the myofiber is a multinucleate structure but there exists a differentially high transcriptional activity of acetylcholine receptor genes at the level of the subjunction of nuclei. In general, understanding protein structure and superstructure requires an understanding of its past evolutionary and developmental histories.

ARE THERE GENERAL LAWS?

In the physical sciences, one is looking for the simplifying concepts embodied in "general laws." The Maxwell equations describe, for instance, the field of electrodynamics in a unique and definitive manner. The question is, can more or less general laws also be formulated for macromolecules and their assemblies? This can be doubted on both philosophical and practical grounds (see, e.g., Parak and Palma, this volume). The concept of "general laws" that rule the universe is rather misleading; the implicit reference to a Platonist world of "Ideas" and "Forms" is no longer acceptable. In general, we refer to more limited and modest "models" which capture regularities in the world on the basis of, for example, a mathematical formalism. No mathematical equation, as "beautiful" as it may be, can pretend to give an exhaustive description of physical reality. We should, therefore, be cautious and humble in our attempt to evaluate regularities and diversities in the world, and in particular, in the world of macromolecules. The eminent evolutionary biologist Ernst Mayr posited that pitfalls might be avoided in the biological sciences by rejecting the so-called "physicist view of the world." We must recognize that living organisms are not homogeneous systems but display an immense structural diversity, unlike systems studied conventionally in physics, and such diversity cannot be understood without rejecting the standard static view of the world and incorporating biological evolution with its very long time scales. (It must be stated that the above paragraph does not reflect the opinion of all authors).

At a more practical level, the question regarding general laws can also be asked in regard to the physical properties of macromolecules. One may ask, for example, if descriptions valid over a wide length scale can be found, e.g., if a conventional full atom molecular dynamics description can be converted into simplified descriptions which include averages over, for example, the 10 to 50 Å scale. This question had been posed in the theory of liquids with the expectation that a full atom molecular dynamics description could be replaced by sequence of pictures which, after coarse graining at a 1000 Å length scale, would assume the well-known forms of hydrodynamics. Despite years of efforts, the theory of liquids did not reach this goal; it turned out to be either extremely difficult or impossible to develop simplified descriptions of liquids valid over the entire length scale of 10 to 1000 Å. Unfortunately, this length scale is that of proteins. One might conclude that similarly it would be hopeless to derive coarse-grained descriptions of proteins. However, certain computationally useful approaches on a variety of length scales are being pursued, such as multiple time step algorithms (Tuckerman et al. 1990; Grubmüller et al. 1991; Humphreys et al. 1995; Zhou and Berne 1995; Littell et al. 1997; Schlick et al. 1997), multipole methods (Appel 1985; Barnes and Hut 1986; Greengard and Rokhlin 1989; Board et al. 1992; Eichinger et al. 1997), lattice sum and grid methods (Luty et al. 1995; Luty and Van Gunsteren 1996), "essential dynamics" (Amadei et al. 1993), "conformational flooding" (Grubmüller 1995), and elastic theory as applied to larger stretches of DNA, both supercoiled and in loops (Schlick 1995; Olson 1996; Balaeff et al. 1999).

With these cautionary tales, we should try to carefully limit our ambition to the recognition of common functional properties (catalysis, cooperativity, signal transduction...) and to the identification of the common structural and dynamical features to which the common functions may be carefully related. This is a rather reasonable program. It is preferable to search for "common rules" rather than to strive for "general laws."

Some Common Rules

We supply a few common rules and invite the reader to add to the list.

1. The function of enzymes, transport proteins, receptors, motor proteins, and many other systems is associated with some structural flexibility. Therefore,

the free energy of structural stabilization is very small and corresponds to 2 to 3 hydrogen bonds only. Exceptions include so-called mechanical proteins, like titin, that are protected against mechanical strain.

2. Proteins possess features such that they fold, function, and evolve. Only defined parts of the molecular structure contribute directly to a specific function (see, however, Martorana et al. 1996, 1998). Other parts may become significant for function upon changes of the environment.
3. Many proteins make clever and highly sophisticated use of the entropic component of free energy. Without, of course, violating thermodynamic laws, they nevertheless appear to come closer to Maxwell's Demon than most other many-body systems. An example is the light-driven proton pump, bacteriorhodopsin, which after light absorption stores only a small fraction of the photon's energy. It apparently primes its pump through photochemical selection of a state with high entropy content. The universality of ATP in driving cell processes likewise seems to be based on the high degree of topochemical specificity (directedness) of its endothermic hydrolysis reaction.
4. Catalysis arises from transition state stabilization rather than from ground state destabilization. It appears that increasing the catalytic rate constant k_{cat} involves electrostatic complementarity between the charge distribution of the transition state and the surrounding protein active site.
5. The reduction of complexity to simplicity in the self-organization of polymers is made easier if polar, directional interactions dominate, as is the case for nucleic acids as opposed to proteins.
6. Assembly of a protein into a finite oligomeric structure confers symmetry properties and cooperative behavior.
7. In the case of metalloproteins, the electronic configuration of the metal has a large number of low energy levels (2–600cm^{-1}) in contrast to inorganic and organic metal components (first excited level $\approx$ 10,000 cm^{-1}). This is necessary for reversibility of binding small molecules or accepting additional electrons under conditions of constant temperature or pressure.

CONCLUSIONS

The examples in this chapter present complexity at a variety of experimental and theoretical levels. The prospects of uncovering "general laws" are apparently still distant, as evidenced by the paucity and relative specificity of even the "common rules", above. Perhaps the biologist Stephen Jay Gould said it best, in a recent essay (Gould 1999): "The fallacies and foibles of human thinking lead us into systematic and predictable trouble when we try to grasp the complexities of external reality. Among these foibles, our persistent attempts to build abstractly beautiful, logically impeccable, and comprehensively simplified systems always lead us astray .... Nature, to cite a modern cliché, always bats last. She will not succumb to the simplicities of our hopes or mental foibles, but she remains eminently comprehensible."

Observations may be relatively simple, but elucidating the connections between the undoubted complexity of macromolecular structure and dynamics and these observations remains a major research challenge.

ACKNOWLEDGMENTS

We thank Hans Deisenhofer and Martin Karplus for their thoughtful comments and contributions.

REFERENCES

Alben, J.O., D. Beece, S.F. Bowne, L. Eisenstein, H. Frauenfelder, D. Good, M.S. Marden, P.P. Moh, L. Reinisch, A.H. Reynolds, and K.T. Yue. 1983. Isotope effect in molecular tunneling. *Phys. Rev. Lett.* **44**:1157–1160.

Amadei, A., A.B.M. Linssen, and H.J.C. Berendsen. 1993. Essential dynamics of proteins. *Proteins* **17**:412–425.

Ansari, A., J. Berendzen, S.F. Bowne, H. Frauenfelder, I.E.T. Iben, T.B. Sauke, E. Shyamsunder, and R.D. Young. 1985. Protein states and protein quakes. *Proc. Natl. Acad. Sci. USA* **82**:5000–5004.

Ansari, A., J. Berendzen, D. Braunstein, B.R. Cowen, H. Frauenfelder, M.K. Hong, I.E.T. Iben, J.B. Johnson, P. Ormos, T.B. Sauke, R Scholl, A. Schulte, P.J. Steinbach, J. Vittitow, and R.D. Young. 1987. Rebinding and relaxation in the myoglobin pocket. *Biophys. Chem.* **26**:337–355.

Appel, A.W. 1985. An efficient program for many-body simulation. *SIAM J. Sci. Stat. Comput.* **6**:85–103.

Åqvist, J., and A. Warshel. 1993. Simulation of enzyme reactions using valence bond force fields and other hybrid quantum/classical approaches. *Chem. Rev.* **93**:2523–2544.

Austin, R.H., K.W. Beeson, L. Eisenstein, H. Frauenfelder, and I.C. Gunsalus. 1975. Dynamics of ligand binding to myoglobin. *Biochemistry* **14**:5355–5373.

Bahnson, B.J., T.D. Colby, J.K. Chin, B.M. Goldstein, and J.P. Klinman. 1997. A link between protein structure and enzyme catalyzed hydrogen tunneling. *Proc. Natl. Acad. Sci. USA* **94**:12,797–12,802.

Bailey, M., U. Schulten, and J.E. Johnson. 1998. The use of solid physical models for the study of macromolecular assembly. *Curr. Opin. Struct. Biol.* **8**:202–208.

Balaeff, A., L. Mahedevan, and K. Schulten. 1999. Elastic rod model of a DNA loop in the lac operon. *Phys. Rev. Lett.* **83**:4900–4903.

Balsera, M.A., W. Wriggers, Y. Oono, and K. Schulten. 1996. Principal component analysis and long time protein dynamics. *J. Phys. Chem.* **100**:2567–2572.

Barnes, J., and P. Hut. 1986. A hierarchical O(N log N) force-calculation algorithm. *Nature* **324**:446–449.

Batey, R.T., and J.A. Doudna. 1998. The parallel universe of RNA folding. *Nature Struct. Biol.* **5**:337–340.

Ben-Nun, M., F. Molnar, H. Lu, J.C. Phillips, T.J. Martinez, and K. Schulten. 1998. Quantum dynamics of retinal's femtosecond photoisomerization in bacteriorhodopsin. *Faraday Soc. Disc*. **110**:447–462.

Bizzari, A.R., O.A. Iakovleva, and F. Parak. 1995. Spin lattice relaxation in Mössbauer spectra of metmyoglobin: Investigation of crystals, water, and water-glycerol solutions. *Chem. Phys.* **191**:185–194.

Board, J.A., J.W. Casey, J.F. Lethrum, A. Windemuth, and K. Schulten. 1992. Accelerated molecular dynamics simulation with the fast multipole algorithm. *Chem. Phys. Lett.* **198**:89–94.

Brandt, U., and B.L. Trumpower. 1994. The protonmotive Q cycle in mitochondria and bacteria. *CRC Crit. Rev. Biochem. Mol. Biol.* **29**:165–197.

Brugé, F., S.L. Fornili, G.G. Malenkov, and M.B. Palma-Vittorelli. 1997. Solvent-induced forces on a molecular scale: Non-additivity, modulation, and causal relation to hydration. *Chem. Phys. Lett.* **254**:283–291.

Buhks, E., and J. Jortner. 1985. The dynamics of carbon monoxide binding to hemoglobin and to cytochrome. *J. Chem. Phys.* **83**:4456–4462.

Bulone, D., V. Martorana, P.L. San Biagio, M.B. Palma-Vittorelli. 1997. Effects of electric charges on hydrophobic forces. *Phys. Rev. E* **56**:R4939–R4942.

Bulone, D., P.L. San Biagio, M.B. Palma-Vittorelli, and M.U. Palma. 1993. On the role of water in hemoglobin function and stability. *Science* **259**:1335–1336.

Changeux, J.-P. 1990. Functional architecture and dynamics of the nicotinic acetylcholine receptor: An allosteric ligand-gated ion channel. Fidia Research Foundation Neuroscience Award Lectures. **4**:21–168.

Changeux, J.-P., and S.J. Edelstein. 1998. Allosteric receptors after 30 years. *Neuron* **21**:959–980.

Chothia, C., S. Wodak, and J. Janin. 1976. Role of subunit interfaces in the allosteric mechanism of hemoglobin. *Proc. Natl. Acad. Sci. USA* **73**:3793–3797.

Cordone, L., P. Galajda, E. Vitrano, A. Gassmann, A. Ostermann, and F. Parak. 1998. A reduction of protein specific motions in CO-ligated myoglobin embedded in a trealose glass. *Eur. Biophys. J.* **27**:173–176.

Dehaene, S., and J.-P. Changeux. 1997. A hierarchical neuronal network for planning behavior. *Proc. Natl. Acad. Sci. USA* **94**:13,293–13,298.

Doudna, J.A., and E.A. Doherty. 1997. Emerging themes in RNA folding. *Folding and Design* **2**:R65–R70.

Duclert, A., and J.-P. Changeux. 1995. Acetylcholine receptor gene expression at the developing neuromuscular junction. *Phys. Rev.* **75**:339–368.

Ebrey, T. 1993. Light energy tranduction in bacteriorhodopsin. In: Thermodynamics of Membranes, Receptors and Channels, ed. M. Jacobson. New York: CRC Press.

Edelstein, S., O. Schaad, E. Henry, D. Bertrand, and J.-P. Changeux. 1996. A kinetic mechanism of nicotinic acetylcholine receptors based on multiple allosteric transitions. *Biol. Cybern.* **75**:361–379.

Eguchi, Y., and J. Tomizawa. 1991. Complexes formed by complementary RNA stem loops: Their formations, structures, and interactions with Col E1 Ram protein. *J. Mol. Biol.* **220**:831–842.

Eichinger, M., H. Grubmüller, and H. Heller. 1995. User manual for EGO_VIII, Release 2.0, electronic access: http://www.mpibpc.gwdg.de/abteilungen/o71/ego.html.

Eichinger, M., H. Grubmüller, H. Heller, and P. Tavan. 1997. FAMUSAMM: A new algorithm for rapid evaluation of electrostatic interactions in molecular dynamics simulations. *J. Comp. Chem.* **18**:1729–1749.

Eiselé, J.L., S. Bertrand, J.L. Galzi, A. Devillers-Thiéry, J.-P. Changeux, and D. Bertrand. 1993. Chimaeric nicotinic-serotonergic receptor combines distinct ligand binding and channel specificities. *Nature* **366**:479–483.

Elber, R. 1996. Reaction path studies of biological molecules. In: Recent Developments in Theoretical Studies of Proteins, ed. R. Elber. Tai Seng, Singapore: World Scientific.

Field, M.J., P.A. Bash, and M. Karplus. 1990. A combined quantum-mechanical and molecular-mechanical potential for molecular-dynamics simulations. *J. Comp. Chem.* **11**:700–733.

Frauenfelder, H., S.G. Sligar, and P.G. Wolynes. 1991. The energy landscapes and motions of proteins. *Science* **254**:1598–1603.

Frauenfelder, H., and P.G. Wolynes. 1985. Rate theories and puzzles of hemeprotein kinetics. *Science* **229**:337–345.

Gabbianelli, R., A. Battistoni, E. Polticelli, B. Meier, M. Schmidt, G. Rotolio, and A. Desideri. 1997. Effect of Lys 175 mutation on structure-function properties of Propionibacterium shermanii superoxide dismutase. *Protein Eng.* **10**:1067–1070.

Garbers, A., F. Reifarth, J. Kurreck, G. Renger, and F. Parak. 1998. Correlation between protein flexibility and electron transfer from QA to QB in PSII membrane fragments from spinach. *Biochemistry* **37**:11,399–11,404.

Garcia, A.E. 1992. Large-amplitude nonlinear motions in proteins. *Phys. Rev. Lett.* **68**:2696–2699.

Getzoff, E.D., D.E. Cabelli, C.L. Fisher, H.E. Parge, M.S. Viezzoli, L. Banci, and R.A. Hallewell. 1992. Faster superoxide dismutase mutants designed by enhancing electrostatic guidance. *Nature* **358**:347–351.

Go, N., and H.A. Scheraga. 1969. Analysis of the contribution of internal vibrations to the statistical weights of equilibrium conformations of macromolecules. *J. Chem. Phys.* **51**:4751–4767.

Gould, S.J. 1999. Branching through a wormhole: Lamarck's ladder collapses. *Nat. History* **108**:24ff.

Greengard, L., and V. Rokhlin. 1989. On the evaluation of electrostatic interactions in molecular modeling. *Chem. Scr.* **29A**:139–144.

Grubmüller, H. 1995. Predicting slow structural transitions in macromolecular systems: Conformational flooding. *Phys. Rev. E* **52**:2893–2906.

Grubmüller, H., H. Heller, A. Windemuth, and K. Schulten. 1991. Generalized Verlet algorithm for efficient molecular dynamics simulations with long-range interactions. *Mol. Simul.* **6**:121–142.

Grubmüller, H., B. Heymann, and P. Tavan. 1996. Ligand binding: Molecular mechanics calculation of the streptavidin-biotin rupture force. *Science* **271**:997–999.

Grubmüller, H., and P. Tavan. 1998. Multiple time step algorithms for molecular dynamics simulations of proteins: How good are they? *J. Comp. Chem.* **19**:1534–1552.

Gullingsrud, J., R. Braun, and K. Schulten. 1999. Reconstructing potentials of mean force through time series analysis of steered molecular dynamics simulations. *J. Comp Phys.* **151**:190–211.

Hayward, S., A. Kitao, and N. Go. 1994. Harmonic and anharmonic aspects in the dynamics of BPTI: A normal mode analysis and principal component analysis. *Physica Scripta* **3**:936–943.

Heidmann, T., and J.-P. Changeux. 1982. Un modèle moléculaire de régulation d'efficacité d'un synapse chimique au niveau postsynaptique. *C.R. Acad. Sci. Paris* **295**:665–670.

Heller, H., H. Grubmüller, and K. Schulten. 1990. Molecular dynamics simulation on a parallel computer. *Mol. Simul.* **5**:133–16.

Herz, J.M., D.A. Johnson, and P. Taylor. 1989. Distance between the agonist and noncompetitive inhibitor sites on the nicotinic acetylcholine receptor. *J. Biol. Chem.* **264**:12,439–12,448.

Heymann, B., and H. Grubmüller. 1999. ANO2-Hapten unbinding forces studied by molecular dynamics AFM simulations. *Chem. Phys. Lett.* **303**:1–9.

Hu, X., T. Ritz, A. Damajanovic, and K. Schulten. 1997. Pigment organization and transfer of electronic excitation in the purple bacteria. *J. Phys. Chem. B* **101**:3854–3871.

Hucho, F., V.I. Tsetlin, and J. Machold. 1996. The emerging three-dimensional structure of a receptor, the nicotinic acetylcholine receptor. *Eur. J. Biochem.* **239**:539–557.

Humphreys, D.D., R.A. Friesner, and B.J. Berne. 1995. Simulated annealing of a protein in a continuum solvent by multiple-time-step molecular dynamics. *J. Phys. Chem.* **99**:10,674–10,685.

Hwang, J.-K., and A. Warshel. 1996. How important are quantum mechanical nuclear motions in enzyme catalysis? *J. Am. Chem. Soc.* **118**:11,745–11,751.

Iwata, S., J.W. Lee, K. Okada, J.K. Lee, M. Iwata, B. Rasmussen, T.A. Link, S. Ramaswamy, and B.K. Jap. 1998. Complete structure of the 11-subunit bovine mitochondrial cytochrome bc1 complex. *Science* **281**:64–71.

Izrailev, S., A.R. Crofts, E.A. Berry, and K. Schulten. 1999. Steered molecular dynamics simulation of the Rieske subunit motion in the cytochrome bc1 complex. *Biophys. J.* **77**: 1753–1768.

Izrailev, S., S. Stepaniants, M. Balsera, Y. Oono, and K. Schulten. 1997. Molecular dynamics study of unbinding of the avidin-biotin complex. *Biophys. J.* **72**:1568–1581.

Kale, L., R. Skeel, M. Bhandarkear, R. Brunner, A. Gursoy, N. Krawelz, J. Phillips, A. Shinozaki, K. Varadarajan, and K. Schulten. 1999. NAMD2: Greater capability for parallel molecular dynamics. *J. Comp. Phys.* **151**:283–312.

Karplus, M., and J.N. Kushick. 1981. Method for estimating the configurational entropy of macromolecules. *Macromolecules* **14**:325–332.

Kim, H., D. Xia, C.-A. Yu, J.-Z. Xia, A.M. Kachurin, L. Zhang, L. Yu, and J. Deisenhofer. 1998. Inhibitor binding changes domain mobility in the iron-sulfur protein of the mitochondrial bc1 complex from bovine heart. *Proc. Natl. Acad. Sci. USA* **95**:8026–8033.

Leggett, A.J., S. Chakravarty, A.T. Dorsey, M.P.A. Fisher, A. Garg, and W. Zwerger. 1987. Dynamics of the dissipative two-state system. *Rev. Mod. Phys.* **59**:2–86.

Littell, T., R. Skeel, and M. Zhang. 1997. Error analysis of symplectic multiple time stepping. *SIAM J. Numerical Analysis* **34**:1792–1807.

Lu, H., B. Isralewitz, A. Krammer, V. Vogel, and K. Schulten. 1998. Unfolding of titin immunoglobin domains by steered molecular dynamics simulation. *Biophys. J.* **75**:662–671.

Luty, B.A., I.G. Tironi, and W.F. van Gunsteren. 1995. Lattice-sum methods for calculating electrostatic interactions in molecular simulations. *J. Chem. Phys.* **103**:3014–3021.

Luty, B.A., and W.F. van Gunsteren. 1996. Calculating electrostatic interactions using the particle–particle particle–mesh method with non-periodic long-range interactions. *J. Phys. Chem.* **100**:2581–2587.

Marrink, S.-J., O. Berger, P. Tieleman, and F. Jähnig. 1998. Adhesion forces of lipids in a phospholipid membrane: Studies by molecular dynamics simulations. *Biophys. J.* **74**:931–943.

Marszalek, P.E., H. Lu, H. Li, M. Carrion-Vazquez, A.F. Oberhauser, K. Schulten, and J.M. Fernandez. 1999. Mechanical unfolding intermediates in titin modules. *Nature* **402**: 100–103.

Martorana, V., D. Bulone, P.L. San Biagio, M.B. Palma-Vittorelli, and M.U. Palma. 1997. Collective properties of hydration: Long range and specificity of hydrophobic interactions. *Biophys. J.* **73**:31–37.

Martorana, V., G. Corongiu, and M.U. Palma. 1996. Correlated solvent-induced forces on a protein at single residue resolution: Relation to conformation, stability, dynamics, and function. *Chem. Phys. Lett.* **254**:292–301.

Martorana, V., G. Corongiu, and M.U. Palma. 1998. Interaction of explicit solvent with hydrophobic/philic/charged residues of a protein: Residue character vs. conformational context. *Proteins* **32**:129–135.

McMahon, B.J., J.D. Müller, C.A. Wraight, and G.U. Nienhaus. 1998. Electron transfer and protein dynamics in the photosynthetic reaction center. *Biophys. J.* **74**:2567–2587.

Molteni, C., and M. Parrinello. 1998. Glucose in aqueous solution by first principles molecular dynamics. *J. Am. Chem. Soc.* **120**:2168–2171.

Monod, J., J.-P. Changeux, and F. Jacob. 1963. Allosteric proteins and cellular control systems. *J. Mol. Biol.* **6**:306–320.

Muegge, I., P.X. Qi, A.J. Wand, Z.T. Chu, and A. Warshel. 1997. The reorganization energy of cytochrome c revisited. *J. Phys. Chem. B* **101**:825–836.

Nienhaus, G.U., K. Chu, and K. Jesse. 1998. Structural heterogeneity and ligand binding in carbonmonoxy myoglobin crystals at cryogenic temperatures. *Biochemistry* **37**:6819–6823.

Novella, M., and K. Schulten. 1991. Molecular dynamics simulation of electron transfer in proteins: Theory and application to QA → QB transfer in the photosynthetic reaction center. *J. Phys. Chem.* **95**:2059–2067.

Oesterhelt, D., J. Titov, and E. Bamberg. 1992. A unifying concept for ion translocation in retinal proteins. *J. Bioenerg. Biomembr.* **24**:181–191.

Olson, W.K. 1996. Simulating DNA at low resolution. *Curr. Opin Struct. Biol.* **6**:242–256.

Olson, J.S., and Q.H. Gibson. 1972. The reaction of n-butyl isocyanide with human hemoglobin. *J. Biol. Chem.* **247**:1713–1726.

Orešic, M., and D. Shalloway. 1994. Hierarchical characterization of energy landscapes using Gaussian packet states. *J. Chem. Phys.* **101**:9844–9857.

Paas, Y. 1998. The macro- and microarchitectures of the ligand-binding domain of glutamate receptors. *Trends Neurosci.* **21**:117–125.

Parak, F., E.N. Frolov, A.A. Kononenko, R.L. Mössbauer, V.I. Goldanskii, and A.B. Rubin. 1980. Evidence for a correlation between the photoinduced electron transfer and dynamic properties of the chromatophore membranes from Rhodospirillum rubrum. *FEBS Lett.* **117**:368–372.

Parrinello, M. 1997. From silicon to RNA: The coming of age of ab initio molecular dynamics. *Solid State Commun.* **102**:107–120.

Parson, W.W., Z.T. Chu, and A. Warshel. 1998. Reorganization energy of the initial electron-transfer step in photosynthetic bacterial reaction centers. *Biophys. J.* **74**:182–191.

Perman, B., V. Srajer, Z. Ren, T.-Y. Teng, C. Pradervand, T. Ursby, D. Bourgeois, F. Schotte, M. Wulff, R. Kort, K. Hellingwerf, and K. Moffat. 1998. Energy transduction on the nanosecond timescale: Early structural events in a xanthopsin photocycle. *Science* **279**:1946–1950.

Perutz, M.F. 1970. Stereochemistry of cooperative effects in hemoglobin. *Nature* **228**:726–739.

Polisky, B. 1998. ColE1 replication control circuiting: Sense from antisense. *Cell* **55**:929–932.

Pyle, A.M., and J.B. Green. 1995. RNA folding. *Curr. Opin. Struct. Biol.* **5**:303–310.

Radmer, R.J., and P.A. Kollmann. 1997. Free energy calculation methods: A theoretical and empirical comparison of numerical errors and a new method for qualitative estimates of free energy changes. *J. Comp. Chem.* **18**:902–919.

Rivetti, C., A. Mozzarelli, G.L. Rossi, E.R. Henry, and W.A. Eaton. 1993. Oxygen binding in single crystals of hemoglobin. *Biochemistry* **32**:2888–2906.

Roitberg, A., R.B. Gerber, R. Elber, and M.A. Ratner. 1996. Anharmonic wave functions of proteins: Quantum self-consistent field calculations of BPTI. *Science* **268**:1319–1322.

San Biagio, P.L., D. Bulone, V. Martorana, M.B. Palma-Vittorelli, and M.U. Palma. 1998. Physics and biophysics of solvent induced forces: Hydrophobic interactions and context-dependent hydration. *Eur. Biophys. J.* **27**:183–196.

Schlick, T. 1995. Modeling superhelical DNA: Recent analytical and dynamic approaches. *Curr. Opin. Struct. Biol.* **5**:245–262.

Schlick, T., E. Bartha, and M. Mandziuk. 1997. Biomolecular dynamics at long timesteps: Bridging the timescale gap between simulation and experimentation. *Ann. Rev. Biophys. Biomol. Struct.* **26**:181.

Schlick, T., R. Skeel, A. Brünger, L. Kale, J. Board, J. Hermans, and K. Schulten. 1999. Algorithmic challenges in computational molecular biophysics. *J. Comp. Phys.* **151**:9–48.

Schulten, K. 1995. Curve crossing in a protein: Coupling of the elementary quantum process to motions of the protein. In: Proc. Ecole de Physique des Houches, ed. D. Bicout and M.J. Field, Les Editions de Physique, pp. 85–118. New York: Springer.

Schulten, K., W. Humphrey, I. Logunov, M. Sheves, and D. Xu. 1995. Molecular dynamics studies of bacteriorhodopsin's photocycles. *Isr. J. Chem.* **35**:447–464.

Shoichet, B.K., W.A. Baase, R. Kuroki, and B.W. Matthews. 1995. A relationship between protein stability and protein function. *Proc. Natl. Acad. Sci. USA* **92**:452–456.

Stanton, R.V., D.S. Hartsough, and K.J. Merz, Jr. 1993. Calculation of solvation free energies using a density functional/molecular dynamics coupled potential. *J. Phys. Chem.* **97**:11,868–11,870.

Treutlein, H., K. Schulten, C. Neidermeier, J. Deisenhofer, H. Michel, and D. Devault. 1988. Electrostatic control of electron transfer in the photosynthetic reaction center of *Rhodopseudomonas viridis*. In: The Photosynthetic Bacterial Reaction Center: Structure and Dynamics, ed. J. Bredon and A. Vermeglio, NATO ASI Series A 149, pp. 139–150. New York: Plenum.

Tuckerman, M.E., G.J. Martyna, and B.J. Berne. 1990. Molecular dynamics algorithm for condensed systems with multiple time scales. *J. Chem. Phys.* **93**:1287–1291.

Warshel, A. 1984. Dynamics of enzymatic reactions. *Proc. Natl. Acad. Sci. USA* **81**:444–448.

Warshel, A. 1991. Computer Modeling of Chemical Reactions in Enzymes and Solutions. New York: John Wiley.

Warshel, A., Z.T. Chu, and W.W. Parson. 1989. Dispersed polaron simulations of electron transfer in photosynthetic reaction centers. *Science* **246**:112–116.

Warshel, A., and J.-K. Hwang. 1986. Simulation of the dynamics of electron transfer reactions in polar solvents: Semiclassical trajectories and dispersed polaron approaches. *J. Chem. Phys.* **84**:4938–4957.

Warshel, A., and M. Levitt. 1976. Theoretical studies of enzymatic reactions: Dielectric, electrostatic and steric stabilization of the carbonium ion in the reaction of lysozyme. *J. Mol. Biol.* **103**:227–249.

Warshel, A., and R.M. Weiss. 1981. Energetics of heme–protein interactions in hemoglobin. *J. Am. Chem. Soc.* **103**:446–451.

Wesolowski, T., R.P. Muller, and A. Warshel. 1996. Ab Initio frozen density functional calculations of proton transfer reactions in solution. *J. Phys. Chem.* **100**:15,444–15,449.

Xia, D., C.A. Yu, J.Z. Xian, A.M. Kachurin, L. Zhang, L. Yu, and J. Deisenhofer. 1997. Crystal structure of the cytochrome bc1 complex from bovine heart mitochondria. *Science* **277**:60–66.

Xu, D., and K. Schulten. 1994. Coupling of protein motion to electron transfer in a photosynthetic reaction center: Investigating the low-temperature behavior in the framework of the spin-boson model. *Chem. Phys.* **182**:91–117.

Zarrinkar, P.P., and J.R. Williamson. 1994. Kinetic intermediates in RNA folding. *Science* **265**:918–924.

Zhang, Z.L., L.S. Huang, V.M. Shulmeister, Y.I. Chi, K.K. Kim, L.W. Hung, A.R. Crofts, E.A. Berry, and S.H. Kim. 1998. Electron transfer by domain movement in cytochrome bc1. *Nature* **392**:677–684.

Zhou, R., and B.J. Berne. 1995. A new molecular dynamics method combining the reference system propagator algorithm with a fast multipole method for simulating proteins and other complex systems. *J. Phys. Chem.* **103**:9444–9459.

16

Why Structure Changes More Slowly than Sequence in Protein Evolution

C. CHOTHIA[1], A.M. LESK[2], I. GELFAND[3] and A. KISTER[3]

[1]MRC Laboratory of Molecular Biology, Hills Road,
Cambridge, CB2 2QH, U.K.
[2]Dept. of Haematology, Wellcome/MRC Building, Cambridge University
Clinical School, Hills Road, Cambridge, CB2 2XY, U.K.
[3]Dept. of Mathematics, Rutgers University,
New Brunswick, NJ 08903, U.S.A.

Keywords: conformational change in proteins, deep structure, surface structure

ABSTRACT

In the evolution of protein families, structure is conserved much more strongly than sequence. This is because the constraints of the preexisting structure strongly favor the acceptance of mutations that, although different to the original residues, are compatible with either the original structure or one close to it. Only the peripheral regions have relatively few constraints and can adapt readily to mutations by changes in conformation. The limited flexibility of the close-packed interior does allow small changes in structures; however, large changes usually require cooperative changes in the conformation of less-constrained peripheral regions against which they pack.

INTRODUCTION

In this chapter, we discuss why structure changes more slowly than sequence in protein evolution. First, we give an overview of the relation observed between the divergence of sequence and structure and briefly list features that constrain changes in protein

Simplicity and Complexity in Proteins and Nucleic Acids
Edited by H. Frauenfelder, J. Deisenhofer, and P.G. Wolynes

structure. Then we describe, in some detail, the nature of the differences and similarities seen in the 5300 different native sequences that are known for immunoglobulin variable domains. Finally, we discuss the implications of the patterns found in the variable domain sequences for the mechanisms of protein evolution.

OVERVIEW OF THE DIVERGENCE OF SEQUENCE AND STRUCTURE IN PROTEIN FAMILIES

The first two protein structures to be determined were sperm whale myoglobin and horse hemoglobin. They had 28% sequence identity. Their structures appeared so similar that Kendrew and Watson (1966) argued that the mutations of buried residues that occur during evolution of protein families are complementary. Thus, if a mutation at one site changed a large residue to a small one, it would be followed at an adjacent site by a mutation in the opposite direction.

Subsequent analysis of the structures of eight members of globin family showed that mutations are not complementary but are accommodated by changes in structure (Lesk and Chothia 1980). For globins with sequence identities of 18%, homologous pairs of helices can differ in their relative positions by 7 Å and in relative orientation by 20°. Shifts and rotations at the different helix interfaces are coupled so that the residues that form the heme-binding pocket maintain their geometrical arrangement .

Analyses of protein families whose structures are built of β-sheets showed that they also respond to sequence divergence by shifts and rotations in the relative geometry of their secondary structures. They also showed local changes in the conformation of peripheral regions, the loops connecting secondary structures and strands on the edges of β-sheets, in response to mutations, particularly those that produced insertions or deletions.

A systematic comparison of 32 pairs homologous proteins from eight protein families gave a general relation between the divergence of sequence and structure (Chothia and Lesk 1986). The structures of an homologous pair of proteins can be divided into two parts: (a) that which has the same fold in both proteins (the common core) and (b) that in which the fold is different. For the regions that have the same fold, changes in sequence produce adjustments in their packing and small differences in conformation. The extent of the changes in structure in the common cores of a pair of homologous proteins can be measured by their optimal superposion and the calculation of the root mean square difference in the position of their main-chain atoms, Δ. Data from the 32 pairs of homologous proteins showed that to a good approximation Δ is related to the extent of the sequence differences by the equation:

$$\Delta = 0.4e^{1.87H} , \qquad (16.1)$$

where Δ is measured in Å and H is the fraction of the residues in the common core that are not identical (see Figure 16.1).

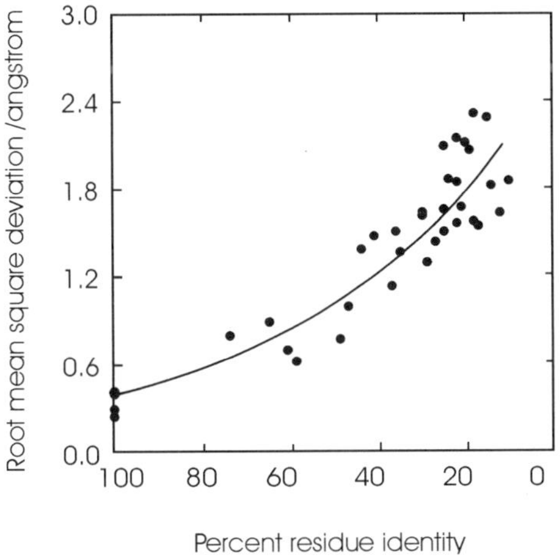

Figure 16.1 The structural differences in the common core of 32 pairs of homologous proteins (as measured by the root mean square deviation of their backbone atoms) are plotted against the residue identity of the common core.

Homologous proteins often differ somewhat in size because of insertions and deletions. If two homologous proteins have n_1 and n_2 residues and c residues in their common core, the fraction of each protein that is in the common core is c/n_1 and c/n_2, respectively. The fraction of each protein that has a different fold is $1 - (c/n_1)$ and $1 - (c/n_2)$. Inspection of homologous proteins with sequence identities of 40% or more shows they have normally 90% or more of their structures in the common core. Going below sequence identities of 40% often produces a rapid drop in the size of the common core: for proteins with 20% identity, it is usual for only half the structure to be in the common core (Figure 16.2).

CONSTRAINTS ON PROTEIN EVOLUTION

Residue packing and hydrogen bonding in the protein interior. Residues in the interior are close packed (Richards 1974): their mean volume is a little less in proteins than in crystals of amino acids (Harpaz et al. 1994). Almost all polar groups buried in proteins form hydrogen bonds.

Low-energy conformations. The very large majority of residues have torsion angles for the main chain and for side chains that have low-energy conformations. This was first demonstrated for structures determined at medium resolution (Ramachandran and Sasisekharan 1968; Janin et al. 1978); the subsequent refinement of structures at high resolution has increased the strength of this generalization.

Free energy of proteins. Although proteins are close packed, have low-energy conformations, and hydrogen bond buried polar groups, the free energy of the native state

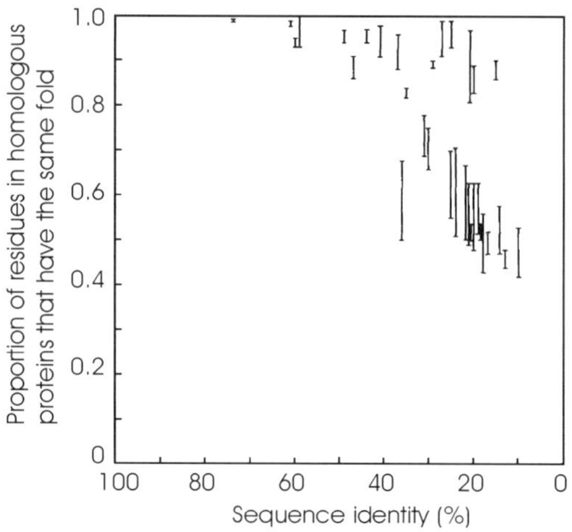

Figure 16.2 The proportion of residues that have same conformation in 32 homologous pairs of proteins (their common cores) is plotted against the sequence identity of the residues in the common cores. Thus, if two proteins with n_1 and n_2 residues respectively have c residues in same conformation, the proportions of residues in their common core are c/n_1 and c/n_2. For each pair we plot these proportions, joined by a line, against the residue identity of the core residues.

is only 5–15 kcal per mole lower than that of the unfolded state (Privalov and Khechinashvili 1974). This implies that an individual mutation is only acceptable if it costs no more than a few kcals. More important, if an "expensive" mutation does occur, the next mutation must either be complementary, which is unlikely, or it must stabilize the structural change induced by the previous mutation.

Intrinsic flexibility of close-packed segments in proteins. The previous paragraphs suggest a rather rigid view of protein structures. In fact, close-packed segments of a polypeptide have an intrinsic flexibility (this is reviewed in Gerstein et al. 1994). For example, close-packed α-helices can readily shift relative to one another by shear motions of 1.5 Å and rotations of 15° (Chothia et al. 1983). The shifts are produced mainly by "rocking" on the side chains that form the interface between the segments. The rocking motion is produced typically by χ-angle changes of 5° to 15° and the occasional flip to an alternative conformation. These types of changes occur when the protein environment changes as, for example, on a binding ligand. Protein engineering studies have shown that local changes of the same kind can, in some cases, accommodate changes in the size of buried residues (see Baase et al., this volume).

Function. Changes in conformation allow individual structures to accept a wide range of mutations, especially those that retain the chemical character of the residue and do not involve large changes in volume in one step. Limitations to this process are found in proteins that have extensive functional requirements. For proteins, such as histones and actin, the functions of which involve extensive interactions with several different proteins and DNA, there are severe limitations and the rate of change in their sequences is about 10% per billion years. Conversely, for certain domains in the immunoglobulin superfamily, where duplicated genes have different functions, about a quarter of domain structure is conserved and the only sequence conservation is the hydrophobic character and approximate size of some 12 residues.

CONSERVATION AND VARIATIONS IN THE SEQUENCES OF VARIABLE DOMAINS

Immunoglobulins and Their Variable Domains

Many of the results described in this section are taken from the recent paper by Chothia et al. (1998). A typical immunoglobulin molecule is built from two heavy chains and two light chains. The light chain folds into two domains: the variable, V_L, and the constant, C_L. The heavy chain folds into four domains: one variable, V_H, and three constant, C_{H1}, C_{H2}, and C_{H3}. The domains are homologous. Different variable domains usually have sequence identities of about 35% or more, and two thirds or more of their structures have the same conformation. However, if variable domains are compared with constant domains, the differences are greater: residue identities that are typically 10% to 20% and the regions with the same fold usually comprise only one third to one half of each domain (Lesk and Chothia 1982).

The data base of Kabat et al. (1991) now contains the sequences of some 5300 different variable domains. The atomic structures of some 140 variable domains have been determined and are available from the Protein Data Bank (Bernstein et al. 1977). These data make the variable domains an excellent subject for an investigation of (a) what residues need to be conserved for a given fold and (b) the mechanisms used to accommodate the changes in sequence that do take place.

The variable domains of heavy (V_H) and light chains (V_L) are built from two β-sheets: one with four strands, A, B, E, and D; the other with six strands, A′, G, F, C, C′, and C″ (Figure 16.3). The binding site for antigens is formed by the interstrand links BC, C′C″, and FG from each domain. Variations in sequence and conformation of these six regions are the major determinants of specificity and affinity for antigens.

The Common Core of Variable Domains

The structure common to a protein family (the common core) is the set of atoms that occupy the same relative positions in space in all members of the family. These positions can be identified by the application of a number of different techniques to the structures known for family members. The application of these methods to the known variable domain structures gives results that have no significant differences. At 76 sites the residues have same conformation: the dispersion of the main-chain atoms of residues at these positions in different structures is very small: usually less than 1 Å and in all cases less than 2.5 Å. This common core of 76 residues comprises about 70% of each variable domain. It includes all, or almost all, of strands A, A′, B, C, C′, D, E, F,

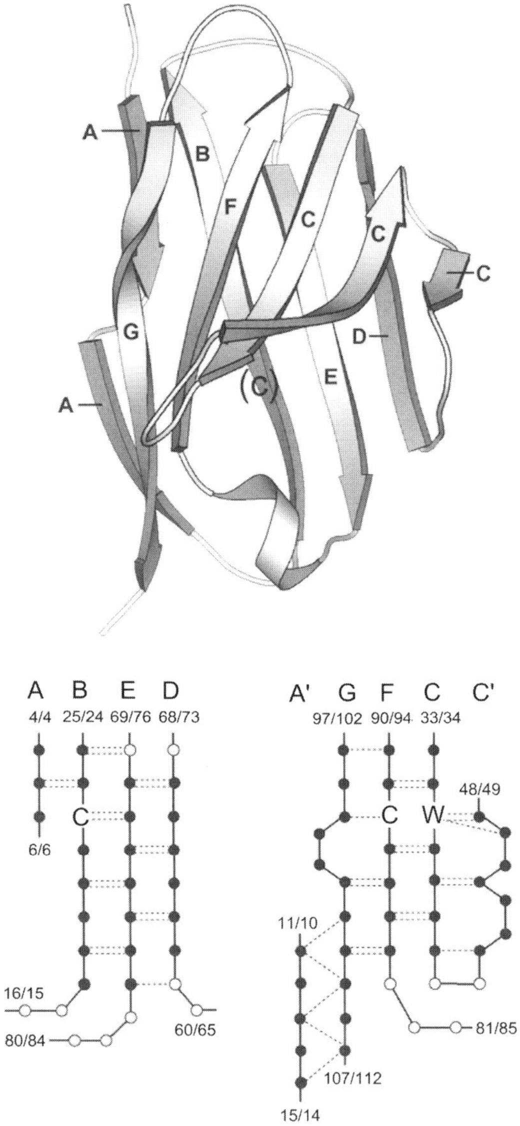

Figure 16.3 Top: The structure of a typical variable domain. β-sheet strands are shown as ribbons and labeled A, A′, B, C, C′, C″, D, E, F, and G. Note how a β-bulge produces the appearance of a rotation in the G-strand. The EF turn contains a short helix. Bottom: A plan of the common core of variable domains. Filled circles represent β-sheet residues and open circles turn residues. This structure is believed to be common to all, or almost all, immunoglobulin variable domains. Different families or classes of variable domains have different conformations for the regions outside this core. On the plan residue sites are numbered *m*/*n* where *m* is the Kabat number of the V_L site and *n* is the Kabat number of the structurally equivalent V_H site.

and G and interstrand links A′B, CC′, and EF. These regions are illustrated in Figure 16.3. The regions that differ in conformation or position are the C″ strand, the other six interstrand regions, and a small number of adjacent residues.

The number of different structures that has been compared is small in comparison with the number of different sequences that are known. The sequences of these structures, however, do cover a large part of the observed range of residues found in the known variable domain sequences. Thus it is expected that all, or almost all, sequences have structures with a core that has a conformation very close to that described here.

Residue Conservation at Core Sites and Its Classification

The frequency of residues at common core sites of variable domains was calculated from sequences in the Kabat data base (Kabat et al. 1991). This contains a complete collection of the known immunoglobulin sequences. It includes the sequences of heavy and light chain variable domains from human, mouse, rat, rabbit, shark, and other species; some 5300 in all. These 5300 sequences were aligned and the frequencies with which different residues are found at each of the 76 core sites determined. Residue variation at the different sites can be described as one of three kinds.

Sites That Conserve One Type of Residue

At this kind of site, only one residue is found in almost all sequences. There are nine sites where a single type of residue is found in 93% to 99% of sequences (Table 16.1). For example, W at the position 35/35 and C at the position 88/92 is found in more than 99% of the sequences. These sites are called invariant residue (IR) sites. (Note that residue sites are numbered here *m*/*n*, where *m* is the Kabat number of the V_L site and *n* is the Kabat number of the structurally equivalent V_H site.)

Sites That Have One of a Few Closely Related Residues

There are 17 sites where one of a set of closely related residues is found in 91% to 99% of sequences. Here similar residues are defined as those that have the same chemical character and volumes that differ by no more than the equivalent of one methylene group. An example of such a site is 21/20, where V, L, I, or M are found in 99% of the sequences (Table 16.1). The side chains of these three residues differ in shape, but all are hydrophobic and occupy similar volumes. This class of sites are similar residue (SR) sites.

Sites That Conserve the Chemical Character of the Residues

Two thirds of the sites are occupied by residues that have a wider range in chemical character and/or volume than that found at SR sites. In nearly all cases, these sites do not contain all types of residues but are limited to certain classes of residues. These sites are residue class (RC) sites.

Table 16.1 Residue conservation and accessible surface areas at the sites that form the common core of the variable domains.

(a) Sites that have an invariant residue or one of closely related residues

SITE	RES	%	ASA	SITE	RES	%	ASA	SITE	RES	%	ASA
Sites with invariant residues (IR)											
23/22	C	99	0	35/36	W	99	1	38/39	Q	93	16
82/86	D	98	5	86/90	Y	99	0	88/92	C	99	0
99/104	G	99	3	101/106	G	99	13	102/107	T	98	5
Sites that conserve closely related residues (SR)											
4/4	LM	95	8	6/6	QE	99	14	16/15	GS	97	58
21/20	VLIM	99	1	22/21	ST	96	42	47/48	VLIM	92	7
61/66	RK	95	41	63/68	ST	91	58	65/70	ST	95	64
73/80	LIMF	97	0	75/82	LIM	93	0	78/82c	VLM	96	3
84/88	GA	97	6	87/91	YF	99	6	98/103	FW	99	18
104/109	VL	99	0	106/111	VLI	99	7				

(b) Sites that conserve residue class(es) (RC sites)

SITE	%	ASA	SITE	%	ASA	SITE	%	ASA	SITE	%	ASA
Sites that conserve hydrophobic and neutral residues (*bn*)											
11/10	75	36	12/11	99	73	13/12	92	12	15/14	99	74
19/18	99	12	25/24	99	11	33/35	99	2	36/37	99	1
44/45	99	7	46/47	94	17	48/49	99	0	62/67	99	10
64/69	99	11	71/78	99	1	83/87	91	58	85/89	93	3
97/102	99	34									
Site that conserves neutral residues (*n*)											
40/41	98	105	72/79	89	36						
Sites that conserve hydrophilic and neutral residues (*sn*)											
5/5	79	85	14/13	99	74	17/16	99	74	18/17	99	116
20/19	99	77	24/23	98	73	34/35	97	8	39/40	92	62
41/42	99	82	42/43	99	95	43/44	97	33	60/65	98	110
66/71	79	73	67/72	97	74	68/71	99	56	69/74	99	46
70/77	93	47	74/81	97	29	76/82a	98	57	77/82b	99	49
79/83	99	82	80/84	97	77	81/85	99	113	89/93	91	1
90/94	96	7	100/105	99	76	103/108	76	96	105/110	98	58
107/112	92	60									
Site that conserves hydrophilic residues (*s*)											
37/38	92	21	45/46	94	89						

The % column indicates the proportion of the ~5000 V domain sequences that conserve the residues or residue types given in the RES column. Mean accessible surface area of each site (ASA) is given in $Å^2$.

Different types of RC sites can be defined by the particular class of residues that they conserve. The classification of these sites is based on two correlated properties: (a) their tendency to be on the surface or in the interior of a protein and (b) their hydrophobic character, as measured by the energy involved in their transfer between water and organic solvent. These two properties are correlated and show that residues can be put into one of three groups (Miller et al. 1987 and references therein). The hydrophobic group contains residues that have a high probability of being buried in the protein interior and we will refer to them as *b* (buried) group residues. The *s* (surface) group contains hydrophilic residues that have a high probability of being on the surface and the *n* (neutral) group contains residues that have a roughly equal chance of being on the surface or in the interior. The residues in each group are:

s:	R,	K,	E,	D,	Q,	and	N;	
n:	P,	H,	Y,	G,	A,	S,	and	T;
b:	C,	V,	L,	I,	M,	F,	and	W.

RC sites in the variable domains, i.e., those sites whose residues are not confined to a single type of residue or to a few closely related types, are classified as *s*, *n*, *b*, *sn*, or *bn*. For example, a site labeled *b* would contain several, but in most cases not all, of the residues from the C, V, L, I, M, F, W group. A site that has residues from both the *s* and *n* groups, or from both the *b* and *n* groups, is labeled *sn* or *bn*, respectively.

At 46 sites the conservation involves the chemical character of the residues (Table 16.1). Of the 46, two sites conserve neutral (*n*) residues (89% and 98% of all sequences); and two conserve hydrophilic residues (92% and 94%). On a broader, but still very significant, level there are 16 sites with *bn* residues in 91% to 99% of sequences and 26 sites with *sn* residues in 91% to 99% of sequences. There are four sites in the variable domains where the conservation of residue type is not much different from a random distribution of residues: 5/5, 11/10, 66/71, and 103/108.

Buried and Surface Sites in Variable Domains

In discussing the role of residues in the structure of a protein, it is useful to know whether their sites are buried in the protein interior, highly exposed on the surface, or in an intermediate position. A good quantitative measure of this property is the area accessible to the solvent (Lee and Richards 1971). The accessible surface area (ASA) of residues in 16 V domains was determined. The ASA of residues occupying structurally equivalent sites in the different domains are similar and, using the data from the 16 structures, the mean accessible surface area of the residues that occupy the 76 common core sites was determined. The values are given in Table 16.1.

The Residues at Sites That Form the Interior of Variable Domains

In the preceding sections, we have described the nature of the residue conservation at the 76 sites in the common core of variable domains and the extent to which these sites are buried or on the surface of the domain. There has been evidence for a considerable time that the major determinant of a protein structure is the residues that form the interior (Perutz et al. 1965, Lesk and Chothia 1980; Alber et al. 1987), and in this section, we discuss the conservation, role, and interactions of residues at the sites that form the interior of variable domains.

The types of residues found in the inner faces of the two β-sheets are described in Figure 16.4. The inner face of the ABED β-sheet has 15 sites while that of the A′GFCC′ β-sheet has 21 sites: 36 in all (Figure 16.3). It is useful to include two sites, 78/82c and 82/86, that are in the EF turn but which are buried and in contact with the interior residues. Of these 38, 19 are IR or SR sites. The other 19 are RC sites: 11 *bn* sites, and 8 *sn* or *s* sites.

The interface between the two β-sheets can be divided into two regions. A central region that contains both hydrophobic and adjacent polar residues forms the "deep" structure of the variable domain fold. It involves a set of 21 residues, almost all of which are strongly conserved. Adjacent to the deep structure there is a "peripheral" region formed by residues at ten sites that are buried but not conserved in size.

Deep Structure Sites: 4/4, 6/6, 19/18, 21/20, 23/22, 35/36, 37/38, 47/48, 61/65, 73/80, 75/82, 78/82c, 82/86, 84/88, 86/90, 88/92, 99/104, 101/106, 102/107, 104/109, and 106/111

Residues at 15 of these sites form a hydrophobic region that fills the center of the interface between the β-sheets (Figures 16.4 and 16.5). Nearly all these sites are very strongly conserved, i.e., IR or SR sites (Figure 16.4). Buried on two sides of this central hydrophobic region are polar residues whose side chains form buried hydrogen bonds: 82/86, 37/38, and 61/65 on one side and 6/6 on the other.

Peripheral Buried Sites: bn sites: 11/10, 13/12, 25/24, 33/34, 48/49, 62/67, 64/69, 71/78, and 97/102, and the sn site 90/94

The preceding paragraph described a set of sites nearly all of which have invariant residues, or closely related residues, and which pack together to form the deep structure of the variable domains. It is surrounded by nine *bn* sites and one *sn* site. Eight are almost totally buried and two are largely buried. All have, within the constraints of their residue classes, a very wide range of residues.

Site	ASA ($Å^2$)	Residues commonly found at the site
11/10	36	E, G, V, L, M
13/12	12	G, A, T, V, L
25/24	11	G, A, S, T, V
33/34	2	A, V, L, I, M, Y, W
48/49	0	G, A, S, V, I
62/67	10	A, T, V, L, I, F
64/69	11	G, V, L, I, M, F
71/78	1	A, V, L, F, Y
90/94	7	R, K, Q, N, H, S, T, L
97/102	34	T, V, I, Y

This set of ten residues forms the "peripheral" buried structure.

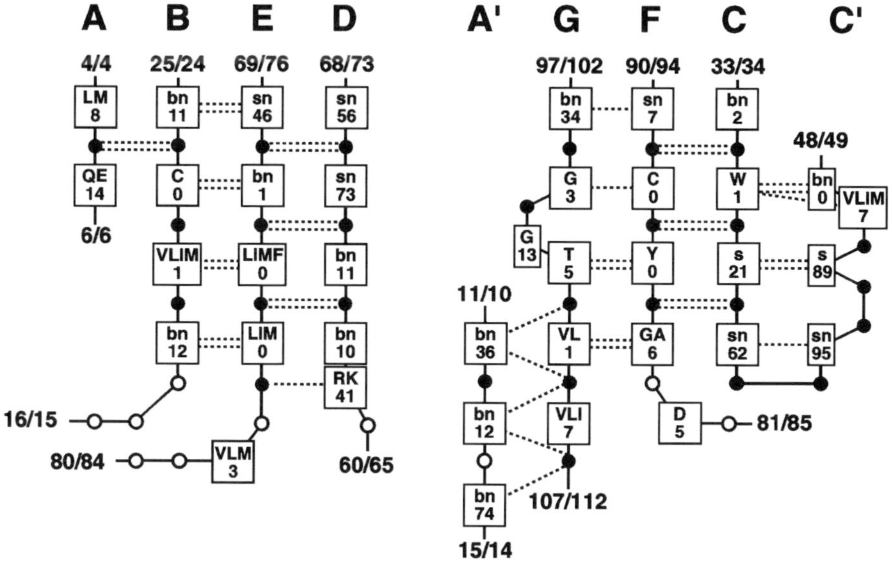

Figure 16.4 Residue conservation and accessible surface area at sites on the inner surface of the ABED β-sheet (left) and of the A′GFCC′ β-sheet (right). For each site we indicate the nature of the conservation in upper case letters for residues at IR and SR sites and in lower case letters for residue classes at RC sites (see Table 16.1). The mean accessible surface area of residues at each site is also given; the units are Å^2.

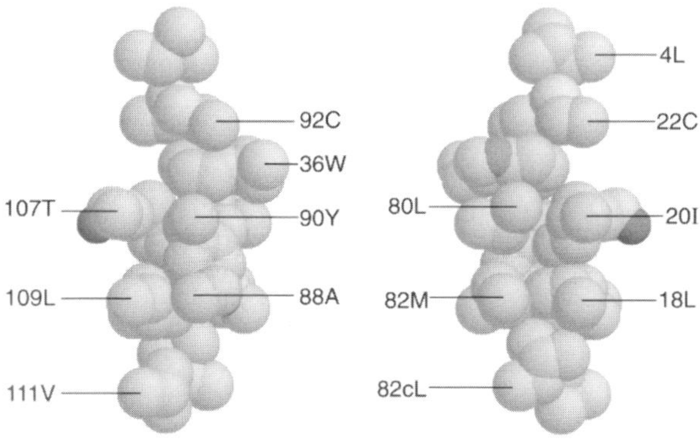

Figure 16.5 The packing in the central buried hydrophobic region. The side chains of the fifteen residues that form the central hydrophobic region are shown in space-filling drawings. On the left we show a "front" view of the residues, i.e., from the point of view used in Figures 16.1 and 16.4. On the right we show a "back" view of the residues. We list the residues commonly found at these sites in other variable domains. At fourteen of the sites the residues listed are almost absolutely conserved. At site 19/18 residues VLIM are found in 85% of sequences (Table 16.1).

How are the variations in the size of the residues in the peripheral structure accommodated? Inspection of the atomic structures of immunoglobulins shows that most of the variations are accommodated by the differences in the conformations and sizes of loop regions that are not part of the common core. Indeed, differences in residues at six of these common core sites play an important role in the different conformations of loops that form the hypervariable regions. To give one example: at site 71/78, V_λ domains usually have Ala but V_κ domains usually have Phe or Tyr. This large difference in volume is accommodated by the BC loops having quite different conformations that move a hydrophobic side chain in the loop close to the Cα atom of the residue at 71/78 in V_λ domains, or further from it in V_κ domains.

Sites 11/10 and 13/12 are in strands on the edge of the β-sheet and increases in volume at these sites can be accommodated by putting the additional atoms on the surface. Indeed, residues such as R, K, and E are found occasionally at these sites. The hydrophobic part of these side chains is packed between the β-sheets, and the charged head group is on the surface.

Saul and Poljak (1993) have described how the effects of sequence differences at position 67 in V_H spread right across the β-sheet. If F replaces L, I, V, T, or A at site 67, the conformation of the neighboring side chain, 82, and its neighbor, 18, are switched. The switch in conformation at 18 is allowed by a change in conformation of the AA′ link.

In summary, the variations in the size of these peripheral buried residues are accommodated by the changes in conformation of those parts of the protein that are not in the common core, small changes in the buried regions themselves, or by the residues extending into the solvent.

(Six *sn* sites, 39/40, 42/43, 45/46, 66/71, 68/73 and 69/76, and one *bn* site, 13/14, are on the corners of the β-sheets. The twist and coil of the β-sheet at the corners exposes these residues to solvent.)

ACCEPTANCE AND NONACCEPTANCE OF SEQUENCE CHANGES IN PROTEIN EVOLUTION

The major requirement for the creation of a stable three-dimensional structure for a protein is a hydrophobic interior that is sufficiently large, close packed and free of strain, and a surface that is sufficiently hydrophilic. These requirements place different constraints upon different sites.

The Deep Structure

At the center of variable domains, 21 residues form the deep structure with a hydrophobic core and two adjacent buried polar regions. These residues are highly conserved. Sites that have residues with very specific properties, such as Cys or Trp, are invariant. Sites with less specific residues, such as those with medium-sized hydrophobic side chains, have one of a closely related set of residues. For these less specific sites, the requirements of close packing and the limitations on strain mean that we should expect the nature and extent of substitutions permitted at different sites to vary and that only particular combinations are allowed. This view is supported by protein engineering experiments on λ-repressor (Lim and Sauer 1989), T4 lysozyme (Karpusas et al. 1989; Gassner et al. 1996), and Barnase (Buckle et al. 1996; Axe et al. 1996).

These results suggest that, in the deep structure, mutations to very different residues are not consistent with the retention of the common core structure of variable domains. If large changes do occur at these sites, within the context of normal free energy of the protein and without a change in the conformation of part of the core structure, the lack of accommodation results in the loss of stability.

Recent experimental evidence supports this conclusion and also explains how large changes do occasionally take place at sites in the deep structure. In the natural antibody ABPC48, the Cys at 92 in V_H, which is part of the conserved disulfide, is replaced by Tyr. Proba et al. (1997) found that the $V_L V_H$ dimer of this antibody is significantly *less* stable than that of the average $V_L V_H$ dimer. They also showed that a mutant, in which the Cys has been restored, has a stability significantly *greater* than that of the average $V_L V_H$ dimer. Thus, in this case, the natural mutation Cys to Tyr at site 92 is made tolerable by the fortuitous evolution of a variable domain whose structure has an exceptionally high stability.

Additional support for this conclusion is given by the distant homologues of variable domains. Two examples of this can be seen in structure of domain 1 of CD2 (Jones et al. 1992), which has 59 residues in a conformation the same as that in variable domains. This smaller common core region is related to difference in residues at sites equivalent to those that form the deep structure of the variable domains. For example, in CD2 the cysteines are replaced by Ile and Val. This increase in volume is accommodated in CD2 by the movement of residues, equivalent to those that form the end of B strand in variable domains, moving away from the D strand into space made free by the absence of the A strand.

The Peripheral Buried Structure

Ten residues form the peripheral buried structure around the edge of the deep buried structure. Eight sites are completely buried and two sites are mostly buried. The variations in the size of residues at these sites are accommodated mostly by conformational differences in the loop regions of the variable domains to which they are adjacent and which are not part of the common core. The different residues of the peripheral buried structure and the different conformations of the associated loops must be the result of long period of coordinated (but not complementary) evolution.

Surface Residues

Except for sites involved in recognition processes, surface sites in variable domains strongly conserve not the identity of residues but their chemical character as hydrophilic or neutral residues. Of the 34 sites that have accessible surface areas of 40 $Å^2$ or more, 28 have hydrophilic or neutral residues in an average of 97% of sequences. Very similar results were found by Bashford et al. (1987) for the globin family.

CONCLUSION

The results described here suggest that during the evolution of proteins, structure is conserved more strongly than sequence because the constraints of the preexisting structure strongly favor the acceptance of those residues that are compatible with it or with one close to it. The limited flexibility of the close-packed interior allows slow changes in structures to occur. However, for these to occur in the buried core requires coordinated changes in the conformation of less constrained peripheral regions.

ACKNOWLEDGMENTS

A.M. Lesk is supported by the Wellcome Trust.

REFERENCES

Alber, T.S., S. Dao-pin, J.A. Nye, D.C. Muchmore, and B.W. Matthews. 1987. Temperature sensitive mutations of bacteriophage T4 lysozyme occur at sites with low mobility and low accessibility in the folded protein. *Biochemistry* **26**:3754–3758.

Axe, D.D., N.W. Foster, and A.R. Fersht. 1996. Active Barnase variants with completely random hydrophobic cores. *Proc. Natl. Acad. Sci. USA* **93**:5590–5594.

Bashford, D., C. Chothia, and A.M. Lesk. 1987. Determinants of a protein fold: Unique features of the globin amino acid sequences. *J. Mol. Biol.* **196**:199–216.

Bernstein, F.C., T.F. Koetzle, G.J.B. Williams, E.F. Meyer, M.D. Brice, J.R. Rodgers, O. Kennard, T. Shimanouchi, and M. Tasumi. 1977. The protein data bank: A computer-based archival file for macro-molecular structures. *J. Mol. Biol.* **112**:535–542.

Buckle, A.M., P. Cramer, and A.R. Fersht. 1996. Structural and energetic responses to cavity creating mutations in hydrophobic cores. *Biochemistry* **35**:4298–4305.

Chothia, C., I. Gelfand, and A. Kister. 1998. Structural determinants in the sequences of immunoglobulin variable domains. *J. Mol. Biol.* **278**:457–479.

Chothia, C., and A.M. Lesk. 1986. The relation between the divergence of sequence and structure in proteins. *EMBO J.* **5**:823–826.

Chothia, C., A.M. Lesk, G.G. Dodson, and D.C. Hodgkin. 1983. Transmission of conformational change in insulin. *Nature* **302**:500–505.

Gassner, N.C., W.A. Baase, and B.W. Matthews. 1996. A test of the jigsaw puzzle model for protein folding by multiple methionine substitutions within the core of T4 lysozyme. *Proc. Natl. Acad. Sci. USA* **93**:12,155–12,158.

Gerstein, M., A.M. Lesk, and C. Chothia. 1994. Structural mechanisms for domain movements in proteins. *Biochemistry* **33**:6739–6749.

Harpaz, Y., M. Gerstein, and C. Chothia. 1994. Volume changes on protein folding. *Structure* **2**:641–649.

Janin, J., S. Wodak, M. Levitt, and B. Maigret. 1978. Conformation of amino acid side chains in proteins. *J. Mol. Biol.* **125**:357–386.

Jones, E.Y., S.J. Davis, A.F. Williams, K. Harlos, and D.I. Stuart. 1992. Crystal structure at 2.8 Å resolution of a soluble form of the cell adhesion molecule CD2. *Nature* **360**:232–239.

Kabat, P.A., T.T. Wu, H.M. Perry, K.S. Gottesman, and C. Foeller. 1991. Sequences of Proteins of Immunological Interest. National Institutes of Health Publ. No. 91–3242 (DHHS, PHS, N.I.H., Bethesda, MD) 5th Ed.

Karpusas, M., W.A. Baase, M. Matsumura, and B.W. Matthews. 1989. Hydrophobic packing in T4 lysozyme probed by cavity filling mutants. *Proc. Natl. Acad. Sci. USA* **86**:8237–8241.

Kendrew, J.C., and H.C. Watson. 1966. Stabilizing interactions in globular proteins. In: Principles of Biomolecular Organization, ed. G.E.W. Wolstenholme and M. O'Connor, pp. 86–100. London: J. and A. Churchill.

Lee, B., and F.M. Richards. 1971. The interpretation of protein structures: Estimation of static accessibility. *J. Mol. Biol.* **55**:379–400.

Lesk, A.M., and C. Chothia. 1980. How different amino acid sequences determine similar protein structures: The structure and evolutionary dynamics of the globins. *J. Mol. Biol.* **136**:225–270.

Lesk, A.M., and C. Chothia. 1982. The evolution of proteins formed by β-sheets: II. the core of the immunoglobulin domains. *J. Mol. Biol.* **160**:325–342.

Lim, W.A., and R.T. Sauer. 1989. Alternative packing arrangements in the hydrophobic core of λ-repressor. *Nature* **339**:31–36.

Miller, S., J. Janin, A.M. Lesk, and C. Chothia. 1987. Interior and surface of monomeric proteins. *J. Mol. Biol.* **196**:641–656.

Perutz, M.F., J.C. Kendrew, and H.C. Watson. 1965. Structure and function of haemoglobin: II. Some relations between polypeptide chain configuration and amino acid sequence. *J. Mol. Biol.* **13**:669–678.

Privalov, P.L. and N.N. Khechinashvili. 1974. A thermodynamic approach to the problem of stabilization of globular protein structure: A calorimetric study. *J. Mol. Biol.* **86**:665–684.

Proba, K., A. Honegger, and A. Plückthun. 1997. A natural antibody missing a cysteine in V_H: Consequences for thermodynamic stability and folding. *J. Mol. Biol.* **265**:161–172.

Ramachandran, G.N., and V. Sasisekharan. 1968. Conformation of polypeptides and proteins. *Adv. Prot. Chem.* **23**:283–437.

Richards, F.M. 1974. The interpretation of protein structures: Total volume, group volume distributions and packing density. *J. Mol. Biol.* **82**:1–14.

Saul, F.A., and R.J. Poljak. 1993. Structural patterns at residue positions 9, 18, 67, and 82 in the V_H framework regions of human and murine immunoglobulins. *J. Mol. Biol.* **230**:15–20.

17

How Much Sequence Variation Can the Functions of Biological Molecules Tolerate?

W.A. BAASE[1], N.C. GASSNER[1], X.-J. ZHANG[2], R. KUROKI[3], L.H. WEAVER[1], D.E. TRONRUD[1], and B.W. MATTHEWS[1]

[1]Institute of Molecular Biology, Howard Hughes Medical Institute and Dept. of Physics, University of Oregon, Eugene, OR 97403, U.S.A.
[2]Crystallography Program, Oklahoma Medical Research Foundation, 825 Northeast 13th Street, Oklahoma City, OK 73104, U.S.A.
[3]Central Laboratories for Key Technology, Kirin Brewery Co., Ltd., 1-13-5, Fukuura, Kanazawa-ku, Yokohama 236, Japan

Keywords: protein simplification, sequence redundancy, information content, T4 lysozyme

ABSTRACT

The analysis of many mutants of T4 lysozyme has made it clear that the protein is remarkably tolerant of changes in its amino acid sequence. This is especially the case for amino acids on the surface of the protein. These can be changed almost at will with little effect on folding, stability, and structure. Such "tolerant" sites constitute over 50% of the overall amino acid sequence. It seems clear that the folding of the protein to its biologically active form is determined primarily by residues within the core; however, the way in which they do so is yet to be determined.

Even within the active site region there is room for change. Alterations in appropriate residues can lead to trapped intermediates, can change the anomeric form of the product, and can introduce nonnative transglycosylase activity. There is only a single residue (Glu11) that seems to be especially important for the biological activity of T4 lysozyme, and even here some variation is allowed.

Simplicity and Complexity in Proteins and Nucleic Acids
Edited by H. Frauenfelder, J. Deisenhofer, and P.G. Wolynes

TOLERANCE TO AMINO ACID SUBSTITUTION

The observation that many amino acids in a protein sequence can be freely replaced with little, if any, effect on protein stability (Matthews 1987; Reidharr-Olson and Sauer 1988) leads one to consider the protein folding problem in a new light. What fraction of the amino acids in a given polypeptide sequence is, in fact, essential for the successful folding of the protein? Could it be as low as, say, 50%, or even 20%?

In an extensive analysis, Poteete and coworkers (Rennell et al. 1991) used suppressor strains to substitute, in turn, 13 different amino acids (Gly, Ala, Leu, Gln, Cys, Ser, Tyr, Phe, His, Pro, Glu, Arg, and Lys) at every site in T4 lysozyme except the N-terminal methionine. The phage containing the mutant lysozyme genes was tested on indicator plates. Plates showing normal halos indicate that the mutant lysozyme is folded and functional, essentially like wild-type. Plates showing small or absent halos indicate that there is reduced lysozyme activity. This might occur because the mutation prevents folding, but could also occur for other reasons, such as a change in a residue that is essential for binding or catalysis. Notwithstanding this ambiguity, the overall results, summarized in Figure 17.1, are still very striking.

Excluding the N-terminal methionine, there are 163 amino acids in T4 lysozyme. At 89 of these sites (55%), 13 different substitutions could be made with little effect on function. There were only 32% of the sites at which one or more of the 13 substitutions caused a major loss in folding and/or function. The remaining 13% of the sites were intermediate in character.

As can be seen in Figure 17.1, the sites that are tolerant of amino acid substitution tend to be mostly on the surface of the protein. These sites also correspond to those side chains that are most mobile, as seen in the crystal structure of the wild-type protein (Figure 17.2). The sites that are least tolerant of change tend to be buried in the core of the protein. Also, in the survey of Rennell et al. (1991) there were a number of intolerant sites in the vicinity of the active site cleft (Figure 17.1). These variants may have affected catalytic activity, rather than folding. Overall, T4 lysozyme is seen to be very tolerant of amino acid substitution.

Studies in our own laboratory, based on purified mutant lysozymes, are very much in agreement with the survey of Rennell et al. (1991) which, as noted, used an *in vivo* assay.

CAN ONE SIMPLIFY THE PROTEIN FOLDING PROBLEM?

If, in fact, proteins are so tolerant of change, could one take a known protein and replace a large number of "nonessential" residues with a single amino acid such as alanine, yet still have a folded functional protein? In such a "polyalanine protein" all nonessential side chains would be deleted. This would allow one to focus on those parts of the amino acid sequence that are critical for the folding process. (Another approach might be to try to construct a folded, functional protein from a very limited

(a)

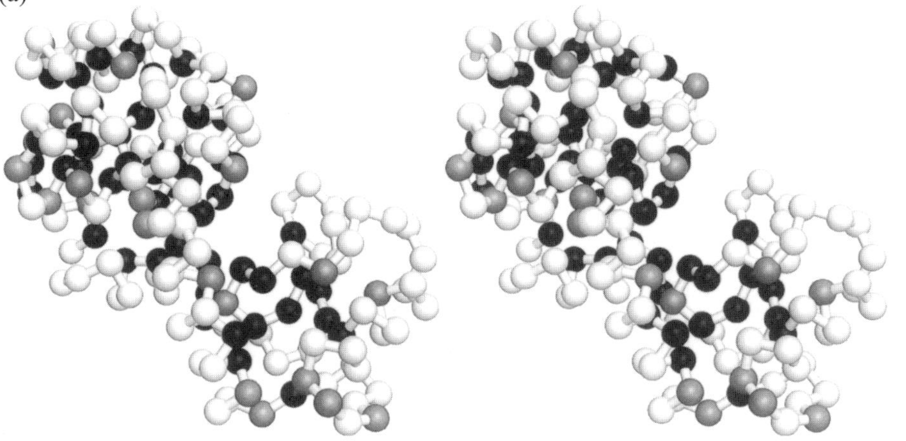

(b)

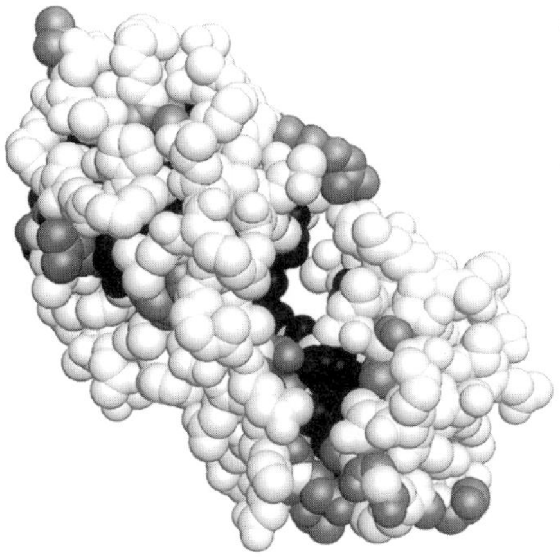

Figure 17.1 (a) Backbone trace of T4 lysozyme, drawn as a stereo pair, showing the tolerance of different residues to change, as inferred from the survey of Rennell et al. (1991). Thirteen different amino acid substitutions were made at every site in the protein except at the N-terminal methionine. The sites in the figure shaded off-white were most tolerant of change. Here, none of the 13 substitutions gave a negative phenotype. The sites shaded black were least tolerant in that one or more of the 13 substitutions resulted in a negative phenotype. The gray sites were intermediate in character. Here one or more of the 13 substitutions gave a "+/–" phenotype.
(b) The same data as (a) are shown but the figure is drawn in a space-filling representation, including all atoms in the T4 lysozyme molecule.

subset of amino acid types; cf. Kamtekar et al. [1993].) To test this idea, we have made single and multiple alanine replacements in various parts of T4 lysozyme. The results for α-helices 39–50, 115–123, and 126–134 (Zhang et al. 1991, 1992; Heinz et al. 1992; Blaber et al. 1995; Figures 17.3 and 17.4) give very consistent results. In general, the substitution of alanine at solvent-exposed sites either *increases* stability slightly (consistent with alanine being a helix-favoring residue) or has very little effect on stability. (Note, however, that the stabilization resulting from the replacement Ser117 → Ala is more substantial. This seems to be due to the elimination of an excessively short hydrogen bond [Blaber et al. 1995].) Not altogether surprisingly, the only

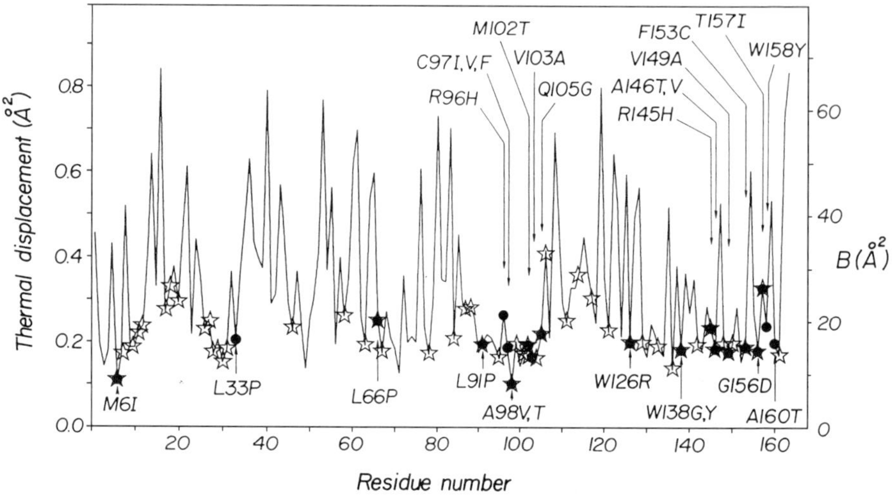

Figure 17.2 Locations of substantially destabilizing mutations in T4 lysozyme. The vertical axis shows the average "thermal motion" of each side chain in the crystal structure of wild-type T4 lysozyme. The solid circles indicate the locations of temperature-sensitive mutants of T4 lysozyme, tabulated by Alber et al. (1987), and identified on the figure. Destabilizing substitutions are only observed in the more rigid parts of the structure. The open stars show sites at which substitution with one or more of the amino acids Gly, Ala, Leu, Gln, Cys, Ser, Tyr, Phe, His, Pro, Glu, Arg, or Lys gives a negative suppression pattern at 37°C or 25°C (Rennell et al. 1991). From Matthews (1993).

cases in which alanine substitution substantially reduces stability is in the case of buried or largely buried bulky residues. For α-helix 126–134 there are two such residues, Trp126 and Leu133, and there are also two such residues, Lys43 and Leu46, for α-helix 40–49 (Figure 17.4).

It also appears that individual "polyalanine" helices that are far apart in the protein can be successfully combined together, with the effects on stability being essentially additive. This is illustrated by the combination of mutant A40–49 ($K^{43}L^{46}$) with A127–134 (L^{133}) to give a combined mutant that is actually more stable than wild-type at pH 3 (Figure 17.3) (X.-J. Zhang, W.A. Baase, and B.W. Matthews, unpublished). (In this mutant Lys43, Leu46, and Leu133, each of which has been identified as "critical" for stability, are retained as in wild-type. All other residues within the segments 40–49 plus 127–134 are alanine.) The structure of this mutant is very similar to the wild-type protein. This suggests that it should, in principle, be possible to extend the "simplification" approach to larger parts of the protein.

The polypeptide segments 40–49 plus 127–134 include a total of 18 residues of which six are alanine in the wild-type protein. In the mutant described above, nine

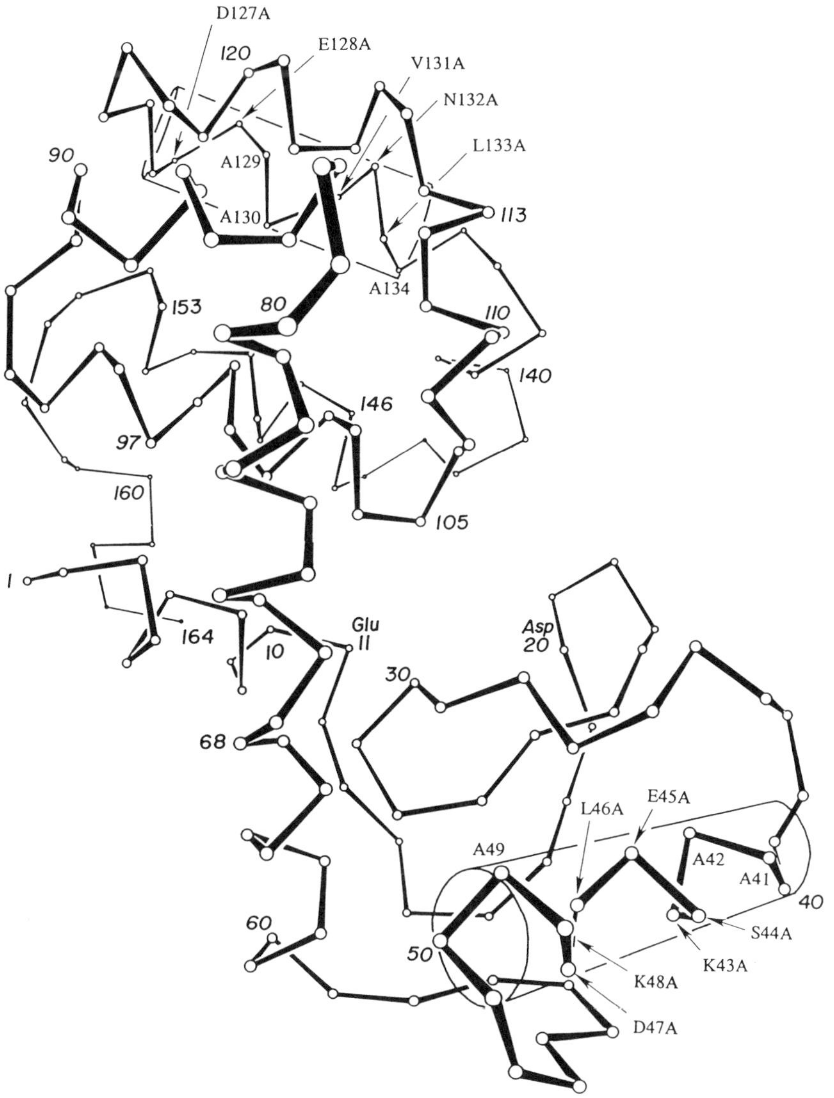

Figure 17.3 Backbone structure of T4 lysozyme showing two of the α-helices within which polyalanine mutagenesis has been tested. The combination mutant in which all of residues 40–49 plus all of residues 127–134 are alanine except Lys43, Leu46, and Leu133 has a melting temperature at pH 3.0, which is 5.5°C above wild-type.

residues have been replaced by alanine (i.e., 50% of the total residues). This result, while preliminary, suggests that the fraction of residues in T4 lysozyme that define its overall structure could be 50% or even less.

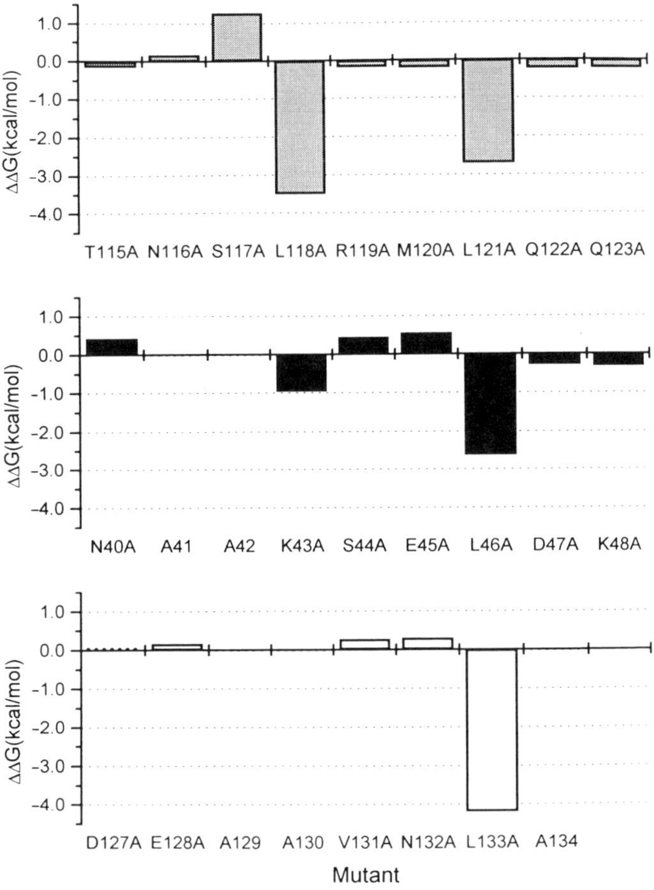

Figure 17.4a Changes in stability for individual alanine mutations within three amphipathic α-helices of phage T4 lysozyme. The single-letter code is used to indicate the wild-type residue at the given position. ΔΔG is the free energy of unfolding so that positive values indicate mutants that are more stable than wild-type. Helix 115–123 (gray), helix 39–50 (black), helix 126–134 (white).

STRUCTURAL STRAIN AND STABILITY COMPENSATION IN CORE MUTANTS

Support for the idea that core residues are less tolerant of change than surface residues was found for mutations at site 98 (Dao-pin et al. 1991). Significant strain resulted from close steric contacts, and attempts to compensate with adjacent mutations were unsuccessful (Dao-pin et al. 1991). Genetic selection (Baldwin et al. 1993) and computational design (Hurley et al. 1992) both showed that alternative core sequences with changes at up to five adjacent sites could be identified which resulted in viable

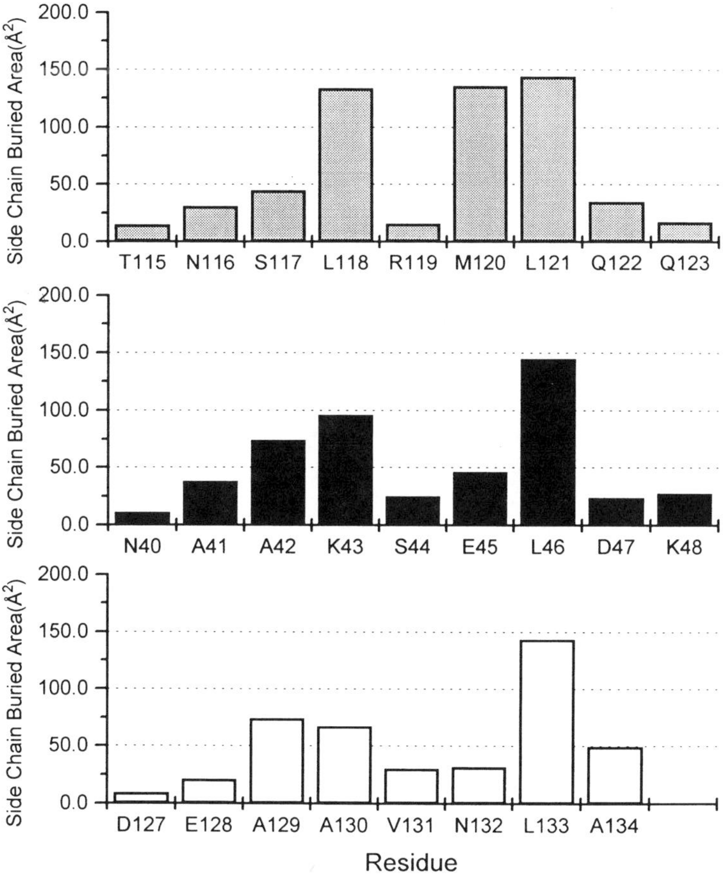

Figure 17.4b Hydrophobic side-chain surface area buried in the folded structure of WT* lysozyme for residues in the α-helices shown in Figure 17.4a.

proteins. These variants were of somewhat lower stabilities than wild-type and had distortions of up to about 0.7 Å in the main-chain backbone. The stabilities of the variants with multiply mutated cores were higher than the sums of the individual mutants would suggest, indicating some compensation.

The role of structural and stability compensation was investigated in more depth by "size-switch" mutations in which a pair of adjacent core residues, one large and the other small, were interchanged (Baldwin et al. 1996). The general scheme of the experiment is shown in Figure 17.5a. If the core is tolerant of change, then the substitution of the small-plus-large pair should compensate for the original large-plus-small pair. As shown in Figure 17.5b, the mutants with the combinations small-plus-small or large-plus-large are substantially less stable than wild-type. This is as expected, in the

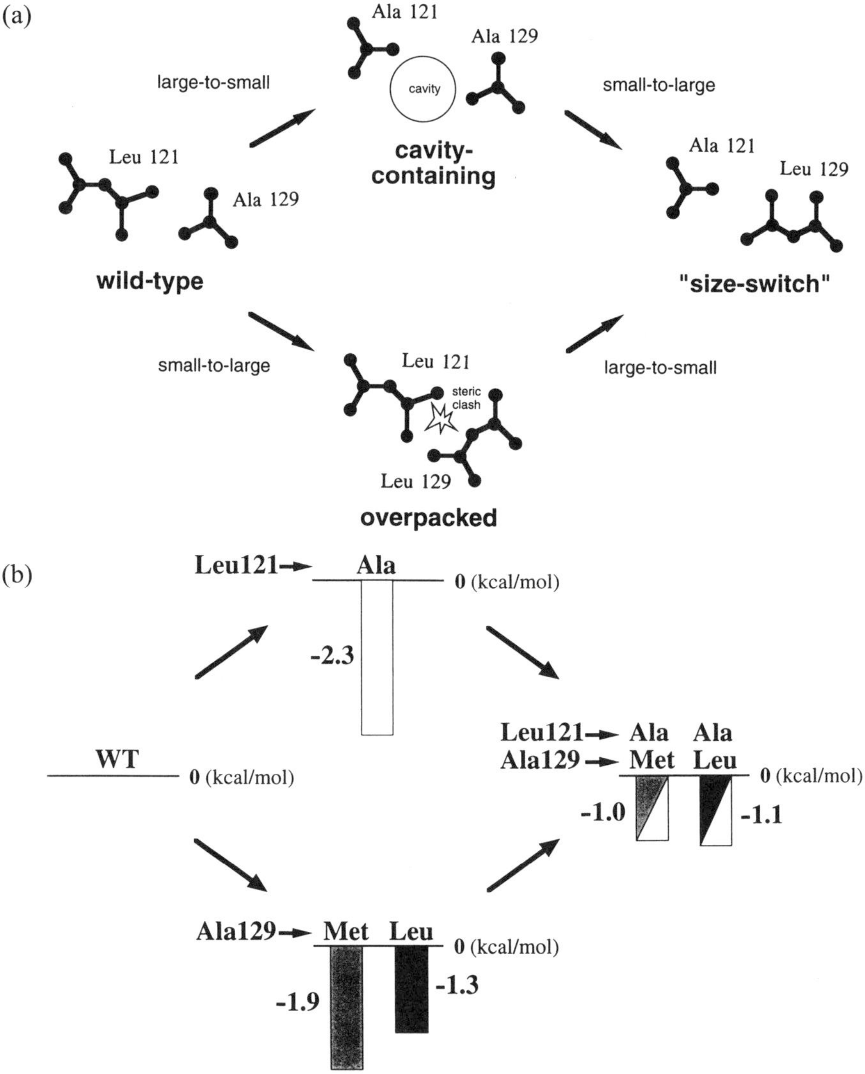

Figure 17.5 (a) Schematic showing how the combination of Leu121 with Ala129 is expected to be roughly equivalent to the size-switch mutant in which Leu121 is replaced by Ala and Ala129 is replaced by Leu. (b) The changes in unfolding free energy associated with the steps shown in (a). Data are also included for the combination of a methionine residue at site 129 with an alanine residue at site 121.

first case because of the loss of hydrophobic stabilization plus cavity formation and, in the second case, as a result of steric interference. Stability is somewhat regained in the small-plus-large double mutant but still remains less than the wild-type protein.

Parallel behavior is also observed in the structural response to these mutations. The introduction of the single mutations "large → small" or "small → large" both perturb the structure of the protein (Baldwin et al. 1996). To some extent these structural changes are alleviated in the double (small plus large) mutant but residual, noncompensated, perturbations persist (Baldwin et al. 1996). Thus, changes of adjacent residues within the core of the protein show only limited thermodynamic compensation. They also do not show complete structural compensation.

THE "JIGSAW-PUZZLE" MODEL FOR PROTEIN FOLDING

To examine the role of side-chain flexibility and to determine in what manner the assembly of a core is akin to a jigsaw puzzle, Gassner et al. (1996) replaced up to ten residues within the core of T4 lysozyme with methionine. In contrast to the rigid pieces of a jigsaw puzzle, a methionine side chain is flexible and, rather than rigidly predefining the space that it has to occupy, will tend to adjust to fill whatever volume happens to be available.

A number of multiple-methionine variants were constructed and found to be active and to fold cooperatively, although with reduced stability. The structure of a seven-methionine variant was determined crystallographically and found to be very similar to wild-type. The mobilities of the side chains in the core of the mutant protein were also found to be quite similar to those in the wild-type protein. This result shows that the interactions between the core residues are not strictly comparable with the precise spatial complementarity of the pieces of a jigsaw puzzle. Rather, there is a certain amount of give and take in forming core structure such that replacement of about a third of the protein core sequence with a single type of amino acid is possible, although with significant loss of stability. Thus, as with multiple alanine replacements made on the surface of T4 lysozyme, simpler core sequences are also possible.

Furthermore, this study illustrates how the conformation that is adopted and the space that is occupied by a given core residue can be specified by the conformations of the surrounding ones. This reemphasizes the complexity of the protein folding problem. It may be difficult to dissect out the role in folding played by a single residue. Rather, the role played by one residue may be dependent on the roles played by many other residues in the protein.

TOLERANCE TO INSERTIONS AND DELETIONS

It is generally thought that insertions within β-sheets and α-helices of proteins are less well tolerated than point substitutions. In a number of different tests, however, it has been found that the loss of stability associated with insertions is comparable with large-to-small or small-to-large substitutions within the core of the protein (e.g., Sondek and Shortle 1992; Heinz et al. 1993, 1994; Vetter et al. 1996).

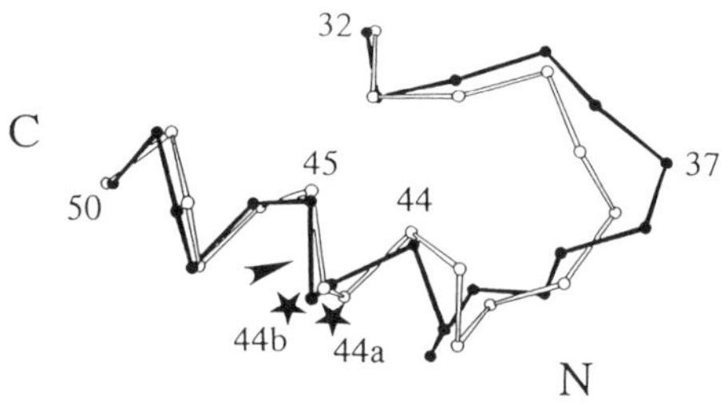

Figure 17.6a Schematic figure showing the structural response to the insertion of two amino acids (Ala–Ala) within an α-helix in T4 lysozyme (from Heinz et al. 1993). In the wild-type structure (open C^{α}–C^{α} bonds) there is a surface loop (residues 32–38) followed by an α-helix (residues 39–50). In the mutant lysozyme (solid bonds), which crystallizes in a different crystal form, the α-helix is preserved and the effect of the mutation is propagated into the surface loop which becomes two amino acids longer. The site of the mutation is indicated by the arrowhead and the two inserted alanines (residues 44a and 44b) are indicated by stars.

Structural analysis shows that an insertion within the middle of an α-helix is often accommodated by a translocation in which the effect of the insertion is, in effect, moved from the middle of the helix to a loop region at one end or the other. "Bulging" or "looping out" within the middle of a helix, while not impossible, is rarely observed.

A typical response is observed for mutant S44-[AA] of T4 lysozyme, in which two alanines were inserted following Ser44 (Heinz et al. 1993, 1994). The mutation causes T4 lysozyme to crystallize in a different crystal form. In the mutant structure, however, the α-helix that contains residue Ser44 is conserved, and the loop at the N-terminus of the helix becomes two residues longer (Figure 17.6a).

A rather different situation, however, occurs following the insertion of three alanines following Ala73 (mutant A73-[AAA]) (Vetter et al. 1996). In this case the long α-helix (helix α_C) that includes the insertion undergoes a translocation and becomes three residues longer (Figure 17.6b). This follows the normal pattern of behavior. Unexpectedly, however, helices α_F and α_G, which are essentially at right angles in the parent structure, reorganize in the mutant to form a single straight helix, $\alpha_{F\text{–}G}$. Although α-helices α_F and α_G are close to helix α_C in the folded structure, they are removed in the sequence of the protein by more than 30 residues from the site of the insertion. Thus, this mutant provides a rather dramatic example of a case in which an insertion in one part of the amino acid sequence of a protein can leave the local structure relatively unchanged, yet induce a much larger change at a site elsewhere in the protein at which the amino acid sequence is unchanged. It shows that all the amino acids in a protein contribute in a synergistic way to the determination of the final fold. Indeed, it is in part for this very reason that protein structures are tolerant of localized

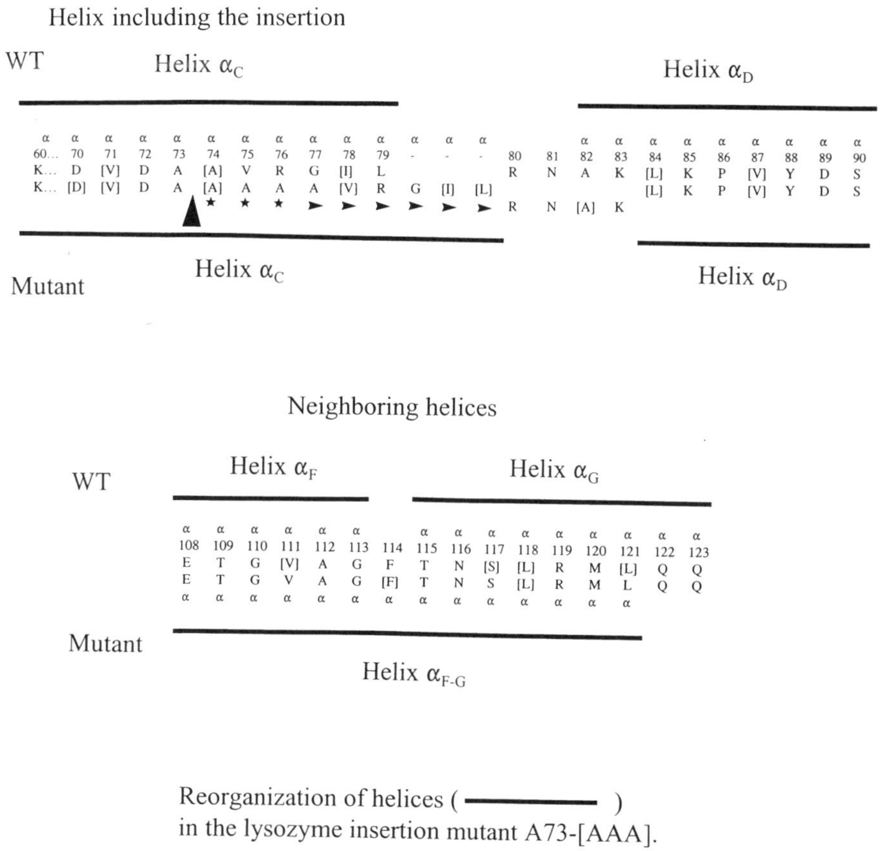

Figure 17.6b Structural consequences of the insertion of three alanines within the long α-helix, α_C (Figure 17.3), that connects the two domains of T4 lysozyme (Vetter et al. 1996). The sequence of the wild-type protein is shown above, with the structurally corresponding amino acid in the mutant shown underneath. The α-helices in the respective structures are indicated by the solid lines. The vertical arrowhead shows the site of the insertion with the three added alanines indicated with stars. Sideways arrows indicate residues translocated in the mutant structure. Residues that are largely buried in the respective structures are indicated in square parentheses. In the mutant, there is a translocation (as in Figure 17.6a) such that helix α_C, the site of the insertion, becomes three residues longer. At a site that is remote in the sequence, however, there is an additional major change such that helices α_F and α_G, which form an almost right-angle bend in the wild-type structure, reorganize in the mutant to form a single essentially straight helix, $\alpha_{F\text{–}G}$.

substitutions. The folding of a protein may not rely exclusively on the information coded by any single amino acid. To the extent that this is true, it will complicate attempts to simplify the protein folding problem in the way envisaged at the outset of this article.

TOLERANCE FOR CATALYSIS

The theme of this chapter is that the folding of T4 lysozyme, and, by inference, proteins in general, is very tolerant of amino acid substitution. What about catalytic activity?

The glutamic acid at site 11 has been identified as the most important catalytic residue in T4 lysozyme. This is the counterpart to Glu35 in hen egg white lysozyme (HEWL). There is a second acidic group, Asp20, in the active site region; however, it does not structurally correspond to Asp52 of HEWL. Rather, from a structural point of view, it is Thr26 of T4 lysozyme that aligns best with Asp52 of HEWL (Weaver et al. 1985).

Rather unexpectedly, it was found that the mutation of Thr26 to glutamic acid led to the formation of a covalent adduct in which an extended cell-wall fragment was bound to T4 lysozyme (Kuroki et al. 1993). Further studies (Kuroki et al. 1995) showed that the substitution of histidine at the same site gave an enzyme in which the product of hydrolysis was the β-anomer rather than the usual α-anomer (Table 17.1; Figure 17.7). It has also been shown that the same mutant T4 lysozyme (Thr26→His) has transglycosidase activity (Kuroki et al. 1998, 1999).

These results, taken together, illustrate both the versatility and the adaptability of the active site region of T4 lysozyme. In the wild-type enzyme, Glu11 is presumed to donate a proton to the glycosidic oxygen and a water molecule, bound by Asp20 and Thr26 on the opposite side of the substrate, is thought to attack the C-1 carbon, inverting the configuration in a single displacement reaction. The mutant with Glu11 replaced with an aspartate gives the same product as the wild-type enzyme and presumably maintains the same catalytic activity, albeit with a loss of efficiency (Table 17.1). This is also true for the mutant in which Asp20 is replaced with glutamic acid (Table 17.1). In the Thr26→His mutant, however, it is presumably the histidine that

Table 17.1 Catalytic activity of some representative active-site mutants of T4 lysozyme.

Lysozome Variant	Active-site Residue			K_{obs} ($M^{-1}min^{-1}\times10^{-4}$)	Anomeric Form of Product
WT	Glu11	Asp20	Thr26	8680	α
T26E	Glu11	Asp20	Glu26	covalent adduct	
T26D	Glu11	Asp20	Asp26	0.04	no data
T26H	Glu11	Asp20	His26	15	β
E11D	Asp11	Asp20	Thr26	2.1	α
D20E	Glu11	Glu20	Thr26	2.4	α

The locations of the three active-site residues are shown schematically in Figure 17.7. Data are from Kuroki et al. (1995).

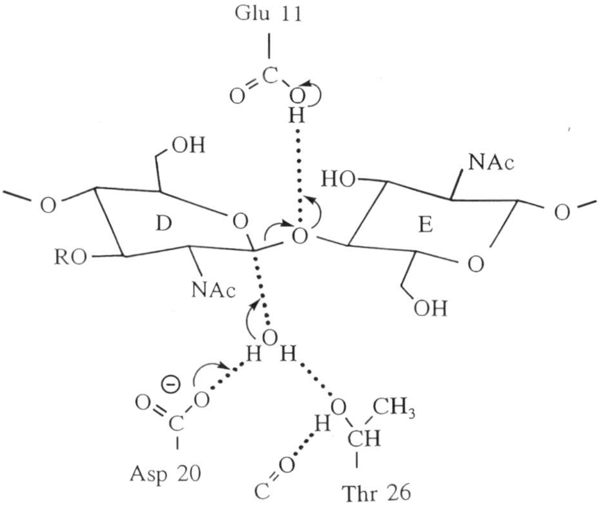

Figure 17.7 Schematic illustration of the key residues in the active site of T4 lysozyme and the presumed mechanism of action of the wild-type protein (see text). After Kuroki et al. (1995).

attacks the C-1 carbon to give a covalent enzyme-substrate adduct, which decomposes via attack of a water molecule from the same side as Glu11 (Figure 17.7). This double displacement reaction necessarily retains the chirality of the substrate, giving the β-anomeric product.

These somewhat surprising results illustrate that not only folding is tolerant of amino acid substitution, but, to some degree, so is catalytic activity. As far as is known to date, there is no amino acid in T4 lysozyme that is indispensable for biological function.

ACKNOWLEDGMENTS

We gratefully acknowledge the key contributions of a number of colleagues including Enoch Baldwin, Michael Blaber, Dirk Heinz, Joel Lindstrom, Ingrid Vetter, Joan Wozniak, and Jian Xu. This work was supported in part by NIH grant GM21967 to B.W. Matthews.

REFERENCES

Alber, T., S. Dao-pin, J.A. Nye, D.C. Muchmore, and B.W. Matthews. 1987. Temperature-sensitive mutations of bacteriophage T4 lysozyme occur at sites with low mobility and low solvent accessibility in the folded protein. *Biochemistry* **26**:3754–3758.

Baldwin, E.P., O. Hajiseyedjavadi, W.A. Baase, and B.W. Matthews. 1993. The role of backbone flexibility in the accommodation of variants that repack the core of T4 lysozyme. *Science* **262**:1715–1718.

Baldwin, E., J. Xu, O. Hajiseyedjavadi, W.A. Baase, and B.W. Matthews. 1996. Thermodynamic and structural compensation in "size-switch" core repacking variants of T4 lysozyme. *J. Mol. Biol.* **259**:542–559.

Blaber, M., W.A. Baase, N. Gassner, and B.W. Matthews. 1995. Alanine scanning mutagenesis of the α-helix 115–123 of phage T4 lysozyme: Effects on structure, stability and the binding of solvent. *J. Mol. Biol.* **246**:317–330.

Dao-pin, S., T. Alber, W.A. Baase, J.A. Wozniak, and B.W. Matthews. 1991. Structural and thermodynamic analysis of the packing of two α-helices in bacteriophage T4 lysozyme. *J. Mol. Biol.* **221**:647–667.

Gassner, N., W.A. Baase, and B.W. Matthews. 1996. A test of the "jigsaw puzzle" model for protein folding by multiple methionine substitutions within the core of T4 lysozyme. *Proc. Natl. Acad. Sci. USA* **93**:12,155–12,158.

Heinz, D.W., W.A. Baase, F.W. Dahlquist, and B.W. Matthews. 1993. How amino-acid insertions are allowed in an α-helix of T4 lysozyme. *Nature* **361**:561–564.

Heinz, D.W., W.A. Baase, and B.W. Matthews. 1992. Folding and function of a T4 lysozyme containing 10 consecutive alanines illustrate the redundancy of information in an amino acid sequence. *Proc. Natl. Acad. Sci. USA* **89**:3751–3755.

Heinz, D.W., W.A. Baase, X.-J. Zhang, M. Blaber, F.W. Dahlquist, and B.W. Matthews. 1994. Accommodation of amino acid insertions in an α-helix of T4 lysozyme. Structural and thermodynamic analysis. *J. Mol. Biol.* **236**:869–886.

Hurley, J.H., W.A. Baase, and B.W. Matthews. 1992. Design and structural analysis of alternative hydrophobic core packing arrangements in bacteriophage T4 lysozyme. *J. Mol. Biol.* **224**:1143–1159.

Kamtekar, S., J.M. Schiffer, H. Ziong, J.M. Babik, and M.H. Hecht. 1993. Protein design by binary patterning of polar and nonpolar amino acids. *Science* **262**:1680–1685.

Kuroki, R., K. Morimoto, and B.W. Matthews. 1998. Converting T4 phage lysozyme into a transglycosidase. *Ann. NY Acad. Sci.* **864**:352–365.

Kuroki, R., L.H. Weaver, and B.W. Matthews. 1993. A covalent enzyme-substrate intermediate with saccharide distortion in a mutant T4 lysozyme. *Science* **262**:2030–2033.

Kuroki, R., L.H. Weaver, and B.W. Matthews. 1995. Structure-based design of a lysozyme with altered catalytic activity. *Nature Struct. Biol.* **2**:1007–1011.

Kuroki, R., L.H. Weaver, and B.W. Matthews. 1999. Structural basis of the conversion of T4 lysozyme into a transglycosidase by re-engineering the active site. *Proc. Natl. Acd. Sci. USA*, in press.

Matthews, B.W. 1987. Genetic and structural analysis of the protein stability problem. *Biochemistry* **26**:6885–6888.

Matthews, B.W. 1993. Structural and genetic analysis of protein stability. *Ann. Rev. Biochem.* **62**:139–160.

Reidhaar-Olson, J.F., and R.T. Sauer. 1988. Combinatorial cassette mutagenesis as a probe of the informational content of protein sequences. *Science* **241**:53–57.

Rennell, D., S.E. Bouvier, L.W. Hardy, and A.R. Poteete. 1991. Systematic mutation of bacteriophage T4 lysozyme. *J. Mol. Biol.* **222**:67–87.

Sondek, J., and D. Shortle. 1992. Structural and energetic differences between insertions and substitutions in Staphylococcal nuclease. *Proteins* **13**:132–140.

Vetter, I.R., W.A. Baase, D.W. Heinz, J.-P. Xiong, S. Snow, and B.W. Matthews. 1996. Protein structural plasticity exemplified by insertion and deletion mutants in T4 lysozyme. *Protein Sci.* **5**:2399–2415.

Weaver, L.H., M.G. Grütter, S.J. Remington, T.M. Gray, N.W. Isaacs, and B.W. Matthews. 1985. Comparison of goose-type, chicken-type, and phage-type lysozymes illustrates changes that occur in both amino acid sequence and three-dimensional structure during evolution. *J. Mol. Evol.* **21**:97–111.

Zhang, X.-J., W.A. Baase, and B.W. Matthews. 1991. Toward a simplification of the protein folding problem: A stabilizing polyalanine α-helix engineered in T4 lysozyme. *Biochemistry* **30**:2012–2017.

Zhang, X.-J., W.A. Baase, and B.W. Matthews. 1992. Multiple alanine replacements within α-helix 126–134 of T4 lysozyme have independent, additive effects on both structure and stability. *Protein Sci.* **1**:761–776.

18

Is Preservation of Diversity a Constraint on Molecular Evolution?

P. SCHUSTER

Institut für Theoretische Chemie und Molekulare Strukturbiologie der Universität Wien, Währingerstr. 17, 1090 Vienna, Austria

Keywords: error threshold, molecular phenotype, neutral evolution, punctuation, RNA structure, secondary structure

ABSTRACT

Populations are nature's genetic reservoirs from which selection chooses better adapted variants. In asexually replicating species, diversity is created mainly by mutations. Recombination, obligatory with sexual reproduction of higher organisms, introduces an additional potential of diversity and allows variation to boil down to the level of individual genes or "loci" in sufficiently large populations. The concept of "molecular quasi-species," as stationary mutant distributions were called, provides a natural measure of diversity for populations of asexual species in the absence of neutrality. Error rates are confined between two limits: the lower limit is given by the maximal accuracy of replication while the upper limit results from error propagation. Neutrality with respect to selection introduces a new feature into evolution: variation of genotypes at constant phenotypes. Then, diversity of genotypes is not an appropriate measure for the evolutionarily effective variability. A new concept of molecular evolution is introduced which extends conventional population genetics by explicit considerations of the migration of populations through sequence space and the mapping of genotypes into phenotypes. The concept is exemplified by means of the RNA model, which consists of a simplified but nevertheless realistic simulation of RNA evolution *in vitro*. Mapping of RNA genotypes into phenotypes, which are represented by RNA secondary structures, revealed several results that are relevant for evolutionary optimization. These coarse-grained structures fall into two classes: common ones, which are accessible from almost everywhere in sequence space within a sphere with a

Simplicity and Complexity in Proteins and Nucleic Acids
Edited by H. Frauenfelder, J. Deisenhofer, and P.G. Wolynes

computable radius, and rare ones which, in essence, cannot be found by populations *in vivo* and *in vitro* and hence play only a very limited or no role in evolution. A typical secondary structure is formed by a large number of sequences which constitute a neutral network that represents the pre-image of the corresponding phenotype in sequence space. An evolutionarily relevant, probabilistic notion of nearness between phenotypes has been derived by considering accessibility through point mutations in the boundary of a neutral network. Global and local neighborhoods are inspected by considering structures accessible through single point mutations. Evolutionary optimization in populations occurs in steps: Fast adaptive periods of great improvement are interrupted by long quasi-stationary phases of almost constant fitness. During neutral evolution on fitness plateaus, the populations show either changing genotypes forming the same phenotype or changing phenotypes of the same fitness which are accessible through "continuous" transitions to globally frequent neighbor phenotypes. At the end of each fitness plateau, a "discontinuous" transition is observed that leads to a locally frequent but globally rare neighbor. The validity of results obtained within the frame of the RNA model for evolution in general is discussed.

INTRODUCTION

The paradigm of Darwinian evolution is based on the balance of two classes of counteracting processes: (a) mutation and recombination increasing diversity through production of new genotypes, and (b) selection reducing diversity through elimination of less fit variants. Hence, actual diversity of genotypes in populations is a result of both, variation caused by mutation and recombination and selection driven by fitness differences. Even when selection has come to an end in constant environments there will be a remaining diversity because mutation and recombination will not stop. This minimal diversity is a straightforward consequence of limited reproduction accuracy. In variable environments, however, selection does not cease to operate, and a certain diversity exceeding the minimum is then necessary for evolution. How much diversity is needed for survival and what determines the variations observed in natural populations?

This chapter aims at an analysis of variability in populations and the role of genotype and phenotype diversity in evolution. An important but hard problem to analyze is the relation and interaction between genotype and phenotype diversity. Variation can only be discussed properly at the molecular level, and hence we are heading for a molecular description of evolutionary phenomena. Since phenotypes, and their unfolding from genotypes encoding them, are extremely complex in nature as well as in artificial evolution, the evolutionary systems have to be reduced as much as possible to find suitable candidates for successful studies. The simplest conceivable case is evolution of RNA molecules in the test tube (Spiegelman 1971; Biebricher and Gardiner 1997; Strunk and Ederhof 1997). Here, genotype and phenotype are two features of the same molecule, sequence, and structure, respectively. We shall primarily consider RNA evolution and simplify further by considering a coarse-grained notion of structure known as secondary structure of polynucleotides. Eventually, we shall try to generalize to more complex cases.

Optimization of RNA structures and functions through variation and selection shares the following features with folding of biopolymers: The numbers of possible genotypes like the numbers of possible configurations are of combinatorial complexity, which implies that they increase exponentially with the chain length n ($N = \kappa^n$). Inspection of the local environment of an individual genotype reveals ruggedness and complexity just as the surrounding of a biopolymer structure in configuration space does. The evolutionary process, however, finds optimal solutions with astonishing ease as a biopolymer molecule usually finds its way from a random coil configuration to the native structure in an amazingly short period of time. New approaches to this puzzle of simultaneous simplicity and complexity, based on the paradigm of landscapes, seem to provide convincing answers in both cases.

THE ORIGIN OF DIVERSITY

Creation of diversity in nature is commonly attributed to two different mutually exclusive sources: genetic and epigenetic variation. Being interested primarily in evolutionary phenomena, we shall concentrate on genetic variation since this part is transmitted from generation to generation through an already understood mechanism of inheritance. This does not imply, however, that considering genotypes, which apparently dominates present-day molecular genetics, is sufficient for understanding evolution. On the contrary, it is the variation of phenotypes induced by genotypic changes that provides the diversity on which selection acts. Indeed, the attainability of phenotypes through accessible genetic variation determines the course of evolution. Accordingly, development of phenotypes through unfolding of genotypes is a core issue of evolutionary biology.

Two major sources of natural variation of genotypes are commonly distinguished: mutations or replication errors and recombination (see, e.g., Figure 18.1). Mutations come in three classes: (a) point mutations, (b) deletions, and (c) insertions. Point mutations are single digit exchanges that occur at constant chain length. In deletions, part of the sequence is omitted during replication and the new molecules are shorter than the parents. Insertions are produced by double or multiple replication of part of the sequence and thus lead to offspring with longer chain lengths. Recombination requires two genotypes that are split into parts and joined to yield new combinations. Recombination may occur at constant size of polynucleotides but may also produce offspring of different chain lengths (Figure 18.1).

In addition to mutation and recombination, diversity of genotypes in artificial evolution and evolutionary biotechnology can be created by a third process: random synthesis of oligonucleotides, which is readily achieved by means of present-day automata. A powerful technique currently used is based on an insertion of random stretches into polynucleotides of already known structures and functions (Bartel and Szostak 1993). From these partially random sequences, Bartel and Szostak derived catalytically active RNA molecules, so-called ribozymes, with new functions.

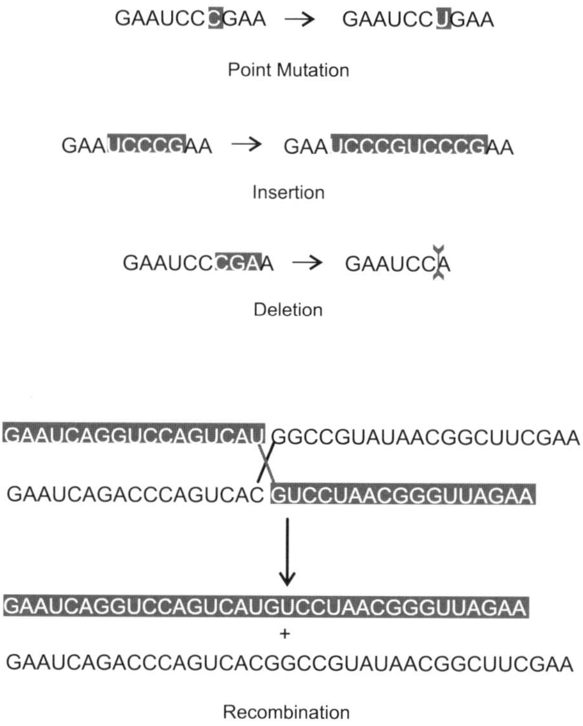

Figure 18.1 Natural variation of genotypes through mutation and recombination. Three classes of mutations are distinguished: (a) point mutations, (b) insertions, and (c) deletions. Recombinations involve two genotypes. They are split into two (or more) parts and recombined in different order. Mutation is a (more or less) local event in sequence space in the sense that properly aligned sequences have small Hamming distances whereas recombination is nonlocal since it transfers both genotypes into new positions of sequence space thereby bridging distances that are the larger the more the two recombined sequences differ.

No physical process can occur with ultimate accuracy, and polynucleotide replication is no exception in this respect. A certain minimum frequency of errors is thus unavoidable. The evolution of the biosphere can be seen as a fight against replication errors. Increasing accuracy of replication is a *conditio sine qua non* for growing sizes of genotypes (see section on **ERROR THRESHOLDS AND POPULATION DIVERSITY**) and thus, in essence, also for increasing biological information (at least up to the level of multicellular organisms[1]). The hitherto best understood cases are (a) viral RNA replication in infected host cells using a single enzyme and (b) prokaryotic DNA

[1] In a recent publication, Wolfgang Wieser (1997) shows convincingly that an appropriate measure of biological complexity is obtained when genome size is replaced by the number of cell types for multicellular organisms and, later on, by the relative size of the central nervous systems for higher mammals.

replication, which makes use of a machinery consisting of approximately twenty different enzymes. The two processes differ strongly with respect to their intrinsic accuracy limits. RNA replication is hardly more accurate than one error in 10^4 nucleotides (Drake 1993) whereas DNA replication with "proof-reading" easily achieves a precision of about one error in 10^8 nucleotides. Replication errors produce new biopolymer sequences by local changes.

Recombination, as many biologists think, is a unavoidable by-product of sexual reproduction because intertwined DNA strands cannot be disentangled in meiosis without breaking and recombining them. Thus, we encounter a situation analogous to mutation: a certain amount of recombination is obligatory and the question is whether and how nature drives enhanced recombination in order to increase diversity. The mechanism of recombination has a fundamentally different characteristic compared to mutation: it is a "nonlocal" process on the genotype sequence (Figure 18.1) and more recombination does not produce a larger number of different genotypes than mutation but more different genotypes.

Although it is an obligatory by-product of meiosis in sexual reproduction, recombination is a much less frequent event in bacterial reproduction and occurs very rarely indeed with viruses and *in vitro* evolution. These earlier forms of recombination, often called para-sexuality, use special mechanisms and are less general (hence we shall not consider them here further). They are, however, not less important since bacteria exchange factors for resistance against antibiotics by such processes. The reason for recombination being "unpopular" in the evolution of simple replicating systems is twofold: (a) recombination requires a fairly elaborate machinery operating on two nucleic acid molecules at a time or has to depend on rare events, and (b) recombination is a powerful tool only when it works on hierarchically ordered structures by exchanging modules. Otherwise, it is more likely to destroy already formed phenotypic structures than to create new ones. Recombination occurs in test tube evolution of RNA and a straightforward mechanism has been suggested for it which is based on a "jump" of the replicase from one template strand to another (Biebricher and Luce 1992). In case the recombination event does not destroy the enzyme recognition site on the RNA, the recombined molecules are "viable" in the sense that are accepted by the replicase and replicated. Nonviable molecules arise from recombination events interfering with the replication initiating structure. Recombination in virus reproduction is generally based on multiple infection of single host cells. Influenza A virus is an excellent example for the exchange of modules in a primitive system: the genome is segmented into eight independent RNA molecules (each carrying one or more genes) and the genome is recruited by "drawing" from a pool of genes. Double infection leads to a mixture of genes from both "parent" variants, and viable virions arise from reassembling complete sets of genes, which either come completely from one of the two parents or from both. Recombination in bacteria occurs by several more elaborate mechanisms which guarantee the transfer of entire genes. Examples are gene transfer mediated by phages, transfer of plasmids, or recombination of bacterial DNA.

SEQUENCE SPACE, FITNESS LANDSCAPES, AND BIOPOLYMER DIVERSITY

The discussion of diversity in biology has been turned into a problem that is accessible to quantitative mathematical analysis by the introduction of a genotype space. Sewall Wright (1932) was the first to use a hypercube for an ordered representation of genotypes. He also conceived the idea of fitness landscapes guiding evolving populations of genotypes towards (local) fitness optima. Fitness is the scalar quantity that determines the outcome of natural selection in the sense that it is a measure of the fertile progeny transmitted by a genotype into future generations. In natural ecosystems it is a highly complicated function of (many) properties of phenotypes in the context of an environment that contains also other species. In evolution in the test tube, the environment is constant and the complexity of fitness is reduced such that it becomes a measurable and or computable quantity. Then (and only then), the notion of a fitness landscape is more than an illustrative metaphor.

It is also important to clarify the notion of genotype space. In essence, we are concerned here with the simplest example of genotype space, known as sequence space. It is based on point mutations being the exclusive source of variation. Then, the number of positions in which sequences differ, known as Hamming distance, is tantamount to the minimal number of point mutations that are required to convert one sequence into the other. It induces a metric in sequence space. Similarly, it is straightforward to define a metric space that is based on recombination being the fundamental genetic operation (Stadler and Wagner 1998). Insertions and deletions are more difficult to incorporate efficiently into these formal concepts because they lead inevitably to changes in the chain lengths of genotypes. In the same way as in sequence alignment (Waterman 1995), a dummy gap symbol can be introduced to compensate for different lengths of polynucleotides.

Sequence space is of combinatorial complexity in the sense that letters chosen from an alphabet are combined to strings or sequences: an alphabet of κ letters allows the formation of κ^n different strings. Sequence space is a discrete space ($\mathcal{S}$; h), in which every genotype is represented by a single point, with the Hamming distance h, counting the number of positions in which two sequences differ, as metric. In sequence space, nearest neighbors[2] have Hamming distance one. To turn sequence space into a geometric object, a graph precisely, all nearest neighbors are connected by straight lines. In the case of binary sequences, the graph obtained is a hypercube of the same dimension as the chain length of the polynucleotides (n). Sequence spaces built upon other alphabets (e.g., the natural four letter alphabet: **AUGC**) are more complex but

2 Genotypes that are separated by a single operation or "move," either by a mutation or a recombination event, are considered nearest neighbors in genotype space. It is important to distinguish between genotype space defined for genetic operations occurring in evolution form and sequence space, which is based on point mutations and thus implies usage of the Hamming distance as a metric.

follow very simple construction principles. A sequence of chain length n has $3n$ nearest neighbors since every nucleotide can be exchanged by three other nucleotides, for example **A→U, A→G,** or **A→C**. Accordingly, we have $\binom{n}{2}3^2$ next nearest neighbors accessible by two point mutations and, in general, $\binom{n}{2}(\kappa-1)^k$ k-error mutants of a sequence in an alphabet with κ letters. A very important but very difficult to visualize property of sequence space is its high dimensionality. For the purpose of illustration we assume a set of N points in an arrangement of dimension n. Then, the maximal "city-block" distance[3] is easily computed:

$$d_{\max} = n\sqrt[n]{N} - n \,. \tag{18.1}$$

As an example, we consider $N = 1024$ points. In a one-dimensional array, the maximum distance is 1023; on a square lattice, $d_{\max}$ is reduced to 62; on the hypercube, $N = 2^n$, we obtain a distance maximum of $d_{\max} = n = 10$.

In the last two decades a novel approach was introduced to facilitate classification and handling of combinatorial optimization problems. The term "landscape" was coined for mapping from a space of combinatorial complexity into the real numbers (for reviews on the usage of the landscape paradigm in combinatorial optimization and molecular evolution, see Schuster and Stadler 1994; Stadler 1996; Schuster 1997a). As an example, we consider the mapping from genotypes into fitness values which is basic to evolutionary dynamics (Figure 18.2). In nature and in most models, the assignment of fitness to individual genotypes is done in two separate steps[4]:

genotype ⇒ phenotype ⇒ fitness.

The first step, genotype–phenotype mapping, maps sequence space into another space of nonscalar objects:

$$\Sigma\colon\ (\mathcal{S}; h) \Rightarrow (\mathcal{Y}; \eta), \tag{18.2}$$

and has been characterized as a *combinatory map* in order to indicate that it is not a proper landscape. Here, the metric space of phenotypes or *shape space* is denoted by $(\mathcal{Y}; \eta)$, where we assume the existence of a distance between phenotypes (η), which measures their relatedness. One particularly simple example of such a distance between phenotypes can be derived for RNA in test tube evolution (see section on **RNA GENOTYPES AND PHENOTYPES**). Fitness values are functions of the evolutionarily relevant properties of phenotypes, and a fitness landscape, accordingly, is a mapping from shape space into the real numbers:

$$\Lambda : (\mathcal{Y}; \eta) \Rightarrow \mathbf{R}_1 \,. \tag{18.3}$$

---

3 The "city-block" distance counts the number of blocks one has to walk from one corner to another with diagonal short cuts forbidden.

4 A few models related to spin glasses (e.g., Kauffman's NK model [Kauffman 1993], omit the explicit consideration of a phenotype and assign fitness values directly to genotypes.

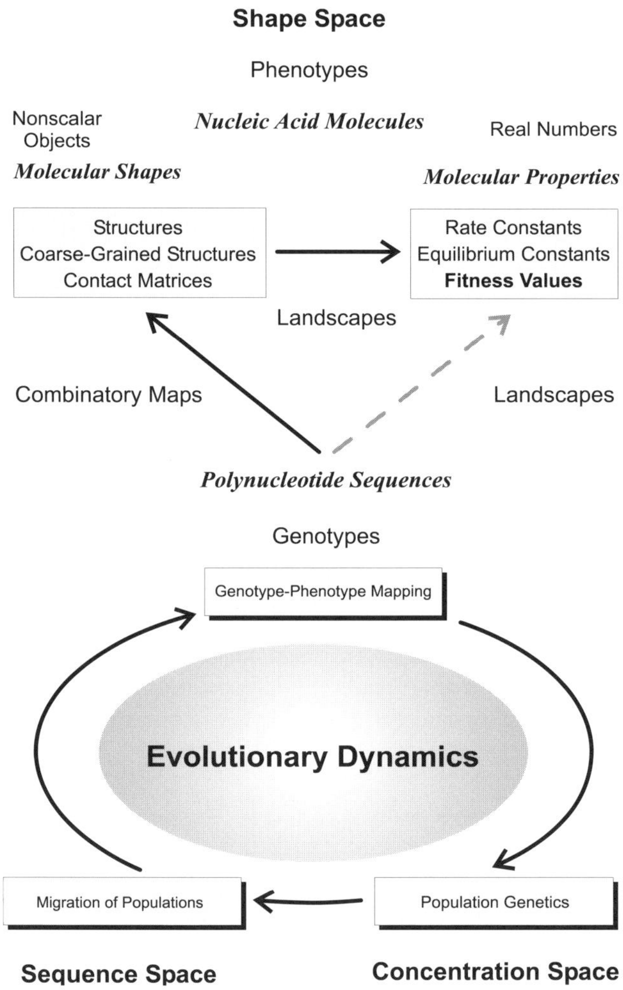

Figure 18.2 Evolutionary dynamics. Evolution in its full complexity is partitioned into three simpler phenomena that can be studied separately: (a) population genetics, (b) migration of populations or population support dynamics in sequence space, and (c) genotype–phenotype mapping. In molecular evolution, population genetics is tantamount to chemical reaction kinetics of replication, mutation, and selection. Population support dynamics describes migration of populations in sequence space. Genotype–phenotype mapping deals with the unfolding of biological information stored in polynucleotide sequences. Two classes of mappings are properly distinguished: (a) combinatory maps from genotype space into another vector space of no-scalar objects, for example structures, and (b) landscapes which map genotype space into the real numbers. In molecular evolution landscapes provide rate constants, equilibrium constants, and composite scalar properties of phenotypes, for example fitness values. The landscapes are commonly constructed in two steps: (a) a mapping of polynucleotide sequences into molecular structures, and (b) an evaluation of the structures to yield the (scalar) molecular properties.

What can we gain from such a formal representation of the evolutionary optimization problem? First, the use of definitions with a firm mathematical background provides a basis for classification. Statements or theorems are based on rigorous proofs and do not change when more data become available. What changes with more knowledge is the interpretation of the empirical results and the range of validity of the model. (It may be personal taste and preference, but some researchers feel more comfortable with an admittedly often small part of their model built upon certainty.) Second, only clear definitions allow assumptions to be made precisely and their consequences to be discussed. Third, the landscape paradigm provides a new access to sequence-structure relations of biopolymers beyond the conventional "one-sequence-one-structure paradigm."

ERROR THRESHOLDS AND POPULATION DIVERSITY

To describe molecular evolution as kinetics of replication and mutation constrained by suitable environmental conditions, Manfred Eigen made use of the sequence space concept (Eigen 1971). These investigations led to the idea of a molecular quasi-species (Eigen and Schuster 1977; Eigen et al. 1989), which is a stationary mutant distribution consisting of a most frequent master genotype or *master sequence* and its closely related mutants of sufficient fitness. A typical quasi-species is shown in Figure 18.3. The existence of such stationary distributions has been studied by analyzing the kinetic differential equations for replication and mutation in populations of constant size (Eigen et al. 1989) as well as stochastic models (Demetrius et al. 1985). It depends on mutation rates as well as fitness values. Under certain simplifying assumptions, the condition of stationarity can be expressed by a simple relation stating that the accuracy of replication has to exceed the critical value of an *error threshold*:

$$Q_{mm} > Q_{min} = \sigma_m^{-1} \tag{18.4}$$

The replication accuracy Q_{mm} is the fraction of correct replicas that are synthesized as copies of the master genotype (error-free copying implies $Q_{mm} = 1$ whereas no correct copy is tantamount to $Q_{mm} = 0$). The minimal replication accuracy, Q_{min} , is equal to the reciprocal value of the superiority of the master genotype, σ_m , which represents the fitness of the master against the weighted rest of the population expressed as the ratio of fitness values f_m and $\bar{f}_{-m}$:

$$\sigma_m = \frac{f_m}{\bar{f}_{-m}};\ \bar{f}_{-m} = \frac{\sum_{k,k\neq m} \bar{x}_k f_k}{\left(1-\bar{x}_k\right)};\ \sum_{k,k\neq m} \bar{x}_k = 1-\bar{x}_k\ . \tag{18.5}$$

The stationary frequencies of individual genotypes, I_i, are denoted by $\bar{x}_i$, and the corresponding nonstationary values of the variables by $x_i(t)$. As in population genetics we make use of normalized variables:

$$\sum_{i=1}^{n} \bar{x}_i = \sum_{i=1}^{n} x_i(t) = 1\ . \tag{18.6}$$

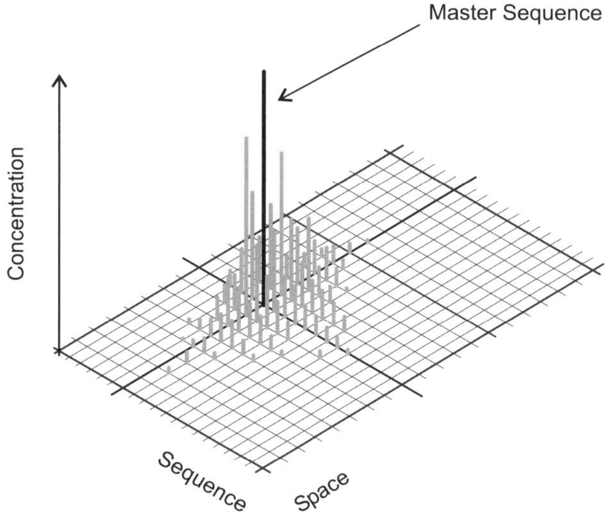

Figure 18.3 The molecular quasi-species in sequence space. The quasi-species is a stationary mutant distribution centered around a (fittest and most frequent) master sequence. The frequencies of individual mutants in the quasi-species are determined by their fitness values and by the Hamming distance from the master. A quasi-species occupies some region in sequence space called the population support. In the nonstationary case the (population) support migrates through sequence space.

Since the master genotype has the largest fitness value, the superiority σ_m is always larger than one by definition[5]. If fitness is only expressed in terms of replication rate constants, the superiority is the ratio in rate constants of the master and the rest of the population. Quantitative analysis of *in vitro* evolution of RNA in the $Q\beta$ assay has shown, however, that fitness is in reality a more complex function of rate and equilibrium constants (Biebricher and Gardiner 1997).

Replication accuracy below the threshold value ($Q_{mm} < Q_{min}$) does not sustain stationary populations. Instead, one observes a random drift phenomenon: all genotypes have finite lifetimes and the population migrates through sequence space (Demetrius et al. 1985; Eigen et al. 1989; Fontana and Schuster 1987; Huynen et al. 1996). For most distributions of fitness values and, in particular, for the realistic ones, the transition from the stationary regime to migrating populations at Q_{min} is very sharp and is reminiscent of cooperative transitions in biopolymers.[6] The consequences of the error threshold is easily illustrated by means of the uniform error rate model which assumes that mutation rates are independent of the individual nucleotide (**A**, **U**, **G** or **C**) and the

5 The limit $\sigma \rightarrow 1$ is the case of selective neutrality: the master genotypes and the mutatnts present in the population have identical fitness values. The conventional quasi-species concept is unable to deal with this situation and is discussed later in this section.

6 For the sake of completeness it is mentioned that there are also flat fitness landscapes on which the transition from the quasi-species to migrating populations is smooth.

position on the sequence. Then, mutation frequencies can be expressed in terms of the single digit replication accuracy q (which is simply one minus the error rate per site and replication act, $q = 1 - p$) and the Hamming distance between template and mutant, d_{ij} :

$$Q_{ij} = q^n \left(\frac{1-q}{q} \right)^{d_{ij}} . \tag{18.7}$$

The error threshold is easily expressed in terms of the single digit accuracy and can be used to define a maximum chain length n_{max} of RNA molecules which can form a quasi-species at constant single digit accuracy:

$$q_{\min} = \sqrt[n]{\frac{1}{\sigma_m}} \text{ and } n_{\max} = \frac{ln\sigma_m}{lnq} \cong \frac{ln\sigma_m}{1-q} = \frac{ln\sigma_m}{p} . \tag{18.8}$$

The maximum chain length is thus of the same order of magnitude as the reciprocal error rate, which means on the average approximately one error per replication of the whole genome. Comparison of replication accuracy (Drake 1993) and genome lengths in RNA viruses has shown that reproduction in these parasites operates indeed very near the error threshold.

In Figure 18.4, population diversity expressed as the fraction of mutants is plotted against the error rate. The domain that is accessible in evolution reaches from the physical accuracy limit, which is given by the properties of the replication machinery, to the error threshold. At the error threshold, the diversity of the population reaches the highest value at which error propagation is still compatible with a stationary mutant distribution. The case of the RNA viruses mentioned above is an indication for an evolutionary demand of maximal diversity. These viruses have genome lengths that are consistent with the replication accuracy of their specific replicases and show vastly heterogeneous populations (Domingo et al. 1995). We may interpret this finding as the result of a condition for variation that is imposed on the parasite by the defense mechanism of the host, for example, by the immune system of higher vertebrates.

So far we have excluded the case of neutrality against selection. We discuss neutrality now in the (more restrictive but easier to analyze) form of redundancy in sequence-structure maps: the same structure is formed by more than one sequence and this implies identical evolutionary properties and identical fitness of several genotypes. Thus, the superiority of the master sequence is one and formally the minimum accuracy of replication becomes unity, $Q_{min} = \sigma_m^{-1}$, implying that any small error rate is incompatible with a stationary mutant distribution. Changing genotypes, however, need not imply impossibility of evolutionary optimization. The question we must ask is: Can stationary phenotype distributions exist despite changing genotypes? To find the answer, the kinetic model has to be reformulated such that variables now describe frequencies of phenotypes (S_k) rather than genotypes (I_i): all genotypes forming the same phenotypes are lumped together:

$$y_k(t) = \sum_{i \in \Sigma_k} x_i(t) , \tag{18.9}$$

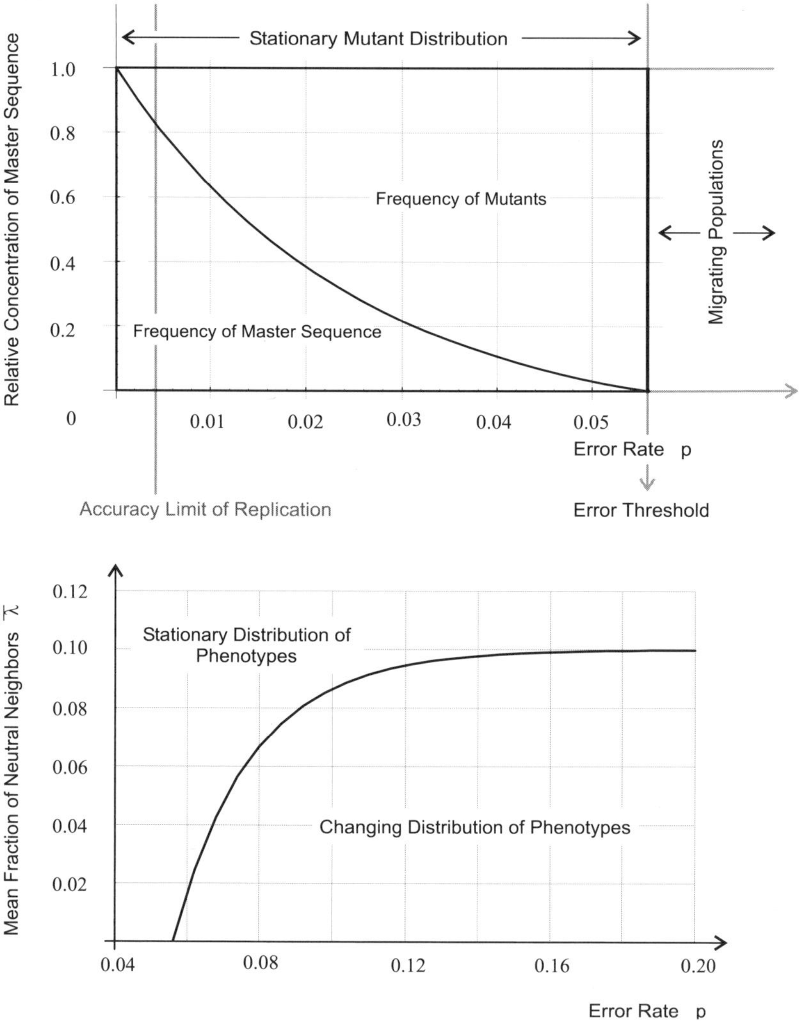

Figure 18.4 Error thresholds. The upper part of the figure shows the fraction of the master sequence in a quasi-species as a function of the error rate p (the fraction of mutants is the complement to one of this quantity). Natural populations are confined between two bounds: a lower bound for mutation rates that is determined by the maximal accuracy of the replication machinery and an upper bound being the genotypic error threshold above which populations start to migrate through sequence space (and thus loose the fittest genotypes). Neutrality with respect to selection as expressed by the mean fraction of neutral neighbors, $\bar{\lambda}$, allows more replication errors under the condition of a stationary phenotype distribution to be tolerated. The lower part of the figure shows the value of the phenotypic error rate as a function of the degree of neutrality. It is interesting to note that there is a limiting value, $\bar{\lambda}_{cr} = \sigma_m^{-1}$, above which the error rate is unable to destabilize the stationary distribution. The parameter values used in the computation were: $\sigma_m = 10$ and $n = 40$.

where Σ_k is the set of all genotypes forming the phenotype S_k. With this definition it is straightforward to extend the quasi-species concept to phenotype distributions in populations. Mathematical analysis reveals again the existence of an error threshold, the so-called *phenotypic error threshold*:

$$Q_{mm} + \bar{\lambda}_m (1 - Q_{mm}) = \tilde{Q}_{mm} > Q_{\min} = \sigma_m^{-1}, \tag{18.10}$$

where $\bar{\lambda}_m$ represents the mean fraction of neutral neighbors of the master phenotype. It is obtained by computing the fraction of neutral neighbors in the one-error neighborhoods of sequences forming the master phenotype and then averaging over the entire neutral network:

$$\bar{\lambda}_m = \frac{\sum_{I_i \in \Sigma_m} \lambda_m(I_i)}{N(\Sigma_m)}, \tag{18.11}$$

where Σ_m is the set of all $N(\Sigma_m)$ sequences (I_i) forming the master phenotype. The equation for the application accuracy is easily modified to yield the minimum single digit accuracy, q_{min}, for the phenotypic error threshold within the uniform error rate model:

$$q_{min}\left(\sigma_m, \bar{\lambda}_m\right) = \sqrt[n]{\frac{1 - \bar{\lambda}_m \sigma_m}{\left(1 - \bar{\lambda}_m\right)\sigma_m}}. \tag{18.12}$$

As is to be expected for an extension of the genotypic error threshold, the genotypic expression is obtained in the limit, $\lim \bar{\lambda} \rightarrow 0$. Neutrality allows higher error rates to be sustained: larger fractions of mutations can be tolerated with an increasing mean fraction of neutral neighbors. Interestingly, we find a critical value, $\left(\bar{\lambda}_m\right)_{cr} = \sigma_m^{-1}$, above which populations can replicate with arbitrarily high mutation rates without loosing the master phenotype.

Diversity of genotypes is an unavoidable by-product of molecular replication. To become relevant for evolution, it has to be transformed into diversity of phenotypes on which selection can act. Relations between sequences and structures, in particular the degree of neutrality $\left(\bar{\lambda}\right)$ and the distribution of neutral sequences in sequence space, are moderating and modifying genotype diversity in evolution. Knowledge on genotype–phenotype mapping is thus indispensable for an understanding of the evolutionary mechanisms.

DYNAMICS OF MOLECULAR EVOLUTION

Population genetics describes the time dependence of genotype (or allele) distributions in (commonly sexually reproducing) populations. It does not, however, deal explicitly with the scenario of migrating populations in sequence space nor does it consider the interplay between genotype–phenotype mapping and selection. Nothing can be said therefore about the accessibility of phenotypes and the efficiency of optimization. To be able to account for these highly relevant aspects of evolution, a new concept has been developed that extends conventional population genetics and it is particularly suitable for modeling molecular evolution (Schuster 1997b). The complex process is seen as the result of three phenomena (Figure 18.2): (a) population genetics, (b) migration of populations described by support dynamics[7] in sequence space, and (c) genotype–phenotype mapping. As indicated in Figure 18.2, the three processes are linked by a circular relationship in the sense that each process provides the input for the next one in the cycle. Genotype–phenotype mapping creates the phenotypes and thus produces the fitness relevant properties, which enter as parameters into the equations of population genetics. Population genetics determines the fate of individual genotypes, when they are created through mutation and when they disappear, and thus provides the input for support dynamics. Support dynamics in turn decides on the regions in sequence space where populations migrate. It describes the area where new genotypes come from and thereby defines the input for genotype–phenotype mapping. Evolutionary dynamics is represented by a kind of circular causality which is typical for self-organizing systems.

Genotype–phenotype mapping is the true source of complexity in biology (Schuster 1996). The current perception gained from chemical and biochemical models as well as from the fossil record reveals complexity increasing in steps through major evolutionary transitions (Szathmáry and Maynard Smith 1995), where each steps opens up an additional hierarchical level: RNA world → DNA-and-RNA-and-protein world → prokaryotic cells → eukaryotic cells → multicellular organisms → animal societies → human society. Although a mechanism for major evolutionary transitions based on molecular symbiosis was proposed twenty years ago (Eigen and Schuster 1978, 1982), no experimental assay (comparable to the RNA system for *in vitro* Darwinian evolution) has yet been developed that would allow the study of cooperative interactions in populations. (For a very promising development along this line see, for example, McCaskill [1997].) Nevertheless, there is general agreement that the Darwinian mechanism is not sufficient to guide evolution from a hierarchical level to the next higher one.

The enormous complexity of the relations between genotypes and phenotypes, however, is almost always prohibitive for systematic investigations. The only currently known exception is *in vitro* evolution of molecules. According to Spiegelman (1971), genotype and phenotype are two features of the same molecule: sequence and structure, respectively. The development of the phenotype from the genotype is reduced to folding the RNA sequence into a three-dimensional structure, which determines the rate and equilibrium constants for all steps involved in the replication process (Biebricher and Gardiner 1997) and thus provides all quantities that enter the expression of the fitness.

---

7 Support is a notion from the theory of distributions. The support of a population is the area in sequence space which is occupied by the actually present genotypes irrespective of their frequency.

RNA GENOTYPES AND PHENOTYPES

The simplest conceivable case of genotype–phenotype relations is encountered in RNA evolution in the test tube. As mentioned above, in this case the genotypes are the sequences and the phenotypes are the structures of RNA molecules, and genotype–phenotype mapping amounts to sequence–structure relations. Although prediction of biopolymer structures from known sequences is a particularly hard problem, it is by far simpler than any other relation between a genotype and a phenotype. Moreover, it is often not necessary to use structures at atomic resolution. Appropriately coarse-grained structures or "shapes" serve most purposes. In the special case of RNA, a coarse-grained notion of structure, the so-called secondary structure,[8] can be used for systematic studies because it is simple enough for computer-assisted investigations yet sufficiently realistic to serve as a model for RNA evolution. Secondary structures also permit us to derive straightforward and simple distance measures η (cf. section on **SEQUENCE SPACE, FITNESS LANDSCAPES, AND BIOPOLYMER DIVERSITY**): the structures may be encoded symbolically by assigning dots to unpaired bases and parentheses for base pairs; then the Hamming distance between two encoded structures can be used as structure distance.

Mappings of RNA sequences into secondary structures with minimal free energies have been studied extensively by means of a mathematical model based on random graph theory (Reidys et al. 1997), by means of exhaustive folding of all sequences of given chain length (Grüner et al. 1995), as well as by statistical analysis of appropriately chosen samples (Fontana et al. 1993; Schuster et al. 1994). The condition of minimal free energy is used here for convenience. It is not critical for the statistical results reported; a systematic study (Tacker et al. 1996) showed that essentially the same findings are obtained with the maximum matching criterion, with kinetically folded RNA molecules as well as with an algorithm computing partition functions. The investigations on RNA secondary structures revealed four regularities:

1. The numbers of sequences exceed the numbers of structures by several orders of magnitude and hence, sequence-structure maps are many to one.
2. Relatively few common structures are contrasted by many rare structures which usually play no role in evolution.[9] In the limit of long chains ($\lim n \rightarrow \infty$) almost all sequences fold into a tiny subfraction of all structures.

---

8 The secondary structure of a single-stranded RNA molecule is defined as a listing of Watson–Crick (**AU** and **GC**) and **GU** base pairs which fulfil the following conditions: (a) base pairs are not allowed between neighbors in the sequences $[i,i+1]$, and (b) if $[i,j]$ and $[k,l]$ are two base pairs then (apart from permutations) only two arrangements along the sequence are acceptable: $[i<j<k<l]$ and $[i<k<l<j]$, respectively. Condition (b) guarantees that the formed structures are free of knots and pseudoknots.

9 Both properties, "few" and "many," refer here to relative amounts. The numbers of common and the numbers of rare structures are exponentially growing quantities ($N_k = a_\kappa \times \alpha_k{}^n$) but the values of the basis (α_k) are smaller or larger, respectively.

3. To find at least one sequence (that folds into it under the minimum free energy criterion) for each of the common structures, one does not need to explore whole sequence space. It is sufficient to search a relatively small spherical neighborhood of an arbitrarily chosen sequence (*shape space covering*; see Figure 18.5 and Schuster et al. [1994]).
4. Common structures are characterized by a high degree of neutrality expressed by the mean fraction of nearest neighbors in sequence space, $\bar{\lambda}$, which form the same structure. The sets of sequences folding into a given structure (called pre-images in the theory of distributions) form extended *neutral networks* in sequence space (Figure 18.6).

These four properties are particularly important for evolution. Evolutionary optimization, for example, is facilitated substantially because of the redundancy in sequence–structure relations. It becomes even more simple since, in essence, only the common structures are the targets of evolution. Rare structures are often formed by a single sequence or only a few sequences, and to find them in an evolutionary search has almost zero probability. Rare structures, in addition, are unstable against mutation as almost every point mutation will change the minimal free energy structure. All structures of small RNA molecules ($n < 100$) found in nature fall into the class of common ones. (This is also presumably true for larger natural RNA structures but has not yet been confirmed.) It is also possible to guess from a given structure whether or not it is common; however, the current experience is not yet sufficient for faithful predictions. Shape space covering provides a powerful tool for the design of protocols for efficient searches in sequence space (Schuster 1995). The existence of neutral networks, finally, is essential for the efficiency of optimization (see section on **EVOLUTION OF**

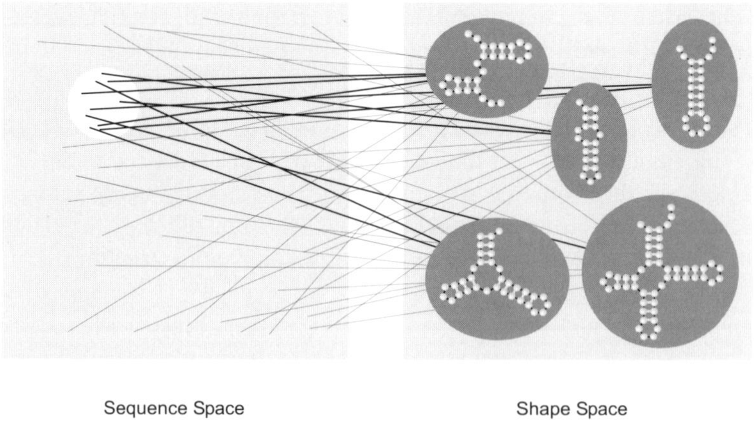

Figure 18.5 Shape space covering. To find at least one RNA sequence for each of the common structures, which forms it under the specified conditions, it is not necessary to explore the whole sequence space. Searching a (relatively small) spherical environment around any arbitrarily chosen reference sequence is sufficient. The radius of this shape space covering sphere, r_{cov}, can be readily estimated from properly chosen samples of structures.

Phenotypes) because it helps populations to escape from local fitness optima representing a kind of "evolutionary traps."

It is also worth mentioning that the results reported here represent by no means a special property observed only in RNA secondary structures. For example, a recent work on lattice models of protein structures report a very similar partitioning of folds into rare and common ones (Li et al. 1996). Results derived from simple off-lattice protein models by Nelson and Onuchic (1998) indicate, however, that shape space covering might not be fulfilled by proteins; they found foldable molecules in parts of sequence space that were separated by gaps. It should be noted, however, that these

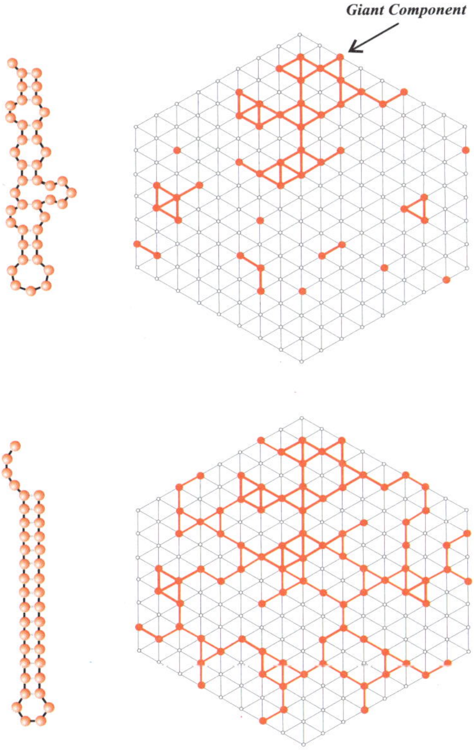

Figure 18.6 Neutral networks in sequence space. The lower structure forms a connected neutral network, spanning the whole sequence space as shown in the two-dimensional sketch. This class of network is typical for common structures. The upper part of the figure is an example of a rare structure forming a disconnected network in sequence space. Such a network consists typically of a giant component and many small components. Connectivity of neutral networks depends on the mean fraction of neutral neighbors ($\bar{\lambda}$) of the structure in sequence space. Random graph theory (Reidys et al. 1997) provides an analytical expression for the critical degree of neutrality which has to be exceeded for the connectivity condition: $\bar{\lambda} > \lambda_{cr}(\kappa) = 1 - \kappa^{-1/(\kappa-1)}$, where κ is the size of the alphabet.

searches covered only cross sections of sequence space, which must be extended to higher dimensions before definite conclusions can be drawn.

EVOLUTION OF PHENOTYPES

Computer simulation of the full, comprehensive dynamics illustrated in Figure 18.2 essentially serves two purposes: (a) the results on genotype–phenotype mapping of RNA reported in the previous section are tested with respect to their relevance for evolution, and (b) the analysis of the recorded data provides molecular insights into the evolutionary process which can be used for predictions and in the design of new experiments. Below, the results of recent computer models are reported which, in addition, initiated a new approach towards an evolutionarily relevant notion of nearness between phenotypes that is based on a kind of statistical topology induced by RNA secondary structures into sequence space (Fontana and Schuster 1998a,b).

The model system discussed here consists of samples of a thousand and more RNA molecules which replicate in a flow reactor. By means of a dilution flux that compensates for the excess production of molecules through elimination of randomly chosen individuals, the population size is constrained to be essentially constant (within fluctuations of $\sqrt{N}$ size around the mean value N). In other words, the molecules have infinite lifetimes and selection is achieved by means of the dilution flux. All RNA sequences are folded to yield secondary structures with minimum free energies. The structures are evaluated according to predefined rules in order to derive numerical values of replication rate constants (k_α is the rate constant of the structure S_α independent of the particular sequence folding into it; the selective value, the quantity that is relevant for selection, is then: $w_i = k_\alpha Q_{ii}$, where "i" refers to the particular sequence I_i that forms the minimum free-energy structure S_α). Structure optimization was performed by assigning a rate constant of the form $k_\alpha = (\delta + \eta(\alpha, \tau) / n)^{-1}$, where δ is some constant, n the chain length of the RNA, and $\eta(\alpha,\tau)$ the distance between structure S_α and the target structure S_τ. Most of the evolutionarily relevant results, however, were found to be independent of specific choices of constants and the detailed analytical expression used for the fitness function. In recent computer simulations (Fontana and Schuster 1998a, b), the initial population consisted of one thousand RNA molecules being identical copies of an arbitrary sequence with chain length $n = 76$. The population at time $t = 0$ was thus homogeneous in phenotype, the start shape, and genotype. A tRNA cloverleaf was chosen to be the target structure. The results of this simulation are shown in Figure 18.7. They are characteristic for a great number of similar computer experiments performed with different initial conditions and different target structures.

The computer simulations on RNA optimization showed a number of interesting features independent of the particular initial conditions and target structures:

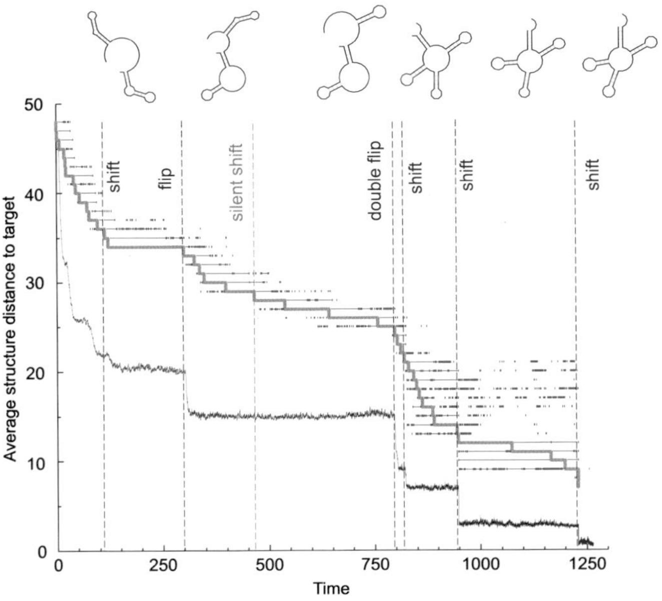

Figure 18.7 Transitions in a computer simulation of RNA optimization towards a tRNA shape. The figure shows how much optimization has progressed at the macroscopic level by plotting the population average of the Hamming distance to the target structure. The fitness curve is superimposed by the *relay trace* showing the flow of causality from start shape to target (see text for definition). The approach to the target occurs in 41 steps. Seven discontinuous transitions or major changes are marked by vertical lines. The corresponding generalized shifts are named (see Figure 18.8), and the shapes before and after the transition are shown (except for the first standard shift in order to avoid congestion of the figure). All other transitions after the first shift are continuous in the sense that they involve minor changes and occur within statistical neighborhoods. Horizontal lines before and after the occurrence of a shape in the relay series indicate periods when the shape is present in the population. The flow reactor was stochastically constrained to maintain an average of 1000 molecules of chain length $n = 76$ and the error rate was 0.001 per nucleotide.

1. The approach towards the target structure occurs in steps. Periods of fast decrease in the distance to the target are interrupted by long quasi-stationary phases or "plateaus" of essentially constant average fitness.
2. Every optimization run is initiated by a relatively short period of many successive improvements, reflecting the fact that initial distance to the target is very large; hence almost every change in structure reduces the distance to the target.
3. After the initial phase, the analysis of the succession of shapes shows always the same regular scenario: a plateau is reached through a rapid succession of readily occurring transitions between phenotypes. For reasons discussed below, we shall call these transitions *continuous*. On the plateau we observe steadily changing genotypes whereas the phenotype is either constant as with

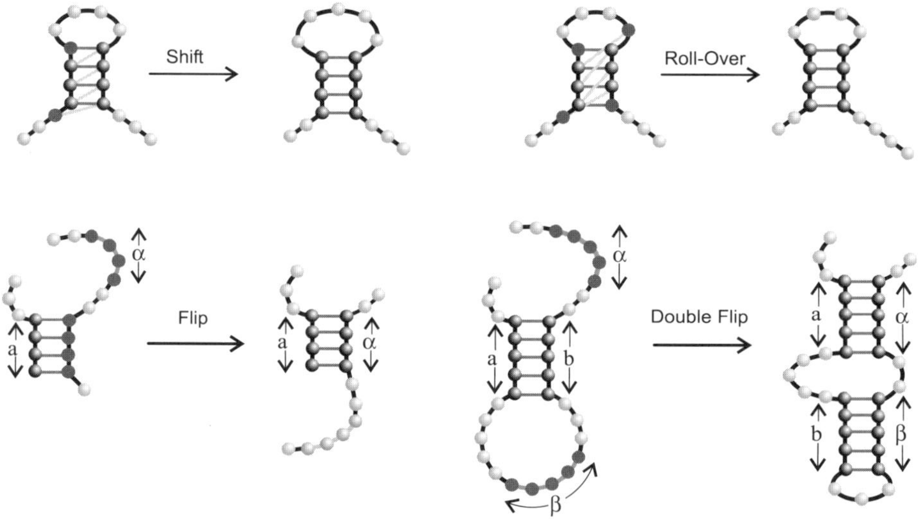

Figure 18.8 Discontinuous transitions between RNA secondary structures. Here four classes of major changes or generalized shifts in RNA secondary structures are shown which dislocate four or more base pairs. Nucleotides that change from paired to unpaired and vice versa are shown in dark gray. Examples for all four classes denoted as *shifts* (in the narrow sense), roll-overs, flips, and double-flips have been observed at the ends of quasi-stationary phases of optimization runs (see, e.g., Figure 18.7).

neutral evolution in the more restricted sense or phenotypes of equal fitness change more or less randomly through continuous transitions. At the end of each plateau we observe a rare transition which we call *discontinuous*.

4. Continuous transitions represent minor changes in the phenotype,[10] they occur readily and lead to globally frequent structures at the boundary of the neutral network of the current phenotype.
5. Discontinuous transitions represent large structural changes (see Figure 18.8). They occur rarely and lead to globally rare but locally frequent structures in the boundary of the neutral network of the current phenotype.

Computer runs have been repeated several times (with different seeds of the random number generator) to learn about reproducibility of results in evolution. Despite some scatter in the time required to reach the target structure, the number and the nature of the discontinuous transitions (five shifts, one flip, and one double flip in Figure 18.7) but not their sequence was found to be fairly constant. Comparison of entry and exit sequences of the shapes in the relay series (these are the first and last sequence along the

[10] Small changes are either the addition of a base pair to an already existing stack or the removal of a base pair from one. In addition, the opening of an entire stack that is under especially strong energetic constraints like, for example, the closing stack of a multiloop, were also found to occur readily.

relay series which fold into the shape considered) revealed an interesting tendency. In most but not all cases, structures become more stable in the sense of lower minimal free energy during evolution on a fitness plateau. This is presumably a secondary effect mediated by the structural details of neutral networks: sequences have higher fitness when they have more neutral neighbors in sequence space, and sequences with more neutral neighbors tend to have lower minimal free energies.

An important question deals with the difference between target-oriented searches (like the one described above or *in vitro* RNA optimization experiments) and evolutionary processes optimizing nothing but fitness. In an earlier paper (Fontana and Schuster 1987), it was shown by computer simulation that both types of evolutionary processes are very similar with respect to the overall appearance (stepwise approach based upon continuous and discontinuous transitions). Repeated untargeted searches yielded, as expected, different structures meeting the optimal fitness criterion.

The regularities observed in computer simulations of RNA optimization call for an interpretation by means of a concept that is generalizable to other evolutionary systems. The meaning of continuity in shape space becomes clearer when one considers the one-error boundaries of neutral networks in sequence space. The full boundary is formed by folding all one-error mutants of all sequences belonging to the neutral network. (It is sufficient, however, to consider a representative random sample as a statistical approximation to the full set.) Evaluation of single point mutation boundaries of neutral networks suggested the development of a nearness relation between phenotypes. Inspection of these one-error boundaries showed that it is straightforward to define frequent shapes. In the rank-ordered distribution of probabilities of occurrence, only a few shapes qualify as frequent neighbors that are clearly separated from the rest by a gap in frequencies. The proposed statistical neighborhood relation truncates the distribution at this gap. Nearness of phenotypes is a property of entire networks and hence it is of global nature.[11] Continuous transitions occur towards shapes within the statistical neighborhood. Every sequence has its own specific single point mutation boundary that contains shapes at substantial frequencies which are not in the statistical neighborhood of the neutral network to which it belongs. Such shapes are locally frequent but globally rare. All discontinuous transitions observed in computer simulations proceeded to such shapes (of course, with the additional constraint of increasing fitness).

In summary, continuous transitions occur within the statistical neighborhood whereas a transition is discontinuous when it leaves the neighborhood. Consequently, continuous transitions occur everywhere on the neutral network and thus readily. The occurrence of discontinuous transitions imposes (severe) constraints on the sequence which can only be fulfilled locally. It is worth mentioning that the differences in

[11] Shapes in the statistical neighborhood relation need not commute: S_α near S_β does not imply S_β near S_α. Assume, for example, two neutral networks of very different size, $\#(S_\alpha) = \{$number of sequences forming $S_\alpha\} >> \#(S_\beta)$, and the shape corresponding to the large network being a frequent neighbor of the small network. Then, the shape of the small network may be only a rare neighbor for the large network: $\{S_\alpha$ near $S_\beta\}$ but $\{S_\beta$ not near $S_\alpha\}$.

frequencies of shapes between local and global boundaries of neutral networks correspond well with the correlation length derived from the autocorrelation function of the sequence-structure map (Fontana et al. 1993).

WHAT THE RNA MODEL TELLS ABOUT EVOLUTION

The subtle details of genotype–phenotype mapping discussed in the last two sections provide a straightforward explanation for the occurrence of punctuated rather than continuous evolution. Target shapes or optimal phenotypes are commonly not accessible through continuous transitions. To bring about the required discontinuous transitions, suitable sequences must be prepared. Populations approach these special points in sequence space through random drift on neutral networks, which manifests itself in the fitness plateaus of the optimization runs (for a sketch, see Figure 18.9).

Stepwise optimization in constant environments is not only observed in computer simulations of RNA evolution. It seems to be a generic feature of genetic algorithms (van Nimwegen et al. 1997) and, even more interesting, it is commonly encountered in experimental recordings. Analysis of the early Spiegelman experiments (Spiegelman 1971), for example, shows a punctuated increase in the rate of RNA synthesis (Fontana and Schuster 1998a). Still more striking are steps in the increase of *Escherichia coli* cell size observed under the constant conditions of a chemostat (Elena et al. 1996). The straightforward explanation of the fitness plateaus is that the population is waiting for a rare event. This may be a rare mutation in absence of neutrality or a special position that allows a switch from one neutral network to another by means of a discontinuous transition.

The RNA model described here certainly represents a drastic reduction of the real biological phenomenon, and it is a legitimate question to ask whether and under which aspects the results apply to more complex systems. The neighborhood relations in sequence and shape space are indeed only based on the degree of neutrality, and their derivation does not make use of any RNA specific property. It is straightforward to attribute general validity to them. To check this conjecture for more complex systems than molecules evolving *in vitro* is certainly very hard. The rapidly increasing information on viral and bacterial genomes indicates promising possibilities in the not too distant future.

Returning to the initial question of diversity in evolution, I summarize the conclusions concerning genotypes and phenotypes:

1. Genotypic diversity is required for change, and it is unavoidable because of limitations in replication accuracy. Error propagation sets an upper bound. How much diversity is required is a matter of environmental variation. RNA viruses represent an extreme example since they are apparently driven to maximal diversity by the defense mechanisms of their hosts. The reproduction of higher organisms does not operate close to error threshold conditions, indicating that they are not forced by their environments to vary as much as possible.

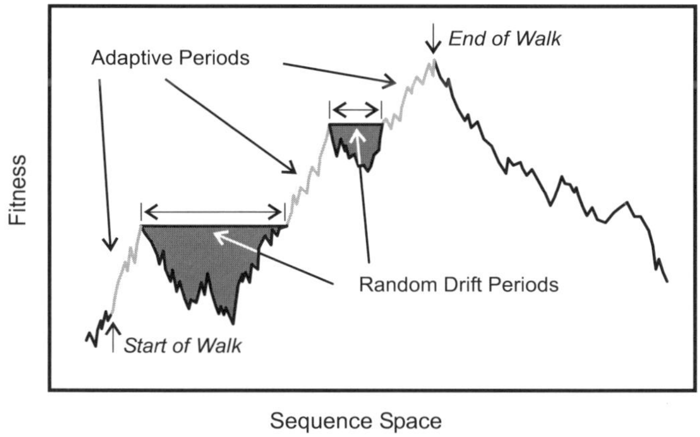

Figure 18.9 A sketch of optimization through adaptive walks of populations. Adaptive walks allow to choose the next step arbitrarily from all directions where fitness is (locally) nondecreasing. Because of the quasi-species-like mutant distributions (Figure 18.3) populations can bridge over narrow valleys with widths of a few point mutations. Major fitness peaks, however, are evolutionary traps in which optimization ends unless the population can escape through random drift along neutral networks. Populations on rugged landscapes with sufficiently high degree of neutrality evolve by a combination of adaptive walks and random drift (at essentially constant fitness) along the neutral networks. When networks span the entire sequence space populations may eventually reach the global maximum of the fitness landscape. Optimization occurs on two time scales: fast adaptive phases interrupted by long quasi-stationary periods of random drift or neutral evolution.

2. Phenotypic diversity is the true target of selection. It is trivially related to genotypic diversity only in the absence of neutrality. Properties of biopolymers and the usability of their sequences for the reconstruction of phylogenetic trees, however, tell us that we are dealing with high degrees of neutrality in nature. Then, genotypic diversity *may, but need not, reflect* variation in the properties of phenotypes. Relations between genotypes and phenotypes are shaping the evolutionary process. Continuity in evolution is brought in context to neighborhood relations in sequence and shape space. The attainability of phenotypes acts as a filter on genetic variation.

In the light of genotype–phenotype relations, evolutionary optimization appears to be exceedingly complex. At the same time, however, it is apparently simple since it occurs spontaneously in nature and in various experimental assays used in test tube evolution and in evolutionary biotechnology. The concept of selection is straightforward and it works so well. Neighborhood relations in sequence space and their interpretation in terms of continuity and discontinuity provide the link between apparent simplicity and overwhelming complexity in the molecular details.

ACKNOWLEDGMENTS

The work reported here has been supported financially by the Austrian Fonds zur Förderung der wissenschaftlichen Forschung (Projects No. 10578-MAT and No. 11065-CHE), by the Commission of the European Community (CEC Contract Study PSS*0884), and by the Santa Fe Institute. Fruitful discussions with Drs. Walter Fontana, Christian Forst, Ivo Hofacker, Christian Reidys, and Peter Stadler are gratefully acknowledged.

REFERENCES

Bartel, D.P., and J.W. Szostak. 1993. Isolation of new ribozymes from a large pool of random sequences. *Science* **261**:1411–1418.

Biebricher, C.K., and W.C. Gardiner. 1997. Molecular evolution of RNA *in vitro*. *Biophys. Chem.* **66**:179–192.

Biebricher, C.K., and R. Luce. 1992. *In vitro* recombination and terminal elongation of RNA by $Q\beta$ replicase. *EMBO J.* **11**:5129–5135.

Demetrius, L., P. Schuster, and K. Sigmund. 1985. Polynucleotide evolution and branching processes. *Bull. Math. Biol.* **47**:239–262.

Domingo, E., J.J. Holland, C.K. Biebricher, and M. Eigen. 1995. Quasispecies: The concept and the word. In: Molecular Basis of Virus Evolution., ed. A. Gibbs, C.H. Calisher, and G. García-Arenal, pp. 171–180. Cambridge, U.K.: Cambridge Univ. Press.

Drake, J.W. 1993. Rates of spontaneous mutations among RNA viruses. *Proc. Natl. Acad. Sci. USA* **90**:4171–4175.

Eigen, M. 1971. Self-organization of matter and the evolution of biological macromolecules. *Naturwissenschaften* **58**:465–523.

Eigen, M., J.S. McCaskill, and P. Schuster. 1989. The molecular quasispecies. *Adv. Chem. Phys.* **75**:149–263.

Eigen, M., and P. Schuster. 1977. The hypercycle. A principle of natural self-organization. Part A: Emergence of the hypercycle. *Naturwissenschaften* **64**:541–565.

Eigen, M., and P. Schuster. 1978. The hypercycle. A principle of natural self-organization. Part C: The realistic hypercycle. *Naturwissenschaften* **65**:341–369.

Eigen, M., and P. Schuster. 1982. Stages of emerging life. Five principles of early organization. *J. Mol. Evol.* **19**:47–61.

Elena, S.F., V.S. Cooper, and R.E. Lenski. 1996. Punctuated evolution caused by selection of rare beneficial mutations. *Science* **272**:1802–1804.

Fontana, W., D.A.M. Konings, P.F. Stadler, and P. Schuster. 1993. Statistics of RNA secondary structures. *Biopolymers* **33**:1389–1404.

Fontana, W., and P. Schuster. 1987. A computer model of evolutionary optimization. *Biophys. Chem.* **26**:123–147.

Fontana, W., and P. Schuster. 1998a. Continuity in evolution. On the nature of transitions. *Science* **280**:1451–1455.

Fontana, W., and P. Schuster. 1998b. Shaping space. The possible and the attainable in RNA genotype–phenotype mapping. *J. Theor. Biol.* **194**:491–515.
(Also published as Santa Fe Institute Preprint No. 97–11–081.)

Grüner, W., R. Giegerich, D. Strothmann, C. Reidys, J. Weber, I.L. Hofacker, P.F. Stadler, and P. Schuster. 1995. Analysis of RNA sequence-structure maps by exhaustive enumeration. *Monatshefte Chemie* **127**:339–397.

Huynen, M.A., P.F. Stadler, and W. Fontana. 1996. Smoothness within ruggedness. The role of neutrality in adaptation. *Proc. Natl. Acad. Sci. USA* **93**:397–401.

Kauffman, S.A. 1993. The Origins of Order. Self-organization and Selection in Evolution. New York: Oxford Univ. Press.

Li, H., R. Helling, C. Tang, and N. Wingreen. 1996. Emergence of preferred structures in a simple model of protein folding. *Science* **273**:666–669.

McCaskill, J.S. 1997. Spatially resolved *in vitro* molecular ecology. *Biophys. Chem.* **66**:145–158.

Nelson, E.D., and J.N. Onuchic. 1998. Proposed mechanism for stability of proteins to evolutionary mutations. Preprint.

Reidys, C., P.F. Stadler, and P. Schuster. 1997. Generic properties of combinatory maps. Neutral networks of RNA secondary structure. *Bull. Math. Biol.* **59**:339–397.

Schuster, P. 1995. How to search for RNA structures. Theoretical concepts in evolutionary biotechnology. *J. Biotech.* **41**:239–257.

Schuster, P. 1996. How does complexity arise in evolution. Nature's recipe for mastering scarcity, abundance, and unpredictability. *Complexity* **2(1)**:22–30.

Schuster, P. 1997a. Genotypes with phenotypes. Adventures in an RNA toy world. *Biophys. Chem.* **66**:75–110.

Schuster, P. 1997b. Landscapes and molecular evolution. *Physica D* **107**:351–365.

Schuster, P., W. Fontana, P.F. Stadler, and I.L. Hofacker. 1994. From sequences to shapes and back. A case study in RNA secondary structures. *Proc. Roy. Soc. Lond. B* **255**:279–284.

Schuster, P., and P.F. Stadler. 1994. Landscapes: Complex optimization problems and biopolymer structures. *Computers Chem.* **18**:295–314.

Stadler, P.F. 1996. Landscapes and their correlation functions. *J. Math. Chem.* **20**:1–45.

Stadler, P.F., and G. Wagner. 1998. The algebraic theory of recombination spaces. *Evol. Comp.* **5**:241–275.

Spiegelman, S. 1971. An approach to the experimental analysis of precellular evolution. *Qtly. Rev. Biophys.* **4**:213–253.

Strunk, G., and T. Ederhof. 1997. Machines for automated evolution experiments *in vitro* based on the serial-transfer concept. *Biophys.Chem.* **66**:193–202.

Szathmáry, E., and J. Maynard Smith. 1995. The major evolutionary transitions. *Nature* **374**:227–232.

Tacker, M., P.F. Stadler, E.G. Bornberg-Bauer, I.L. Hofacker, and P. Schuster. 1996. Algorithm-independent properties of RNA secondary structures. *Eur. Biophys. J.* **25**:115–130.

van Nimwegen, E., J.P. Crutchfield, and M. Mitchell. 1997. Finite populations induce metastability in evolutionary search. *Phys. Lett. A* **229**:144–150.

Waterman, M.S. 1995. Introduction to Computational Biology. Maps, Sequences, and Genomes. London: Chapman and Hall.

Wieser, W. 1997. A major transition in Darwinism. *Trends Ecol. Evol.* **12**:367–370.

Wright, S. 1932. The roles of mutation, inbreeding, crossbreeding, and selection in evolution. In: Intl. Proc. 6th Intl. Congress on Genetics, ed. D.F. Jones, vol. 1, pp. 356–366. Ithaca, N.Y.

19

Chance and Necessity in Biomolecular Chemistry

Is Life as We Know It Universal?

S.A. BENNER[1] and C.Y. SWITZER[2]

[1]Departments of Chemistry and Anatomy and Cell Biology, University of Florida, Gainesville, FL 32611, U.S.A.
[2]Department of Chemistry, University of California, Riverside, CA 92502, U.S.A.

Keywords: Darwinian evolution, biopolymer models, indicators of life

ABSTRACT

Life is defined as a self-sustaining chemical system capable of undergoing Darwinian evolution. The structures of biopolymers that might support life so defined are discussed in this chapter with an eye towards identifying features of these biopolymers that might be found in any life form, regardless of its genesis, and inaccessible (or improbable) as the product of any nonliving chemical system. Also explored is whether the structure of contemporary life is more or less complex than it needs to be, and whether systems with greater complexity existed earlier in the history of the planet. We then consider why some theories are successful at guiding the design and manipulation of complex organic molecules, while others are not.

INTRODUCTION

Unless it is intelligent enough to speak to us directly, extraterrestrial life will ultimately be recognized by some distinctive character of its chemistry. Chemistry may be even more useful than physiology (or fossil remnants of physiology) for this purpose, as the recent discussion of the "fossils" from Martian meteorites has shown (see

Simplicity and Complexity in Proteins and Nucleic Acids
Edited by H. Frauenfelder, J. Deisenhofer, and P.G. Wolynes

review in Kerr 1997). This discussion has been largely uninformed by relevant information from contemporary bio-organic chemistry, which must ultimately be brought to bear on the search for evidence of life on Mars. Nevertheless, it is clear from the published literature that the microstructures found in the Allan Hills 84001 meteorite might have been formed by nonliving processes as well as living processes. Thus, such microfossils cannot be taken as reliable indicators of life.

The same can be said, of course, for many chemicals that might be viewed as signatures of life. Fifty years ago, the discovery of adenine or alanine or ribose — three chemical building blocks of nucleic acids, proteins, and carbohydrates, respectively — in extraterrestrial samples might have been viewed as evidence for extraterrestrial life. No longer. "Prebiotic chemistry," which seeks to create abiotically in the laboratory the chemical components of terrestrial living systems, has produced adenine, alanine, and ribose derivatives, not only without the intervention of enzymes and cells, but also under conditions that (with a little imagination) are conceivable at least somewhere in the galaxy. While successful prebiotic experiments increase the possibility of spontaneous generation of life, every successful prebiotic experiment also diminishes the number of molecules that can be used as reliable signatures of life.

The problem is even more complicated, however. Samples where extraterrestrial life perhaps *used* to exist but exists no more (samples from the surface of Mars, for example) will undoubtedly be obtained by the NASA planetary exploration program before samples where extraterrestrial life perhaps *presently* exists (the oceans of Europa or the subsurface of Mars). Detecting the chemical and structural signatures of expired life is more difficult than identifying those of extant life, as degradation of carbon-containing molecules from life yields products that resemble even more the products of nonlife. For example, "polycyclic aromatic hydrocarbons" (PAHs) are molecular fragments of graphite. Such fragments are found in cigarette smoke and coal (indisputably derived from organic matter) and in interstellar space, presumably not arising from living matter (Li and Greenberg 1997). They will undoubtedly be found on Mars, provided that we look hard enough. From a chemical point of view, it will be difficult to tell whether these arose from carbon in living systems or carbon in nonliving systems. Likewise, in the evidently oxidizing top surface of Mars (Zent and McKay 1994), organic matter is likely to be degraded to carbon dioxide, also available from nonbiological sources.

Thus we are left with a conundrum. If structures that look like fossils of cells can be generated by nonlife, and if molecules that look like biochemicals (or their degradation products) can be generated by nonlife, how are we to identify life (or its remnants) on Mars, Europa, or elsewhere?

This conundrum can be solved if we can come to understand what chemical features are *universal* in living systems (that is, found in life regardless of its genesis) and *unique* to living systems (that is, not generated by any nonlife processes). To begin, we adopt the working definition of "life" proposed by a committee that addressed this problem for NASA in 1994 (Joyce et al. 1994), modified as suggested by Luisi (1998) to state explicitly certain aspects of the definition that NASA implied but did not state:

Life is a chemical system that is self-sustaining via the internal synthesis of components from external energy and nutrients, and capable of undergoing Darwinian evolution (reproduction, mutation, and natural selection).

The definition makes it evident that biology is not a discipline distinct from chemistry (a point perhaps obvious in 1999, when it is difficult to find a biologist who is not attempting to put a molecular face on the biological phenomena under study). The definition also captures the fundamentals of our understanding of how life works. Very few mechanisms have been proposed for how molecular systems could emerge with the high information content needed to be self-sustaining in a chemical sense:

1. Creationism, generation of heritable information in a living system by another system carrying still more information.
2. Lamarkianism, feedback of information from the environment to be encoded in a heritable form.
3. Darwinianism, the generation of heritable information by a process of random variation followed by natural selection.

Of these, Darwinianism is the only mechanism known to function in the natural world. Biology can therefore be defined as the subdiscipline of chemistry that studies molecular systems which undergo Darwinian evolution, that is, can reproduce, can suffer mutation, and evolve under the pressures of natural selection.

THE TWO-BIOPOLYMER MODEL

What features are universal in molecular systems that can undergo Darwinian evolution? One must begin with the only example of molecular systems capable of Darwinian evolution known to us: life on Earth. While one has little choice but to start here, we recognize the prejudice created by our knowledge of terrestrial life. A rather comprehensive knowledge of chemistry is required to go even one step beyond these prejudices (and even then, it is very difficult). Sociological problems are also confounding. "Chemistry for biologists" is frequently one of the worst courses in the academic curriculum. Conversely, most organic chemists do not involve themselves in issues central to defining organic reactivity that is general to life; many consider this type of research "flaky." Thus, it is difficult to match the problem with the expertise needed to solve it.

The one self-sustaining chemical system indisputably known to undergo Darwinian evolution is the "two-biopolymer" system built from nucleic acids and their encoded proteins of contemporary terrestrial life.[1] Proteins and DNA might be sought as

---

[1] Terrestrial life is actually a three-biopolymer system: DNA, RNA, and proteins. Molecular biologists often treat RNA and DNA as "the same," but they are not, and need not have been as similar as they actually are, given what we know about how small perturbations in backbone structure change the properties of oligonucleotides.

extraterrestrial markers of life. Indeed, if they were to be found together on Mars or Europa (and if the possibility of recent terrestrial contamination could be excluded), they would be strong indicators of extraterrestrial life in those locales.

However, before planetary expeditions to Mars or Europa are designed from this optimistic scenario, we must recognize that this starting point carries with it the prejudices noted above. We know that a two-biopolymer system works on Earth, but is the system universal? In particular:

1. Is it universal in its details, (i.e., will all life on all planets use proteins and nucleic acids with the same twenty amino acids and the same four heterocyclic nucleobases, which we will call *standard* biopolymers)?
2. Is it universal without its details, (i.e., will all life use proteins and nucleic acids, with the amide linkages and phosphodiester linkages as the universals, but with different amino acid side chains and the nucleobases)
3. Is the two-biopolymer system universal only in its broad outlines, (i.e., will all life use two biopolymers, one as an information repository, the other for catalysis, with the first encoding the other) but based on biopolymers other than those joined by peptide (amide) or phosphodiester bonds?

Considerable experimental work now shows that Pt. 2 is chemically possible. Nucleic acids with twelve distinct bases joined as base pairs via six different patterns of hydrogen bonding are competent as components of a genetic system (Switzer et al. 1989; Piccirilli et al. 1990). These can be used to expand the number of amino acids in encoded peptides (Bain et al. 1992), and technology has been developed by Hecht, Schultz, Chamberlin, and others that synthesize polypeptides with nonstandard amino acids by recruiting the stop codon (for a review, see Benner 1994). These experimental results rule out Pt. 1 as a defensible assumption when seeking signs of extraterrestrial life. Even if a two-biopolymer model for life is assumed to be universal, the scope of the search must extend past the chemistry of terrestrial biopolymers to at least Pt. 2.

THE ONE-BIOPOLYMER MODEL

Even the terrestrial version of the "two-biopolymer" model contains a well-recognized paradox, one that relates to "origins." On Earth, the two biopolymers (proteins and nucleic acids) are specialized to perform two functions (catalysis and information repository). The sequence of one is related to the sequence of the other by an encoded–encoder relationship. It is highly improbable, however, that both biopolymers arose spontaneously and simultaneously in an environment lacking life, and astronomically improbable that they arose in this way with an encoded–encoder relationship.

A variety of "single-biopolymer" hypotheses have been proposed to address this paradox. Such hypotheses hold that life originated as a single biopolymer that performed both catalytic and repository roles. A variety of these are based on "genetic takeover" models, where one biopolymer presumed to be easy to form under prebiotic conditions eventually templates the standard biopolymers known in contemporary terrestrial life.

The most popular model today is the hypothesis that the first (or perhaps an early) molecular system capable of Darwinian evolution was based on RNA, a hypothesis that dates back to seminal papers by Alex Rich (1962), Woese (1967), Orgel (1968), and Crick (1968), Usher and McHale (1976), White (1976), and Visser and Kellogg (1978). The phenomenal discoveries by Cech and Altman (Cech et al. 1981; Zaug and Cech 1986), that RNA performs catalytic functions in contemporary organisms, has made the "RNA world" a part of the culture in contemporary molecular biology (Watson et al. 1987).

Even simple "single-biopolymer" models have relevance to the search for extraterrestrial life. For example, biologists have taken pains to point out that the microfossils in the Allan Hills meteorite, which are as small as 20–100 nanometers across, are too small to be living cells (Kerr 1997). After all, the argument is made, the ribosome is 25 nm across, and ribosomes are a basic requirement for life.

Actually, ribosomes are a basic requirement for life based on *two* biopolymers. If a single biopolymer (such as RNA) can serve both genetic and catalytic functions, then ribosomes are not needed for life. Indeed, much of the metabolism of contemporary cells (aminoacyl tRNA synthetases and amino acid biosynthesis enzymes, for example, which comprise more than half of what is believed to be the "core metabolism" [Benner et al. 1993]) is also not needed for life. A cell based on a "single-biopolymer" genetic system can be far smaller than one based on two biopolymers. This means that the "fossils" in the Martian meteorite are *not* too small to be remnants of a single-biopolymer form of life. Conversely, if the meteorite structures are indeed fossils, then they almost certainly are fossils of an organism that used only a single biopolymer as its molecular system capable of Darwinian evolution.[2] A similar analysis can be applied to any structures that might be encountered in planetary exploration missions.

Because they do not require the invention of translation, we might naively view single-biopolymer life forms to be the most likely form of life to be encountered extraterrestrially. In fact, this view requires that three hypotheses be true:

1. Single biopolymers capable of self-sustenance and Darwinian evolution must exist (the "existence" hypothesis).

---

[2] If one accepts the notion that life on Earth began with an organism that used a single biopolymer such as RNA, one might ask whether there might be environments where a single-biopolymer organism would remain even after the invention of translation. Normally, the superior power of proteins as catalysts provides presumptive arguments that the older form of life was extinguished by the derived form, and none remains on Earth. The fact that cells containing single-biopolymer life can be much smaller than two-biopolymer cells suggests, however, that the one-biopolymer life might survive where small size offers a selective advantage. In subterranean matrices, e.g., formations can have pore sizes that are too small to permit a two-biopolymer organism to live, but might permit a single-biopolymer cell to reside free from competition from its more adept protein-using cousins.

2. The fitness of such biopolymers must be reasonable (the "fitness" hypothesis).
3. The "breakthrough" (Benner, Ellington et al. 1989) to translation must be chemically challenging.[3]

If these hypotheses are true, then we expect to encounter single-biopolymer systems in the galaxy with higher probability than two-biopolymer systems. If intrinsic chemical contradictions of a single biopolymer attempting to both store information and catalyze reactions (see below) render them intrinsically unfit, then single-biopolymer systems that fail to invent translation rapidly are expected to be easily extinguished, and the ratio of two-biopolymer living systems to single-biopolymer systems will be proportionately higher (at the same time as implying that the total number of living systems in the galaxy will be fewer).

IN SEARCH OF UNIVERSAL FEATURES OF A SINGLE-BIOPOLYMER LIFE

What then are expected to be molecular features both universal in and unique to single-biopolymeric systems capable of undergoing Darwinian evolution? What structures bring together replicatability, mutability, and adaptive function[4] (e.g., catalytic activity) in one biopolymer that can evolve in response to natural selection? As with any research, one cannot ask such questions in a vacuum. "Structure space" is far too large; there are simply too many polymers that might be made to undertake a research project to make them one at a time and inspect their properties, although this approach has certainly been tried (Freier and Altmann 1997). Instead, the search must be rationally constrained. Several constraints being used by many workers are briefly summarized below.

Constraint 1: Attempt to limit the range of possibilities using prebiotic chemistry

One of the successes of exobiology has been the generation of organic molecules characteristic of living systems under prebiotic conditions. If only a few organic molecules

3 None of these points are free of controversy. For example, dozens of laboratories have attempted to identify an inherent chemical relationship between the DNA/RNA codon and the encoded amino acid ever since the genetic code was established. Such efforts are ongoing in many laboratories (we cannot cite all of these papers; two recent papers with leading references are DiGiulio [1997] and Alberti [1997]). If an intrinsic link exists between the chemistry of proteins and nucleic acids, then inventing translation becomes proportionately easier, which in turn increases the ratio of two-biopolymer living systems to single-biopolymer living systems in the galaxy, at the same time as increasing the likelihood that those two biopolymers are indeed proteins and nucleic acids. Indeed, it would argue that the genetic code is truly universal (in the cosmologic sense of the term).

4 The phrase "adaptive function" is redundant in the Darwinian view, which holds that function can arise only through selection placed atop random variation. We use the redundancy here because so many biochemists confuse "function" with "behavior," "structure," or other observables that may or may not be adaptive.

could have emerged spontaneously under conditions inferred for primitive Earth, prebiotic chemistry would be a powerful guide in our search for universal structures in life (assuming that wherever life emerged, the same prebiotic chemistry prevailed). It would force us to consider these (and only these) few molecules as the starting point to obtain Darwinian systems. Conversely, if many organic molecules could have been generated under conditions inferred for primitive Earth (or if many conditions can be imagined, each leading to different sets of prebiotic chemicals), then prebiotic chemistry is a less useful guide for research in Darwinian chemistry.

Unfortunately, prebiotic studies have been so successful that the latter is more the case than the former. An example is found in the "tholins" of the Cornell group, which have the deep red-brown color known to those who have cleaned out a chemical stockroom (Khare et al. 1993). Even when generated from simple organic compounds such as ethane (and more so if ammonia is added, and still more so if additional compounds are added to the mix), tholins contain so many compounds that it is difficult to exclude any specific compound in particular from the prebiotic environment, at least in small amounts. Further, the conditions under which these mixtures are formed (e.g., ethane in water with irradiation), appear to be quite accessible on both early and modern planets.

Indeed, an entirely new problem has emerged from prebiotic research: the problem of complexity. For example, template-directed synthesis of chiral compounds can be inhibited by building blocks of the opposite chirality (Orgel 1992; Schmidt et al. 1997). This is one example of a well-known problem relating to the chemistry of complex mixtures: one component of a prebiotic mixture can inhibit as well as facilitate a desirable reaction by another component of the same mixture. Chemists have long known that it helps to isolate compounds in relatively pure form before trying to crystallize them. Combinatorial chemistry, a new approach in medicinal chemistry derived by analog to evolution in natural systems (Thompson and Ellman 1996), is troubled by similar problems.

One task for the next phase of prebiotic/combinatorial chemistry is to define how complex mixtures might be resolved to obtain mixtures with useful complexity (but not greater). Indeed, at this stage in the development of prebiotic chemistry as a field, it would be more useful to ask how mixtures of organic chemicals might be fractionated on a primitive planet. Life may have been guided not by what organic molecules were synthesized in large quantities in a prebiotic setting, but by what compounds were efficiently concentrated by prebiotic chemical and mechanical processes.[5]

The second reason limiting prebiotic chemistry as a guide for Darwinian chemistry is the converse of the first. Even while many compounds are now known to be generated abiotically, there is no guarantee that a compound *not* presently known to be a product of abiotic synthesis will not become "prebiotic" by some future abiotic experiment. This has led to a sociological problem in the field, the propensity of prebiotic

---

[5] We do not regard this problem as hopeless. An active planet has abundant opportunity to fractionate, crystallize, and separate, as crystalline mineral deposits illustrate.

chemists to pronounce a compound as "not prebiological" simply because it is not (yet) known to be generated abiologically.[6]

Constraint 2: Attempt to limit the range of possibilities by identifying molecular systems that undergo template-directed reproduction

Reproduction is, of course, the first property that a molecular system must display to partake in Darwinian evolution. Many molecular systems make copies of themselves, of course, including some in quite trivial ways. For example, a crystal of sodium chlorate ($NaClO_3$), if fragmented, will nucleate (or "seed") the formation of more crystals of $NaClO_3$. The surface of the crystal is a template for the self-assembly of freely wandering ions into organized matter. In this particular case, the system spontaneously generates homochirality, a property widely believed to be unique to living systems (Macdermott et al. 1996). Rectangular prisms of $NaClO_3$, in either the right-hand or left-hand form of the chiral space group P2(1)3, form from the solutions of the achiral ions. If the solution is stirred, only right-handed or only left-handed crystals are formed (McBride and Carter 1991). If the solution is *not* stirred, crystals of both enantiomorphic forms precipitate in equal amounts. This highly reproducible result is a consequence of a "self-reproduction." Stirring breaks apart the first crystal formed with random chirality, whose fragments then seed the formation of many "daughter" crystals with the same chirality.

Templated replication is widespread in chemistry, and many examples are now known in bio-organic chemistry as well. Rebek (Wintner et al. 1994; cf. the challenge by Menger et al. [1995] over the validity of Rebek's interpretation and Rebek's rebuttal [Wintner et al. 1995]), Ghadiri (Lee et al. 1997), and von Kiedrowski et al. (1991) have shown how small molecules, peptides, and oligonucleotides undergo template-directed synthesis. It seems not to be excessively difficult to design such systems, and many are being examined in many laboratories.

Just as the success of prebiotic chemistry diminishes its value to constrain the search for structures that are universal molecular features of life, so has the widespread success of experiments to obtain templated reproduction. If self-templating systems made with nucleic acids work, with peptides, with salts, and with small organic molecules of many shapes and sizes, no single system has the "universality" attribute that demands its selection as a system for further study, let alone as a system that might be sought on Mars or Europa.

---

6 Our favorite example of this is the discovery by Albert Eschenmoser of a relatively efficient abiological synthesis of ribose-2,4-diphosphate in the context of a literature that had all but excluded ribose as a "prebiotic" substance. Whether the Eschenmoser synthesis was in fact used prebiotically remains uncertain. However, the point remains: just because we do not know of a "prebiotic" synthesis of a compound does not mean that one does not exist.

Constraint 3: COSMIC-LOPER biopolymers

Mutation is the second property that a molecular system must display to partake in Darwinian evolution. In some usage, "mutation" denotes any structural change. Thus, a molecular system of $NaClO_3$ might be said to "mutate" if the sodium is replaced by potassium to yield $KClO_3$.

In practice, mutation cannot be this drastic if it is to be effectively coupled with natural selection. This is both a practical matter and a theoretical one. At the very least, mutation in a molecular system that can self-replicate must not change the physical properties of the molecules involved in a way that obviates mechanisms used for self-replication. The replacement of Na by K in the example above does.

To support a self-sustaining chemical system capable of undergoing Darwinian evolution, a biopolymer must be capable of searching mutation-space independent of concern that it will lose properties essential for replication. We will designate biopolymers that have this property as COSMIC-LOPER biopolymers (the acronym stands for "Capable Of Sustaining Mutation Independent of Concern over Loss Of Properties Essential for Replication").

The need for a single-biopolymer form of life to be based on a COSMIC-LOPER biopolymer is nearly axiomatic. If a substantial fraction of the mutations encompassed within an encoding system cause a biopolymer to precipitate, unfold, form a new folded structure and therefore no longer be recognized by a catalyst for replication, or acquire some other physical trait that makes it unable to guide template-directed replication, the biopolymer is not suited for Darwinian chemistry. Thus, the COSMIC-LOPER property is expected to be a universal in single-biopolymer living systems. We might conjecture that two-biopolymer forms of life must contain one COSMIC-LOPER polymer.

In contemporary terrestrial life, it is clear that proteins are not COSMIC ("Capable of Sustaining Mutation Independent of Concern") polymers — even in cases where they can direct template-based replication (as the splendid work of Ghadiri has found). The physical properties of proteins can change dramatically upon point mutation within the mutation space allowed by the twenty standard amino acids. The textbook case is sickle-cell hemoglobin, where a single amino acid substitution creates a variant that undergoes partial precipitation.

It is well known to those who work with peptides that precipitation is a common property; a peptide of random sequence chosen from the twenty standard amino acid is most likely to be insoluble, a property also known to those who boil eggs (after heating to escape the metastable state where the proteins are soluble). Thus, it is not remarkable that sickle-cell hemoglobin precipitates; what is remarkable is that native hemoglobin is soluble at the nearly gram-per-milliliter concentrations found in red blood cells. One of the most characteristic features of the divergent evolution of the hemoglobin gene family is the extent to which substitution has been constrained to avoid insolubility.

Experimental work illustrates the non-COSMIC properties of peptides. Small (14 amino acid) polypeptides that formed nearly exclusively a four-helix bundle have been designed (Johnsson et al. 1990, 1993) and their structure determined by NMR in solution, one of the first times that this was done for a designed peptide. The identical peptide was then prepared, but with an N-terminal acetyl group. The small change was sufficient to cause the peptide to aggregate. Still other point mutations caused the helix not to form. If solubility and/or helix formation are essential to the replicatability of a peptide template, a large range of plausible mutation would destroy it, at least in this system.

Synthetic organic chemists have long been aware of the anti-COSMIC behavior of many small molecules. The most common situation is encountered in natural product synthesis. Frequently, synthetic efforts on the natural product are preceded by exploratory work with a "model system," a simpler molecule that is more accessible than the natural product itself, but which is presumed to be representative of the kinds of problems that will be encountered when the natural product is synthesized. As often as not, reactivities of the model and the real natural product are sufficiently different that chemistry developed on the first is defeated by the second. A single methyl group can make the difference, with the practical implications of real chemistry unanticipated by models.

The molecular biologist relies on the fact that natural oligonucleotides with different sequences nevertheless behave similarly. Every (or almost every) oligonucleotide will precipitate in ethanol. Every (or almost every) oligonucleotide will bind to its complement in a rule-based fashion. Every (or almost every) oligonucleotide will be a template for a polymerase. Every (or almost every) oligonucleotide will migrate as expected on an electrophoresis gel. This regularity is normal for oligonucleotides but is exceptional for virtually every other class of molecule.

Even standard nucleic acids are not entirely capable of searching mutation space independent of concern that they will lose properties essential to replication. Best known to the molecular biologist are RNA molecules that have G-rich sequences. These adopt tertiary structures around a G-quartet, and these structures often disrupt the templating ability of an RNA sequence (Wang and Patel 1994). This means that if an RNA molecule searches G-rich regions of sequence space, it runs the risk that it will lose certain properties essential for replication. In comparison with virtually every other class of organic molecule, however, nucleic acids are the most COSMIC-LOPER polymer known.

The notion of COSMIC-LOPER behavior is virtually absent from theoretical models of self-replicating systems. These nearly always simply assume that some molecular system (generally with undefined structure) exists that can be a template, a catalyst, and undergo COSMIC mutation to improve fitness, all at the same time. In the real world with real molecules displaying real reactivity, it is not so simple. Indeed, many theoretical discussions of self-replication can be criticized because they are disconnected from the chemical realities of real organic molecules.

In "What if?" and "Why not?" studies, some of the structural features of nucleic acids that give them their COSMIC-LOPER properties have been defined. One of the most important of these studies involved the generation of oligonucleotide analogs with nonionic backbones. One of the best studied of these took a small step from the natural backbone, replacing the phosphate diester linkers in DNA and RNA by nonionic dimethylenesulfone linking units. The sulfone group is an "isosteric" and "isoelectronic" replacement for a phosphate. The sulfone functional group is dipolar but nonionic, intrinsically soluble in water, stable to alkaline degradation, and not stereogenic (Richert et al. 1996).

Nevertheless, oligomers built from these units display some remarkable properties. First, they fold: for example, the octamer $A_{SO_2}U_{SO_2}G_{SO_2}G_{SO_2}U_{SO_2}C_{SO_2}A_{SO_2}U$ folds in solution to give a folded form in water having a high melting temperature (ca. 87°C). More remarkably, a synthetic intermediate leading to this oligosulfone was found to be a "catalyst" for a self-debenzoylation reaction (Richert et al. 1996). Still more remarkably, such behavior is not a general property of oligosulfones. For example, the sequence $d(T_{SO_2}T_{SO_2}T_{SO_2}C_{SO_2}T_{SO_2}T_{SO_2}T_{SO_2}T)$ displays in water no particularly stable folded conformation, judging by a multidimensional NMR spectrum (Greenbaum, Eschgfäller, Benner, unpublished). The sequence $G_{SO_2}C$ in the crystal gives a antiparallel duplex approximately isomorphous with the analogous RNA. The $A_{SO_2}T$ gives an open structure with no Watson–Crick pairing in the crystal. The $U_{SO_2}C$ dinucleotide gives a complex featuring backbone-to-backbone and backbone-to-nucleobase hydrogen bonds (Steinbeck and Richert, unpublished).

Even within a relatively small search of mutation space, these nonionic oligonucleotide analogs retained no conformational or physical property that would offer a common basis for replication. Indeed, the concern that the next variant with a new sequence would display some unusual physical or catalytic behavior remains the central concern of the graduate students working with these molecules. In this respect, oligosulfone analogs of DNA and RNA have much the same behavior as peptides and small organic molecules, not the nucleic acids upon which they are modeled.

These results show that the need for COSMIC-LOPER behavior is a strong constraint on what types of biopolymers need to be considered as the basis for single-biopolymer life. They also suggest that a polyelectrolyte (polyanion or polycation) structure is important for the regular, largely sequence-independent behaviors that we see in nucleic acids. In particular, the anionic phosphate groups confer several properties important for COSMIC-LOPER behavior independent:

(a) Phosphate groups force the interaction surface between strands as far distant from the backbone as possible. This is the Watson–Crick "edge" of the nucleobases. As a consequence, sugar–sugar interstrand interactions, sugar–backbone interstrand interactions, interactions between the sugar and backbone groups of one strand and the Hoogsteen edge of the nucleobases on the other, Hoogsteen–Hoogsteen interstrand interactions, Watson–Crick–Hoogsteen interstrand interactions, all become important without interstrand phosphate–phosphate repulsion. Non-ionic oligonucleotide analogs should (and do) have rich intermolecular conformational properties (just like peptides).

(b) Phosphates keep the DNA molecule from folding on itself, allowing it to act as a template. The statistical mechanical theory of polymers suggests that the polyanionic backbone will cause natural oligonucleotides to adopt an extended structure (Flory 1953; Brant and Flory 1965). Nonionic oligonucleotide analogs should (and do) have rich *intra*molecular conformational properties. They should (and do) fold just like peptides. The repeating charge in the backbone is responsible for the rule-based pairing of complementary DNA strands.

(c) Electronic distribution in a molecule is described as an infinite series (monopole + dipole + quadrapole + ...). The first nonvanishing term dominates the behavior of the molecule, when viewed at a distance. The repeating monopole (charge) in DNA renders dipolar interactions (hydrogen bonding) secondary. The physical behavior of a DNA molecule is the same regardless of its sequence, to a first approximation. An encoding molecule needs physical properties that are largely independent of its sequence. One does not want to mutate a gene to get a better protein, only to discover that the mutant DNA gene precipitates.

From these experiments has come a working hypothesis that may guide our search for universal chemical structures in single-biopolymer systems, and in encoding biopolymers in all forms of life:

Hypothesis 1: As a universal chemical characteristic, living systems must contain at least one biopolymer having a repeating charge, either polyanionic or polycationic,[7] as these are structures that are mutable without creating dysfunctional physical properties (such as precipitation). This biopolymer will perform both catalytic and repository roles (in a one-biopolymer system) and the information repository role (in a two-biopolymer system).

This hypothesis constrains our studies from the universe of all molecules that are known to undergo template-directed replication (including salts, small organic molecules, peptides, and nucleic acids) to a subset of these, those with repeating charges. This directs us to nucleic acids at a first level, to nucleic acid analogs at the next, to cationic analogs at the next (Dempcy et al. 1994), and then to other polycationic and polyanionic molecules. While an extraordinarily large number of structures remain, the scope of exploration is limited.

This hypothesis is particularly useful because if true, it suggests a way to detect chemical remnants of life in extraterrestrial samples: One searches for biopolymers or their fragments with regularly spaced positive or negative charges. This turns out to be a rather simple structural feature to search for experimentally. A chip having an

7 In water.

absorbent with a regularly spaced negative or positive charge will bind the polycation or polyanion (respectively) more tightly than competing monocations or anions.

A molecular system also exists that can challenge the hypothesis. Peptide-linked Nucleic-acid Analogs (PNAs) are a nonionic oligonucleotide analog that is known to support rule-based molecular recognition of oligonucleotides following Watson–Crick rules. Structural studies with PNAs suggest a novel type of interstrand interaction that makes these molecules behave so differently from other nonionic oligonucleotide analogs. Miller (1997) recently argued on prebiotic grounds that PNAs might be a predecessor to RNA as the first living biopolymer. PNA is therefore a likely candidate for a biopolymer that violates Hypothesis 1.

Those who work experimentally with PNAs recognize signs of anti-COSMIC behavior. Their propensity to self-aggregate is recorded in the literature (Egholm et al. 1992), as is self-structure that appears to interfere with PNA–DNA duplex formation (Dueholm et al. 1994). G-rich sequences of PNAs are again problematic, although this phenomenon is not well explored. Many subtle modifications on the backbone destroy Watson–Crick like pairing. Not surprisingly, modification of the ends on the PNA polymer that might improve the COSMIC properties of PNAs involve the addition of charges (see, for example, Gangamani et al. 1997).

Constraint 4: Demands placed on a biopolymer under direct selection pressure for fitness

Perhaps the most useful constraint in the search for universal and unique structural features of Darwinian biopolymers arises from their need to contribute fitness. In a single-biopolymer system, fitness is more than replicatability and mutability. Fitness also requires a chemical feature that confers self-sustaining properties on the system, in particular, to enable the internal synthesis of components from external energy and nutrients. Most simply, fitness can arise if the biopolymer can have catalytic power, and this will be the primary focus of the work proposed here. A biopolymer that can catalyze reactions essential to its own self-reproduction at a higher rate or with greater efficiency is presumably fitter than a competing biopolymer.

Following the work of Cech and Altman, which demonstrated the catalytic power of RNA in contemporary terrestrial organisms (work foreshadowed by reports by Usher and others of RNA catalysis in a variety of model systems [Usher and McHale 1976]), RNA became the system of choice for seeking a single-biopolymer system capable of Darwinian evolution. Coupling affinity separations with PCR amplification led to the development of *in vitro* selection in the pioneering work of Joyce, Szostak, Cech, Gold, and Ellington. This technology will come, we believe, to be recognized as one of distinctive scientific advances of the late twentieth century.

At one level, these experiments have been fantastically successful. Aptamers, catalysts, ligands, and receptors have all been prepared. At another, these experiments have raised serious doubts that catalytic RNA formed the first living system. Some carefully quantitative *in vitro* selection experiments have shown that RNA displays notably poor fitness as a catalytic biopolymer relative (for example) to proteins (Bartel and Szostak 1993).[8]

One measure of the intrinsic fitness of a biopolymer for a particular role is given by the fraction of the sequence space that performs this role to a defined level of specification. For example, to have a population of antibodies with a 50% chance of containing one that has an affinity for a specified ligand with a disassociation constant of approximately 1 mM, approximately 1 million antibody molecules must be prepared. This gives one a measure of the intrinsic ability of the antibody scaffolding to be a receptor. If the population size needed to be larger to have the same result, the scaffolding would be poor.

In vitro selection experiments are capable of measuring fitness in this way with some accuracy. For example, elegant studies by Bartel and Szostak (1993) have shown that to find a single RNA molecule that increases the rate of a templated ligation (a simple reaction) by a modest four orders of magnitude, one must examine ca. 2×10^{13} random RNA sequences. This number must be normalized over the size of the sequence space possible (only a small fraction of the total sequence space is searched in the Bartel–Szostak experiment, of course).

This measure of intrinsic fitness can be contrasted with the intrinsic capacity of a peptide for catalysis. For example, short peptides designed to catalyze the decarboxylation of oxaloacetate produced a rate enhancement of between three and four orders of magnitude for the rate-determining step. In this case, the sequence space contained approximately 10^{18} molecules. A combination of design (Johnsson et al. 1990, 1993), combinatorial experiments (Perezpaya et al. 1996), and refinement (Baltzer, unpublished) were performed, with the estimate that only 10^7 random sequences must be searched to obtain this degree of rate enhancement (Johnsson et al. 1990, 1993). This suggests that peptides as a biopolymer is intrinsically 1 to 10 *million*-fold fitter as a catalyst than RNA.

This comparison is "unfair" in a variety of ways. The reactions being compared are different, each better suited to its own catalyst. The nucleic acid is, in fact, not alone the catalyst; it requires divalent metal ion as a cofactor. Nevertheless, the point is clear: nucleic acids are orders of magnitude worse catalysts than proteins.

---

[8] This conclusion has been clouded by literature that paradoxically refers to RNA as a "perfect" catalyst (Herschlag and Cech 1990). The argument was based on the fact that once bound, substrate nearly always turns over to product. This was proposed as a criterion for perfection in a catalyst based on a paper of Albery and Knowles. In fact, the paper of Albery and Knowles (which itself is flawed; see Ellington and Benner 1987) does not assign an enzyme the status of "perfect" if every substrate molecule, once bound, turns over to product. If substrate binding is very tight, then unbinding is very slow. Through Watson–Crick interactions, the binding of RNA molecules to other RNA molecules can be fantastically tight. Thus, very slow RNA catalysts are easily viewed as "perfect," catalysis is very, very slow. However, the release of bound substrate is still slower. So every molecule bound eventually goes to product.

The superiority of proteins as catalysts compared with RNA reflects (at the very least) the fact that proteins have more building blocks and more functional groups available to serve catalytic roles than RNA. Further, it is clear that the increased catalytic fitness of proteins as a biopolymer more than compensates for the increased size of protein sequence space (for a given molecular weight).

This has consequences for how we view the origin of terrestrial Darwinian chemistry (Benner 1987; Benner et al. 1987). Prebiotic conditions could conceivably generate a library of 10^7 random peptide sequences. It is, however, more difficult to imagine a collection of 10^{13} random RNA sequences emerging under abiotic conditions.

More important to the search for extraterrestrial life is what these and analogous experiments imply about the robustness of single-biopolymer systems when challenged to improve their fitness under Darwinian selection pressure. Joyce and his coworkers have done a fascinating series of experiments to explore the mutation space of the *Tetrahymena* ribozyme challenged in various ways (Beaudry and Joyce 1992; Tsang and Joyce 1994; Lehman and Joyce 1993). Remarkable about these studies is the breadth of the sequence space explored (5% mutagenesis of the catalytic core) and the improvement in "fitness" (substrate binding, for example, by 25-fold; catalytic rate by 50-fold; metal ion dependency by 170-fold), when compared to analogous experiments in directed protein evolution *in vitro* (Johnsson et al. 1993; Perezpaya et al. 1996), by phage display, or *in vivo* (a particularly relevant case being the adaptation of HIV reverse transcriptase when challenged with azidothymidine). By any measure, the proteins are better — much better.

A decade ago, Benner et al. (1987) discussed the intrinsic limitations of standard nucleic acids as a matrix for obtaining functional behavior under conditions of Darwinian selection. These observations generated the second hypothesis:

Hypothesis 2: If RNA is to serve as a scaffolding for a single-biopolymer chemical system capable of undergoing robust Darwinian evolution, it must have more functionality than it presently has.

This functionality can be delivered by either extra letters in the genetic alphabet or by cofactors designed to bind to oligonucleotides. Interestingly, Miller has recently shown that 5-position functionalized pyrimidines are accessible under "prebiotic" conditions (Miller 1997). Further, functionalized ribonucleosides found in tRNA are very ancient and carry much of the functionality needed for catalysis (reviewed in Benner, Ellington et al. 1989).

The need for the polymer in a single-biopolymer Darwinian system to be both capable of suffering mutation independent of concern over the loss of properties essential for replication (COSMIC-LOPER) and to confer fitness through catalytic activity generates competing (and occasionally contradictory) demands on the structure of the biopolymer. Specifically:

1. A biopolymer specialized to be a catalyst must have many building blocks, so that it can display a rich versatility of chemical reactivity; a biopolymer

specialized to store information must have few building blocks, as a way of ensuring faithful replication. The inverse relation between fidelity and the number of building blocks is seen in both theory (Szathmary 1992) and experiment (Lutz et al. 1996).

2. A biopolymer specialized to be a catalyst must fold easily so that it can form an active site; a biopolymer specialized to store information does not fold easily, so that it can serve as a template. This is one of the clearest contradictions between the demands for a catalyst and the demands for an information repository.
3. A biopolymer specialized for catalysis must be able to change its physical properties rapidly with few changes in its sequence, enabling it to explore "function space" during divergent evolution; a biopolymer specialized to encode information must have physical properties largely unchanged even after substantial change in its sequence, so that the polymer remains acceptable to the enzymes required for replication (the COSMIC-LOPER property).

The contradiction between the structures/properties required for catalysis and those required for information storage creates problems. At the very least, any biopolymer forming a single-biopolymer Darwinian system must be a compromise between these goals. Under this model, the advantage of a two-biopolymer system is that it allows the two biopolymers to specialize for genetic and catalytic roles, respectively. Thus, if an RNA world existed and if it had more than the four standard building blocks, one would have expected the number of building blocks to have been reduced after the breakthrough to translation to allow specialization of nucleic acids for encoding function. Further, once invented, proteins would have been expected to acquire functionality rapidly to match that not already available in RNA cofactors. The "palimpsest" of modern metabolism can be read exactly as such, if one is inclined to do so (Benner, Glasfeld, et al. 1989).

Yet no law requires that any real compound be able to make this compromise in a satisfactory way. The demands for functional diversity, folding, and rapid search of function space *might be* so stringent, and the demands for few building blocks, templating ability, and COSMIC-LOPER ability so stringent, that *no* biopolymer structure could achieve a suitable compromise. This would make the single-biopolymer model for the origin of life unavailable as a solution to the "chicken-or-egg" paradox in the origin of two-biopolymer systems. Life would be scarce in the universe, and if a single-biopolymer system did arise, it would be poorly adaptable, not robust, and easily extinguished.

Conversely, it is conceivable that many biopolymeric systems could make an acceptable compromise between the demands of catalysis and the demands of genetics. If so, life would be abundant in the universe, and in many forms. There would be a less pressing need to invent two-biopolymer systems once a one-biopolymer system was established. Small cells would flourish with many different types of biopolymers, with only a few making the breakthrough to two-biopolymer systems.

Organic chemistry seems, in reality, to offer neither extreme. Hypothesis 1 holds that the information repository molecule must be a polyelectrolyte (polyanionic or polycationic). It is clear that some (albeit poor, and not very diverse catalysis) can emerge from RNA and DNA as a representative polyelectrolyte. It is also clear that the number of building blocks in DNA and RNA can be increased without destroying COSMIC-LOPER behavior (Switzer et al. 1989; Piccirilli et al. 1990). Thus, the problem becomes experimentally interesting. The more we can do in the laboratory to increase the structural diversity of polyanionic biopolymers (such as RNA and DNA) without destroying their COSMIC-LOPER properties, the more likely life is to emerge under the single-biopolymer model, the more likely it is to have been robust after it emerged, and the more likely we are to encounter it in planetary exploration.

It could be that many biopolymer systems retain COSMIC-LOPER behavior with a wide range of building blocks. If so, we may return to prebiotic analysis. With "What if?" and "Why not?" experiments showing the capability of a wide range of structures to be self-templating, and prebiotic chemistry experiments showing a wide range of structures accessible abiologically, the next phase of exobiological research must learn whether or not these chemical contradictions are resolvable by any real biopolymer (the "existence" question) and whether the fitness of this biopolymer is in any way sufficient to provide "self-sustaining" properties required by life.

With these thoughts in mind, we discuss briefly possible chemical embodiments of a possible biopolymer that combines the replicatability of nucleic acids with the functional group diversity found in proteins, without obviously destroying the COSMIC-LOPER properties necessary for the biopolymer to be a substrate for Darwinian evolution. One embodiment is based on the recognition that the Watson–Crick base pair could easily accommodate 12 bases forming 6 distinct base pairs, rather than the two found in natural RNA, by exploiting more of the hydrogen bonding patterns conceivable with three hydrogen bond acceptor/donor groups (Figure 19.1) (Switzer et al. 1989; Piccirilli et al. 1990). Additional letters in the genetic alphabet could carry a richer diversity of functionality; indeed (Figure 19. 2), one could readily conceive of a new type of biopolymer, one carrying functionalization like proteins but that can be copied like nucleic acids (Kodra and Benner 1997).

Some details of modern biochemistry support the view that life on earth might have at one point early in its history exploited an expanded alphabet of more than four nucleobases. Furthermore, the RNA world might have expanded its functional diversity by incorporating modified standard nucleobases. Contemporary tRNA and rRNA are known to contain much of the functionality lacking in contemporary encoded RNA, including amino, carboxylate, and aliphatic hydrophobic groups (Figure 19.3) (Limbach et al. 1994). Some of these might even be placed by parsimony in the protogenome, the reconstructable genome at the three fold trifurcation point joining the archaebacterial, eubacterial, and eukaryotic kingdoms (Benner, Ellington et al. 1989). Could these be vestiges of a functionalized RNA world?

To obtain information relating to this question, experiments must be done. First, we must learn whether the structures drawn on paper (Figures 19.1 and 19.2) indeed can

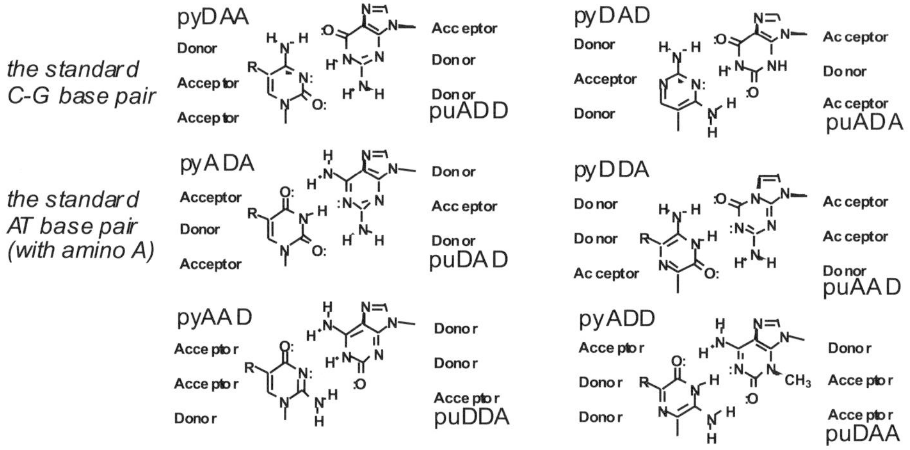

Figure 19.1 Twelve bases that are possible in a DNA- or RNA-based "alphabet" within the constraints of the Watson–Crick base pair geometry. Pyrimidine base analogs are designated by "py," purine by "pu." The upper case letters following the designation indicate the pattern of hydrogen-bonding acceptor (A) and donor (D) groups. Thus, cytosine is pyDAA, guanine is puADD, adenine is puDA- (diaminopurine, puDAD, completes the Watson–Crick base pair, and thymine is pyADA. The remainder of the base pairs are joined by *nonstandard* hydrogen-bonding schemes.

serve as effective components of a genetic information system. Then, we must ask whether *in vitro* selection based on an expanded genetic alphabet indeed improves the catalytic versatility of RNA. Only if both answers are affirmative would the RNA world have the motive to use an expanded genetic alphabet.

Then, we must ask whether the RNA world had the opportunity to use an expanded genetic alphabet. If the components of an expanded genetic alphabet are accessible

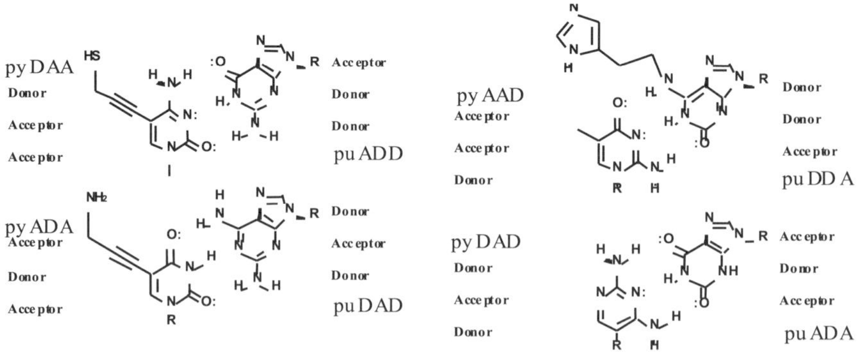

Figure 19.2 Nonstandard and standard nucleobases with functionality. Note that the pyDAD nucleobase can be protonated below pH 7 ($pK_a = 7.4$)

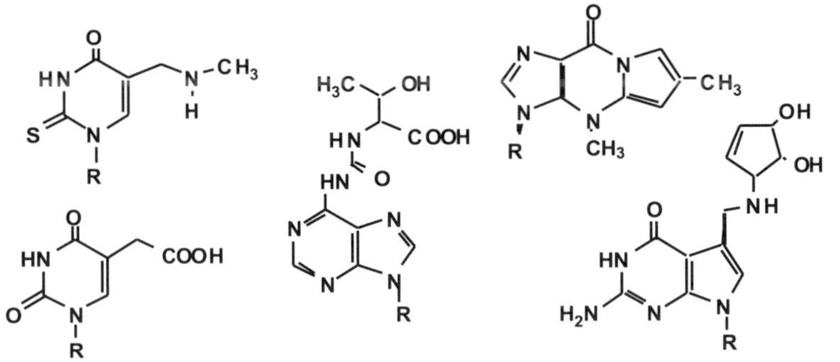

Figure 19.3 Transfer RNA contains a rich collection of functionalized standard nucleobases, created by post-transcriptional modification, that deliver functional groups (amino groups, carboxylic acid groups, aliphatic hydrophobic groups) not found within unmodified RNA. Could these be vestiges of functionalized RNA originating in the RNA world?

under abiological (prebiotic) conditions reconstructed in the laboratory, then it becomes more plausible that the RNA world exploited them.

Organic synthesis generated the nonstandard nucleobases by synthesis (Switzer et al. 1989; Piccirilli et al. 1990; Vögel, Altorfer, et al. 1993; Vögel and Benner 1994; Kodra and Benner 1997; Tarasow et al. 1997). Their structures were optimized for stability and pairing properties (Piccirilli et al. 1990; Vögel, von Krosigk, et al. 1993). New protecting group chemistry was developed to permit automated synthesis of oligonucleotides containing them (Huang and Benner 1993; von Krosigk and Benner 1995). Polymerases were found that would catalyze their incorporation into oligonucleotides by the polymerase chain reaction (Horlacher et al. 1995; Lutz et al. 1996).

Once these nonstandard nucleotides were in hand, it was possible to show that the RNA world had both the motive and the opportunity to exploit nonstandard and functionalized nucleobases. Three key advances have supported this conclusion:

(a) The first case where functionalized standard RNA has been exploited to generate a catalyst was reported by Tarasow et al. (1997). These researchers used Selex experiment with oligonucleotide libraries that incorporated a functionalized standard pyADA nucleobase (Figure 19.4). The result was a ribozyme that catalyzes a Diels-Alder reaction. As noted above, earlier Selex work with libraries that did not contain functionalized nucleotides had failed to generate a catalyst (Morris et al. 1994). For this reason, it is difficult to state quantitatively how much better this particular functionalized RNA is than standard RNA as a matrix for generating catalysis for this particular reaction. However, it appears as if the improvement is by several orders of magnitude.

(b) The first experiments where functionalized standard DNA was used to generate receptors were recently completed (Battersby et al. 1999). A functionalized 2′-deoxyuridine (trivially designated "J", Figure 19.4) was incorporated into an *in vitro* selection experiment that followed a recipe developed by Huizenga

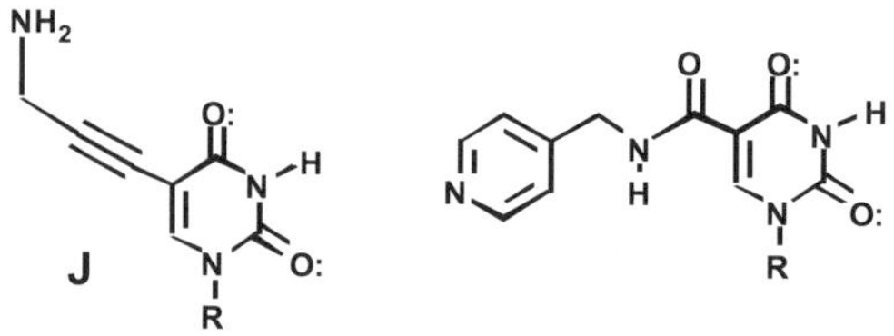

Figure 19.4 Functionalized standard bases that have been used in *in vitro* selections.

and Szostak (1995), who selected for adenosine receptors using a standard DNA library.

The results were intriguing. Most of the J-containing sequences were not analogous to those obtained from a standard DNA library, and presumably adopted a fold different from the fold adopted by the standard DNA receptors. A few of the sequences emerging from the J-containing library appeared to be analogous to the sequences obtained from a standard DNA library, however, and may fold analogously as well. Further, when one of the J-containing sequences was resynthesized with T replacing J, it still bound adenosine derivatives, only more weakly. Thus, the functionalization contributed by J influences modestly, rather than dramatically, the affinity-structure landscape, at least for receptors that bind adenosine derivatives selected under non-demanding conditions. More recent experiments (Ang, Westerman-Clark, and Benner, unpublished) show a larger impact of the functionality under more demanding conditions, especially in catalysis.

Careful analysis by capillary electrophoresis of the affinity of the J-receptors for adenosine showed that both the natural and the functionalized motifs bound not one, but two ATP molecules cooperatively. Likewise, the affinity of the receptors for ATP had converged; in both cases, the receptors are 50% at the 3 mM concentrations of ATP presented during the selection. The convergence of phenotype suggests that the outcome of this selection experiment was determined by features of the environment during the selection, in particular, a highly loaded affinity resin used in the selection step, and that an optimal molecular phenotype has been achieved by both selections for the selection conditions. This interplay between environment demanding a function from a biopolymer and the ability of the biopolymer to deliver that function is strictly analogous to that observed during natural selection.

The third advance comes from the field of "prebiotic chemistry," which seeks to discover ways by which the components of living systems might have emerged in the early Earth. Robertson and Miller (1995) found that the intrinsic nucleophilicity of the 5-position of pyrimidines such as uracil can be exploited to generate functionalized uracil derivatives that carry positive charges at the 5-position under abiological conditions (Robertson and Miller 1995). Analogous chemistry can be used to generate other functionalized derivatives. The products resemble the amino group functionalized uracils found in some tRNA molecules (Figure 19.3). This suggests that the RNA world may have had the opportunity to, indeed, may have needed to, use some functionalized nucleosides when life first emerged on Earth.

A "SECOND GENERATION" MODEL FOR NUCLEIC ACIDS

It is intriguing to note that the "first generation" theory of nucleic acid structure proposed nearly fifty years ago by Watson and Crick permitted the design of alternative nucleobases, but failed to support the design of altered backbones. The Watson–Crick model is only barely reductionist, at least compared to quantum mechanics, molecular mechanics, or other models that are used to interpret the result of chemical modification. Nevertheless, it supports vast amounts of research effort in molecular biology, as well as the modest efforts to redesign genetic biopolymers discussed here.

There is clearly the opportunity for the development of a "second generation" model for nucleic acids, one that is supported by the synthesis of alternative nucleic acid structures. The model would certainly incorporate a more substantial role for the backbone, and especially the negative charge in the backbone. It would also incorporate the growing understanding of the flexibility of base pairing in strand–strand recognition process.

Above all, the second generation model for nucleic acid structure must incorporate our growing understanding of the ability of nucleic acids to have selectable function (phenotype) as well as performing genetic tasks. In particular, we must reconsider more seriously whether a single polymer can robustly carry information and catalyze reactions, in light of the contradicting chemical demands for performing each role. Chemical experiments attempting to resolve these contradictions in single-biopolymer systems will, we believe, provide the most important results in the coming decade to understand life's origins on Earth and how best to find life elsewhere.

ACKNOWLEDGMENTS

Work described in this paper was supported in part by the Office of Naval Research (N00025-96-1-0362), the National Aeronautics and Space Administration, and the National Institutes of Health (GM 54048).

REFERENCES

Alberti, S. 1997. The origin of the genetic code and protein synthesis. *J. Mol. Evol.* **45**:352–382.

Bain, J.D., A.R. Chamberlin, C.Y. Switzer, and S.A. Benner. 1992. Ribosome-mediated incorporation of non-standard amino acids into a peptide through expansion of the genetic code. *Nature* **356**:537–539.

Bartel, D.P., and J.W. Szostak. 1993. Isolation of new ribozymes from a large pool of random sequences. *Science* **261**:1411–1418.

Beaudry, A.A., and G.F. Joyce. 1992. Directed evolution of an RNA enzyme. *Science* **257**:635–641.

Benner, S.A. 1987. Redesigning life. Organic chemistry and the evolution of proteins. *Chimia* **41**:142–148.

Benner, S.A. 1994. Expanding the genetic lexicon. Incorporating non-standard amino acids into proteins by ribosome-based synthesis. *Trends Biotech.* **12**:158–163.

Benner, S.A., R.K. Allemann, A.D. Ellington, L. Ge, A. Glasfeld, G.F. Leanz, T. Krauch, L.F. MacPherson, S.E. Moroney, J.A. Piccirilli, and E.G. Weinhold. 1987. Natural selection, protein engineering, and the last riboorganism: Rational model building in biochemistry. *Cold Spring Harbor Symp. Quant. Biol.* **52**:53–63.

Benner, S.A., M.A. Cohen, G.H. Gonnet, D.B. Berkowitz, and K. Johnsson. 1993. Reading the palimpsest: Contemporary biochemical data and the RNA world. In: The RNA World, ed. R. Gesteland and J. Atkins, pp. 27–70. Cold Spring Harbor Laboratory Press, N.Y.

Benner, S.A., A.D. Ellington, and A. Tauer. 1989. Modern metabolism as a palimpsest of the RNA world. *Proc. Natl. Acad. Sci. USA* **86**:7054–7058.

Benner, S.A., A. Glasfeld, and J.A. Piccirilli. 1989. Stereospecificity in enzymology. Its place in evolution. *Top. Stereochem.* **19**:127–207.

Brant, D.A., and P.J. Flory. 1965. The configuration of random polypeptide chains. I. Experimental results. *J. Am. Chem. Soc.* **87**:2788–2791.

Cech, T.R., A.J. Zaug, and P.J. Grabowski. 1981. *In vitro* splicing of the ribosomal RNA precursor of Tetrahymena. Involvement of a guanosine nucleotide in the excision of the intervening sequence. *Cell* **27**:487–496.

Crick, F.H.C. 1968. The origin of the genetic code. *J. Mol. Biol.* **38**:367–379.

Dempcy, R.O., O. Almarsson, and T.C. Bruice. 1994. Design and synthesis of deoxynucleic guanidine. A polycationic analog of DNA. *Proc. Natl. Acad. Sci. USA* **91**:7864–7868.

DiGiulio M. 1997. On the origin of the genetic code. *J. Theor. Biol.* **187**:573–581.

Dueholm, K.L., K.H. Peterson, D.K. Jensen, P.E. Nielsen, M. Egholm, and O. Buchardt. 1994. Peptide nucleic acid (PNA) with a chiral backbone based on alanine. *Biomed. Chem. Lett.* **4**:1077–1081.

Egholm, M., O. Buchardt, P.E. Nielsen, and R.H. Berg. 1992. Peptide nucleic acids. Oligonucleotide analogs with an achiral peptide backbone. *J. Am. Chem. Soc.* **114**:1895–1897.

Ellington, A.D., and S.A. Benner. 1987. Free energy differences between enzyme bound states. *J. Theor. Biol.* **127**:491– 506.

Flory, P. 1953. Principles of Polymer Chemistry. Ithaca, N.Y.: Cornell Univ. Press.

Freier, S.M., and K.H. Altmann. 1997. The ups and downs of nucleic acid duplex stability, structure-stability studies on chemically modified DNA, RNA duplexes. *Nucl. Acids Res.* **25**:4429–4443.

Gangamani, B.P., V.A. Kumar, and K.N. Ganesh. 1997. Spermine conjugated peptide nucleic acids (spPNA), UV and fluorescence studies of PNA–DNA hybrids with improved stability. *Nucl. Acids Res.* **240**:778–782.

Herschlag, D., and T.R. Cech. 1990. Catalysis of RNA cleavage by the Tetrahymena thermophila enzyme. *Biochemistry* **29**:10159–10171.

Horlacher, J., M. Hottiger, V.N. Podust, U. Hübscher, and S.A. Benner. 1995. Expanding the genetic alphabet. Recognition by viral and cellular DNA polymerases of nucleosides bearing bases with non-standard hydrogen bonding patterns. *Proc. Natl. Acad. Sci. USA* **92**:6329–6333.

Huang, Z., and S.A. Benner. 1993. Selective protection and deprotection procedures for thiol and hydroxyl groups. *Syn. Lett.*:83–84.

Huizenga, D.E., and J.W. Szostak. 1995. A DNA aptamer that binds adenosine and ATP. *Biochemistry* **34**:656–665.

Johnsson, K., R.K. Allemann, and S.A. Benner. 1990. Designed enzymes. New peptides that fold in aqueous solution and catalyze reactions. In: Molecular Mechanisms in Bioorganic Processes, ed. C. Bleasdale and B.T. Golding, pp. 166–187. Cambridge: Royal Society of Chemistry,.

Johnsson, K., R.K. Allemann, H. Widmer, and S.A. Benner. 1993. Synthesis, structure and activity of artificial, rationally designed catalytic polypeptides. *Nature* **365**:530–532.

Joyce, G.F. 1994. Forward. In: Origins of Life, The Central Concepts, ed. D.W. Deamer and G.R. Fleischaker. Boston: Jones and Bartlett.

Kalamaraki, M., G.M. McGeehan, and S.A. Benner. 1991. Site-directed mutagenesis of bovine pancreatic ribonuclease A. Serine 16, threonine 17, alanine 19 and alanine 20. In: Structure, Mechanism, and Function of Ribonucleases, Proc. 2nd Intl. Conf. on Ribonuclease, ed. C.M. Cuchillo, R. de Llorens, M.V. Nogués, and X. Parés, pp. 145–150. Barcelona: IBF Publications.

Kamaike, K., M. Takahashi, K. Utsugi, K. Tomizuka, and Y. Ishido. 1995. An efficient method for the synthesis of [4–^{15}N]cytidine and [6–^{15}N]adenosine derivatives from uridine and inosine. *Tetrahedron Lett.* **36**:91–94.

Kerr, R.A. 1997. Ancient life on Mars? Putative Martian microbes called microscopy artifacts. *Science* **278**:1706–1707.

Khare, B.N., W.R. Thompson, L. Cheng, C. Chyba, C. Sagan, E.T. Arakawa, C. Meisse, and P.S. Tuminello. 1993. Production and optical-constants of ice tholin from charged-particle irradiation of (1/6) C_2H_6/H_2O at 77 K. *Icarus* **103**:290–300.

Kodra, J., and S.A. Benner. 1997. Synthesis of an N-alkyl derivative of 2'-deoxyisoguanosine. *Syn. Lett.* :939–940.

Lee, D.H., K. Severin, Y. Yokobayashi, and M.R. Ghadiri. 1997. Emergence of symbiosis in peptide self-replication through a hypercyclic network. *Nature* **390**:591–594.

Lehman, N., and G.F. Joyce. 1993. Evolution *in vitro* of an RNA enzyme with altered metal dependence. *Nature* **361**:182–185.

Li, A.G., and J.M. Greenberg. 1997. A unified model of interstellar dust. *Astron. Astr.* **323**:566–584.

Limbach, P.A., P.F. Crain, and J.A. McCloskey. 1994. The modified nucleosides of RNA. *Nucl. Acids Res.* **22**:2183–2196.

Luisi, P.L. 1998. About various definitions of life. In: Origins Life, in press.

Lutz, M.J., S.A. Benner, S. Hein, G. Breipohl, and E. Uhlmann. 1997. Recognition of uncharged polyamide-linked nucleic acid analogs by DNA polymerases and reverse transcriptases. *J. Am. Chem. Soc.* **119**:3177–3178.

Lutz, M.J., H.A. Held, M. Hottiger, U. Hübscher, and S.A. Benner. 1996. Differential discrimination of DNA polymerases for variants of the non-standard nucleobase pair between xanthosine and 2,4-diaminopyrimidine, two components of an expanded genetic alphabet. *Nucl. Acids Res.* **24**:1308–1313.

Macdermott, A.J., L.D. Barron, A. Brack, T. Buhse, A.F. Drake, R. Emery, G. Gottarelli, J.M. Greenberg, R. Haberle, et al. 1996. Homochirality as the signature of life. The Seth cigar. *Plan. Space Sci.* **44**:1441–1446.

McBride, J.M., and R.L. Carter. 1991. Spontaneous resolution by stirred crystallization. *Angew. Chem. Intl. Edn.* **30**:293–295.

Menger, F.M., A.V. Eliseev, N.A. Khanjin, and M.J. Sherrod. 1995. Evidence for an alternative mechanism to a previously proposed self-replicating system. *J. Org. Chem.* **60**:2870–2878.

Miller, S.L. 1997. Peptide nucleic acids and prebiotic chemistry. *Nature Struct. Biol.* **4**:167–169.

Morris, K.N., T.M. Tarasow, C.M. Julin, S.L. Simons, D. Hilvert, and L. Gold. 1994. Enrichment for RNA molecules that bind a Diels-Alder transition-state analog. *Proc. Natl. Acad. Sci. USA* **91**:13028–13032.

Orgel, L.E. 1968. Evolution of the genetic apparatus. *J. Mol. Biol.* **38**:381–393.

Orgel, L.E. 1992. Molecular replication. *Nature* **358**:203–209.

Perezpaya, E., R.A. Houghten, and S.E. Blondelle. 1996. Functionalized protein-like structures from conformationally defined synthetic combinatorial libraries. *J. Biol. Chem.* **271**:4120–4126.

Piccirilli, J.A., T. Krauch, S.E. Moroney, and S.A. Benner. 1990. Extending the genetic alphabet, Enzymatic incorporation of a new base pair into DNA and RNA. *Nature* **343**:33–37.

Rich, A. 1962. On the problems of evolution and biochemical information transfer. In: Horizons in Biochemistry, ed. M. Kasha and B. Pullman, pp. 103–126. New York: Academic.

Richert, C., A.L. Roughton, and S.A. Benner. 1996. Nonionic analogs of RNA with dimethylene sulfone bridges. *J. Am. Chem. Soc.* **118**:4518–4531.

Robertson, M.P., and S.L. Miller. 1995. Prebiotic synthesis of 5-substituted uracils. A bridge between the RNA world and the DNA-protein world. *Science* **268**:702–705.

Schmidt, J.G., L. Christensen, P.E. Nielsen, and L.E. Orgel. 1997. Information transfer from DNA to peptide nucleic acids by template-directed syntheses. *Nucl. Acids Res.* **25**:4792–4796.

Switzer, C.Y., S.E. Moroney, and S.A. Benner. 1989. Enzymatic incorporation of a new base pair into DNA and RNA. *J. Am. Chem. Soc.* **111**:8322–8323.

Szathmary, E. 1992. What is the optimum size for the genetic alphabet? *Proc. Natl. Acad. Sci. USA* **89**:2614–2618.

Tarasow, T.M., S.L. Tarasow, and B.E. Eaton. 1997. RNA-catalyzed carbon–carbon bond formation. *Nature* **389**:54–57.

Thompson, L.A., and J.A. Ellman. 1996. Synthesis and applications of small molecule libraries. *Chem. Rev.* **96**:555–600.

Tsang, J., and G.F. Joyce. 1994. Evolutionary optimization of the catalytic properties of a DNA-cleaving enzyme. *Biochemistry* **33**:5966–5973.

Usher, D.A., and A.H. McHale. 1976. Hydrolytic stability of helical RNA, A selective advantage for the natural 3',5'-bond. *Proc. Natl. Acad. Sci. USA* **73**:1149–1153.

Visser C.M., and R.M. Kellogg. 1978. Biotin. Its place in evolution. *J. Mol. Evol.* **11**:171–178.

Vögel, J.J., M.M. Altorfer, and S.A. Benner. 1993. The donor-acceptor-acceptor purine analog. Transformation of 5-aza-7-deaza-*iso*guanine to 2′-deoxy-5-aza-7-deaza-*iso*-guanosine using purine nucleoside phosphorylase. *Helv. Chim. Acta* **76**:2061–2069.

Vögel, J.J., U. von Krosigk, and S.A. Benner. 1993. Synthesis and tautomeric equilibrium of 6-amino-5-benzyl-3-methylpyrazin-2-one. An acceptor-donor-donor nucleoside base analog. *J. Org. Chem.* **58**:7542–7547.

Vögel, J.J., and S.A. Benner. 1994. Non-standard hydrogen bonding in duplex oligonucleotides. The base pair between the acceptor-donor-donor pyrimidine analog and the donor-acceptor-acceptor purine analog. *J. Am. Chem. Soc.* **116**:6929–6930.

von Kiedrowski, G., B. Wlotzka, J. Helbing, M. Matzen, and S. Jordan. 1991. Parabolic growth of a self-replicating hexadeoxynucleotide bearing a 3'.5'-phosphoramidate linkage. *Angew. Chem.* **30**:423–426.

von Krosigk, U., and S.A. Benner. 1995. pH-independent triple helix formation by an oligonucleotide containing a pyrazine donor-donor-acceptor base. *J. Am. Chem. Soc.* **117**:5361–5362.

Wang, Y., and D.J. Patel. 1994. Solution structure of the Tetrahymena telomeric repeat DT2G44 G-tetraplex. *Structure* **2**:1141–1156.

Watson, J.D., N.H. Hopkins, J.W. Roberts, J.A. Steitz, and A.M. Weiner. 1987. In: Molecular Biology of the Gene, p. 1115. Menlo Park: Benjamin Cummings.

White III, H.B. 1976. Coenzymes as fossils of an earlier metabolic state. *J. Mol. Evol.* **7**:101–104.

Wintner, E.A., M.M. Conn, and J. Rebek, Jr. 1994. Autocatalysis and the generation of self-replicating systems. *Accts. Chem. Res.* **27**:198–203.

Wintner, E.A., B. Tsao, and J. Rebek, Jr. 1995. Evidence against an alternative mechanism for a self-replicating system. *J. Org. Chem.* **60**:7997–8001.

Woese, C.R. 1967. The Genetic Code. The Molecular Basis for Genetic Expression. New York: Harper and Row.

Zaug, A.J., and T.R. Cech. 1986. The intervening sequence RNA of Tetrahymena is an enzyme. *Science* **231**:470–475.

Zent, A.P., and C.P. Mckay. 1994. The chemical-reactivity of the Martian soil and implications for future missions. *Icarus* **108**:146–157.

Standing, left to right: Cyrus Chothia, José Onuchic, Thomas Kiefhaber, Maurizio Brunori, Steven Benner
Seated, left to right: Zan Luthey-Schulten, Peter Schuster, Steve Boxer, Brian Matthews
Not pictured: Rudolf Rigler

20

Group Report: What Are the Constraints on the Evolution and Diversity of Biopolymers?

S.G. BOXER, Rapporteur
S.A. BENNER, M. BRUNORI, C. CHOTHIA, T. KIEFHABER,
Z. LUTHEY-SCHULTEN, B.W. MATTHEWS, J.N. ONUCHIC,
R. RIGLER, AND P. SCHUSTER

Keywords: sequence space, exons/introns, selection, adaptation

INTRODUCTION

During a heated discussion at this meeting Martin Karplus said: "The fundamental experimental result in the protein folding field is that proteins fold." Most of the discussions at this conference began with this seemingly obvious result and asked how something that is known to fold actually folds, what methods can be used to characterize this process, and what general and/or detailed models can be used to systematize the results and make predictions. But there is another question: by studying existing proteins, we know for sure that folding occurs; however, is this a rare event in that only a few sequences or a few regions of sequence space have proteins that fold? If so, how were these found? Or, is folding a common feature of proteins, with foldable sequences across sequence space?

If someone handed you a Pentium III processor and you had no prior knowledge of its purpose, careful analysis would lead you to conclude it is mostly sand. Conversely, and more in keeping with the times, if someone handed you a pile of sand and proposed to construct something that would have the capabilities of the Pentium III processor, would you invest? Would it help if you knew about the existence of historical relics like the Pentium I, 486, 386, or 286 chips or contemporary advanced species like the Power PC chip? Certainly if you were asked to invest it would be comforting to know

Simplicity and Complexity in Proteins and Nucleic Acids
Edited by H. Frauenfelder, J. Deisenhofer, and P.G. Wolynes © 1999 Dahlem University Press

that something related had been made before, but would these help you to understand the internal design?

What *is* clear is that it would help to know something about transistors, capacitors, and resistors, and more deeply about solid-state physics, before starting the design process. Yet would knowledge of these provide a route to uncovering the function of the existing device? This was really the premise of this meeting with its strong flavor of physics and physical chemistry applied to the complex structures and functions in biological systems. The laws governing the behavior of electrons in semiconductors are well understood and can be tested precisely on independent systems. It's all in the physics. And these devices were built with a clear purpose. It may not be evident at all just looking at the device, but the purpose is essential to the design.

When we try to make the connection to biology, we must also consider random variations and natural selection, the cornerstones of evolution. One could certainly argue that semiconductor processor design was subject to rules akin to survival of the fittest in the market place, but it is difficult to discern anything like random chance (not to be confused with accidental discovery or good luck). If we start our thinking with a protein that is known to fold, then the goal for understanding is quite clear, even though the process of understanding may be tortured as described in the other sections of this meeting. If we broaden our thinking to all possible protein or nucleic acid sequences, subject to random variations and natural selection, will traditional physical thinking provide the needed insight?

Our group discussed protein and nucleic acid structure and function from this evolutionary perspective. Vast amounts have been written on this subject, some very provocative, some nonsense. Some of our group members are intensely involved in this debate, some not at all. Interesting problems like the origin of life, artificial life, and computer models for life were not discussed extensively (see Benner, this volume). A key issue that came up in discussion is how to understand the emergence of different structural types, e.g., α-helices or β-sheet proteins. Further, how can we move among even modest variations in structure by taking small random steps without passing through a sequence so unstable or unsuitable that it could not possibly confer any selective advantage on the organism? How do the constraints we identify with the selection process actually feed back to the synthesis machinery? Those who come from a protein design viewpoint look to variables related to their experiments: temperature, pH, function, etc. However, in a living system, fitness must be considered collectively for the components that interact physically or in chemical networks, rather than individually, even though the feedback mechanism operates individually.

The following sections summarize compactly the current state of understanding in very selected areas of interest to the group. This begins with a *manifesto* describing our perspective on the problem, some comments on basic chemical constraints (often not considered in detail by either physicists or biologists), followed by a series of approaches with a few results. One of the most widespread approaches is the analysis of sequence data. As more and more sequence data become available, the power of statistical arguments based on this data base becomes increasingly compelling. Within a few years, this data base will be overwhelming. Furthermore, the number of three-dimensional structures, coupled with the analysis of folds, is growing to represent what is believed to be a sizable fraction of possible structure types. Another approach is to change the primary sequence using the powerful tools of molecular biology and protein overexpression, followed by detailed structure and function analysis, to learn the limits and possibilities of small to modest sequence variations within the framework of existing folded proteins. A powerful exploration of sequence space is possible using *in vitro* selection. Simple physical models, coupled to sophisticated simulations, can be used to explore what is possible and might be relevant. Each of these approaches has successes and failures, and directions for future work are described.

MANIFESTO

It is clear from nearly every discussion that different communities of scientists have very different views of what it means to "understand" a phenomenon. These different levels of rigor can coexist, but our group was charged with taking an evolutionary perspective. Much of the discussion at this workshop centered on conformational energy landscapes, dynamic behavior, and other physical properties that can be probed in great detail with contemporary experimental methods and modeled by contemporary theories. Efforts were made to persuade other sections to exploit the value of a Darwinian perspective in guiding and simplifying the experimental and theoretical tasks presented by biological systems. Interestingly, these efforts were resisted. On several occasions, the assertion was voiced that one can (if one wishes) describe biological systems in great detail without *any* reference to biological function, that is, without any concern for which elements of the description were adaptive and which were not. This view raises a more general issue concerning the selection of a system to study, and we were sufficiently surprised by the resistance that we felt it would be helpful at the outset to discuss this further.

What is needed in chemistry first and foremost is a clear and rigorous understanding of fundamental aspects of molecular structure, dynamics, and reactivity. Clearly, more problems arise in systems containing more atoms than in systems containing fewer. One chooses to study the reaction of $H_2 + H$ only when our understanding of the dynamic behavior of molecules undergoing reaction is in its infancy. Conversely, if our understanding of the dynamic behavior of molecules undergoing reaction *is* in its infancy, it is imprudent *not* to study the $H_2 + H$ reaction. Let us stipulate that the $H_2 + H$ reaction is now fully understood. It is clear that because of its simplicity, the $H_2 + H$ reaction is not trivially generalized to examine issues that arise in larger and more complex systems (including proteins). From a purely pragmatic perspective, it would be appropriate to now find the *simplest* system where those issues manifest themselves, and to apply the full power of contemporary theory and experiment to address the issues in those systems, at least until they are understood in their context at least as well as the $H_2 + H$ reaction is in its context.

Other groups as well as ours discussed at length a variety of issues concerning energy surfaces, reaction trajectories, and various "metalanguages" that attempt to capture some important feature of these without needing to describe all of their mathematical intricacy. It was clear that many issues raised under these rubrics were not yet solved for proteins. However, it is equally clear that they are not solved for much smaller molecular systems, systems that would be more amenable both to experimental study and theoretical modeling precisely because they are smaller. These discussions underscore the issue: *Why is all of this theoretical and experimental talent being thrown at protein systems, which by their size are difficult to model and difficult to study experimentally, when many of these issues are just as real in simpler molecules that are easier to model and (frequently) easier to study?*

One answer to this question is that it is easier to raise money and to publish dynamic studies on proteins than on smaller molecules! In fact, a rigorous understanding of solute–solute and solute–solvent interactions in small molecules, if properly done, might contribute more to our understanding of the behavior of proteins than a semi-rigorous study of proteins themselves. The other possibility is that the talent *is* interested in biology. One feature of myoglobin, for example, that is *not* shared by small organic molecules is its history: it is the product of natural selection superimposed upon random variation. This means that certain of its behaviors have been optimized in a way unlikely to be paralleled in laboratory-designed molecules, and these behaviors (but not others) will be revealing about underlying reactivity in a way that is not necessarily to be found in small molecular systems. It is, we believe, reasonable to sacrifice a degree of rigor in both the model and the experiment to explore complex biomolecules if one has the chance of exploiting the Darwinian history of the molecules as a sort of compensation for the loss of completeness and rigor. However, to achieve this compensation, one must actively address, not deny, the question: *What in the behavior of biomolecules that we measure is likely to be adaptive in the Darwinian sense?*

WHAT IS SEQUENCE SPACE?

Central to this discussion is the concept of sequence space. With 20 amino acids available as building blocks (see the final section), the number of possible sequences in a protein of length n is 20^n, i.e., there are 400 possible dipeptides, 8000 possible tripeptides, 160,000 possible tetrapeptides, and so on. Even short (from a biological perspective) proteins 100 amino acids long give rise to an astronomical number of possible protein sequences (20^{100} or about 10^{130}), far larger than the number of atoms in the galaxy. This collection of protein sequences can be represented as a "space" in n dimensions. A relatively simple metric describes the distance between any two points in that space: the number of mutations needed to obtain one sequence from another. The maximum distance in this case is 100 single amino acid mutations (steps). There are, however, a large number of paths between two points. For example, if two proteins differ at 20 positions in their amino acid sequences, there are 20! or about 2×10^{18} direct paths between them.

The effectiveness of the search for fitness that has been made by natural selection depends on the nature of the surface, as well as on geological parameters and mechanisms that allow the space to be sampled. It appears that accidental gene duplication must be essential so that existing function can be retained while new functions emerge. Although single amino acid mutation is simplest to understand (and simulate by experiments in the laboratory), insertions and deletions on a much larger scale also occur.

In Vitro RNA Evolution Explores Sequence Space

Although most of our discussion concerned proteins, a fascinating array of new experiments have been developed that are uniquely able to explore nucleic acid diversity and sequence space. Any simple calculation shows that the numbers of different nucleic acid sequences are so large that only a tiny fraction of all the possible ones could have been explored so far. A few results derived from experiments and computer studies on RNA molecules show that evolutionary processes can nevertheless be successful and suggest why this is so.

1. The design of small RNA molecules, so-called aptamers, which bind preferentially to predefined target molecules, has been very successful with respect to two aspects: the selection of molecules binding with (relatively) large binding constants and high specificity in the sense of discrimination between chemically closely related targets (for example, caffeine and theophylline) (Ellington 1994; Jenison et al. 1994). These experiments use huge pools of random sequences, a mechanism for selection (in this case binding to a substrate), and some mechanism for amplifying the selected sequences (including some error to broaden the pool). This process is repeated until functionally relevant sequences are selected. Structure determination of aptamer target complexes by NMR and crystallography reveal the nature of these complexes. In one particularly well-studied example (Jiang et al. 1997), the aptamer is an RNA with chain length $n = 27$ and has a structure that allows it to achieve high specificity primarily through binding to a bulge and a hairpin loop. An RNA of this length can exist in about 2×10^{16} different sequences. A typical sample pool size for one of these selection experiments is 10^{15} random sequences, and thus we conclude that selection from a random pool of sequences leads to something that is selectable with a high probability (at least in case of the smallest molecules). It is worth mentioning, as an aside, that the smallest molecules that were replicated successfully by virus specific RNA replicases (Zamora et al. 1995) are of roughly the same chain length.
2. Experiments aiming at the design and selection of ribozymes with specific catalytic functions, for example ligase activity, were successful even when they started from pools of RNA molecules with lengths up to 220 with random

sequences (Bartel and Szostak 1993). The chance of finding a given sequence in the initial population of 10^{15} different molecules is, in this case, 10^{-117}. Nevertheless, molecules with weak but distinct activity were isolated and improved by mutation and selection to yield catalysts with catalytic enhancements of 10^6 over background. The success of such experiments is plausible only if certain requirements are fulfilled by the relationships between RNA sequences, RNA structures, and catalytic functions: (a) the (weak) catalytic function is a common property of RNA molecules, (b) many sequences give rise to the same structure and the same function, and (c) these sequences are (more or less) uniformly distributed in sequence space such that a search starting from any arbitrarily chosen reference sequence can be successful when it covers only a (relatively small) fraction of sequence space. Computer searches based on RNA molecules of chain lengths up to $n = 100$ have shown that conditions (b) and (c) are indeed fulfilled with RNA secondary structures (see Schuster, this volume).

3. Computer simulations of evolution experiments with populations of 1000 to 10,000 RNA molecules have shown that it is fairly easy to reach tRNA-like cloverleaf structures in reasonably short times. These computer experiments retrieve pathways leading to target structures at molecular resolution and provide an explanation why optimization occures in stepwise manner and not continuously as is often assumed (Fontana and Schuster 1998). Although all successful trajectories were different in detail, they have characteristic and reproducible features in common which constitute the basis of evolutionary dynamics (Fontana and Schuster 1998; Schuster, this volume). It is worth mentioning that no successful trajectory has led so far to a cloverleaf from a GC-only sequence, although we know from other searches that such molecules do exist. This observation suggests that evolutionary optimization in a four-letter alphabet with two base pairs may be more effective than the analogous searches in a universe of structures formed by a two-letter (GC) alphabet.

How to Explore Sequence Space for Proteins?

There is little solid information for proteins at the level described above for nucleic acids. The mechanisms that make *in vitro* evolution experiments so useful for nucleic acids do not apply to proteins. This is an interesting area for the development of clever new methods. Samples containing random sequences of proteins contain many aggregates (nucleic acids being poly-anions tend not to aggregate), and there is no direct mechanism for amplifying the population (see, e.g., Smith 1985; Scott and Smith 1990; Devlin et al. 1990; Gesareni 1992; Burritt et al. 1996).

Nonetheless, there are strong indications that the situation for proteins is not as hopeless as it might seem. In the early days of site-directed mutagenesis, everyone was concerned that individual amino acid changes would lead to measurable structural changes, but this almost never proves to be the case. To the contrary, starting with a folded protein, one can change nearly any amino acid on the surface to any other without loss of structure. The interior, being more compact, is more susceptible to disruption, but many changes can be tolerated. A particularly interesting effect is described in the background paper by Baase et al. (this volume). Several laboratories have studied the structural and functional consequences of insertions of amino acids into α-helices. Remarkably, such insertions can be tolerated, most often by accommodating the added sequence in loop regions at the ends of the helices, albeit with some overall reduction in stability. In one case, however, a much longer-range structural effect was observed: two short α-helices that were joined at approximately right angles were transformed into a single extended α-helix. This is an example of one clear structural type being transformed into another with only a relatively small change in primary sequence. The most ambitious work randomizing larger segments of sequence likewise suggests remarkable tolerance (Plaxco et al. 1998). Some simplified models have been developed that provide an estimate for the number of compact, folded-like structures expected within the entire pool of random sequences. These models suggest that such compact structures may comprise a measurable fraction of all sequences (Lau and Dill 1990; Chan and Dill 1991; Davidson and Sauer 1994; Socci and Onuchic 1995; Li et al. 1996). Compact structures are not functional units, of course, but it is certainly plausible that random mutations subject to natural selection within these extended basins of molten globules could lead to the evolution of functional proteins. Finally, there is an important connection with the folding problem. Many proteins exhibit complex folding kinetics, suggesting significant population of local minima along the folding pathway. These rather different metastable species offer a plausible model for transitions among different structural types as discussed further below. In summary, the current view, although largely untested by experiment, is that there are measurable regions of sequence space for proteins in which reasonably folded structures are present.

CONSTRAINTS

It is clear that the search for fitness is constrained by physics and chemistry. There are certain things that physical laws exclude from a living system, regardless of how much more fit they would make an organism. More generally, the intrinsic reactivity of the chemical species at hand constrains the search for fitness. The limited solubility of carbon dioxide in water, the poor electrophilicity of the carbon atom in bicarbonate, and the abundance of carbonate as a planetary source of carbon are all constraints that direct the evolution of biomolecules.

Perhaps more interesting are the constraints on the fitness of biomolecules that arise from the structure of the biomolecules themselves. Once proteins and nucleic acids (as we know them in terrean environments) have been chosen as the biopolymers that will serve as catalysts and information storage molecules, the quest for fitness is constrained by the chemical properties of these molecules. This gives rise to a hierarchy of questions that can be raised about any physical or chemical property of the biopolymer, but are raised here with respect to the topic of "conformational landscapes" in proteins, heavily discussed in other group sessions at this Dahlem Workshop (Wolynes et al. 1995; Bryngelson et al. 1995; Onuchic et al. 1997b). Do proteins as a biopolymer have conformational landscapes that are qualitatively different (or quantitatively different, to the extent that the concept of a "landscape" can be quantitatively defined) from those of polymers generally? If so, are these differences such that proteins are uniquely suitable as a matrix for supporting life and its adaptation? Are there functional features of biopolymer structure and dynamics that determine the selection criteria for biopolymers in a biological environment? Biopolymers seem to have evolved to maximize robustness towards a large variety of environmental influences. This is discussed by Baase et al. and Chothia et al. (both this volume), who show that proteins can tolerate a considerable number of mutations without losing the ability to fold to the same stable and functional structure. Additional selection criteria for protein structures are (a) stability, (b) cooperative unfolding transitions, (c) solubility, and (d) interactions with other proteins, cofactors, substrates, etc. It should be emphasized that there is no need to optimize these features fully, but to sufficiently optimize them to the given biological environment. Proteins, for example, have not evolved to be maximally stable, but show a stability that allows them to function in their biological niche. The need for protein turnover in regulation of cellular processes and functionally essential motions places an upper limit on stability of enzymes. On the other hand, structural proteins like extracellular matrix proteins and eye-lens proteins often exhibit remarkable thermodynamic and kinetic stabilities towards unfolding. These examples illustrate that proteins have evolved to be as stable as necessary for their particular function.

In the discussion of selection criteria for dynamic behavior, the question of conserved folding pathways was addressed. Some theoreticians argued for funnel-like folding landscapes as an important selection criterion. Such landscapes lead not only to fast folding but also to robustness towards sequence variations. This would lead to a need for optimizing a measure like T_f/T_g for a given protein, which quantifies the ratio between the energy bias to the folded state and the ruggedness of the landscape. The role and the origin of local energy minima in the energy landscape, which leads to the population of partially folded states during refolding of most proteins, was discussed. Such states might be connected to the evolution of new protein structures as mentioned above. In evolving from one fold to another by small changes in the protein sequence, the signature of the initial fold might still be present as a local energy minimum on the energy landscape of the new structure. For two-state folders, typically the case for small proteins (of the order of 80 residues or fewer), the energy landscape of the new fold might be relatively easy to optimize and thus get rid of local minima. Larger proteins like lysozyme and barnase might be unable to evolve to such a smooth energy landscape.

We also debated at considerable length whether specific dynamic behaviors could be adaptive. It was agreed that they could be and a recent example can illustrate this point. In myoglobin, the rate at which oxygen is released in the tissues is a parameter that is almost certainly adaptive. If this rate is too slow, the protein is "unfit." An origin of this fitness was explored by the following experiments coupled with molecular dynamics simulations (Brunori and Gibson, unpublished). With the globin from Ascaris, the rate is slow, 10^{-2} to 10^{-3} per second, much slower than with sperm whale myoglobin. In the protein from Ascaris, Tyr and Glu are on the distal (oxygen binding) side of the heme. When these residues were placed into the sperm whale myoglobin framework by site-directed mutagenesis, and the structures were solved, the distances between the heme-bound oxygen and the residues in the distal pocket are essentially the same as in the Ascaris protein. The rate constants for oxygen release differ, however, by a factor of 500. To understand this difference, simulations of picosecond motions and geminate behavior were examined (specifically for NO binding, but this is analogous to oxygen). From the simulation, it appears that the difference arises because the ligand finds a pathway to a secondary site, another pocket behind an Ile which undergoes a dynamic motion to make a space for the departing ligand. In the Ascaris globin, Phe is present instead of Ile, and the Phe does not move out of the way on a picosecond time scale, accounting for the overall slower release of the ligand. While this result is preliminary and subject to interpretation, it clearly offers a model for how dynamic behavior on a very short time scale might influence an adaptive behavior, and thereby itself be under Darwinian selective pressure.

What Are the Structural Elements in Protein Evolution?

One of the most interesting questions to ask from an evolutionary perspective is: What is the connection between sequence and structure, and what are the structural building blocks of proteins? The complex relation between sequence and structure would be easier to understand if the tertiary structure could be broken into meaningful parts. The structures of many of the larger proteins appear to be concatenations of domains whose sizes are comparable to small, foldable proteins. There are many ways to view this problem, and several that emerged during the discussion are highlighted below.

The search for the smallest meaningful units of structure has a long and rich history. Already in the early 1970s, Wetlaufer (1973) posited the existence of small modules capablc of folding independently as a solution to the Levinthal Paradox. The idea of modular assembly was made attractive by the discovery of exons and introns, and this led Gilbert (1978) and Blake (1978) to suggest that exons encode structural and functional modules. With this idea in mind, Rose (1979), Go (1981), Ikura et al. (1993), and several other groups developed algorithms for dividing proteins into structural modules based on geometrical criteria. In these geometrical approaches, the role of the sequence variation is to a large degree suppressed so that all analogous structures from homologous proteins have the same structural modules. Using an energetic criterion, specifically the ratio of a contiguous segment's energetic stability gap to the energy variance of that segment's molten globule states, Wolynes and coworkers recently used energy landscape theory to predict kinetically competent, independently folding units or foldons for 50 proteins from various structural classes (Panchenko et al. 1996,

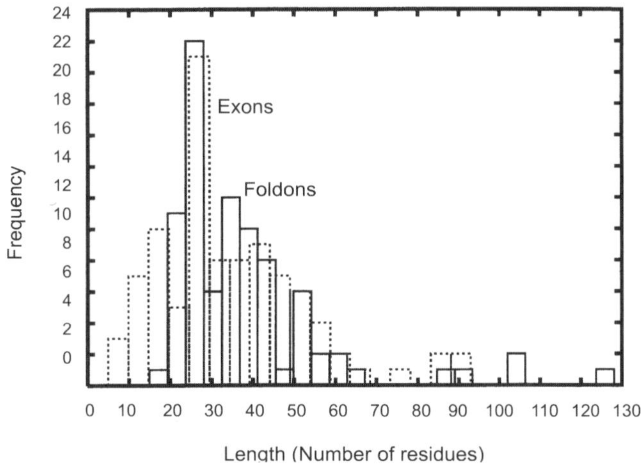

Figure 20.1 Length distribution of foldons and exons for the proteins studied in Panchenko et al. (1996, 1997). If the small exons (less than 15 residues) were omitted, the data sets would have the same mean value of 38 residues.

1997). While the length distribution of the foldons is similar to that of the exons, a significant overlap exists only for the more ancient proteins, as shown in Figure 20.1.

Comparisons between the foldons and the recently published categories of structures proposed by Gilbert (Desouza et al. 1996, 1997, 1998) are being carried out, and the existence of these folding units is being tested experimentally in several laboratories.

Experimental Evidence for Exons as Folding Units

Among the protein families studied by M. Go, myoglobin (and O_2 carriers in general) provided a test for the hypothesis that exons code for functionally independent units. The most remarkable success of Go (1981) was the prediction of a third intron in the O_2 carrier's gene, which was subsequently discovered (Jensen et al. 1981). Support for the proposal that the central exon of myoglobin has all the structural requirements for an "ancestral" O_2 carrier was probed by the following experiment. The preparation of a proteolytic fragment from horse apo-myoglobin comprising residues 32 to 139 seems to be consistent with the hypothesis of a functional central exon. This mini-myoglobin (De Sanctis et al. 1988, 1994) was shown to (a) fold correctly, (b) bind the heme stoichiometrically, and (c) bind O_2 and CO reversibly with the "physiologically" correct constants. Although mini-myoglobin is larger than the central exon (by 13 residues), it still seems the best example for the globin family, and supports the idea that exons 1 and 3 were "subsequently" added to build up more complex functions (assembly, control, etc.), as suggested by analysis of hemoglobin mutants (Eaton 1980). In this perspective, the data are consistent with the hypothesis of variation discussed in the section after the next.

A Critique of the Exon Hypothesis

The exon hypothesis is in fact two hypotheses. One is a hypothesis of origins, describing the initial formation of proteins through a combination of smaller peptides, each of which is some sort of structural or functional unit. The other is a hypothesis of variation within existing sets of functioning proteins during contemporary evolution. Originally, the hypothesis was presented as the latter, where introns were viewed as increasing the genetic distance between the exons, permitting a higher frequency of recombination between them.

There are a number of arguments against the "origins" hypothesis that are, for some of us, compelling. These include:

1. The number and pattern of introns in related proteins can be very different. An example of this is found in phosphofructokinase. The bacterial enzyme is some 220 residues in length, while in higher organisms, the protein contains approximately 450 residues. The longer protein has arisen by evident duplication of the shorter gene followed by fusion to give a protein with two homologous halves. There are introns in both halves: twelve in the first and nine in the second. None of the introns in the first half lies in a position equivalent to its position in the second half, implying that the introns in the first half and the introns in the second half arose *following* the gene duplication.
2. A large fraction of introns appears to be recent in origin by other criteria. For example, tubulin phylogenetic trees can rationalize the very different intron patterns observed in homologous tubulins (Dibb and Newman 1989). This suggests no role for these introns in the origins of these proteins.
3. The correlation between exons and structure units is statistically very weak (Stolzfus et al. 1994).
4. Adjacent introns within domains often have different phases (positions of the intron within the codon), which would imply that if exons did combine to form a domain, then they would have frame shift changes upon their recombination (Patthy 1991).

Introns are almost certainly important for the evolution of multiple domain proteins in metazoa, for example, those involved in "higher" metazoan function, such as cell adhesion molecules and cell-surface receptors. These proteins are frequently built from domains taken from a small repertoire of families. In all, or nearly all, cases these domains are separated by introns in the same phase. This implies that these proteins have arisen from gene duplication of the domains and intronic recombination (Patthy 1994).

Current Estimates of the Number of Protein Families

The number of three-dimensional structures in the Protein Data Bank is now close to 7,000. Many of these, however, are variants of a smaller set of unique proteins:

variants that have, for example, bound substrates or engineered mutations. Nearly all these proteins of known structure are members of one of a number of different families whose members are descended from a common ancestor. Proteins are usually assigned to a family on the basis of general similarities in their amino acid sequences (see Chothia et al., this volume). However, their evolution can proceed beyond the point where simple sequence comparisons can reveal their relationships. In these cases evolutionary relationships are determined using information on their structure, sequence features, and function.

In Table 20.1, we list the number of different protein families formed by the protein structures known in the years 1989 to 1995, when it rose from 103 to 405. This large rise is the result of rapid progress in structure determination by X-ray crystallography and NMR spectroscopy. An examination of the structures available in November, 1997, showed that they form 527 different families. Each year protein structures are determined whose sequences have no simple similarity to the sequences of previous known structures. These have, therefore, the potential of being the first known structure of a new family. On determination of their structures a proportion of these are indeed shown to be the first of a new family. For others, their structure, together with features of their sequence and function, shows that they are members of a previously known family. Between 1989 and 1995 the proportion in this second category rose from 17% to 45%.

If we make the assumption that the proteins that are potential members of a new family are a random selection of proteins, the data in Table 20.1 can be used to make a rough estimate of the total number of protein families. For example, if we know 500 families from previous structures, and half of these proteins turn out to belong to one of these 500, this implies that the total number of families is $2 \times 500 = 1000$. Carrying out this calculation using the data for each of the years 1989 to 1995 gives a fairly consistent set of results (fourth column of Table 20.1) and suggests that the total number of

Table 20.1 Number of proteins formed by proteins of known structure 1989–1995 and estimates of the total number of protein families.

Year	Number of known protein families	Percentage of potential members of a new family that belong to an old family (%)	Estimate of the number of the total number of protein families
1989	103	14	740
1990	129	17	760
1991	160	22	730
1992	194	34	570
1993	240	39	620
1994	297	42	710
1995	405	45	900

Data in the table is a simplified version of that given in Brenner et al. (1997).

protein families is somewhat less than 1000. The assumption made for this calculation, that the new protein structures are random selections of all proteins, is not accurate. Certain classes of proteins, such as integral membrane proteins, have very few representatives thus far. Many membrane proteins, however, come from a small number of sequence families, for example, the ATP-binding cassette proteins, which form the largest family of proteins in bacteria. Thus, this calculation surely underestimates to some extent the total number of protein families; however, it does clearly indicate that most proteins are members of one of approximately 1000 different families.

Probing Ancient Sequence Connections

Only two criteria for demonstrating (inferring) homology and common ancestry are available to us. The first is sequence comparison; the second is conformational analogy (similarity in overall fold). The second, obviously, cannot be used as a criterion for showing that proteins with two different folds arose from a common ancestor, nor can the first if sequences have diverged sufficiently to permit a protein to fold into two forms as they will have lost sufficient similarity to directly infer the fact of their common ancestry (see Chothia et al., this volume). It appears to be true that recent history has seen the origin of only a few folds. There are a variety of mechanisms to explain this: (a) There already are enough folds so that if new function is needed, an existing fold can be recruited, and there is no selective pressure favoring the creation of new folds. (b) The environmental conditions have changed so that new folds cannot be formed. For example, the level of oxygen in the atmosphere has risen dramatically in the last 2 billion years, still more dramatically in the last 600 million years. (c) Mechanisms to conserve existing function might make it difficult to create new folds. The fidelity of polymerases, which must be high in organisms with large genomes, might prevent the level of mutation that is necessary to generate new folds, for example.

What is clear, however, is that short of creating new folds, the geological time scale is sufficient to create most other types of function through recruitment of existing folds. It has taken approximately half of the lifetime of the star to create a molecular system that is capable of cognition (4.5 billion years). Gene duplication is quite rapid on this time scale (Jermann et al. 1995).

Intensive discussion then occurred as to whether a multiplicity of pathways were available to permit divergence of function, or whether only few were. Matthews noted that many of the point mutations that disrupt packing in lysozyme disrupt the stability of the folded form, and therefore would (presumably) not be accepted by natural selection. Benner noted on the other hand that explicit reconstruction of evolutionary intermediates in digestive ribonucleases found pathways that did involve the change of core residues that had no impact on folded stability, while others off of the path evolution actually followed did not (Stackhouse et al. 1990).

The prediction was made that of the 17 changes separating the α-subunits of human and horse hemoglobin 10 if introduced individually would have no impact on the stability of the protein, but that all of the others would lead to a less stable protein. It is clear, however, that horse and human hemoglobin are homologs; they are descendants of a common ancestor, and this ancestor can be reconstructed. This implies that there must be at least one pathway from one to the other that leads via functionally satisfactory intermediates. Recognizing that there are 20! or 2×10^{18} direct evolutionary pathways between the two hemoglobins, it is clear that a more comprehensive search for alternative paths (as well as a study of the actual path followed, which can be reconstructed) would be a productive research project.

How Does Sequence Ordering Influence Structure?

Towards the goal of understanding the constraint between sequence and structure, Nelson and Onuchic use ideas that come from energy landscape theory and apply them to sequence space instead of configuration space (Nelson and Onuchic 1998). One of the interesting theoretical concepts is that different sequence orderings can lead to different fold motifs. Each optimal target structure is associated with a class of minimally frustrated sequences that will fold to the target, and each class is characterized by a particular tendency for ordering the residues called "modes of design." A minimalist off-lattice model using a binary code with hydrophic (*H*) and polar (*P*) residues which allows two kinds of interactions, local and nonlocal, between hydrophobic monomers can be used to test this hypothesis (Nelson et al. 1997). Simulations with these simplified sequences are already able to illustrate such a behavior. Two minimally frustrated sequence classes are induced by modes of design that favor either the cross-chain (nonlocal) or chain-bonded (local) interactions in a sequence. Because these two sequence design modes oppose each other in the way that the *H* monomers are distributed in sequence, a substantial number of exchange mutations are required to evolve from one to the other folding motif. An example of these two design modes is:

P-P-P-P-H-H-H-H-H-H-H-P-P-P-P	*H-P-H-P-H-P-H-P-H-P-H-P-H-P-P*
local	nonlocal

This means that it is "impossible" to evolve between these different minimally frustrated regions by single mutations without crossing a region of frustrated (physiologically unacceptable with low T_f/T_g) slow folding sequences (Onuchic et al. 1997; Nymeyer et al. 1998).

Despite the fact that this hypothesis has been demonstrated only for this minimalist model, the possibility that a similar mechanism may exist for real proteins is intriguing. Guided by these conclusions, we can now speculate on real problems:

1. The model suggests that a similar mechanism may be responsible for selecting the possible folds for real proteins. Although real proteins have a much more diverse set of interactions, the general principles should hold. Some protein motifs should be more common than others and a selection mechanism similar to the one described here should identify the possible ones.
2. Has evolution already identified all possible folds? If not, how could sequences evolve from their present folds and find new ones?

Why These Twenty Amino Acids?

We conclude with some informed speculation on the choices of chemical species, specifically the amino acids. The twenty encoded amino acids found in most contemporary proteins provide a not unreasonable collection of chemical diversity. To cover chemical diversity space one requires positively charged, negatively charged, hydrophilic uncharged, aliphatic hydrophobic, and aromatic hydrophobic functionalities. The selection found in the standard encoded amino acids is not implausible. Compared to the characteristics found in the U.S. Pharmacopoeia, there appears to be an overrepresentation of aliphatic hydrophobic groups and an underrepresentation of heterocycles.

Instead of the structural terms outlined above, one might alternatively consider amino acid side chains in terms of elements of reactivity. The standard amino acids have acids and bases (including imidazole, an acid/base), a thiol, an alcohol (three), a carboxylic acid, and an amine. Some important functionalities are missing. There is no electrophilic carbon (e.g., an aldehyde) available in the amino acid side chain, no redox active side chain (other than cysteine), and no side chain capable of "Umpolung" chemistry. Cofactors, obtained from vitamins, provide the missing functionality for these proteins. Pyridoxal (vitamin B6) provides the electrophilic carbon. Flavin and nicotinamide provide redox active functionality. Thiamine provides the "Umpolung" reactivity.

In some cases, a particular reactivity is excluded from an amino acid side chain because of underlying problems in reactivity. Thus, an amino acid cannot plausibly carry an aldehyde group in its side chain as the aldehyde would react with the amino group. Thus, if a protein is to have access to aldehyde functionality, it must get it via a cofactor: pyridoxal provides this functionality. It is less clear why an amino acid does not exist carrying a redox active functionality (such as nicotinamide adenine dinucleotide or flavin adenine dinucleotide). Two explanations — one historical, the other functional — are conceivable. The vast majority of cofactors are "RNA cofactors"; they carry a piece of RNA attached. This implies that they originated in an "RNA world," an episode of life on Earth where RNA was the only genetically encoded component of biological catalysis. Proteins came later, it is argued, and emerged in a bio-environment already rich in redox active cofactors (Benner et al. 1989).

The functional argument notes that a typical dehydrogenase requires only one redox active amino acid. Considering whether there is a "cost" associated with maintaining an amino acid within the "translation repertoire," one must also maintain the corresponding tRNA, the aminoacyl tRNA synthetase, and must reserve codons that can be used elsewhere. One might argue from a functional perspective that it is cheaper to introduce a reactivity into a protein via a cofactor (rather than via an encoded amino acid) if one is going to use it only infrequently. This also permits much greater diversity in redox potential. Recent work has shown limited roles for radicals of Tyr and Trp and possibly His in a number of metalloproteins (see, e.g., Hoganson et al. 1997 and references therein).

The origins of translation are lost in the evolutionary mist prior to the divergence of the three primary kingdoms. There are, for example, two classes of aminoacyl tRNA synthetases. Amino acid specificities are divided between the two classes in an intriguing way. For example, aspartate falls in one class, while glutamate falls in the other; lysine falls in one, and arginine falls in the other. Each member of a pair of chemically similar amino acids is handled by a different family of tRNA synthetases. This pattern can, it appears, be reconstructed in the protogenome. This implies that aminoacyl tRNA synthetases emerged at a time where the alphabet already contained twenty amino acids, and where errors in tRNA charging would be lethal.

The standard model to explain this behavior is that protein-based aminoacyl tRNA synthetases arose in organisms where proteins were already being biosynthesized via translation, but with RNA molecules serving as the catalysts to charge the tRNA. Under such circumstances, it appears as if the protein-based aminoacyl tRNA synthetases already were built from twenty amino acids, implying that all (or at least most) of the standard amino acids had already been chosen, even before the divergence of the three principal kingdoms of life.

This implies that the origin of translation cannot be approached by the direct reconstructive processes that are so useful in other areas of metabolism. It is now clear (by experiment) that RNA molecules are capable of catalyzing amide bond synthesis; there is little doubt that translation *could* have emerged using RNA catalysts. Likewise, it is clear (again from experiment) that function can be obtained from a smaller alphabet of amino acids. For example, an oxaloacetate decarboxylase was prepared from helices constructed from just three different amino acids, Lys, Ala, and Leu (Johnsson et al. 1993). Biochemical pathways are known that use amino acid-RNA conjugates for functions other than translation (e.g., in chlorophyll biosynthesis, Kannangara et al. 1988), suggesting a possible way in which selective value could be had by intermediates along the path to translation (but short of translation itself). Thus, it is conceivable that translation could have emerged by Darwinian processes, starting from a smaller number of amino acids. However, in the absence of direct reconstructive approaches, the specific process by which translation actually arose in historical reality is likely to remain obscure.

REFERENCES

Bartel, D.P., and J.W. Szostak. 1993. Isolation of new ribozymes from a large pool of random sequences. *Science* **261**:1411–1418.

Benner, S.A., A.D. Ellington, and A. Tauer. 1989. Modern metabolism as a palimpsest of the RNA world. *Proc. Nat. Acad. Sci. USA* **86**:7054–7058.

Blake, C.C.F. 1978. Do genes-in-pieces imply proteins-in-pieces? *Nature* **273**:267.

Brenner, S.E., C. Chothia, and T.J.P. Hubbard. 1997. Population statistics of protein structures: Lessons from structural classifications. *Curr. Opin. Struct. Biol.* **7**:369–376.

Bryngelson, J.D., J.N. Onuchic, N.D. Socci, and P.G. Wolynes 1995. Funnels, pathways and the energy landscape of protein folding: A synthesis. *Proteins* **21**:167–195.

Burritt, J.B, W.W. Bond, K.W. Doss, and A.J. Jesaitis, 1996. Filamentous phage display of oliogopeptide libraries. *Analyt. Biochem.* **238**:1–13.

Chan, H.S., and K.A. Dill, 1991. "Sequence Space Soup" of proteins and copolymers. *J. Chem. Phys*. **95**:3775.

Davidson, A.R., and R.T. Sauer, 1994. Folded proteins occur frequently in libraries of random amino acid sequences. *Proc. Natl. Acad. Sci. USA* **91**:2146–2150.

De Sanctis, G., F. Ascoli, and M. Brunori. 1994. Folding of apominimyoglobin. *Proc. Natl. Acad. Sci. USA* **91**:11,507–11,511.

De Sanctis, G., G. Falcioni, B. Giardina, F. Ascoli, and M. Brunori. 1988. Mini-myoglobin: The structural significance of haem-ligand interactions. *J. Mol. Biol.* **200**:725–733.

Desouza, S.J., M. Long, R.J. Kleln, S. Roy, S. Lin, and W. Gilbert. 1998. Toward a resolution of the introns early/late debate — Only phase zero introns are correlated with the structure of ancient proteins. *Proc. Natl. Acad. Sci. USA* **95**:5094–5099.

Desouza, S.J., M. Long, L. Schoenbach, S.W. Roy, and W. Gilbert. 1996. Intron positions correlate with module boundaries in ancient proteins. *Proc. Natl. Acad. Sci. USA* **93**:14,632–14,636.

Desouza, S.J., M. Long, L. Schoenbach, S.W. Roy, and W. Gilbert. 1997. The correlation between introns and the three-dimensional structure of proteins. *Gene* **205**:141–144.

Devlin, J.J., L.C. Pangaiban, and P.E. Devlin. 1990. Random peptide libraries. A source of specific protein binding molecules. *Science* **249**:404–406.

Dibb, N.J., and A.J. Newman. 1989. Evidence that introns arose at proto-splice sites. *EMBO J.* **8**:2015–2021.

Eaton, W.A. 1980. The relationship between coding sequences and function in haemoglobin. *Nature* **284**:183–185.

Ellington, A.D. 1994. RNA selection. Aptamers achieve the desired recognition. *Curr. Biol.* **4**:427–429.

Fontana, W., and P. Schuster. 1998. Continuity in evolution. On the nature of transitions. *Science* **280**:1451–1455.

Gesareni, G. 1992. Peptide display on filamentous phage capsids. A new powerful tool to study protein–ligand interaction. *FEBS Lett.* **307**:66–70.

Gilbert, W. 1978. Why genes in pieces? *Nature* **271**:501.

Go, M. 1981. Correlation of DNA exonic regions with protein structural units in haemoglobin. *Nature* **291**:90–92.

Hoganson, C.W., and G.T. Babcock. 1997. A metalloradical mechanism for the generation of oxygen from water in photosynthesis. *Science* **277**:1953–1956.

Ikura, T., N. Go, D. Kohda, F. Inagaki, H. Yanagawa, M. Kawabata, S. Kawabata, S. Iwanage, T. Noguit, and M. Go. 1993. Secondary structural features of modules M2 and M3 of barnase in solution by NMR experiment and distance geometry calculation. *Proteins* **16**:341–356.

Jenison, R.D., S.C. Gill, A. Pardi, and B. Polisky. 1994. High-resolution molecular discrimination by RNA. *Science* **263**:1425–1429.

Jensen, E.O., K. Paludan, J.J. Hylding-Nielsen, P. Jorgensen, and K.A. Marcker. 1981. The structure of a chromosomal leghaemoglobin gene from soybean. *Nature* **291**:677–679.

Jermann, T.M., J.G. Opitz, J. Stackhouse, and S.A. Benner. 1995. Reconstructing the evolutionary history of the artiodactyl ribonuclease superfamily. *Nature* **374**:57–59.

Jiang, L., A.K. Suri, R. Fiala, and D.J. Patel. 1997. Saccharide-RNA recognition in an aminoglycoside antibiotic-RNA aptamer complex. *Chem. Biol.* **4**:35–50.

Johnsson, K., R.K. Allemann, H. Widmer, and S.A. Benner. 1993. Synthesis, structure and activity of artificial, rationally designed catalytic polypeptides. *Nature* **365**:530–532.

Kannangara, C.G., S.P. Gough, P. Bruyant, J.K. Hoober, A. Kahn, and D. von Wettstein. 1988. tRNA(Glu) as a cofactor in delta-aminolevulinate biosynthesis. Steps that regulate chlorophyll biosynthesis. *Trends Biochem. Sci.* **13**:139–143.

Lau, K.F., and K.A. Dill. 1990. Theory for protein mutability and biogenesis. *Proc. Natl. Acad. Sci. USA* **87**:638–642.

Li, H., R. Helling, and C. Tang. 1996. Emergence of preferred structures in a simple model of protein folding. *Science* **273**:666–669.

Nelson, E.D., and J.N. Onuchic. 1998. Proposed mechanism for stability of proteins to evolutionary mutations. *Proc. Natl. Acad. Sci. USA* **95**:10,682–10,686.

Nelson, E.D., L.F. Ten Eyck, and J.N. Onuchic. 1997. Thermodynamics and symmetry of protein-like heteropolymers. *Phys. Rev. Lett.* **79**:3534–3537.

Nymeyer, H., A.E. Garcia, and J.N. Onuchic. 1998. Folding funnels and frustration in off-lattice minimalist protein landscapes. *Proc. Natl. Acad. Sci. USA* **95**:5921–5928.

Onuchic, J.N., Z. Luthey-Schulten, and P.G. Wolynes. 1997. Theory of protein folding:The energy landscape perspective. *Ann. Rev. Phys. Chem.* **48**:539–594.

Panchenko, A., Z. Luthey-Schulten, R. Cole, and P.G. Wolynes. 1997. The foldon universe: A survey of structural similarity and self-recognition of independently folding units. *J. Mol. Biol.* **272**:95–105.

Panchenko, A., Z. Luthey-Schulten, and P.G. Wolynes. 1996. Foldons, protein structural modules, and exons. *Proc. Natl. Acad. Sci. USA* **93**:2008–2013.

Patthy, L. 1991. Modular exchange principles in proteins. *Curr. Opin. Struct. Biol.* **1**:351–361.

Patthy, L. 1994. Introns and exons. *Curr. Opin. Struct. Biol.* **4**:383–392.

Plaxco, K.W., D.S. Riddle, V. Grantcharova, and D. Baker 1998. Simplified proteins: Minimalist solutions to the "protein folding problem." *Curr. Opin. Struct. Biol.* **8**:80–85.

Rose, G.D. 1979. Hierarchic organization of domains in globular proteins. *J. Mol. Biol.* **134**:447–470.

Scott, J.K., and G.P. Smith. 1990. Seaching for peptide ligands with an epitope library. *Science*. **249**:386–390.

Smith, G.P. 1985. Filamentous fusion phage. Novel expression vectors that display cloned antigens on the virion surface. *Science* **228**:1315–1317.

Socci, N.D., and J.N. Onuchic. 1995. Kinetic and thermodynamic analysis of protein like heteropolymers: Monte Carlo histogram technique. *J. Chem. Phys.* **103**:4732–4744.

Stackhouse, J., S.R. Presnell, G.M. McGeehan, K.P. Nambiar, and S.A. Benner. 1990. The ribonuclease from an extinct bovid. *FEBS Lett.* **262**:104–106.

Stoltzfus, A., D.F. Spencer, M. Zuker, J.M. Logsdon, and W.F. Doolittle. 1994. Testing the exon theory of genes: The evidence from protein structure. *Science* **268**:1367–1369.

Wetlaufer, D.B. 1973. Nucleation, rapid folding, and globular intrachain regions in proteins. *Proc. Natl. Acad. Sci. USA* **70**:697–701.

Wolynes, P.G., J.N. Onuchic, and D. Thirumalai. 1995. Navigating the folding routes. *Science* **267**:1619.

Zamora, H., R. Luce, and C.K. Biebricher. 1995. Design of artificial short-chained RNA species that are replicated by Qb replicase. *Biochemistry* **34**:1261–126.

Author Index

Austin, R.H. 179–191

Baase, W.A. 297–311
Benner, S.A. 339–363, 365–382
Boxer, S.G. 365–382
Brunori, M. 365–382

Changeux, J.-P. 255–280
Chothia, C. 281–295, 365–382
Crothers, D.M. 255–280

Deisenhofer, J. 1–3, 81–106
Dill, K.A. 81–106
Dobson, C.M. 15–37, 179–191

Eaton, W.A. 117–138, 179–191

Fleming, G.R. 81–106
Frauenfelder, H. 1–3, 81–106

Garel, T. 107–116
Gassner, N.C. 297–311
Gelfand, I. 281–295
Gerwert, K. 81–106
Geyer, M. 59–79
Grubmüller, H. 255–280

Halle, B. 179–191

Karplus, M. 139–177, 179–191
Kiefhaber, T. 365–382
Kister, A. 281–295
Kuroki, R. 297–311

Lesk, A.M. 281–295
Luthey-Schulten, Z. 365–382

Matthews, B.W. 297–311, 365–382
McCammon, J.A. 81–106, 193–198
Michel, H. 81–106
Moffat, K. 255–280

Nienhaus, G.U. 39–57, 255–280

Onuchic, J.N. 365–382
Orland, H. 107–116, 179–191

Palma, M.U. 213–225, 255–280
Parak, F.G. 213–225, 255–280

Richards, F.M. 179–191
Rigler, R. 365–382

Schlichting, I. 59–79, 179–191
Schulten, K. 227–253, 255–280
Schuster, P. 313–337, 365–382
Switzer, C.Y. 339–363

Tronrud, D.E. 297–311

Warshel, A. 199–211, 255–280
Weaver, L.H. 297–311
Wolynes, P.G. 1–3, 5–13, 179–191

Zhang, X.-J. 297–311

Subject Index

accessible surface area (ASA) 289, 291
acetylcholine receptor 271
acetylcholinesterase (AchE) 88, 195
actin polymerization 197
activated dynamics method 143–146, 157
activation barrier 142, 146, 150, 152, 159, 162, 199, 208, 209
  in enzymes 200, 204
  requirement for 144
  in TIM 203
activation free energy 200, 207
adaptive dynamic behavior 372, 373
aggregation 15, 32, 34, 101, 102, 222
  partially folded states in 34, 35
  prevention of 30
  slow 31
alanine "tetrapeptide" 162
all-atom model 159
allostery 59, 117, 123–127
  two-state model 124
alphabet, genetic 355, 370
  RNA-based 356
α-domain 20, 25, 27, 28, 31
α-helices 112, 114, 188
α-lactalbumins 19, 26–29, 31
Alzheimer's disease 30
amino acid
  displacement of residues 193
  substitution 298, 299, 306–309
amyloid fibrils 30–34, 189
anisotropic displacement parameters 63
annealing processes 67, 68
antenna systems, light-harvesting 229, 249, 250
  anti-COSMIC behavior 348, 351
aptamers 369
aromatic ring flips 97, 157, 158
Arrhenius behavior 42, 47, 160
Arrhenius equation 145, 146, 152, 154
Arrhenius law 12, 145, 162, 169, 170
Arrhenius-like temperature dependence 139, 145, 153, 182
assemblies 255, 270, 273
  macromolecular 267, 268, 271
  neuronal 268, 269
A-state 26–29, 31
  hydrogen exchange experiments 27
  proteolytic cleavage in 28
atomistic dynamics calculations 11
ATP 96, 100, 101
  synthase 196
attraction, dynamical basins of 6, 7
autocorrelation of energy gap 203, 204, 207
averaging 213–223
  conceptual 255
  fluctuational 258
  thermodynamic 215–218, 221, 256–258

backbone 110–112, 157
  defects 107
  dipolar effects 115
  negative charge in 359
bacteriochlorophylls (Bchls) 230, 231, 234
  –carotenoid interactions 240
  circular aggregates 241, 242
  role in photon fueling 246, 247
  triplet states 230, 247, 249

bacteriorhodopsin (bR) 40, 55, 95, 258, 263
photoisomerization in 205, 206
barriers in landscapes 22–24, 34, 39, 43, 87, 143, 144, 168
activation 142, 146, 150, 152, 159, 162, 199, 208, 209
enthalpy 42, 50, 54, 215
free-energy 7, 162, 166, 187, 204
high-energy 108
inner 53, 54
Gaussian distribution of 49, 50, 62, 85
barrier crossing 148–151, 157, 194
simulated 53
Na^+ 150
baseline distortions 67, 68
base stacking 82
interaction with base pairing 260
β-domain 20, 28, 31
β-sheets 112, 114, 188, 189
B-factor 62–64
bioelectronic systems 229
biopolymers 193, 194, 196
assemblies of 267
COSMIC-LOPER 347–351, 353
diversity 318–321, 365–380
fitness of 344, 351, 352
one-biopolymer model 342, 350
properties of 81–83
two-biopolymer model 341, 342, 350, 354
Bohr effect 125, 127, 214
bond angles 68
bond lengths 110
Bonnet transformation 115
Born–Oppenheimer approximation 29, 263
bovine pancreatic trypsin inhibitor (BPTI) 49, 150, 157, 158, 194, 220, 221
Bragg reflections 60, 62, 65
Bragg spacing 61, 90
Brownian dynamics 82, 159, 262

carbon monoxide-binding kinetics 214
carbon monoxide stretch bands 53
carbonmonoxy myoglobin (MbCO) 39, 52, 96, 214–216, 222
energy landscape analysis in 41–50
molecular dynamic simulations of 53
structure of 45
X-ray diffraction experiments 46
carotenoids 229–233, 238
–BChl interactions 240
role in photon fueling 246, 247
Car–Parrinello methods 101, 263
Cartesian displacements 206
catalysis 308, 309
cavity formation 304
chain connectivity 11
chaperonins 196
charge 350, 351
transfer processes 207
chemical kinetics 170, 183
chemistry
combinatorial 345
prebiotic 340, 344–346, 355, 358
chirality 109–112, 345
chlorophylls 229, 247
Chou/Fasman rules 84
chromophores 249, 250
arrangement of in PSU 237–241
circular dichroism (CD) 18, 20, 25, 27, 28
classical chemical phenomenology, limits of 6, 8
cluster physics 109
coagulation 223
"cold" neutrons 91
collective coordinates 181
combinatorial chemistry 345
combinatory map 319, 320
common core of variable domains 285, 287, 292
complexity 107–115
combinatorial 314, 319
definition of simple/complex behavior 141, 142, 170, 180
complex reaction profile 256
computer simulation 101, 102, 179, 183, 199, 267
minimalist models 181–184
RNA 330, 331, 333
conceptual averaging 255
conformational changes (see also *folding, unfolding reactions*) 59–61, 67, 100, 122, 130, 139–171, 186, 222
biopolymer 196

hinge bending 140, 157, 159
large-scale 102
lid opening/closing 157
low-energy 283
quaternary 118, 129, 133, 135, 160, 179
sequence-dependent 82
studies of 156–161
tertiary 133, 134, 160, 179
conformational entropy 85, 188
conformational heterogeneity 55
conformational relaxation 131, 132, 166, 208
conformational substates (CS) 39, 40, 51, 89, 188, 194, 215, 216, 266
continuous relaxation spectrum 256, 258
continuous transitions 331–334
cooperative mechanism 119–122, 125–128, 134, 135, 269
core 86
common core of variable domains 285, 287, 292
hydrophobic 85, 86, 113, 115
mutants 302–305
residue 297
structure 305
correlated landscape theory 182
correlation function analysis 150
COSMIC-LOPER biopolymers 347–351, 353
Coulomb interactions 110
Coulomb mechanism 230, 247
couplings
anisotropic nuclear 97
Coulomb 230
dipolar 66, 96
J-coupling 60, 66, 67
scalar 96
spin-spin constant 66
Zeeman 96
crambin 42, 90
creationism 341
Creutzfeldt-Jakob disease 30, 31
crystallization 62, 109
phase transition 109
prerequisite for 92
prevention of 7
crystallographic temperature (B-factor) 62–64
crystallography (see also *X-ray*) 63, 83, 95, 96, 228
crystals 59–65
disordered 142
lattice 63, 64
formation 91
model 109, 163–165, 168, 329
c-type lysozymes 19, 27, 34
cytochrome c 92, 262, 268

Darwinianism 341
Debye–Waller factor (DWF) 62–65, 85
decoupling 18
deep structure 290, 292, 293
definition of simple/complex behavior 141, 142, 170, 180
degrees of freedom 259, 260
deletions 305–307, 369
demixing 220–222
density functional methods 101
deoxyhemoglobin 122, 126–129
deoxymyoglobin 217
deutrons 94, 97
Dexter mechanism 230
Diels-Alder reaction 357
diffusion
-collision model 167
constant 184
processes 200, 216
dihedral angles 66, 68, 110, 157, 181, 194
pseudo 159
dihydrofolate reductase 196
dihydroxyacetone phosphate (DHAP) 140, 146, 147, 151
dimension reduction 259
dipolar backbone effects 115
dipolar chiral chain 109
dipolar coupling 66, 96
dipolar interactions 112, 188
dipolar relaxation rate matrix 68
dipole orientation 219, 220
discontinuous transitions 331–334
disease 16, 30, 33, 35
Alzheimer's 30
Creutzfeldt-Jakob 30, 31

disease (continued)
intervention 81
sickle cell 122, 189, 347
distance geometry (DG) 67
disulfide activation parameters 158
disulfide bridges 19, 27, 28, 49
diversity 325
biopolymer 318–321, 365–380
origin of 314–317
phenotypic 325, 335
sequence diversity 113, 115, 282, 283
structural 282, 283
downfield shift 66
double helix structure 82
Watson–Crick 89, 260
DNA catalysts 82
DNA library 358
DNA photolyase 96
dynamical basins of attraction 6, 7
dynamical effects 199–209, 260
meaning of 200
dynamics (see also *molecular dynamics*) 1, 214–218, 256
activated dynamics method 143–146, 157
adaptive dynamic behavior 372, 373
Brownian dynamics 159, 262
criteria for behavior 372
intramolecular 68, 149, 216
of taxonomic substates 46–50

effective Hamiltonian 243–246
elastic constants 108
electron density map 121, 235, 237
electron transfer processes 55, 206–208
electrostatic effects 203, 264, 266, 268, 277
empirical valence bond (EVB) 203
energy basins 6, 7, 24
energy landscape 5–12, 20–22, 39, 54, 55, 89, 140, 161, 179–189, 219, 257–259
barriers in 22–24, 34, 39–43, 87, 143, 144, 168
in carbonmonoxy myoglobin 41–50
exploration of 1, 15
fitness in 318–321, 335
for folding reactions 22–26
funnel 10, 11, 86, 169, 183, 250
global 9, 12
goal of using 5, 6
how to predict 2
intermolecular 31
mapping 15, 25, 30
role of 180
rugged (rough) 9, 11, 23, 27, 40, 47, 49, 194
statistical energy landscape theory 181–185
energy surface (see *energy landscape*)
ensembles of proteins, heterogeneous 258
enthalpy 109, 158, 161, 169
barrier 42, 50, 54, 215
–entropy compensation 187
surface entropy 112
entropy 85, 87, 111, 158, 161, 169, 261
bottleneck 183
configurational 187
conformational 85, 188
surface 112
environmental properties of molecules 265, 266
enzymes
catalysis 139, 140, 171, 199, 200, 209
converting states of 8
dihydroxyacetone phosphate (DHAP) 140, 146, 147, 151
"gated" 195
glyceraldehyde-3-phosphate (GAP) 140, 146
hyperthermophilic 91
modeling chemical processes in 258
reactions in 200–204
triosephosphate isomerase (TIM) 146, 147, 150, 151, 157–160, 195, 203
equilibrium fluctuations, function of 50, 54
equine lysozyme 25
equine variant 24, 27
error propagation 321–325, 334
Eschenmoser synthesis 346
Escherichia coli 62, 196
excitation funnel system 229, 232, 241–244, 250
effective Hamiltonian for 244–246
excitation transfer 230, 232, 239

exciton band system 240, 242–245
exons 122, 374, 375
exponential time dependence 139, 160, 162, 170
extraterrestrial life 340, 343, 350, 353
evolution 193, 195, 223, 228, 314–317, 365–380
  constraints in 283–285
  Darwinian 339, 343, 351, 352
  divergence in 232, 249, 354
  dynamics of 325, 326
  phenotype 330–334
  protein 281–294
  RNA 334, 335, 369
  sequence changes in 292–294
  structural elements of 372–374
evolutionary optimization 314, 328, 335

Ferry relation 47, 49
field cycling (FC) 97, 98
fitness 351, 352, 369, 371
  biopolymer 344, 351, 352
  landscapes 318–321, 335
flash photolysis 214, 216
flavoprotein 96
Flory-Huggins best fit 222
fluctuational averaging 258
fluorescence energy transfer 19
folding (see also *energy landscape*) 15–35, 85–87, 140, 161–171, 180, 183, 185, 223, 260, 298, 307–309, 365
  code for 16
  conserved 372
  fast 24
  features of 18
  fundamental nature for 16
  intermolecular interactions in 30–33
  lysozyme 23, 25, 30, 33
  model 156
  problems with 298–301
  rate 162, 166–170
  reactions 22–26, 162, 165, 166
  reversible 86, 107
  time dependence in 162–167
  transition state 22, 26, 34, 108–115
  traps 23
  wild-type 30
foldons 372, 374
Förster mechanism 230, 246
Fourier transformation 62
free energy 85, 86, 199, 283
  activation 200, 207
  barriers 7, 162, 166, 187, 204
  landscapes 101
  minima 108
  perturbation method 261
  profile 181, 182
  surface 258
"frozen bath" assumption 148, 149, 152
frustration mechanism 9
FTIR spectroscopy 46, 95, 98, 102
function 82, 117, 179–189, 214, 261, 284
  biological 109–209
  cellular 228
  relation to exons 122
  relation to motion 152, 193–197
  space 354
functional relaxation 87
funnel 10, 11, 86, 169, 183, 223
  excitation system 229, 232, 241–244, 250
future research 83, 89, 93, 102, 181, 355
  development of X-ray lenses 91

gas-phase reactions 142–144
"gated" enzymes 195
gating motion 195, 196
Gaussian distribution
  of barriers 49, 50, 62, 85
  of substate energies 131
geminate rebinding 131, 132
generalized random energy models (GREM) 114
"general" laws 271–273
genetic alphabet 355, 356, 370
Gibbs–Helmholtz relation 142
glasses 5, 8, 39, 47, 142, 215
  spin model 9, 39, 108
glass transition 7, 11, 89, 182–184, 187
  "slaved" 265
  temperature 7, 89, 266
glutamate receptor 269
glyceraldehyde-3-phosphate (GAP) 140, 146

glycine residues 107, 109, 112
glycosylation 151
green fluorescent protein 96
ground state 108, 198, 112, 163, 188

Hamiltonian
effective 243–246
path 108, 109, 112
CHARMM 110, 111
Hamming distances 316, 318, 322, 323, 327
helical secondary structure 66
helium 7
hemagglutinin HA2 196
heme 40, 43, 44, 50–55, 87, 96, 215, 216
–heme interaction 121
–histidine–CO complex 154
ligand binding 258
position of 154
hemoglobin 117–135, 179, 189, 214, 218, 223, 256, 263, 266
deoxyhemoglobin 122, 126–129
at equilibrium 123–127
historical background 119–123
horse 282
kinetics 127–130
oxygenation 119, 120
quaternary transition of 157, 160, 161
schematic structures of 118
sickle-cell 347
Henderson–Hasselbalch relations 44, 52
heterogeneous ensembles of proteins 258
heterogeneous model 87, 88, 89
heteropolymers 107–110, 114
folding surfaces of 164, 165
random 166
hetertropic effects 122, 124, 132, 134
hierarchical energy landscape 42, 89, 194
hierarchical organization 216, 217, 269
hinge bending 140, 157, 159
histidine 44, 51, 113
residues 46
hole burning experiments 97
homogeneous model 87
homopolymer 10, 108, 109, 114, 187–189
models of 111, 112
homotropic effects 122, 124, 133, 134
hydrogen
bond 19, 32, 44, 61, 66, 67, 82, 85, 86, 109–112, 188, 283
nonstandard 356
–deuterium exchange 91, 92
exchange 20, 23, 27
$H + H_2$ paradigm 9
hydrophobic cores 85, 86, 113, 115
hydrophobic interactions 85, 218, 221, 222
temperature dependence in 182
hydrophobic residues 27, 34
hydrophobic stabilization 304
hyperthermophilic enzymes 91

immunoglobulins 285–292
INDO-CIS method 242–244
induced dipole–induced dipole interaction 229, 242
induced fit 59
infrared spectroscopy (IR) 35, 43–46, 53, 94, 95, 171
insertions 305–307, 369
intermolecular interactions 30–32, 34, 68, 189
intramolecular dynamics 68, 149, 216
intrinsic tryptophan fluorescence 21
intron 374, 375
ion channel 269, 270
isotopic labeling 95–98
isotropic analysis 63

J-coupling 60, 66, 67
"jigsaw-puzzle" model 305

Kabat data base 287
Karplus equation 66, 67
kinetic bottlenecks 195
"kinetic rule" 131
KNF model 122, 127
Kramers equation 161

lactate dehydrogenase 196
Lamarkianism 341
Landau continuum approach 111
Larmour precession frequency 66
laser excitation 135

lattice simulations 109, 163–165, 168–170, 182–185
Laue method 60, 65, 90, 92, 96
Le Chatelier's principle 121
Lennard–Jones potential 67
Lennard–Jones spheres 218–220
levels of organization 268–271
Levinthal Paradox 16, 17, 108, 161, 373
lid opening/closing 157–160
life
  artificial 366
  extraterrestrial 340, 343, 350, 353
  indicators of 340, 342
  origins of 366
  universal structure in 345
ligand binding 50, 117, 140, 151, 153, 171, 194–196, 214, 218, 269
  to biopolymers 196
  electronic barrier to 153
  "gated" diffusional 195
  to heme proteins 258
  intrinsic binding constant 120
  by myoglobin 139
  rate 195
ligand rebinding 129, 130, 215
  after Mb photolysis 140, 151–158, 164, 170, 185
ligation state 123, 128, 129, 132, 134
light-harvesting systems 227–249, 267
  antenna 229, 249, 250
  complex I (LH-I) 231, 233
  complex II (LH-II) 231–235, 237, 240, 247
  complex III (LH-III) 233
  function of 229–231, 234
  physical properties of 229–231
  in purple bacteria 230–233
light-induced charge separation 49
light-induced photobiological reactions 199, 200, 209
linear dichroism 94, 95
linear free-energy relation 132, 183
linear response approximation 202
lipidic bicontinuous cubic phases 92, 93
localization 16
loop reorientation 157, 195
low-temperature electrophoresis experiments 133
lysozomes 23, 157, 201
  c-type 19, 27, 34
  equine 25
  folding 23, 25, 30, 33
  hen egg white 222, 308
  T4 lysozyme 297–301, 306–308

macromolecular assemblies 267, 268, 271
mapping
  of energy landscapes 15, 25, 30, 35
  genotype–phenotype 313–320, 325–327, 330, 334
maps
  combinatory 319, 320
  electron density 121, 235, 237
Marcus theory 208, 264
Martian meteorites 339
master sequence 321
Maxwell fields 110, 147
mean-field theory 9
mean-square deviations 85
  Debye–Waller factor (DWF) 62–65
mean-square displacement 62, 216, 217
mechanisms 82, 86–89
memory effects 260, 270
methione 298, 299, 304, 305
Michaelis complex 141
microfossils 340, 343
microwave absorption techniques 216
modes of regulation 255
molecular chaperones 15, 16, 30, 86
molecular clock 196, 197
molecular dynamics (MD) (see also *dynamics*) 95, 108, 268
  restrained (rMD) 67
  simulations 41, 53, 98, 135, 152, 154, 159, 179, 185–187, 195, 197, 202, 219, 221, 223, 259, 267
molecular graphics 267
molecular mechanics (MM) force field 146
molecular motions 193, 194
molecular orbital QM/MM method 146
molecular "quasi-species" 313, 321–325, 335
molecular tweezers 101
molten globule 26

monoclonal antibodies 92
Monod–Wyman–Changeux (MWC) model 120, 122, 124, 125, 127–130, 135, 160, 269
extended kinetic model 132–134
–PSK model 123–125, 127, 134, 135
monomer sequence 82
Monte Carlo methods 24, 101, 108, 133, 167, 185, 202, 260
Moore's law 101
Mössbauer spectroscopy 64, 216, 217
multiple anomalous diffraction (MAD) 62, 89, 91
multiple isomorphous replacement (MIR) method 62
multisubunit proteins 119, 179
mutagenesis experiments 30
mutations 30, 33, 313–317, 326, 335, 347, 369
"size-switch" 303
space 353
myoglobulin (Mb) 1, 2, 8, 40, 42, 60, 130, 132, 140, 154, 184, 187, 195, 214–218, 263, 374
conformational relaxation in 166
deoxymyoglobin 217
function of 50
photodissociation 155, 156
relaxation phenomena in 153
sperm whale 282, 373
structure 41, 54

nanosecond-resolved spectroscopy 129
native fold, characteristics of 20
native structure 15, 81, 82
characteristics of 18–20
stabilizing nature 16, 23, 26, 31
neighborhoods, statistical 314, 331, 333
neural network 314, 328, 329, 333, 334, 335
neuronal assemblies 268, 269
neurotransmitter proteins 257, 265, 269
receptors 270, 271
neutrons
"cold" 91
diffraction 91, 92
neutron scattering experiments 152, 216

Newtonian paradigm 6
nonadditivity 220
nonexponentional relaxation 152
nonradiative transitions 229, 263
normal modes
analysis of 63, 216, 259, 266
simulation of 65
Nose–Hoover thermostat 187
novel pulse sequences 94
nuclear magnetic relaxation dispersion (NMRD) 49, 186
NMR spectroscopy 16, 23–30, 40, 60, 61, 67–71, 83, 93–98, 171, 194, 228, 261
data on ring flip 157
protein structure determination 65, 66
solid state deuterium 159
structure determination of aptamer 369
nuclear Overhauser effect (NOE) 19, 24, 25, 60, 61, 68, 93, 186
spectroscopy (NOESY) 66, 68
nuclear spin relaxation times 96
nucleobases, nonstandard 357

one-biopolymer model 342, 350
optical rotation signal (ORD) 222
optical spectroscopy 40
organizational levels 268–271
osmotic pressure measurements 119

P68 protein 120
Paracoccus denitrificans 92
Peptide-linked Nucleic-acid Analogs (PNAs) 351
peripheral buried sites 290, 293
Perutz stereochemical mechanism 123–128, 135
pH 107, 115, 122, 124, 129, 134
phase space 107, 108
phenotype 314, 315, 327–329
diversity of 325, 335
evolution of 330–334
shape space 319, 320, 328, 335
phosphate groups 349, 350
photoactive yellow protein 90, 96, 263
photodissociation 151–156
photoisomerization 205, 206, 263
photolysis 185, 214

photosynthesis 206–208
light harvesting in 231–233
photosynthetic bacteria 227
purple 227, 229–233, 241, 249
photosynthetic light-harvesting protein 88
photosynthetic reaction centers (RC) 40, 42, 55, 96, 197, 229, 231, 234, 237, 258
photosynthetic unit (PSU) 229, 231–233, 247–249
architecture of 231
arrangement of chromophores 237–241
excitation transfer 239
structural model of 233–237
photosystem II of spinach 266
Poisson process 7
Poisson–Boltzmann equation 262
polyalanine hypothesis 112
polyalanine protein 298, 300
polycyclic aromatic hydrocarbons (PAHs) 340
polymers 12, 257
synthetic 39
polynucleotides, secondary structure 314
porins 92, 93
potassium channel 171
potential-of-mean force (PMF) 147–150, 180
prebiotic chemistry 340, 344–346, 355, 358
pre-exponential factor 167
PROCHECK 68
proline residues 107, 109, 112
protein amide exchange experiments 66
proteinase inhibitor IIA 60
Protein Data Bank (PDB) 83–85, 102, 285, 375
Kabat data base 287
protein conformations 128, 180, 185, 187
protein disorder 63, 64
protein engineering studies 22, 180, 284
protein families 83, 84, 375–377
protein-specific motions 216
proteolytic cleavage in A-state 28
protomers 269
protonation 46, 52
proton transfer 151, 200, 201
pump-probe experiments 95
purple bacteria 227, 229–233, 241, 249
antenna systems of 249
light-harvesting in 230–233
Rhodobacter sphaeroides 233–237, 240, 245, 246, 247
Rhodopseudomonas (Rps.) acidophila 231, 235, 247, 249
Rhodospirillum (Rs.) molischianum 231, 235–237, 240–243, 245, 247, 249

quantum mechanics 264, 265
consistent force-field/π-electron (QCFF/PI) approach 243
description 229, 241–244, 262, 263
perturbation theory 229
quaternary structures 121
changes 139, 160, 181
conformational changes 118, 129, 133, 135, 160, 179
R- 121, 124, 129, 130
T- 121–126, 129

Raman spectra 65, 95
Ramachandran
plot 68, 188
surface 16, 17
random charged (RC) chain 113, 114
random coil (state) 16, 17, 27, 32, 34, 189
chemical shift 28
values 66
Random Energy Model (REM) 9, 110, 113, 114, 156, 166, 182
random graph theory 329
random hydrophilic–hydrophobic (RHH) chain 113, 114
random synthesis of oligonucleotides 315
rate constant 208, 257, 258
criterion for existence 143–145
reaction centers 49, 207, 208
efficiency of energy transfer in 266
photosynthetic 40, 42, 55, 96, 197, 229, 231, 234, 237, 258
reaction coordinates 18, 143–148, 154, 157, 165, 166, 184, 185, 215, 256
reaction energy landscape 51–53
reaction flux method 146
reactive fluctuations 200, 201

rebinding 129, 130
barrier to 154
geminate 131, 132
myoglobin after photolysis 140, 151–156, 158, 164, 170, 185
recombination 313–317
DNA technologies 62
relaxation 153, 156, 197, 223
after perturbation 142
of autocorrelation function 207
in Mb 153, 166
behavior 156
conformational 131, 132, 208
dipolar relaxation rate matrix 68
functional 87
NMR studies 186
nonexponentional relaxation 152
rate 153
tertiary relaxation 130–132
thermal 243
relaxation methods 46, 49, 93, 97
competitive 67
flash photolysis kinetics 46, 47, 48, 52
functional 87
pressure relaxation 46, 48
spin 97
relaxation spectrum 256, 258
relaxation times 8, 96, 129, 143, 256, 258
reorganization energy 197, 204, 207, 268
replication, template-directed 346, 347, 350
residues
core 297
conservation of 287–289, 291
glycine 107, 109, 112
histidine 46
hydrophobic 27, 34
packing of 283
proline 107, 109, 112
stability of 26, 29, 30
surface 294
use of site-directed mutagenesis 26
restrained molecular dynamics (rMD) 67
R-factor 61, 62, 64, 68
"rf pulse labeling" experiment 25
Rhodobacter sphaeroides 233–237, 240, 245–247
Rhodopseudomonas (Rps.) acidophila 231, 235, 247, 249
rhodopsin 205, 206, 263
Rhodospirillum (Rs.) molischianum 231, 235–237, 240–249
Rhodospirillum rubrum 266
ribosome 15, 16, 32, 82, 86, 343
ribozymes 315, 353, 369
RNA
computer simulation 330–333
evolution *in vitro* 313, 317, 322, 326, 351, 352
folds of 82
optimization 330–333
secondary structures 313, 327–330
root mean square (RMS) deviation 67, 68, 283
rotational correlation times 66
"rough" energy surface 9, 11, 23, 27, 40, 47, 49, 165, 183, 184, 194, 196, 216, 219
R-quaternary structure 121, 124, 129, 130
deoxyhemoglobin in 122, 127

salt bridges 85, 122–125, 127, 134
salt concentration 107
scalar couplings 96
scattering 18, 92
small angle X-ray scattering 18, 23
Schrödinger's cat 99
secondary structures 109, 111, 112, 115
helical 66
polynucleotides 314
prediction of 84
selectivity 194–196
self-aggregation 247, 248
self-replication 345, 348
sequence diversity 113, 115, 282, 283
sequence space 257, 260, 269, 314, 318–322, 325–327, 334, 335, 368–371, 378
neural networks in 314, 328, 329
sequence variation 297–309
shape space 319, 320, 328, 333
side chains 107, 108, 113–115
disorder of 27
signal-to-noise ratios 93, 99

single-molecule experiments (see also *hemoglobin*) 8, 99–101, 151, 156, 194
singular-value-decompostion (SVD) analysis 130
site-directed mutagenesis 26, 44, 95, 98, 135, 183
"size-switch" mutations 303
small angle X-ray scattering 18, 23
solvation effect 149
  nonequilibrium 203, 204
solvents 186, 187, 218, 220, 265, 266
  accessible surface area (ASA) 289
  -induced force (SIF) 218, 220–223, 266, 267
  –protein system 187, 201
  role of 214
spectral density function (SDF) 97, 98
spin glass model 9, 39, 108
spin systems 10, 66
spongiform encephalopathies 30, 31
stability 86, 257
  compensation 302–305
  gap in 188
  hydrophobic 304
statistical energy landscape theory 181–185
statistical neighborhoods 314, 331, 333
statistical substates 39, 50
stereochemical mechanism 123
steric interactions 86–88
structural biology 117, 118, 123
structural diversity 282, 283
structural strain 302–305
structure prediction 12, 89
structure space 344
substates 102, 259
  A-state 26–29, 31
  conformational (CS) 39, 40, 51, 89, 188, 194, 215, 216, 266
  hierarchy in 89
  statistical 39, 50
  taxonomic substates 39, 43–44, 55
  transitions 89
subtilisin 203, 204
sulfone group 349
superoxide dismutase (SOD) 261, 262
synchrotron X-ray sources 60, 90, 98
T4 lysozyme 297, 298, 306, 307
  backbone trace of 299, 301
  mutations in 300, 308
taxonomic substates 39, 43–44, 55
  dynamics 46–50
temperature dependence 141–145, 162, 182, 216
  of folding rate 167–170
  of rebinding 152, 215
template-directed reproduction 346
tensorial-order parameter 112
tertiary relaxation 130–132
tertiary structures 82, 109, 133
  conformational changes 134, 160, 179
tetrapide 181
thermal diffuse scattering (TDS) 60, 62, 65
thermal fluctuations 200, 202
thermal motion/vibrations 65, 264, 300
thermodynamic averaging 215–218, 221, 256–258
thermodynamics 85, 163, 182, 217
  phases 108
  "sink" in energy landscape 33
  specificity 88
θ point 109, 111
tholins 345
time dependence 141, 197, 201, 206, 209
  energy gap in 264
  of folding 162–167
  of motion 193
  transmission coefficient ($\kappa(t)$) 144
torsion angles 61, 67, 93
  determination of 66
T-quaternary structure 121–126, 129
  low affinity of 134
  transition state ensemble 180, 183
transition states 18, 22, 26, 34, 40, 47, 48, 51, 165, 166, 183, 204, 260, 261
  conformational 269, 270, 271
  continuous 331–334
  desensitization 270
  discontinuous 331–334
  folding 22, 26, 34, 108–115
  stabilization of 264
transition state theory (TST) 49, 144, 148, 200, 202
  value 149, 150

translation-libration-screw (TLS) model 63
translocation 16, 306, 307
transmission coefficient 144–146, 149–151
calculation of 148
"transmition" factor 200–203, 207–209
triosephosphate isomerase (TIM) 146, 147, 150, 151, 157–160, 195, 203
tunneling 262
two-level tunneling system (TLS) 216
two-biopolymer model 341, 342, 350, 354
"two-state" behavior 18, 22, 23, 26
folding reactions 34
protonation model 52
two-state quantum system 263

unfolding 141, 157, 159–162, 171
barrier for 162
free energy of 170
rate 168
simulation 166, 169
thermal 31
universal biomolecule 2
universal structure in life 345
upfield shift 66
urea concentrations 16, 28, 29
UV circular dichroism (CD) 18, 19, 23

valence angles 110
valence bond 101
van der Waals interactions 6, 110, 188
radii 67
variable domains 286
common core of 285, 287, 292
interior of 289–292
vibrational motions 6, 262

water 214, 218, 220
–protein interface 186
Watson–Crick base pair 349, 355, 356
Watson–Crick double helix 89, 260
What_Check 68
wild-type protein 30, 31

X-ray crystallography 59–65, 83, 89–91, 93, 102, 125, 228, 261
development of 121
structure determination of aptamer 369
time-resolved 135, 171
X-ray diffraction 46, 60

Zeeman couplings 96
zero point quantum effects 262
zooming-in 213–223